VOLUME SIX HUNDRED AND NINE

METHODS IN ENZYMOLOGY

Enzyme Nanoarchitectures: Enzymes Armored with Graphene

METHODS IN ENZYMOLOGY

Editors-in-Chief

ANNA MARIE PYLE
Departments of Molecular, Cellular and Developmental Biology and Department of Chemistry
Investigator, Howard Hughes Medical Institute
Yale University

DAVID W. CHRISTIANSON
Roy and Diana Vagelos Laboratories
Department of Chemistry
University of Pennsylvania
Philadelphia, PA

Founding Editors

SIDNEY P. COLOWICK and NATHAN O. KAPLAN

VOLUME SIX HUNDRED AND NINE

METHODS IN ENZYMOLOGY

Enzyme Nanoarchitectures: Enzymes Armored with Graphene

Edited by

CHALLA V. KUMAR
Department of Chemistry;
Department of Molecular and Cell Biology,
University of Connecticut,
Storrs, CT, United States

ELSEVIER

ACADEMIC PRESS
An imprint of Elsevier

Academic Press is an imprint of Elsevier
50 Hampshire Street, 5th Floor, Cambridge, MA 02139, United States
525 B Street, Suite 1650, San Diego, CA 92101, United States
The Boulevard, Langford Lane, Kidlington, Oxford OX5 1GB, United Kingdom
125 London Wall, London, EC2Y 5AS, United Kingdom

First edition 2018

ISBN: 978-0-12-815240-9
ISSN: 0076-6879

Publisher: Fiona Pattison
Acquisition Editor: Ashlie Jackman
Editorial Project Manager: Shellie Bryant
Production Project Manager: Divya Krishna Kumar
Cover Designer: Christian Bilbow

Typeset by SPi Global, India

CONTENTS

CONTRIBUTORS

Lital Alfonta
Department of Life Sciences and Ilse Katz Institute for Nanoscale Science and Technology, Ben-Gurion University of the Negev, Beer-Sheva, Israel

Murali Anuganti
Department of Chemistry, University of Connecticut, Storrs, CT, United States

Palaniappan Arumugam
CSIR-Central Electrochemical Research Institute, Karaikudi, Tamil Nadu, India

Shamoon Asmat
Department of Biochemistry, Faculty of Life Sciences, Aligarh Muslim University, Aligarh, India

Omar Azzaroni
INIFTA Instituto de Investigaciones Fisicoquímicas Teóricas y Aplicadas (INIFTA)—Departamento de Química, Facultad de Ciencias Exactas, Universidad Nacional de La Plata—CONICET, La Plata, Argentina

Sheela Berchmans
CSIR-Central Electrochemical Research Institute, Karaikudi, Tamil Nadu, India

Christina Bliem
AIT Austrian Institute of Technology GmbH, Biosensor Technologies, Vienna, Austria

Paulina Bolibok
Faculty of Chemistry, Physicochemistry of Carbon Materials Research Group, Nicolaus Copernicus University in Toruń, Toruń, Poland

Alexandra V. Chatzikonstantinou
Department of Biological Applications and Technology, Laboratory of Biotechnology, University of Ioannina, Ioannina, Greece

Anil Kumar Chhillar
Centre for Biotechnology, Maharshi Dayanand University, Rohtak, Haryana, India

Cecilia Cristea
Department of Analytical Chemistry, Faculty of Pharmacy, "Iuliu Haţieganu" University of Medicine and Pharmacy, Cluj-Napoca, Romania

Debika Debnath
Department of Biomedical Engineering, University of Bridgeport, Bridgeport, CT, United States

Bhushan Dharmadhikari
Department of Computer Science and Engineering, University of Bridgeport, Bridgeport, CT, United States

Nalok Dutta
ICMR-National Institute of Cholera and Enteric Diseases, Kolkata, West Bengal, India

Luminita Fritea
Department of Preclinical Disciplines, Faculty of Medicine and Pharmacy, University of Oradea, Oradea, Romania

Elena Gkantzou
Department of Biological Applications and Technology, Laboratory of Biotechnology, University of Ioannina, Ioannina, Greece

Dimitrios Gournis
Department of Materials Science & Engineering, University of Ioannina, Ioannina, Greece

Qayyum Husain
Department of Biochemistry, Faculty of Life Sciences, Aligarh Muslim University, Aligarh, India

Zhongyi Jiang
Key Laboratory for Green Chemical Technology of Ministry of Education, School of Chemical Engineering and Technology, Tianjin University; Collaborative Innovation Center of Chemical Science and Engineering (Tianjin), Tianjin, China

Ankarao Kalluri
Department of Biomedical Engineering, University of Bridgeport, Bridgeport, CT, United States

Maryam Khan
Department of Biochemistry, Faculty of Life Sciences, Aligarh Muslim University, Aligarh, India

Wolfgang Knoll
AIT Austrian Institute of Technology GmbH, Biosensor Technologies, Vienna, Austria

Challa V. Kumar
Department of Chemistry; Department of Molecular and Cell Biology, University of Connecticut, Storrs, CT, United States

Melissa R. Limbacher
Department of Chemistry, University of Connecticut, Storrs, CT, United States

Yao Lin
Department of Chemistry; Polymer Program, Institute of Materials Science, University of Connecticut, Storrs, CT, United States

Jagriti Narang
Amity Institute of Nanotechnology, Amity University, Noida, Uttar Pradesh, India

Michaela Patila
Department of Biological Applications and Technology, Laboratory of Biotechnology, University of Ioannina, Ioannina, Greece

Prabir Patra
Department of Biomedical Engineering; Department of Mechanical Engineering, University of Bridgeport, Bridgeport, CT, United States

Esteban Piccinini
INIFTA Instituto de Investigaciones Fisicoquímicas Teóricas y Aplicadas (INIFTA)—Departamento de Química, Facultad de Ciencias Exactas, Universidad Nacional de La Plata—CONICET, La Plata, Argentina

Megan K. Puglia
Department of Chemistry, University of Connecticut, Storrs, CT, United States

Chandra S. Pundir
Department of Biochemistry, Maharshi Dayanand University, Rohtak, Haryana, India

Caterina M. Riccardi
Department of Chemistry, University of Connecticut, Storrs, CT, United States

Katarzyna Roszek
Department of Biochemistry, Faculty of Biology and Environmental Protection, Nicolaus Copernicus University in Toruń, Toruń, Poland

Malay K. Saha
ICMR-National Institute of Cholera and Enteric Diseases, Kolkata, West Bengal, India

Robert Sandulescu
Department of Analytical Chemistry, Faculty of Pharmacy, "Iuliu Haţieganu" University of Medicine and Pharmacy, Cluj-Napoca, Romania

Orr Schlesinger
Department of Life Sciences and Ilse Katz Institute for Nanoscale Science and Technology, Ben-Gurion University of the Negev, Beer-Sheva, Israel

Jiafu Shi
School of Environmental Science and Engineering, Tianjin University, Tianjin; Guangdong Provincial Key Laboratory of New and Renewable Energy Research and Development, Guangzhou, China

Asieh Soozanipour
Department of Biotechnology, Faculty of Advanced Sciences and Technologies, University of Isfahan, Isfahan, Iran

Haralambos Stamatis
Department of Biological Applications and Technology, Laboratory of Biotechnology, University of Ioannina, Ioannina, Greece

Yiying Sun
Key Laboratory for Green Chemical Technology of Ministry of Education, School of Chemical Engineering and Technology, Tianjin University; Collaborative Innovation Center of Chemical Science and Engineering (Tianjin), Tianjin, China

Asghar Taheri-Kafrani
Department of Biotechnology, Faculty of Advanced Sciences and Technologies, University of Isfahan, Isfahan, Iran

Mihaela Tertis
Department of Analytical Chemistry, Faculty of Pharmacy, "Iuliu Haţieganu" University of Medicine and Pharmacy, Cluj-Napoca, Romania

Manju Venkatesan
CSIR-Central Electrochemical Research Institute, Karaikudi, Tamil Nadu, India

Chirnajeevi Srinivasa Rao Vusa
CSIR-Central Electrochemical Research Institute, Karaikudi, Tamil Nadu, India

Marek Wiśniewski
Faculty of Chemistry, Physicochemistry of Carbon Materials Research Group, Nicolaus Copernicus University in Toruń, Toruń, Poland

Neelam Yadav
Centre for Biotechnology, Maharshi Dayanand University, Rohtak, Haryana, India

Shaohua Zhang
Key Laboratory for Green Chemical Technology of Ministry of Education, School of Chemical Engineering and Technology, Tianjin University; Collaborative Innovation Center of Chemical Science and Engineering (Tianjin), Tianjin, China

Jingjing Zhao
Key Laboratory for Green Chemical Technology of Ministry of Education, School of Chemical Engineering and Technology, Tianjin University; Collaborative Innovation Center of Chemical Science and Engineering (Tianjin), Tianjin, China

PREFACE

This is the third volume in the Nanotechnology Portfolio of Methods in Enzymology and we are proud that numerous authors have been contributing to this monumental work. The first volume of the portfolio was about enzymes protected by nanoparticles, using the most recent methods of nanochemistry and nanobiology and the next volume focused on the armoring of enzymes with synthetic polymers such that these unique nanosystems not only preserved the catalytic properties of enzymes but also stabilized them against denaturation by heat or harsh environments. The current volume is dedicated to the application of graphene in stabilizing sensitive enzymes while preserving their catalytic function. Graphene is a special nanomaterial for a large number of reasons (Kumar & Pattammattel, 2017). One of the attractive attributes of graphene for enzymology applications is that graphene has enormous surface area per unit mass, and thus, a large amount of enzyme can be accommodated on a small amount of graphene. Several examples establishing this advantage are documented in this volume.

Graphene is considered to be the wonder material or the material of the millennium but its use in biology is severely limited. However, recent efforts in making graphene, using biocompatible water dispersion has successfully overcome this issue, and large quantities of this material can be made by using a simple apparatus, in an inexpensive manner, even in resource-limited laboratories. Therefore, this review is timed well to aid enzymologists to exploit the amazing properties of graphene, which are too numerous to mention here. Chapters 1, 10, and 12 highlight the use of graphene and graphene quantum dots for this important application.

One major drawback of pristine graphene is its high level of hydrophobicity which makes it difficult to dissolve or suspend in aqueous media. This problem has been addressed successfully in the above three chapters, but the remaining chapters of this volume address this issue differently. The alternatives of either chemical or physical derivatization of graphene are reported in these other chapters, so that the graphene derivative is more compatible with aqueous media with extraordinary diversity, effectiveness, and utility. Thus, these chapters are equally important in enumerating the attractive attributes of graphene and its chemical derivatives with a variety of enzymes.

The selection of the articles for this volume, we hope, is a good mix while including a diverse set of enzymes, diverse set of graphene and its derivatives, numerous different methods, and very interesting applications, as in the previous volumes.

Easy separation, reaction workup, and recycling of the graphene-based biocatalysts are expected to reduce the process cost, and weight of the biocatalyst, because small amounts of graphene can support high loadings of the enzyme of interest and hence, this kind of a light weight and high surface area biocatalyst could be very important for industrial applications. The layered structure of graphene and its exfoliation makes good access for its enormous interior surface, thus, making these unique attributes possible for the corresponding biocatalysts. We thank the authors for their hard work and making this volume possible.

ACKNOWLEDGMENTS

Author thanks the National Science Foundation for partial financial support of this work (DMR-1401879) and the NASA STTR grant.

Challa Vijaya Kumar
Department of Chemistry, University of Connecticut, Storrs, CT, United States
Department of Molecular and Cell Biology, University of Connecticut, Storrs, CT, United States

REFERENCE

Kumar, C. V., & Pattammattel, A. (2017). *Introduction to graphene: Chemical and biochemical applications*. Amsterdam, The Netherlands: Elsevier.

CHAPTER ONE

Interlocking Enzymes in Graphene-Coated Cellulose Paper for Increased Enzymatic Efficiency

Melissa R. Limbacher*, Megan K. Puglia*, Caterina M. Riccardi*, Challa V. Kumar*,†,1

*Department of Chemistry, University of Connecticut, Storrs, CT, United States
†Department of Molecular and Cell Biology, University of Connecticut, Storrs, CT, United States
[1]Corresponding author: e-mail address: challa.kumar@uconn.edu

Contents

Abstract

A simple method for interlocking glucose oxidase and horseradish peroxidase in a network of cellulose fibers coated with bovine serum albumin (BSA)-exfoliated graphene (biographene) is reported here. The resulting paper reactor is inexpensive and stable. Biographene is expected to function as an electron shuttle, making the reaction between the enzyme and the substrate more efficient, and this hypothesis is examined here. The BSA used to separate the sheets of graphene provides extra carboxylic acid groups and primary amines to help interlock the enzymes and the graphene in between the fibers. The decrease in entropy associated with interlocking the enzymes on a solid support is likely responsible for the increase in enzymatic stability/activity observed.

Methods in Enzymology, Volume 609
ISSN 0076-6879
https://doi.org/10.1016/bs.mie.2018.06.012

Each cellulose disk contained 5.2 mg of enzyme per gram of paper and 93% of the enzyme is retained after washing for 0.5–2 h. This simple methodology provides a low cost, effective approach for achieving high enzymatic activity and good loadings on a benign, versatile support.

1. INTRODUCTION

Attachment of enzymes to porous materials is useful for the fabrication of biosensors, biofuel cells, and diagnostic tools. Restraining enzymes to a solid surface is a technique commonly employed to improve enzyme stability and recycle the attached enzyme multiple times. Both physical and chemical attachment strategies are known to restrain the enzymes in this way. The enzyme of interest is covalently linked to the support after it is loaded onto the surface. Thus, adsorption of the enzyme on the support is essential first step. Physical adsorption is observed when weak interactions exist between the surface and the enzyme (Chiou & Wu, 2004). However, these weak physical interactions can be problematic due to the possibility of the enzyme leaching off the surface during the use. However, the amount of enzyme loaded is limited by the amount of accessible surface area of the supporting material and this issue has been recently addressed by wrapping the enzymes around the cellulose fibers (Riccardi, Kasi, & Kumar, 2017).

1.1 Why Use Cellulose?

Presently, a wide variety of techniques exist for covalent attachment of enzymes to porous solid supports. One method is to use a solid support to trap the enzyme through either covalent attachment or adsorption to an inorganic matrix or a synthetic polymer (Mudhivarthi, Bhanbhani, & Kumar, 2007). Natural polymers are advantageous as supports because they are biocompatible and nontoxic (Mostafavi, 2015). One of the most well-known, and most commonly used natural polymers is cellulose, a robust, chemically inert material (Rojas, 2016; Yang et al., 2017). It is inexpensive and abundant, commonly found in plant-based materials, and naturally synthesized by algae and bacteria. Additionally, cellulose paper has high intrinsic porosity (Wang & Hsieh, 2004) and it is insoluble in most organic solvents (Rojas, 2016). Presently, few examples of graphene derivatives coated on paper for enzyme linking exist in literature. One example uses graphene oxide and horseradish peroxidase (HRP) to electrochemically detect cancer biomarkers (Wu, Xue, Kang, & Hui, 2013). Results showed signal amplification in electrochemical experiments. Additionally, graphene-based biosensor using conducting polymers, graphene, glucose oxidase (GOx), and gold nanoparticles was reported (Gokoglan et al., 2017). Another example is glucose detection with improved biosensor response (Gokoglan et al., 2017). Here, we describe a novel method for enzyme binding to cellulose/graphene composites using our interlocking approach, and our approach requires less preparation and fewer reagents and uses pristine graphene, not graphene oxide or reduced graphene oxide (Scheme 1).

1.2 Why Choose Interlocking?

"Interlocking" enzymes in paper is a novel approach to attach enzymes to the fibrous cellulose matrix without functionalizing the cellulose surface. With this approach, enzyme leaching is prevented, enzyme activity retained and stability enhanced (Riccardi et al., 2016). Interlocking uses carbodiimide (EDC) chemistry to crosslink carboxylic acid groups of one protein molecule to primary amines of another around the fibers of the cellulose matrix. This method crosslinks neighboring protein/enzyme groups at multiple sites to create a

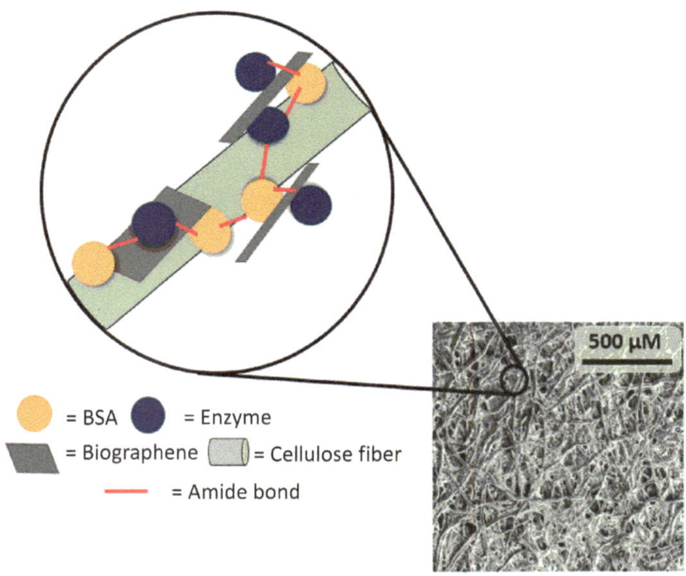

Scheme 1 Molecular view of interlocking of enzymes to biographene in a cellulose matrix through the use of EDC chemistry. The BSA present in the biographene and the enzyme being assayed are represented as *peach* and *blue circles*, respectively. The biographene is represented as a *gray rectangle* and the cellulose fiber as a *gray cylinder*. The *pink lines* indicate the amide bonds formed.

"mesh" around the cellulose fibers. This results in wrapping of the interlocked protein and mechanically interlocks the protein matrix with the cellulose fibers. Thus, no functionalization of cellulose is required.

Meanwhile, the weak interactions of cellulose with the interlocked enzyme are unlikely to perturb the interlocked enzyme to a significant extent. This approach of interlocking differs significantly from covalent crosslinking with cellulose which is generally used for enzyme attachment to solid surfaces, and interlocking provides particular advantages. For example, one step interlocking of any enzyme with a spacer protein such as bovine serum albumin (BSA) promotes interlocking under benign conditions. The numerous amine and carboxyl groups of BSA provide an extensive network to interlock acidic enzymes or basic enzymes. Thus, we expect this simple and novel method to be versatile and nearly universal. Additionally, the interlocking network of proteins such as BSA stabilizes the interlocked enzyme without hindering its biological activity (Riccardi et al., 2017). When interlocking is carried out in the presence of excess BSA, the BSA layer dilutes the enzyme layer, thus separating the enzyme molecules from each other. This dilution inhibits enzyme aggregation and promotes enzyme stability (Riccardi, Kasi, & Kumar, 2018). Therefore, we chose to interlock specific enzymes in the cellulose fiber mesh of an ordinary Whatman filter paper disks for making enzyme/graphene/cellulose composite bioreactors.

One other advantage of the interlocking approach is that it has the potential to stabilize the interlocked protein. Interlocking restricts the conformation of the bound enzyme in a native-like conformation under the mild aqueous reaction conditions by locking (Scheme 2). The multiple covalent links with its neighbors decrease the entropy of the native and denatured states. The conformational restriction due to the links with its immediate neighbors decreases the entropy of denaturation (ΔS, Scheme 2B). This results in an increase in the free energy of denaturation (ΔG) ($\Delta G = \Delta H - T\Delta S$). For enzyme denaturation is always positive and hence, ΔG should increase, assuming that ΔH is constant. This latter assumption is valid because ΔH is largely controlled by the primary sequence of the enzyme, which is unaltered during interlocking. Thus, the transition between the native (folded) state and the denatured (unfolded) state is inhibited by increasing ΔG of denaturation (Mudhivarthi et al., 2007; Riccardi et al., 2017). These favorable thermodynamic changes are proposed to increase the enzyme stability under harsh conditions such as heat or chemical denaturants.

In this chapter, the cellulose filter paper was coated with biographene and interlocked GOx (redox enzyme) and HRP (peroxidase) into the fibers

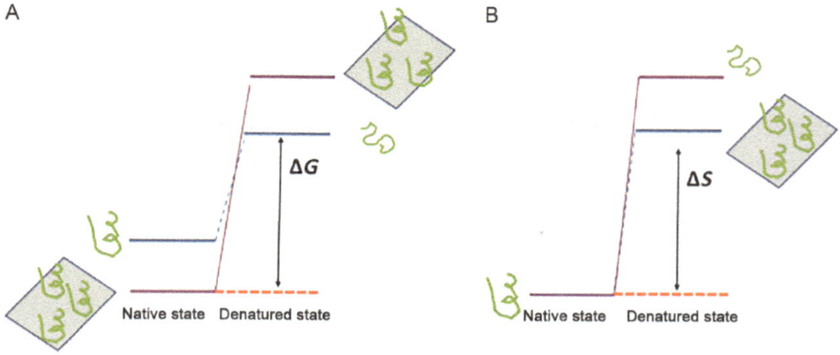

Scheme 2 Energy changes in the system for free and interlocked enzyme. (A) Stabilization of the enzyme by raising the free energy (ΔG) of the denatured state. (B) Decrease in entropy (ΔS) of the denatured state results in an increase in free energy of denaturation. This loss in ΔS is due to locking the enzyme in a specific conformational state by covalent links with its immediate neighbors. Green enzymes on *gray squares* indicate enzymes interlocked in graphene-coated cellulose. Free enzyme is indicated by the *green structures* without the graphene *squares*.

of the cellulose matrix. By coating the cellulose with biographene before interlocking, the efficiency of linking is increased due to the presence of additional BSA coating on the graphene, while numerous amine and carboxyl groups of BSA provide additional anchor points. Using graphene as a bridge, we tested if the biocatalysis of redox enzymes enhances the rates of electron transfer between the active sites of the enzymes and the substrate. Thus, a composite of enzyme/graphene/cellulose is prepared and tested for the retention of enzymatic activity. Biographene is made by the exfoliation of graphite with BSA to delaminate the layers and stabilize them from restacking. Thus, graphene and BSA are expected to enhance interlocking and stabilize the interlocked enzymes.

1.3 Why Use Graphene?

Graphene is being touted as the material of the millennium. Since new materials advanced human civilizations in the human history, such as the stone age, the bronze age, the space age, and so on, current millennium is being dubbed as the graphene age. This is not an exaggeration but its potential for practical applications is not yet realized. Some major assets of graphene include high surface area, high electrical and thermal conductivities comparable to or better than silver, and mechanical strength which is hundred times that of steel (Pattammattel & Kumar, 2017). It is also highly flexible and can be bent over wide angles without breaking or cracking. Single layer graphene is transparent, yet electrically conducting and thus, it is suitable for touch displays. Its high thermal conductivity is being used to fabricate efficient heat radiators for cooling space solar cells designed for outer planetary missions. Its high electrical conductivity is being explored in the current studies to promote enzymatic redox reactions.

Graphene is a two-dimensional layered material that has a multitude of possible biological applications including use in biosensors (Kulia, Bose, Khanra, Kim, & Lee, 2011), biocatalysis (Mostafavi, 2015), supercapacitors (Mosa et al., 2017), or drug delivery vehicles (McCallion, Burthem, Ress-Unwin, Golovanov, & Pluen, 2016; Zhang et al., 2017). The material is made up of single layer of carbon atoms arranged in a honeycomb lattice to form large, flat sheets (Fig. 1) which may bend or curl (Pattammattel et al., 2017).

Graphene, when used in enzymology or other biological systems, can enhance many enzymatic properties. However, there is one major problem that has limited the use of graphene in biological applications. Pristine graphene is hydrophobic and not inherently biocompatible (Pinto, Goncalves, & Magalhaes, 2013; Yang, Asiri, Tang, Du, & Kin, 2013) and it is often made in toxic organic solvents. To increase the biocompatibility of graphene and decrease its hydrophobicity, direct exfoliation of graphene with a protein in water or buffer has been achieved in a kitchen blender (Pattammattel & Kumar, 2015) and by a shear reactor (Pattammattel et al., 2017). In these processes, proteins such as BSA are adsorbed to the graphite surface to facilitate

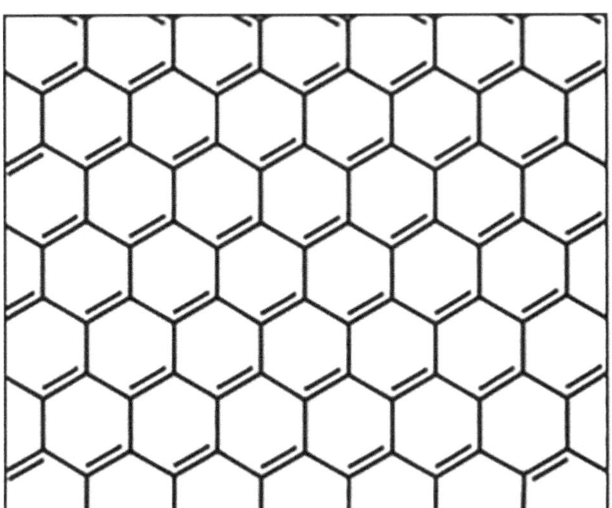

Fig. 1 Schematic of a part of the two-dimensional structure of graphene.

delamination and stabilization of the separated layers of graphene. Exfoliation of graphite with BSA serves to make the graphene more hydrophilic, while also decreasing its cytotoxicity (McCallion et al., 2016; Pattammattel et al., 2017). Adsorption of BSA to the surface of graphene is expected to be via stacking interactions between the aromatic residues of the protein and the sp^2 carbons of graphene, as well as via interactions with the basic residues such as arginine (Gu et al., 2015). Thus, BSA-coated graphene bridge is a good choice to test the hypothesis of increased enzymatic efficiency of redox active enzymes interlocked in cellulose, where the high conductivity and large sheet size of graphene are exploited.

1.4 Why Use BSA?

The use of BSA for enzyme interlocking in graphene-coated paper is advantageous because of the ability of BSA to bind a variety of hydrophobic ligands (Kragh-Hansen, 1981), dyes (Laurence, 1951), drugs (Sulkowska, 2002), lipids (Morrisett, Pownall, & Gotto, 1974), hormones (Kragh-Hansen, 1981), and other substrates. Substrate preconcentration in the BSA layer could enhance the enzymatic rates. In biographene, BSA acts as a stabilizer to keep the layers of graphene separated while at the same time biophylizing the graphene sheets so that they will not denature the delicate enzymes bound to it. BSA also provides extra carboxylic acid and primary amine groups for enzyme interlocking in the cellulose fiber mesh. The BSA layer surrounding the enzyme molecules dilutes the protein layer and inhibits the aggregation of the bound enzyme. For this reason, interlocking with BSA is also expected to increase the thermal stability of the interlocked enzyme (Fig. 2).

BSA is a large (66.4kDa), heart shaped, carrier protein from bovine plasma. It can be readily replaced by other serum albumins when species specificity is required which becomes important for the implant applications of the bioreactors. BSA is abundant, inexpensive, inert, and it is a waste product of the meat industry. It is readily degraded in the environment by proteases which are abundant in the soil. Thus, it is a green, sustainable, and nontoxic material for bioreactor design (Table 1).

1.5 Why Use GOx and HRP?

We chose GOx and HRP as model redox enzymes to test the current hypothesis that graphene may be able to serve as an electron conduit between the active sites of these enzymes and the substrate. Current limitations of their use include instability outside of physiological conditions where bound enzymes often show decreased activities. Thus, while stabilizing the enzymes

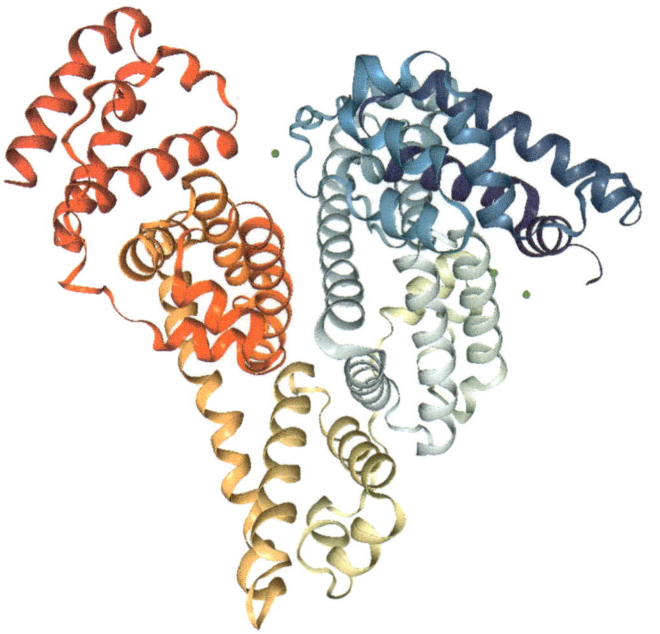

Fig. 2 Structure of bovine serum albumin (BSA) PDB 3V03 (Majorek et al., 2012). *Image from RCSB PDB (www.rcsb.org) of PDB 3V03 (Majorek, K. A., Porebski, P. J., Dayal, A., Zimmerman, M. D., Jablonska, K., Stewart, A. J., et al. (2012). Structural and immunologic characterization of bovine, horse, and rabbit serum albumins.* Molecular Immunology, *52, 174–182).*

Table 1 Key Properties of BSA

Protein	Molecular Weight (Da)	Absorbance Wavelength (nm)	Extinction Coefficient ($M^{-1}cm^{-1}$)	pI	Amine Side Chains	Carboxylic Acid Side Chains
Bovine serum albumin	66,400	280	43,284	5.6	82	99

by interlocking in cellulose matrix, we explored if graphene would serve as an electron bridge and enhance the redox catalysis of these enzymes.

GOx (Table 2) is a large (150–180 kDa), flavoenzyme which is isolated from *Aspergillus niger*. It is strongly acidic and this anionic enzyme has high stability making it useful for fabrication of biosensors to monitor glucose in blood (Ramanavicius et al., 2014). GOx (Fig. 3A, Table 2) is a dimer composed of two identical subunits and catalyzes the oxidation of glucose to gluconate by oxygen while producing the byproduct, hydrogen peroxidase

Table 2 Properties of Enzymes Used in Studies

Enzyme Bound to Biographene Surface	Molecular Weight (Da)	Absorbance Wavelength (nm)	Extinction Coefficient ($M^{-1}cm^{-1}$)	pI	Amine Side Chains[a]	Carboxylic Acid Side Chains[a]
Glucose oxidase	160,000	450	28,200	4.6	72	132
Horseradish peroxidase	40,000	403	100,000	5.1	27	28

[a]All side chain calculations completed using UniProt. HRP P00433, GOx P13006, and Lipase P20261.

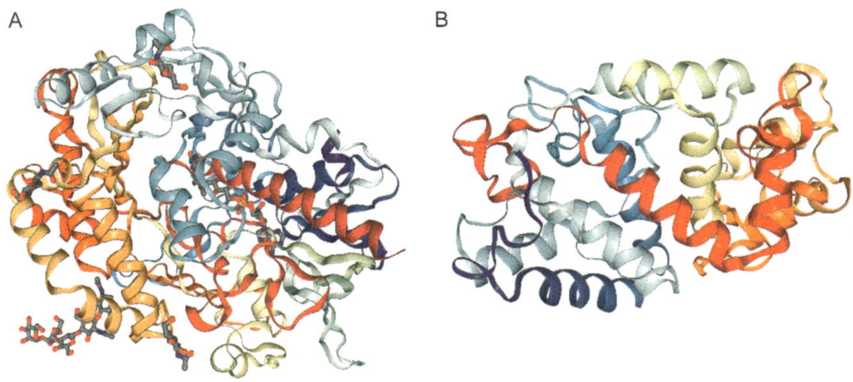

Fig. 3 (A) Structure of monomeric form of glucose oxidase (PDB 1CF3) (Wohlfahrt et al., 1999). (B) Structure of horseradish peroxidase (PDB 1H55) (Berglund et al., 2002). *Panel (A) image from RCSB PDB (www.rcsb.org) of PDB 1CF3 (Wohlfahrt, G., Witt, S., Hendle, J., Schomburg, D., Kalisz, H. M., & Hecht, H.-J. (1999). 1.8 and 1.9 A resolution structures of the* Penicillium amagasakiense *and* Aspergillus niger *glucose oxidases as a basis for modelling substrate complexes.* Acta crystallographica. Section D, Biological Crystallography, *55(Pt 5), 969–977). Panel (B) image from RCSB PDB (www.rcsb.org) of PDB 1H55 (Berglund, G. L., Carlsson, G. H., Smith, A. T., Szörke, H., Henriksen, A., & Hajdu, J. (2002). The catalytic pathway of horseradish peroxidase at high resolution.* Nature, *417(6887), 463–468).*

(Zhang et al., 2017). One method of evaluating GOx enzymatic activity is to estimate the concentration of the byproduct, hydrogen peroxide by a variety of methods. The reported Michaelis constant (K_m) for GOx catalyzed glucose oxidation is $38 \pm 6\,mM$ (Tao, Raffel, & Souid, 2009). We tested here if the activity of GOx is enhanced by graphene.

HRP (Fig. 3, Table 2) is a small (40 kDa) heme-containing enzyme that is isolated from *Armoracia rusticana* (Ramanavicius et al., 2014). This peroxidase uses the hydrogen peroxide produced from the GOx catalysed reaction with glucose, and readily catalyzes the oxidation of ABTS (2,2′-azino-bis (3-ethylbenzothiazoline-6-sulfonic acid). This product is blue-colored that is easily monitored by absorbance at the peak wavelength (Veitch, 2004). The reported K_m for the reaction of H_2O_2 catalyzed by HRP is $2.55 \pm 0.11\,\mu M$ (Zhang, Tsitkov, & Hess, 2016). HRP is widely used commercially in the production of clinical diagnostic kits (McCallion et al., 2016).

2. MATERIALS AND METHODS

2.1 General Equipment

1. *Safety*: Proper PPE such as goggles, lab coat, nitrile gloves, shoes covering the toes
2. *Waste management*: General aqueous waste receptacle
3. Deionized water
4. UV–Vis spectrometer
5. Microcentrifuge tubes
6. Volumetric micropipettes (20, 100, 200, and 1000 μL)
7. Beakers (100–1000 mL)
8. Stir/hot plate
9. Analytical balance
10. pH meter
11. Whatman 1 cellulose filter paper (185 μ*M* diameter, 87 $g m^{-2}$ weight, 185 m*M* diameter, 11 μ*M* pore size)
12. Benchtop centrifuge

2.2 Synthesis of Biographene

1. Graphite powder (MKBW0432V, Sigma)
2. BSA (BAH64-1366, Equitech-Bio, Inc.)

For this approach, biocompatible graphene was produced by exfoliation as described below.

1. Dissolve BSA (20 mg/mL) in distilled water
2. Add graphite (100 mg/mL) and stir at 65°C for 20 days or blend in a kitchen blender for 0.5 h.
3. Separate graphene from graphite by centrifugation at 1500 rpm for 45 min. Characterize the graphene in the supernatant by Raman spectroscopy (Renishaw system 2000 microscope, 1 μ*M* spatial resolution). For more

details, please see Chapter "A simple flow reactor for continuous synthesis of biographene for enzymology studies" by Puglia et al. in this volume.

2.2.1 Characterization of Biographene With Raman Spectroscopy

A common characterization technique for graphene is Raman spectroscopy.

1. Calibrate the spectrometer using a silicon chip with a known Raman peak at 521 cm^{-1}
2. Drop-cast biographene solution on to a Si/SiO_2 microscope slide
3. Excite the sample at 512.5 nm
4. Carry out the analysis to obtain the number of layers and the quality of graphene (Paton et al., 2014; Pattammattel & Kumar, 2015)

Analysis of the Raman data indicated that our biographene had 3–4 layers per sheet (Eqs. 1 and 2) and a lateral sheet size of about 0.4 μm (Eq. 3). The $I_D/I_{D'}$ of the graphene sample was about 2.7, implying that the sheets lack sp^3 defects and not oxidized during the exfoliation process. This ratio of 2.7 suggests fewer edge defects (Paton et al., 2014; Pattammattel & Kumar, 2015). The equations used for the analysis are reported below.

$$\text{Number of layers} = 10^{0.84M + 0.45M^2} \tag{1}$$

$$M = \frac{\left[I_{2D\text{ peak}}/I_{2D\text{ peak}-30\text{ cm}^{-1}}\right]_{\text{graphene}}}{\left[I_{2D\text{ peak}}/I_{2D\text{ peak}-30\text{ cm}^{-1}}\right]_{\text{graphite}}} \tag{2}$$

$$\text{Lateral sheet size} = \frac{k}{[I_D/I_G]_{\text{graphene}} - [I_D/I_G]_{\text{graphite}}} \tag{3}$$

*k was experimentally calculated to be 0.17 using transmission electron microscopy data as detailed in our previously published work (Paton et al., 2014; Pattammattel & Kumar, 2015) (Fig. 4).

2.3 Method for Enzyme Interlocking in Paper

1. GOx (*A. niger*, GOx, EC 1.1.3.4, Sigma-Aldrich)
2. HRP (EC 1.11.1.7, Calzyme)
3. 1 Ethyl-3-(3-(dimethylamino)propyl) carbodiimide (EDC, TCI America)
4. Citrate–phosphate buffer (100 mM citric acid, 200 mM phosphate, pH 4.5)
5. Biographene (synthesized in Section 2.2)

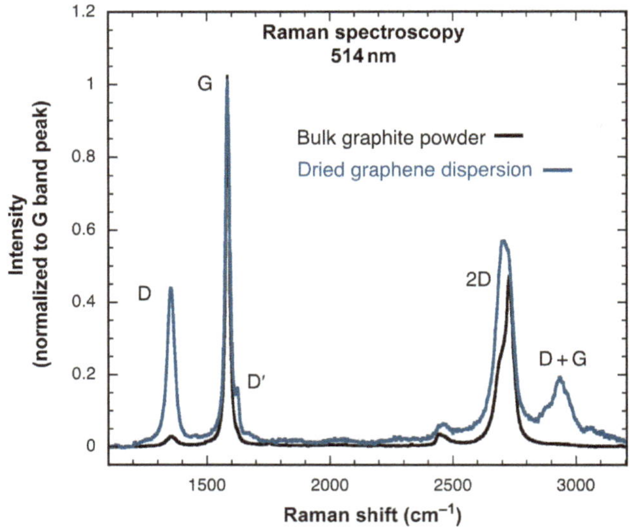

Fig. 4 Raman spectrum for dried graphene sample (*blue*) on glass microscope slide compared to that of bulk graphite powder (*black*). The spectra are normalized to the intensity of the G peak. Excitation wavelength is 514.5 nm.

For this approach, filter paper was coated with biographene first. An enzyme–graphene conjugate was then formed on the surface of the biographene-coated cellulose. Using standard carbodiimide (EDC) chemistry, the amine and carboxylic acid side chains of HRP and GOx were covalently linked to those of BSA in biographene (Scheme 3). The resulting biosensor can be colorimetrically assayed, so that the enzyme activity could be visually confirmed in addition to quantitation by spectroscopy.

1. Load biographene (5 mg/mL graphene, 20 mg/mL BSA) onto a Whatman filter paper, let dry for 30 min.
2. Prepare stocks of GOx (100 μ*M*) and HRP (16 μ*M*) in deionized water.
3. Mix GOx (16 μ*M*), HRP (3.2 μ*M*), and citric acid/sodium phosphate buffer (100 m*M* citric acid, 200 m*M* phosphate, pH 4.5) stir 10 min.
4. Add EDC (100 m*M*) to the GOx/HRP mixture, immediately drop-cast 227 μL onto the biographene/paper disk. *Tip:* EDC is hygroscopic, do not weigh out until ready to use
5. Let the filter paper to air dry at room temperature for 1 h.
6. Wash with deionized water for 30 min to remove any free protein/enzyme, graphene, and EDC urea. Do multiple washes, if needed.
7. Let the washed filter disk to dry at room temperature for 1 h. *Tip:* Save some of the wash to test for unbound enzyme and total protein by Bradford Assay.

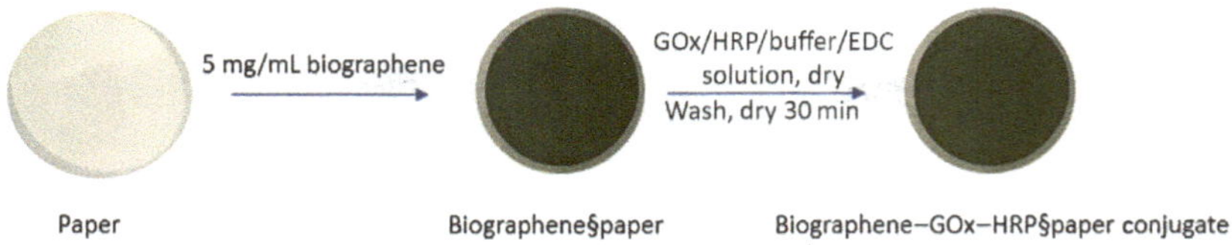

Scheme 3 Synthesis of enzyme interlocked graphene filter paper. Biographene (5 mg/mL) is coated onto cellulose paper and allowed to dry. Next, an enzyme/ 200 m*M* phosphate/100 m*M* citrate buffer, pH 4.5/EDC solution is drop-cast onto the biographene-coated cellulose paper. After drying, the sample is washed in deionized water for 30 min to remove any free enzyme or by-products and allowed to dry. mg/mg ratios are listed in Table 3.

Table 3 The Mass Ratios of Synthesis Components vs That of the Paper Disk

Component	Mass (mg Component/mg Disk)
Biographene	0.51
BSA	3.9×10^{-2}
GOx	5×10^{-3}
HRP	2.4×10^{-4}

2.4 Enzyme Loading on Paper by Bradford Assay

1. Brilliant Blue G-250 (Sigma)
2. Ethanol (Sigma-Aldrich)
3. Phosphoric acid (Sigma-Aldrich)
4. Whatman 1 filter paper, standard grade

To determine the extent of conjugation and whether or not the enzyme and BSA washed off the filter, total protein concentration was calculated by mixing samples with Bradford reagent and measuring the absorbance at 595 nm. A calibration plot was made for the reagent using BSA (Fig. 5) as a standard and used to quantify the concentration of total protein that is released into the wash and hence, not crosslinked around the cellulose fibers.

2.5 Colorimetric Enzymatic Activity Assays

1. ABTS (Roche Diagnostics)
2. Glucose (Sigma)
3. Phosphate buffer (100 m*M*, pH 7.4)
4. Plate Reader (Flex Station 3)

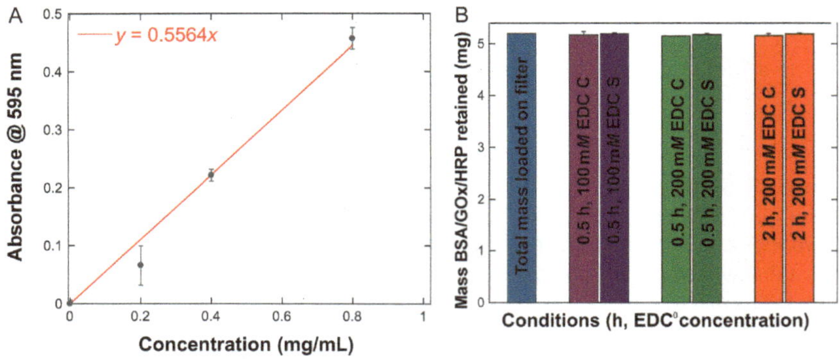

Fig. 5 (A) Bradford assay used to calculate mass of protein retained on the filter disk. (B) Summary of Bradford analysis of sample washes; C and S denote control and graphene samples.

2.5.1 GOx/HRP Activity Assay

To monitor the enzymatic activity of the reactor containing GOx and HRP interlocked on biographene-coated filter paper, the 2,2-azino-bis (3-ethylbenzothiazoline-6-sulfonate) (ABTS) assay was used. This assay (Scheme 4A) utilizes the oxidation of glucose to gluconate and H_2O_2. The H_2O_2 then activates the peroxidase to oxidize ABTS (Scheme 4B) to form ABTS radical cation ($ABTS^{+}\cdot$). Formation of $ABTS^{+}\cdot$ can be seen using absorbance spectroscopy to monitor 420 nm and can be seen colorimetrically by color change from clear to blue (Scheme 4C). Experiments were completed using the procedure described below (Scheme 4D).

1. Mix 20 m*M* phosphate buffer (pH 7.4), 1 m*M* glucose, 1 m*M* ABTS, and deionized water in a glass vial.
2. Add two disks from the Whatman 1 filter paper samples to each vial (each disk will have ~0.588 mg of GOx).
3. Measure the absorbance at 420 nm ($\varepsilon_{420} = 36{,}000\ M^{-1}\ cm^{-1}$) over time to calculate the rate and efficiency of the reaction.

To determine whether or not graphene actively works as an electron shuttle, the enzymatic activities were calculated for each sample by determining the initial rate. To calculate the percent retention, the rate of graphene–GOx–HRP§Paper was compared to BSA–GOx–HRP§Paper. When GOx/HRP is free in a biographene suspension (Fig. 6A), the percent retention increased to 112 ± 12 as compared with GOx/HRP/BSA in suspension in which 100% retention is assumed. Interlocking GOx and HRP into biographene-coated cellulose (Fig. 6B) further increased the percent retention of GOx/HRP to 279 ± 1, a 280-fold increase compared to GOx/HRP

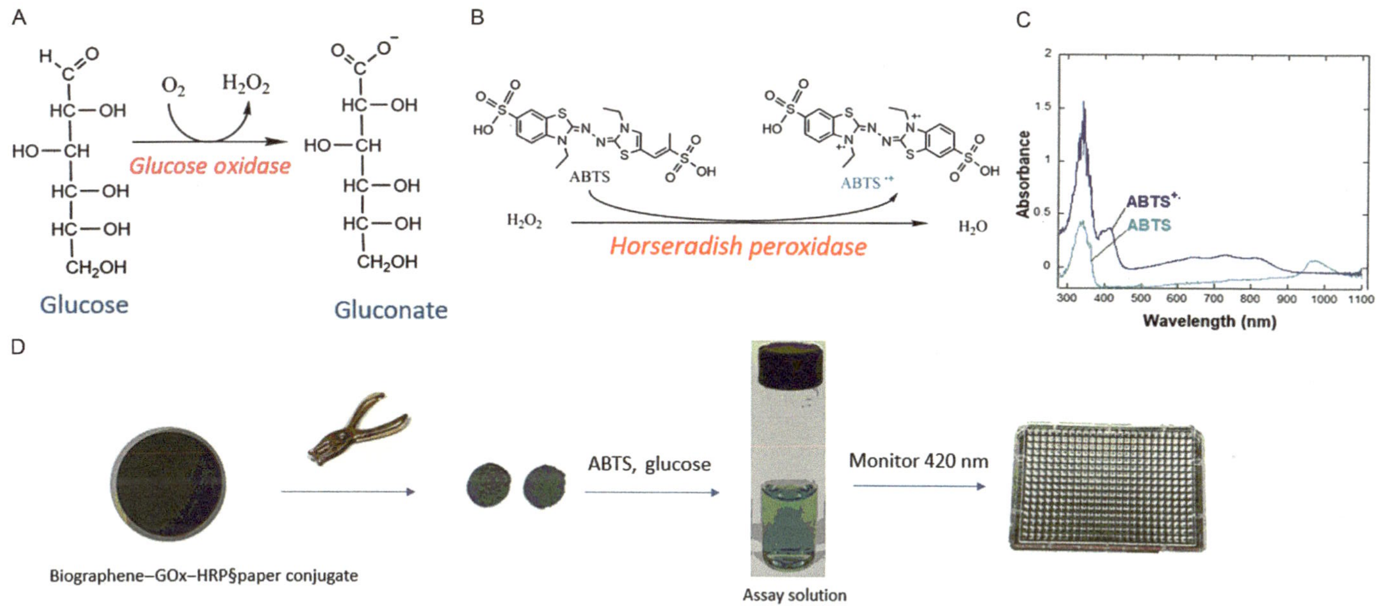

Scheme 4 (A) Mechanism for glucose oxidase assay. (B) Mechanism for horseradish peroxidase assay. (C) Absorbance spectra of ABTS oxidation during enzymatic reaction. (D) Experimental setup of assay.

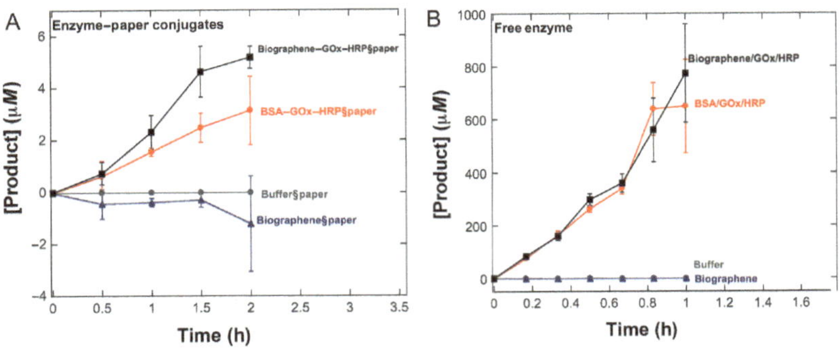

Fig. 6 Biographene-coated sample, enzymatic activity of GOx and HRP as a function of substrate concentration and time. (A) Biographene–GOx–HRP§Paper. (B) Free enzyme.

in suspension with biographene. The increase in rate and percent retention is thought to be due to interlocking enzymes and the graphene in cellulose matrix and graphene may function as an electron conduit.

2.6 Quantification of K_m and V_{max}

To determine the affinity of the substrate to the enzymes in the cellulose matrix, Michaelis–Menten kinetic experiments were carried out at increasing concentrations of the substrate. Tested here are 5 mg/mL graphene-coated (Fig. 7A), 5 mg/mL graphene-coated sample with an extra addition of EDC (Fig. 8B), and 5 mg/mL graphene sample with a 2-h wash time to remove by-products and unbound protein/enzyme (Fig. 7C).

1. Choose concentrations of substrate to test. Here, 0.1, 0.5, 1, and 2 m*M* ABTS/glucose were used.
2. Set up the experiments as shown previously (Section 2.5).
3. Measure product formation at 420 nm for at least 2 h (need to calculate the initial rate for each set of conditions).
4. Plot initial rate against the inverse of the substrate concentration.
5. Fit the initial rates to Michaelis–Menten model.
6. Calculate K_m and V_{max}.

Samples summarized in Table 4 indicate that samples that were coated with $2\times$ excess of graphene and then washed for 30 min have the lowest enzyme affinity for the substrate ($K_m = 3 \pm 3$) while samples that contain one coating of EDC followed by a 30-min wash have the highest enzyme affinity for the substrate ($K_m = 0.4 \pm 0.8$). Samples containing 200 m*M* EDC and a 2-h wash fall in the middle with $K_m = 3 \pm 2.4$. The large amounts of error that are associated with the samples are likely due to light scattering from the paper disk being suspended in the reaction mixture. More work is being done to mitigate this error.

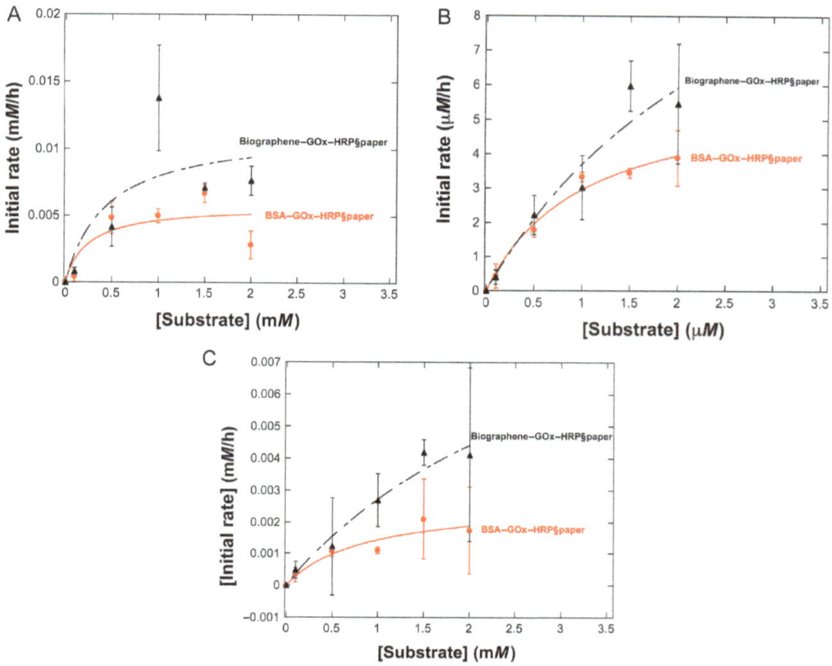

Fig. 7 Michaelis–Menten plots of samples. (A) Samples coated in 5 mg/mL biographene, 100 m*M* EDC, 30-min wash. (B) Samples coated in 5 mg/mL biographene, 200 m*M* EDC, 30-min wash. (C) Samples coated in 5 mg/mL biographene, 200 mm EDC, 2-h wash. All samples contained a 1:1 ratio of ABTS to glucose.

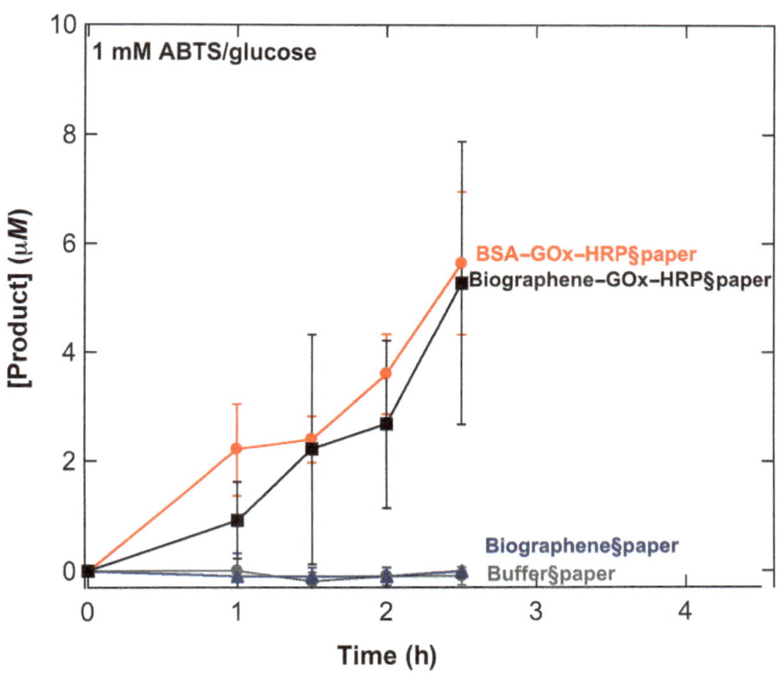

Fig. 8 Activity of GOx and HRP plotted as a function of substrate concentration 2 × excess of EDC.

Table 4 Summary of Michaelis–Menten Kinetic Studies and the Relative Enzymatic Activities of the Samples Interlocked in the Whatman Filter Paper

Sample	Graphene (mg/mL)	EDC (m*M*)	Wash Time (h)	K_m (m*M*)	V_{max} (m*M*/h)	Relative Activity (%)
Sample 1	5	100	0.5	0.4 ± 0.8	$0.01 \pm 5.9 \times 10^{-3}$	279 ± 1
Sample 2	5	200	0.5	3 ± 3	$0.15 \pm 9.9 \times 10^{-3}$	91 ± 37
Sample 3	5	200	2	3 ± 2.4	$0.003 \pm 8.9 \times 10^{-4}$	259 ± 97

Table 5 Summary of Concentration BSA/GOx/HRP Washed Off Filter Paper Under Varied Conditions

Sample	Concentration of EDC (m*M*)	Wash Time (h)	Concentration of BSA/GOx/HRP Washed Off (mg/mL)
BSA/GOx/HRP	100	0.5	0.03 ± 0.08
Biographene/GOx/HRP	100	0.5	0.003 ± 0.04
BSA/GOx/HRP	200	0.5	0.05 ± 0.09
Biographene/GOx/HRP	200	0.5	0.02 ± 0.4
BSA/GOx/HRP	200	2	0.04 ± 0.04
Biographene/GOx/HRP	200	2	0.003 ± 0.007

2.7 Optimization of GOx/HRP/Biographene Interlocking in Paper

2.7.1 Effect of Increased EDC Concentrations for Interlocking

To optimize the amount of BSA, GOx and HRP interlocked in the cellulose paper, the concentration of EDC was increased from 100 to 200 m*M* and the enzymatic activities of the sample tested (Fig. 8).

As EDC concentration is increased, less BSA was present in the wash solution. Less than 1 mg/mL of protein (BSA, GOx, and HRP) was found in the wash for all EDC concentrations tried. The data indicated that 93% of the BSA, GOx, and HRP are bound to the paper (Table 5).

2.7.2 Effect of Wash Time

To better understand the amount of protein that is actually interlocked in the paper after interlocking, experiments were done to test for free BSA, GOx, and HRP concentrations. Instead of a 30-min wash, a 2-h wash was used to

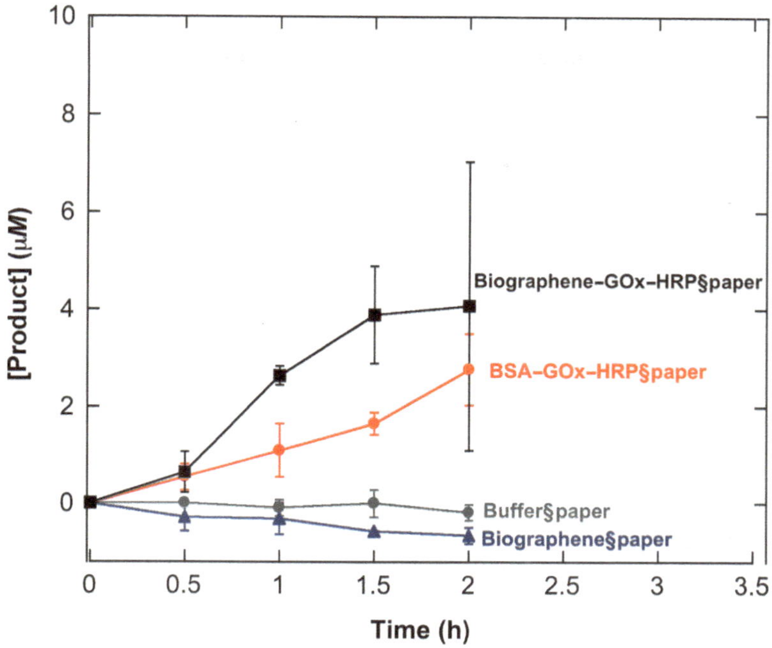

Fig. 9 Enzymatic activity of GOx and HRP after 2-h wash interlocked with 200 m*M* EDC.

confirm that all of the EDC by-products and free enzyme were removed. While the 30-min wash sample showed a product retention that was lower than that of the graphene free sample, the 2-h wash showed significant enhancement in enzymatic activity retention, with a calculated value of 259 ± 97 (Fig. 9).

3. CONCLUDING REMARKS

Interlocking enzymes in cellulose coated with graphene helped enhance the activity of the enzymes used here. By varying the amount of EDC, it is possible to minimize the amount of BSA and enzyme that was retained on the paper and up to 93% of the enzyme has been retained. Additionally, samples made with a 100 m*M* EDC and a 30-min wash and samples made with 200 m*M* EDC and a 2-h wash show an increase in relative enzymatic activity retention with respect to the free enzymes in a biographene suspension. Conversely, using 200 m*M* EDC and washing for 30-min inhibited the activity of the interlocked enzyme. This could be due to EDC by-products or protein clogging the pores of the filter paper, and keeping the substrate from reaching the enzyme active site.

The enhanced activity of the graphene interlocked in paper is likely in part due to the ability of graphene serving as an electron conduit, but further studies are needed to test other redox enzymes. Another possibility is that graphene functions as an artificial peroxidase (Garg, Bisht, & Ling, 2015). Under the conditions used in our experiments, we did notice any product formation without the inclusion of HRP indicating that the rate enhancement is due to the conductivity of graphene.

ACKNOWLEDGMENTS

We acknowledge financial support from the US NASA and NSF EAGER Grant (DMR1401879).

REFERENCES

Berglund, G. L., Carlsson, G. H., Smith, A. T., Szörke, H., Henriksen, A., & Hajdu, J. (2002). The catalytic pathway of horseradish peroxidase at high resolution. *Nature, 417*(6887), 463–468.

Chiou, S.-H., & Wu, W.-T. (2004). Immobilization of *Candida rugosa* lipase on chitosan with activation of the hydroxyl groups. *Biomaterials, 25*, 197–204.

Garg, B., Bisht, T., & Ling, T.-C. (2015). Graphene-based nanomaterials as efficient peroxidase mimetic catalysts for biosensing applications: An overview. *Molecules, 20*, 14155–14190.

Gokoglan, T. C., Kesik, M., Soylemez, S., Yuksel, R., Unalan, H. E., & Toppare, L. (2017). Paper based glucose biosensor using graphene modified with a conducting polymer and gold nanoparticles. *Journal of the Electrochemical Society, 164*(6), G59–G64.

Gu, Z., Yang, Z., Wang, L., Zhou, H., Jimenez-Cruz, C. A., & Zhou, R. (2015). The role of basic residues in the adsorption of blood proteins onto the graphene surface. *Scientific Reports, 5*, 10873.

Kragh-Hansen, U. (1981). Molecular aspects of ligand binding to serum albumin. *Pharmacological Reviews, 33*(1), 17–53.

Kulia, T., Bose, S., Khanra, P., Kim, N. H., & Lee, J. H. (2011). Recent advances in graphene-based biosensors. *Biosensors and Bioelectronics, 26*, 4637–4648.

Laurence, D. J. R. (1951). A study of the adsorption of dyes on bovine serum albumin by the method of polarization of fluorescence. *Journal of Biochemistry, 51*(2), 168–180.

Majorek, K. A., Porebski, P. J., Dayal, A., Zimmerman, M. D., Jablonska, K., Stewart, A. J., et al. (2012). Structural and immunologic characterization of bovine, horse, and rabbit serum albumins. *Molecular Immunology, 52*, 174–182.

McCallion, C., Burthem, J., Ress-Unwin, K., Golovanov, A., & Pluen, A. (2016). Graphene in therapeutics delivery problems, solution and future opportunities. *European Journal of Pharmaceutics and Biopharmaceutics, 104*, 235–250.

Morrisett, J. D., Pownall, H. J., & Gotto, A. M., Jr. (1974). Study of the fatty acid steroid binding sites using spin-labeled lipids. *The Journal of Biological Chemistry, 250*(7), 2487–2494.

Mosa, I. M., Pattammattel, A., Kadimisetty, K., Pande, P., El-Kady, M. F., Bishop, G. W., et al. (2017). Ultrathin graphene–protein supercapacitors for miniaturized bioelectronics. *Advanced Energy Materials*, 7, 1700358–1700370.

Mostafavi, S. M. (2015). 3D Graphene biocatalysts for development of enzymatic biofuel cells: A short review. *NanoAnalysis*, *2*, 57–62.

Mudhivarthi, V. K., Bhanbhani, A., & Kumar, C. V. (2007). Novel enzyme/DNA/inorganic nanomaterials: A new generation of biocatalysts. *Dalton Transactions*, *47*, 5483–5497.

Paton, K. R., Varrla, E., Backes, C., Smitj, R. J., Khan, U., et al. (2014). Scalable production of large quantities of defect-free few-layer graphene by shear exfoliation in liquids. *Nature Materials*, *13*, 624.

Pattammattel, A., & Kumar, C. V. (2015). Kitchen chemistry 101: Multigram production of high quality biographene in a blender with edible proteins. *Advanced Functional Materials*, *25*, 7088–7098.

Pattammattel, A., & Kumar, C. V. (2017). *Chemical and biochemical applications of graphene.* Elsevier.

Pattammattel, A., Pande, P., Kuttappan, D., Puglia, M., Basu, A. K., et al. (2017). Controlling the graphene-bio interface: Dispersions in animal sera for enhanced stability and reduced toxicity. *Langmuir*, *33*, 14184–14194.

Pinto, A. M., Goncalves, I. C., & Magalhaes, F. D. (2013). Graphene-based materials biocompatibility: A review. *Colloids and Surfaces B: Biointerfaces*, *111*, 188–202.

Ramanavicius, A., Voronovic, J., Semashko, T., Mikhailova, R., Kausaite-Minkstimiene, A., & Ramanaviciene, A. (2014). Comparison of glucose oxidases from *Penicillium adametzii, Penicillium funiculosum, and Asperigillus niger* in the design of amperomeric glucose biosensors. *Analytical Sciences*, *30*, 1143–1149.

Riccardi, C. M., Kasi, R. M., & Kumar, C. V. (2017). Nanoarmoring of enzymes by interlocking in cellulose fibers with poly(acrylic acid). *Methods in Enzymology*, *590*, 476–500.

Riccardi, C. M., Kasi, R. M., & Kumar, C. V. (2018). A general method for interlocking enzymes in Whatman paper. *Angewandte Chemie.* https://doi.org/10.1002/anie.201805074. (provisionally accepted).

Riccardi, C. M., Mistri, D., Hart, O., Anguganti, M., Lin, Y., Kasi, R. M., et al. (2016). Covalent interlocking of gluocose oxidase and peroxidase in the voids of paper: Enzyme-polymer "spider webs". *Chemical Communications*, *52*, 2593–2596.

Rojas, O. J. (2016). Cellulose chemistry and properties: Fibers, nanocelluloses and advanced materials. *Advances in Polymer Science*, *271*, 1–52.

Sulkowska, A. (2002). Interaction of drugs with bovine and human serum albumin. *Journal of Molecular Structure*, *614*, 227–232.

Tao, Z., Raffel, R. A., & Souid, A.-K. (2009). Kinetic studies on enzyme-catalyzed reactions: Oxidation of glucose, decomposition of hydrogen peroxide and their combination. *Biophysical Journal*, *96*, 2977–2988.

Veitch, N. C. (2004). Horseradish peroxidase: A modern view of a classic enzyme. *Phytochemistry (Elsevier)*, *6*, 249–259.

Wang, Y., & Hsieh, Y.-L. (2004). Enzyme immobilization to ultra-fine cellulose fibers via amphiphilic polyethylene glycol spacers. *Journal of Polymer Science, Part A*, *42*(17), 4289–4299.

Wohlfahrt, G., Witt, S., Hendle, J., Schomburg, D., Kalisz, H. M., & Hecht, H.-J. (1999). 1.8 and 1.9 A resolution structures of the *Penicillium amagasakiense* and *Aspergillus niger* glucose oxidases as a basis for modelling substrate complexes. *Acta crystallographica. Section D, Biological Crystallography*, *55*, 969–977.

Wu, Y., Xue, P., Kang, Y., & Hui, K. M. (2013). Paper-based microfluidic electrochemical immunodevice integrated with nanobioprobes onto graphene film for ultrasensitive multiplexed detection of cancer biomarkers. *Analytical Chemistry*, *85*, 8661–8686.

Yang, Y., Asiri, A. M., Tang, Z., Du, D., & Kin, Y. (2013). Graphene based materials for biological applications. *Materials Today, 16*(10), 365–373.

Yang, H., Kannappan, S., Pandian, A. S., Jang, J.-H., Lee, Y. S., & Lu, W. (2017). Graphene supercapacitor with both high power and energy density. *Nanotechnology, 28*, 445401.

Zhang, Y., Tsitkov, S., & Hess, H. (2016). Proximity does not contribute to activity enhancement in the glucose oxidase-horseradish peroxidase cascade. *Nature Communications*, 7, 13982.

Zhang, Q., Wu, Z., Li, N., Pu, Y., Wang, B., Zhang, T., et al. (2017). Advanced review of graphene-based nanomaterials in drug delivery systems: Synthesis, modification, toxicity and application. *Materials Science and Engineering C*, 77, 1363–1375.

CHAPTER TWO

Enzyme Multilayers on Graphene-Based FETs for Biosensing Applications

Christina Bliem[*,1], Esteban Piccinini[†], Wolfgang Knoll[*], Omar Azzaroni[†,1]

[*]AIT Austrian Institute of Technology GmbH, Biosensor Technologies, Vienna, Austria
[†]INIFTA Instituto de Investigaciones Fisicoquímicas Teóricas y Aplicadas (INIFTA)—Departamento de Química, Facultad de Ciencias Exactas, Universidad Nacional de La Plata—CONICET, La Plata, Argentina
[1]Corresponding authors: e-mail address: christina.bliem@ait.ac.at; azzaroni@inifta.unlp.edu.ar

Contents

Abstract

Electrochemical sensors represent a powerful tool for real-time measurement of a variety of analytes of much significance to different areas, ranging from clinical diagnostics to food technology. Point-of-care devices which can be used at patient bedside or for online monitoring of critical parameters are of great importance in clinical daily routine. In this work, portable, low-cost electrochemical sensors for a fast and reliable detection of the clinically relevant analyte urea have been developed. The intrinsic pH sensitivity of reduced graphene oxide (rGO)-based field-effect transistors (FETs) was exploited to monitor the enzymatic hydrolysis of urea. The functionalization of the sensor platform using the layer-by-layer technique is especially advantageous for the immobilization of the biorecognition element provided that this approach preserves the enzyme integrity as well as the rGO surface. The great selectivity of the enzyme (urease) combined with the high sensitivity of rGO-based FETs result in the construction of urea biosensors with

Methods in Enzymology, Volume 609
ISSN 0076-6879
https://doi.org/10.1016/bs.mie.2018.06.001

a limit of detection (LOD) of 1 μ*M* and a linear range up to 1 m*M*. Quantification of Cu^{2+} with a LOD down to 10 n*M* was performed by taking advantage of the specific inhibition of urease in the presence of heavy metals. These versatile biosensors offer great possibilities for further development of highly sensitive enzyme-based FETs for real-time, label-free detection of a wide variety of clinically relevant analytes.

ABBREVIATIONS

Γ	mass surface coverage
$\Delta I_{ds}/\Delta V_g$	transconductance
θ_{min}	SPR angle of minimum reflectivity
μ	charge carrier mobility
APTES	(3-aminopropyl)triethoxysilane
C_{EDL}	electrical double-layer capacitance
C_g	top-gate capacitance
C_Q	quantum capacitance
$d\eta/dC$	refractive index increment
FET	field-effect transistor
GO	graphene oxide
HEPES	4-(2-hydroxyethyl)-1-piperazineethanesulfonic acid
I_{sd}	source–drain voltage
L	channel length
LbL	layer-by-layer
PEI	polyethylenimine
pI	isoelectric point
rGO	reduced graphene oxide
SPS	sodium pyrenesulfonate
V_g	gate voltage
W	channel width

1. INTRODUCTION

Due to numerous advantages over traditional analytical methods which include ease of operation, low cost, and compact size, electrochemical sensors have become a prominent topic of interest in industry and academia. In particular, biosensors which are highly selective represent a powerful tool for real-time measurement of a variety of analytes of considerable interest in different areas, such as food safety, environmental monitoring, drug screening, and diagnosis (Mello & Kubota, 2002). Biosensors represent a subgroup of chemical sensors comprising biological host molecules as recognition elements coupled to a chemical or physical transducer

(Hulanicki, Glab, & Ingman, 1991). Nature offers a variety of recognition strategies using, for example, interactions between antibodies and antigens, cell response, protein interaction, DNA targeting, or enzymatic substrate conversion (Wang, Li, Wang, Li, & Lin, 2011).

In this context, enzymes constitute excellent recognition elements due to their specificity and selectivity toward the analyte (Ariga et al., 2013). Owing to their availability and biocompatibility, the integration of enzymes on sensor surfaces offers a plausible approach for the detection of specific substrates, whereas the variety of enzymes, or even enzymatic cascade reactions, makes the approach applicable to a wide range of analytes. Depending on the enzymatic reaction product, different transducing strategies can be chosen, ranging from direct electronic communication (Shao et al., 2010; Zhu et al., 2012), to electrodes selective to the particular reaction product (e.g., NH_3) (Disawal, Qiu, Elmore, & Lvov, 2003; Koncki, Wałcerz, Ruckruh, & Głab, 1996; Stasyuk, Smutok, Gayda, & Gonchar, 2011; Stasyuk et al., 2012; Valle-Vega, Young, & Swaisgood, 1980), or the detection of pH changes using an interface sensitive to protons (Karacaoğlu, Timur, & Telefoncu, 2003; Komaba, Fujino, Matsuda, Osaka, & Satoh, 1998; Piccinini et al., 2017).

With regard to the sensor platform, field-effect transistors (FETs) based on nanomaterials, such as semiconducting nanowires (Cui, Wei, Park, & Lieber, 2001), single-walled carbon nanotubes, and graphene (Fu, Jiang, van Geest, Lima, & Schneider, 2017) hold a leading position in signal transduction, offering outstanding capabilities for label-free, highly sensitive, real-time detection (Zhang & Lieber, 2016). Owing to the possibility of being operated in aqueous solutions applying small voltages (which is crucial for biological applications), relative good biocompatibility (Zhu et al., 2012), and inherent amplification properties (Zhang et al., 2015), graphene-based liquid-gated FETs have gained vast interest in clinical diagnosis (Yan, Zhang, & Li, 2014). Graphene, a 2D zero bandgap semiconducting material, exhibits extraordinary electronic, chemical, and mechanical properties (Balasubramanian & Kern, 2014; Choi & Lee, 2012; Nehra & Pal Singh, 2015). Its ambipolar character, together with a high carrier mobility and a great sensitivity toward changes in environmental conditions, makes graphene perfectly suitable as transducing material for the use in biochemical sensors (Pumera, 2011; Pumera, Ambrosi, Bonanni, Chng, & Poh, 2010).

The most often applied immobilization techniques for integrating enzymes on solid-sensing platforms include entrapment, encapsulation, cross-linking, covalent binding, and adsorption (Amine, Mohammadi, Bourais, & Palleschi, 2006; Mello & Kubota, 2002). As the biosensor performance strongly depends on the characteristics of the interfacial architecture, the functionalization method has to be chosen with great care in order to retain the enzymatic activity as well as the functionality of the transducing element. Covalent attachment of biomolecules may not only induce damage to the sp^2 structure of graphene (Niyogi et al., 2010), but it can also disrupt the folding and thus the activity of the enzyme if essential groups are involved in the immobilization (Scouten, Luong, & Stephen Brown, 1995; Sheldon & Van Pelt, 2013).

In this regard, layer-by-layer (LbL) assembly offers a versatile and simple technique for the noncovalent functionalization of graphene with enzyme layers preserving the bioactive functionality and guaranteeing substrate accessibility. As numerous different types of building blocks can be employed in this process, the electrostatic assembly of oppositely charged layers opens the door to the fabrication of a great variety of functional thin films with high organization at the nanoscale level (Decher & Schlenoff, 2012; Iost & Crespilho, 2012). Using this technique, the layer thickness varies depending on the nature of the adsorbed components and the deposition conditions. In typical polyelectrolyte–enzyme LbL assemblies, the average layer thickness is from a few nanometers up to tens of nanometers (Forzani, Otero, Pérez, Teijelo, & Calvo, 2002; Lvov & Caruso, 2001). It should be noted that in the case of enzyme-based FET biosensors, it is important to locate the enzyme layer close enough to the transducer surface (in our case, graphene). The distance between the enzyme layer and the graphene surface should be closer than hundreds of micrometers to achieve optimized sensing performances (Temple-Boyer et al., 2008). In this regard, the LbL technique fulfills this requirement and allows reaching distances below 10 nm.

In this chapter, a novel approach for designing and constructing biosensors is presented. The approach is based on the intrinsic pH sensitivity of reduced graphene oxide (rGO) FETs as a strategy to detect the enzymatic conversion of the analyte. We describe the fabrication process, the rGO FET characterization, the LbL assembly of urease onto the sensor platform, and the sensing of urea by detecting pH changes upon enzymatic substrate conversion. Finally, taking advantage of the inhibition of urease activity by heavy metal ions, the detection of Cu^{2+} in the n*M* range is described. Fig. 1 gives an overview of the biosensor fabrication process, which spans from the assembly of rGO flakes on interdigitated microelectrodes to the LbL functionalization with enzyme layers, thus resulting in a highly sensitive FET biosensor.

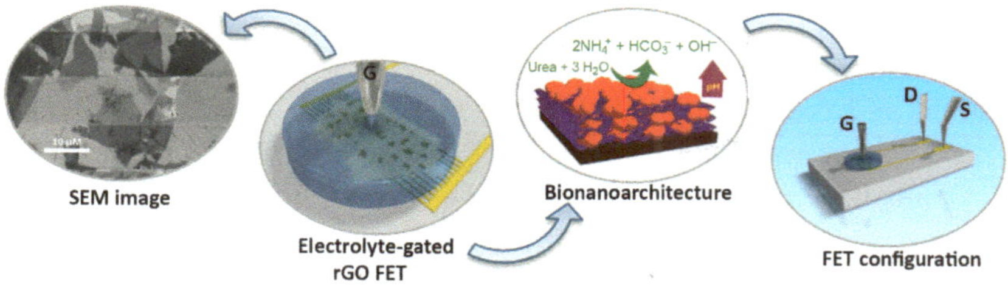

Fig. 1 Overview of the development of an rGO-based enzymatic FET, starting with the assembly of rGO on an interdigitated set of microelectrodes, followed by the functionalization with enzymatic recognition elements using the LbL method.

2. rGO-BASED FETs

In general, a FET consists of a source- and drain electrode connected via a semiconducting channel and a gate electrode. The potential applied between gate and source electrode (V_g) controls the flow of charge carriers between source and drain, which thus determines the flow of current through the channel. The applied gate voltage (V_g) creates an electric field perpendicular to the channel attracting either holes or electrons toward the gate, thus creating a conductive channel between source and drain (Bao & Locklin, 2007; Lee, Kyu Kim, & Kim, 2009; Torsi, Magliulo, Manoli, & Palazzo, 2013).

Due to its ambipolar character, its high conductivity, and the possibility of functionalization, rGO represents an ideal channel material for the fabrication of FETs. Graphene forms a planar sp^2-hybridized honeycomb structure with the delocalized electrons being responsible for the outstanding electronic properties of the material (Wong & Akinwande, 2011). Chemically rGO differs substantially from pristine graphene, owing to its defective structure and remaining functional groups. In spite of the great homogeneity of graphene grown by chemical vapor deposition, chemically derived rGO can be highly advantageous as a sensing material. The defective structure and the remaining functional groups in rGO play a key role not only in surface functionalization (Pumera, 2011) but also in pH-sensing applications (Reiner-Rozman, Larisika, Nowak, & Knoll, 2015; Sohn et al., 2013). The high sensitivity of rGO-based FETs toward small pH changes near the channel surface is caused by the interaction of carboxyl or hydroxyl functional groups at the rGO surface with H^+ ions in the electrolyte (Sohn et al., 2013) giving rise to a change in the surface charge density. The alteration of the surface charge density and the electric double layer leads to electrostatic gating effects (Heller et al., 2010). This intrinsic property of rGO can easily be exploited for biosensing purposes.

Buffers and reagents

- (3-aminopropyl)triethoxysilane (APTES) solution: 2% APTES in EtOH.
- Graphene oxide (GO) solution: A concentrated GO stock solution prepared by a modified Hummer's method was diluted with deionized water to obtain a 18-μg/mL GO solution, according to protocol described by Larisika, Huang, Tok, Knoll, and Nowak (2012).
- Hydrazine monohydrate. Caution: Hydrazine monohydrate is extremely hazardous in case of skin contact (corrosive, irritant), eye contact (irritant), and inhalation. Therefore, it should be handled with extreme care inside the fume hood using gloves, eyes protection, vapor respirator, and complete protection suite.
- Working buffer: 10 m*M* KCl, 0.1 m*M* HEPES in dH_2O (adjusted to the desired pH using KOH or HCl solutions).

Equipment and procedure

- Interdigitated microelectrodes (ED-IDE1-Au Micrux Technologies): The substrate dimensions are 10 × 6 mm
- Semiconductor characterization system (Keithley 4200)
- Electrochemical flow cell (Micrux Technologies)
- Peristaltic pump (Reglo ICC, Ismatec)

1. Incubate the glass substrate for 1 h in the APTES solution.
2. Rinse with EtOH and anneal the APTES layer 1.5 h at 120°C.
3. Drop-cast 30 μL of GO solution onto the interdigitated microelectrodes and incubate for 1 h.
4. Remove surplus GO flakes by rinsing carefully with water.
5. Reduce the attached GO flakes by exposing the electrodes to hydrazine vapor at 80°C overnight (the 80%–90% of the surface should be covered with rGO).
6. Insert the microelectrodes into the flow cell and adjust the buffer flow to 300 μL/min.
7. Record transfer characteristics: Apply a V_{ds} of 0.1 V and sweep the V_g from −0.6 V to 0.6 V at different pH.
8. Record channel current: Apply a V_{ds} of 0.1 V, set the V_g to −0.2 V, and titrate the different pH solutions.

2.1 Fabrication

Wet-chemically synthesized GO can be easily assembled onto a glass substrate using silane chemistry and subsequent reduction by hydrazine with the only disadvantage of the flakes being rather randomly arranged. To overcome this drawback, a novel strategy of FET fabrication was applied by using an interdigitated array of gold microelectrodes on a glass substrate. Despite the random organization of the rGO flakes the bridging between source and drain electrode is assured by the choice of the very narrow channel (10 μm) as well as the large channel area ($0.522\,mm^2$), resulting in a very low ohmic resistance of ~100 Ω between source and drain electrode. The special FET architecture leads to an extremely high transconductance ($\Delta I_{ds}/\Delta V_g$) of up to 800 μS and shows a stable response and a very good device reproducibility, which is crucial for routine applications.

The rGO surface can be subsequently functionalized via π stacking of aromatic molecules onto the graphene plane, which has little impact on the electrical properties of the material (Atta, Galal, & El-Ads, 2015; Loh, Bao, Ang, & Yang, 2010; Singh et al., 2011), see Section 3.

Fig. 2A shows the liquid-gated FET, an enlarged view of the interdigitated microelectrodes as well as a SEM image of the assembled rGO flakes. An ad hoc flow cell (Fig. 2B) together with a peristaltic pump is used to allow precise positioning of the chip with respect to the gate electrode and assure a constant flow rate.

2.2 Characterization

As this type of enzyme sensor relies on the detection of small pH changes during the enzymatic reaction, the transfer characteristics of the transistors in a liquid-gated configuration under different pH values need to be studied.

Linearly sweeping the potential between gate and source (V_g) shifts the Fermi energy from its equilibrium increasing either hole or electron mobility in the channel material, thus changing the drain–source current (I_{ds})

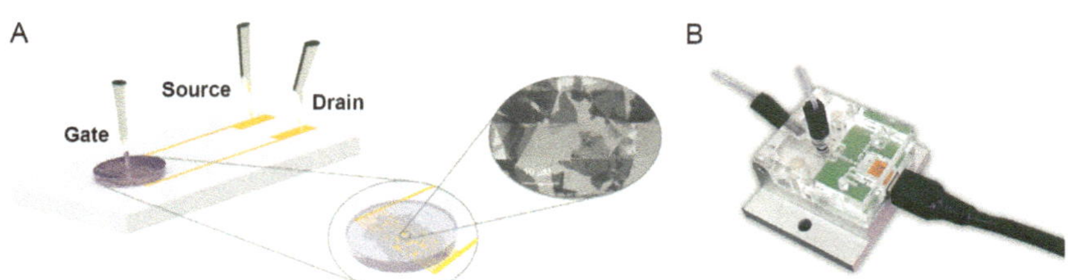

Fig. 2 (A) Illustration of the liquid-gated FET with source, drain, and gate electrodes, an enlarged view of the interdigitated channel and a SEM image of the rGO-modified channel; scale bar = 10 μm. (B) Image of the flow cell (Micrux Technologies).

accordingly. The values of the charge carrier mobility (μ) were calculated as follows (Wang & Burke, 2013):

$$\mu = \left(\frac{L}{WC_g V_{ds}}\right)\left(\frac{\Delta I_{ds}}{\Delta V_g}\right) \tag{1}$$

where L and W are the channel length (10 μm) and the channel width (55.2 mm), respectively. C_g stands for the top-gate capacitance, and $\Delta I_{ds}/\Delta V_g$ represents the rate of change of I_{sd} with respect to V_g, called transconductance. The gate capacitance can be thought as the equivalent capacitance of the quantum capacitance (C_Q) in series with the electrical double-layer capacitance (C_{EDL}) (Ohno, Maehashi, Yamashiro, & Matsumoto, 2009):

$$\frac{1}{C_g} = \frac{1}{C_Q} + \frac{1}{C_{EDL}}$$

Thus, the value of C_g is dominated by the smallest capacitor. For electrolyte solutions, it is generally considered an electrical double-layer thickness given by the Debye–Huckel equation and $C_{EDL} = \epsilon\epsilon_0\kappa$, where κ^{-1} is the Debye screening length. That gives a C_{EDL} value of 23 μF/cm^2 for a 10 mM KCl solution. By using C_Q values previously reported (ranging from 20 up to 1000 nF/cm^2; Du, Guo, Jin, Jin, & Zhao, 2015; Fang, Konar, Xing, & Jena, 2007; Ohno et al., 2009; Wang & Burke, 2013; Xia, Chen, Li, & Tao, 2009), we estimated C_g and then the charge carrier mobility. The estimated value of the μ for holes and electrons is 1.5–72 cm^2 V^{-1} s^{-1}.

The characteristic ambipolar behavior of rGO FETs at pH ranging from 5 to 10 is shown in Fig. 3A. The slope of the two branches remained nearly constant, which indicates a pH independency of the charge carrier mobility. However, a significant shift of the Dirac point (V_i) can be observed, which exhibits a linear pH dependency with a slope of 23 ± 1.8 mV/pH (Fig. 3B). The high pH sensitivity relies on electrostatic gating effects due to the protonation or deprotonation of remaining functional groups at the rGO surface which cause changes in the surface charge density of the material and charge doping effects of adsorbed H^+ or OH^- ions (Heller et al., 2010; Sohn et al., 2013; Wang & Burke, 2014). To assess the reproducibility between rGO FETs, the pH sensitivity was evaluated in five rGO FETs devices. The average pH sensitivity was of 25 ± 4 mV/pH.

Further, the effect of a polyethylenimine (PEI) layer on the pH sensitivity was studied. In order to confer negative charge to the rGO surface, a layer of sodium pyrenesulfonate (SPS) was assembled, followed by the deposition

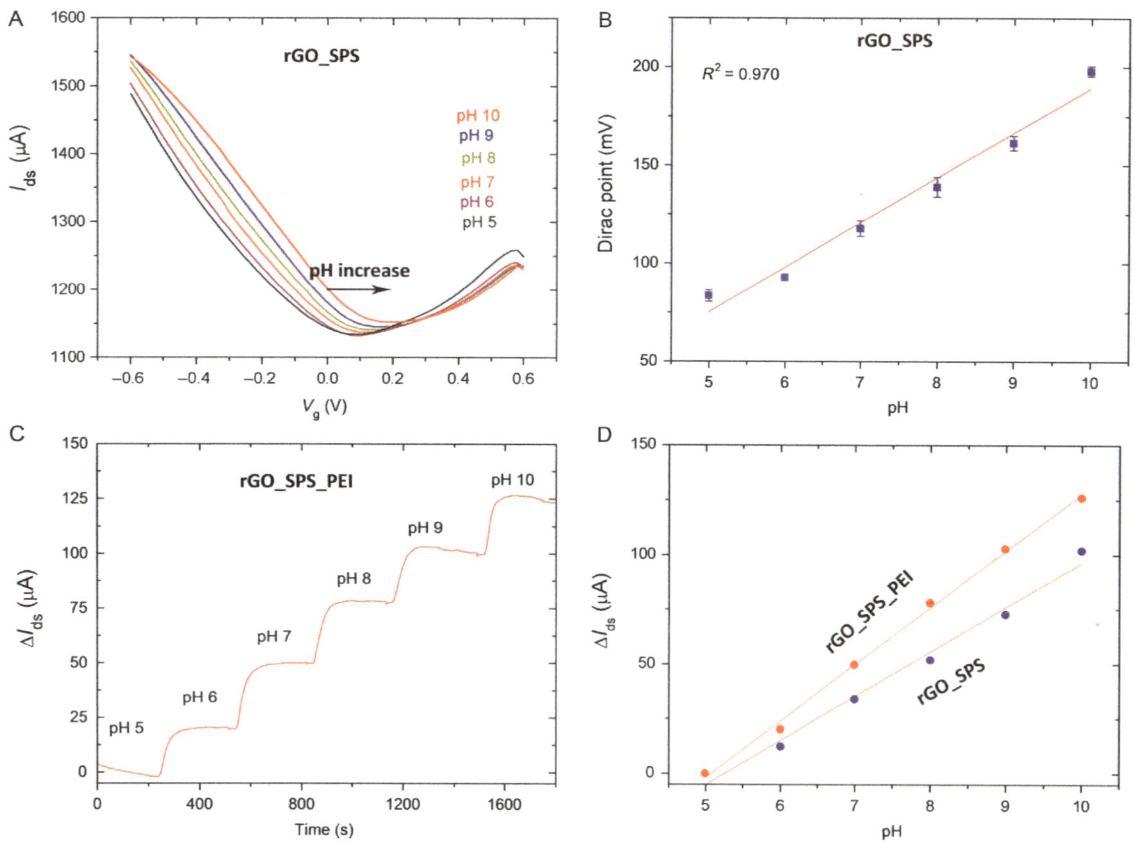

Fig. 3 (A) Transfer characteristics of SPS-modified rGO FETs recorded at pH 5–10. (B) pH dependence of the Dirac point taken from the recorded transfer characteristics. (C) Real-time current response (I_{sd}) of a SPS-PEI-modified rGO FET titrating from pH 5–10. (D) Comparison of the pH dependence of rGO FETs modified with SPS (*blue*) and SPS-PEI (*red*), taken from the real-time current response (I_{sd}). Experimental conditions: in buffer of 10 m*M* KCl and 0.1 m*M* HEPES at a fixed V_g of −0.2 V and V_{ds} of 0.1 V.

of the positively charged polyelectrolyte PEI (for the preparation see Section 3). The application of a constant V_g gives rise to a change in the I_{ds} upon changing the pH of the solution (Fig. 3C). Fig. 3D shows the pH sensitivity of the rGO FET before (blue dots) and after (red dots) the assembly of PEI with a slope of $20.3 \pm 0.6\ \mu A/pH$ and $25.9 \pm 0.6\ \mu A/pH$, respectively, revealing a sensitivity enhancement of 28% by modification with the polyelectrolyte. Since PEI is a weak polycation with a pK_a of 8–9, the degree of protonation of the polymer decreases with increasing pH resulting in a p-doping effect of the transducer by electrostatic gating effects. Therefore, the weak polycation plays a double role: (a) a building block for the LbL assembly and (b) a transducing element amplifying the pH sensitivity of the FET.

3. ENZYME IMMOBILIZATION

As mentioned earlier, FETs can be used as platforms for biosensing by modifying the channel surface with biological recognition elements. The modification technique needs to preserve the functionality of the biological entity as well as that of the transducing element. Further, accessibility of the analyte to the active site of the recognition element has to be guaranteed. The LbL technique poses an ideal means to accomplish the prerequisites. The method relies on the electrostatic interaction of oppositely charged species, thus avoiding covalent attachment, which may not only disrupt the folding and functionality of the enzyme if essential groups are involved in the immobilization (Scouten et al., 1995; Sheldon & Van Pelt, 2013), direct covalent attachment to graphene surfaces can also induce damage to the sp^2 structure and thus impair signal transduction (Niyogi et al., 2010). Another advantage of the LbL technique is the use of mild conditions, like aqueous solutions, which is inevitable to retain protein folding (Rydzek et al., 2015). In spite of the drying steps within the assembling process, entrapped water is still present within the layer, which is of great importance to retain the biomolecule activity. The water content in the film after drying depends on the film components and the relative humidity (RH). For a multilayer film constituted of poly(styrenesulfonate) and poly(diallyldimethylammonium chloride), it was reported a value around 15% water content for 70% RH (De, Cramer, & Schönhoff, 2011). By this means, layered systems with tunable characteristics can be constructed protecting the enzymes within a defined volume (Lisdat, Dronov, Möhwald, Scheller, & Kurth, 2009; Sakr & Borchard, 2013).

An ideal approach for conferring negative charges to the rGO substrate relies on the self-assembly of aromatic molecules via π stacking, which has little impact on the electronic properties of the material (Atta et al., 2015; Loh et al., 2010; Singh et al., 2011).

Buffers and reagents

- Basic piranha solution: H_2O_2 30%, NH_4OH 35% (1:1) in dH_2O
- Cysteamine solution: 5 m*M* in EtOH
- SPS solution: 5 m*M* SPS in DMF
- Enzyme buffer: 10 m*M* HEPES, 10 m*M* KCl in dH_2O adjusted to pH 7.4 (above the isoelectric point (pI) of the enzyme)
- PEI solution: 2 mg/mL PEI in dH_2O (adjusted to pH 8)

Equipment and procedure

- SPR gold substrates (SPR102 AU, BioNavis).
- rGO-functionalized microelectrodes.
- Multiparametric surface plasmon resonance (MP-SPR) instrument (SPR Navi 210A BioNavis).

1. Clean the SPR gold substrate using basic piranha solution (H_2O_2 30% and NH_4OH 35% 1:1) at 60°C for 10 min.
2. Modify the substrate with a self-assembled monolayer, by immersing it in the cysteamine solution for 8 h.
3. To functionalize the SPR substrate with rGO, proceed with steps 3–5 of Section 2, as reported before (Piccinini et al., 2018).
4. Confer negative charge to the rGO surface (of the SPR substrate or the rGO FET, respectively) by submerging the substrate into the SPS solution overnight.
5. Rinse carefully with DMF and water.
6. Assemble the polyelectrolyte layer by submerging the substrate into the PEI solution for 10 min and wash with water.
7. Transfer the substrate to the enzyme solution containing 1 mg/mL of the desired enzyme, incubate for 30 min, and wash with water.
8. Repeat the alternating steps polycation and enzyme adsorption as desired.
9. In order to characterize the assembly, insert SPS-modified SPR gold substrates into the flow cell of the MP-SPR instrument.
10. Follow the adsorption of the polycation and enzyme layers using a flow rate of 10 μL/min with the MP-SPR and a 785-nm laser, wash with water after each assembly step.

3.1 LbL Assembly

LbL assembly represents a versatile bottom-up strategy to fabricate multilayered films on the nanoscale level. Generally, the technique consists of exploiting the electrostatic interaction of oppositely charged species to form ultrathin multilayer films. This technique facilitates the integration of different constituents into the assembly and allows great freedom in the choice of number and sequence of the layers (Ariga, Hill, & Ji, 2007; Tang, Wang, Podsiadlo, & Kotov, 2006). In this work, we used the weak polycation PEI as a positively charged building block to assemble layers of negatively charged enzymes. In principle, the driving force governing the assembly is the increase in entropy release of counter ions and water of hydration from the dissolved polyelectrolyte chains (Bucur, Sui, & Schlenoff, 2006).

In order to confer negative charge to the rGO surface, the FET channel was modified using SPS, which consists of a pyrene anchor and a negatively charged sulfonate group. Substrate immersion into a solution containing a positively charged polyelectrolyte leads to charge reversal due to overcompensation by the charge monomers of the polyelectrolyte (Rydzek et al., 2015). Subsequently, the substrate can be functionalized with a layer of the negatively charged enzyme (at pH > pI). Alternately dipping the substrate into solutions of the weakly charged polycation and the negatively charged enzymes leads to multilayer films with desired structure and thickness (Fig. 4). Loosely attached material is removed by rinsing with water between the consecutive adsorption (Tang et al., 2006). A number of parameters, such as type and molecular weight of the interacting polymers, as well as the pH used for the assembly process define the stability of the multilayers. Therefore, it is crucial to control the pH as well as the ionic strength of the solutions in order to achieve successful alternate adsorption (Decher & Schlenoff, 2012).

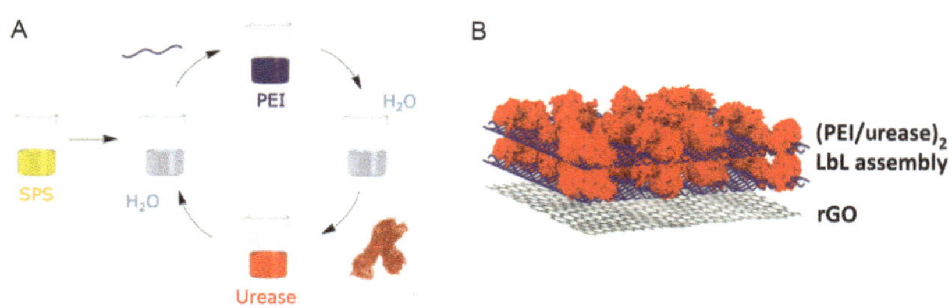

Fig. 4 (A) Schematic of the LbL assembling process of alternating layers of PEI and urease. (B) Representation of the LbL assembly of PEI (*blue*) and urease (*red*) on graphene.

3.2 SPR Characterization

In the assembling process, the pH of the polyelectrolyte and enzyme solution is critical. Enzymes are negatively charged at a pH above their characteristic pI, which is required for their electrostatic adsorption onto the positively charged PEI. Taking into account that the pI of urease is in the range of 5.0–5.2 (Sumner & Hand, 1929), the adsorption process of PEI and urease onto rGO-modified substrates was followed by SPR at enzyme buffer pH 7.4. Fig. 5A shows the reflectivity curves after the adsorption and rinsing for the first three bilayers.

To estimate the surface coverage of the enzyme, the SPR angle of minimum reflectivity (θ_{min}) was measured in situ during the assembly. The shift in the minimum reflectivity angle ($\Delta\theta$) resulting from adsorption of the enzyme was converted into the mass surface coverage (Γ, ng/cm^2) using the following equation (Stenberg, Persson, Roos, & Urbaniczky, 1991):

$$\Gamma = \frac{\Delta\theta k d}{{}^{d\eta}/_{dC}} \quad (2)$$

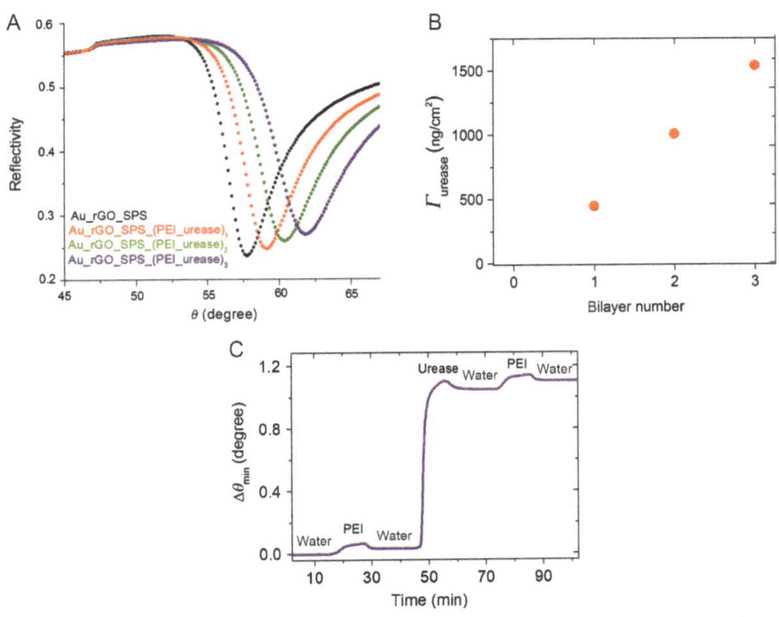

Fig. 5 (A) SPR reflectivity curves of (PEI/urease)$_n$ LbL assemblies prepared on a rGO-modified gold substrate. (B) Enzyme surface coverage as a function of the number of bilayers (n). (C) Time-resolved SPR sensogram of the in situ PEI/urease assembly process.

The parameter k^*d, which depends on the SPR substrate and the laser wavelength, was provided by BioNavis with a value of 1.9×10^{-7} cm/degrees. The refractive index increment ($d\eta/dC$) was considered 1.82×10^{-10} cm^3/ng for urease (Zhao, Brown, & Schuck, 2011).

SPR measurements confirmed the successful assembly of the enzyme and a linear growth behavior resulting in a highly stable supramolecular interfacial architecture. The estimated values of the enzyme surface coverage (Γ) are presented in the Fig. 5B. We obtained an enzyme surface coverage value (per layer) of 515 ng/cm^2. Our result is in excellent agreement with a previously reported value (490 ng/cm^2) for urease/polycation LbL assemblies prepared on polystyrene microparticles (Lvov & Caruso, 2001). To obtain the enzyme surface coverage in mol/cm^2 units, the molecular weight of a urease subunit was considered (91 kDa). It can be seen in Fig. 5C that during the first seconds of the washing step there is a slight desorption of weakly adsorbed enzymes. After that, the interfacial architecture seems to be very stable.

4. ENZYMATIC BIOSENSORS

Owing to their excellent substrate specificity and selectivity, enzymes were chosen as recognition elements onto the FET platform. Especially, enzymes which trigger pH changes at the sensor surface during the enzymatic reaction are of great interest, as no direct electron communication between the active center of the enzyme and the transducing element of the sensor is required. The enzymatic conversion of the analyte can be simply measured by following the pH change at the sensor interface (Sohn et al., 2013; Soldatkin, Montoriol, Sant, Martelet, & Jaffrezic-Renault, 2002).

Urease was chosen as a model system to demonstrate the sensing capabilities of this novel approach based on the integration of enzymes on rGO-based FETs. Urease from Jack bean is a robust enzyme with the function of catalyzing the hydrolysis of urea to generate ammonium and bicarbonate (Krajewska, 2009a). The acid–base equilibrium of the formed products results in a pH increase of the solution. This enzyme presents Ni(II) metal centers in their active sites, whose task is the activation of the substrate and water for the reaction. The enzyme typically forms trimers and hexamers with subunits of 90 kDa. Some heavy metals inhibit the urease activity, being Hg^{2+}, Ag^{+}, and Cu^{2+} the strongest metal inhibitors. This inhibition is ascribed to the reaction of the metal ions with the thiol groups of the enzyme. As we will show in this work, the inhibition can be exploited for the construction of urease inhibition-based sensing systems (Krajewska, 2009b). The detection of urea, the substrate of urease, is of great importance in diagnosis and control of a number of kidney and liver diseases (Carvounis, Nisar, & Guro-Razuman, 2002; Lakard et al., 2011). In particular, in the case of hemodialysis patients, online monitoring of urea may improve the diagnosis of kidney failure as well as prolong the patient's life expectancy (Sant, Temple-Boyer, Chanie, Launay, & Martinez, 2011). The monitoring of urea is mostly performed by means of classical analytical methods, such as fluorescence, colorimetric, and potentiometric methods (Rajesh, Bisht, Takashima, & Kaneto, 2005; Singh, Verma, Garg, & Redhu, 2008). Such measurements imply the use of expensive equipment or laborious procedures and are not suitable for online monitoring. The developed enzyme-based rGO FETs, in contrast, offer many advantages such as easy handling, real-time response, cost-efficiency, operation in aqueous solutions, and electronic readout applying very low voltages (which is elemental for biological sensing) (Zhang et al., 2015).

Buffers and solutions

- Experimental buffer: 10 mM KCl in dH_2O (adjusted to pH 6)

Equipment and procedure

- Semiconductor characterization system (Keithley 4200)
- Electrochemical flow cell (Micrux Technologies)
- Peristaltic pump (Reglo ICC, Ismatec)

1. Place the enzyme-modified rGO FET in the flow cell and adjust the flow rate of the peristaltic pump to 100 μL/min
2. Apply a drain–source bias (V_{ds}) of 0.1 V, a fixed gate voltage (V_g) of −0.2 V and monitor the drain–source current (I_{ds}) using the semiconductor characterization system
3. Titrate the appropriate concentrations of the analyte waiting for an equilibrium to establish at each concentration step
4. In order to apply the senor for heavy metal detection, titrate solutions containing a fixed concentration of urea (0.2 mM) and increasing amounts of Cu^{2+}

4.1 Single-Enzyme Sensor

For the enzyme sensor, urease was assembled onto the rGO FET platform. Fig. 6 shows the hydrolysis of urea catalyzed by the enzyme. The pH change caused by the enzymatic conversion of the substrate can be measured by

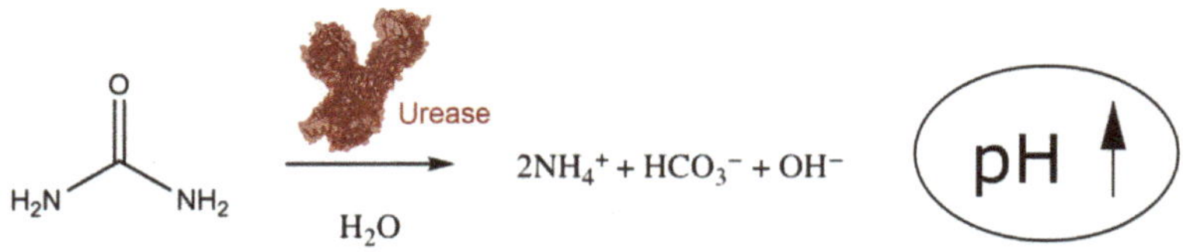

Fig. 6 Enzymatic hydrolysis of urea catalyzed by urease.

exploiting the pH sensitivity of the rGO FET. The response signal can then be correlated to the analyte concentration in solution.

First, the response of the transistor to urea was tested for a (PEI/urease)$_1$ assembly sweeping the V_g from -0.6 to 0.6 V at a constant V_{ds} of 0.1 V. Fig. 7A (top) shows the transfer characteristics in the absence (blue line) and in the presence (red line) of 100 μ*M* urea. A clear shift in the V_i to more positive values was observed in the presence of urea, similar to the shift obtained upon increasing the pH of the solution. This result indicates that the enzyme properly catalyzed the hydrolysis of urea. The hydroxyl ions produced by the enzyme reaction are involved in an acid–base equilibrium, which causes a pH change near the sensor surface as well as an alteration of the protonation of the PEI, resulting in a significant shift in the V_i to a more positive gate voltage.

Further, the real-time response of the sensor toward urea was studied. The sensor modified with a (PEI/urease)$_1$ assembly was exposed to an analyte solution of 100 μ*M* at constant V_{ds} of 0.1 V and a fixed V_g of -0.2 V monitoring the change in I_{ds} (Fig. 7A bottom). After a plateau was reached, referred to as flow response, the flow was stopped resulting in another I_{ds} peak (static response). This sensor response may be explained by the following mass transport phenomena: during flow conditions the flux of the analyte solution rapidly washes away the hydroxyl ions obtained by the hydrolysis of urea by forced convection. The sensor response (I_{ds}) reaches a plateau, when the flow of hydroxyl ions into and out of the film reaches an equilibrium. At static conditions, the outflow of the hydroxyl ions from the film is only governed by diffusion, causing a local increase of hydroxyl ions close to the rGO surface until the urea concentration gradient reaches a maximum value, thus resulting in an I_{ds} peak. When the flow was resumed, the sensor response returned to the flow equilibrium level. These observations are in agreement with the predictions obtained by theoretical studies (Temple-Boyer et al., 2008) for urease–ISFET devices.

Fig. 7B shows the titration of urea increasing the analyte concentration from 1 up to 2500 μ*M* using a (PEI/urease)$_1$ assembly. The I_{ds} increases up to

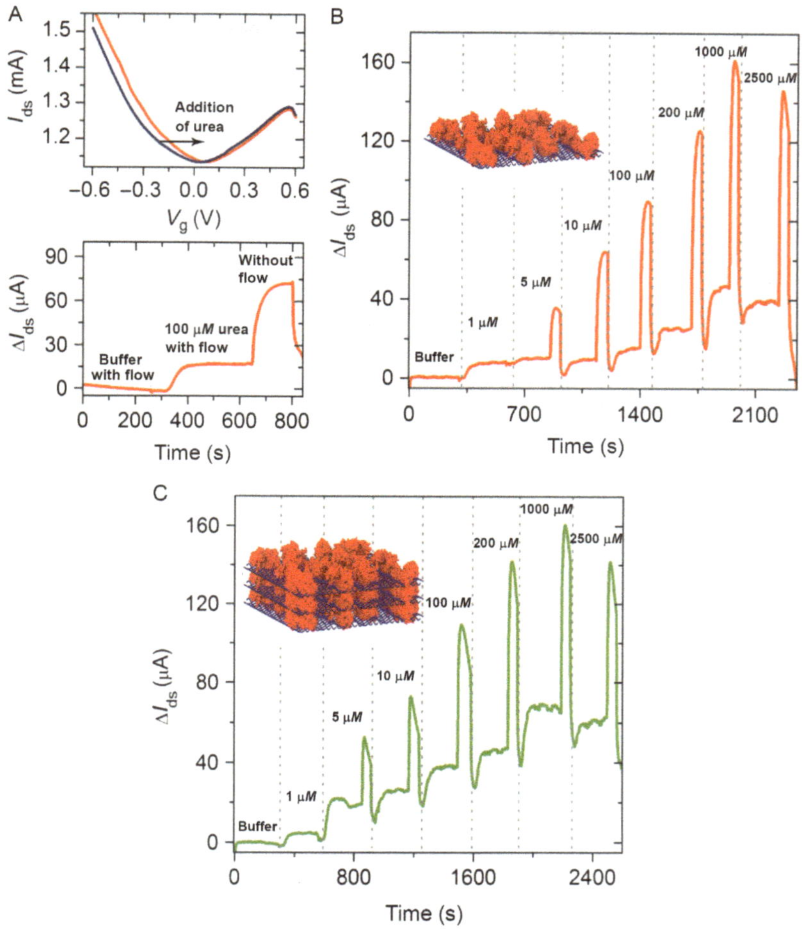

Fig. 7 (A) Transfer characteristics of a (PEI/urease)$_1$-modified rGO FET in the absence (*blue*) and the presence (*red*) of 100 μ*M* urea (*top*). Real-time current response (I_{ds}) of a (PEI/urease)$_1$-modified rGO FET (*bottom*). Current response (I_{ds}) rGO FETs modified with (B) (PEI/urease)$_1$ and (C) (PEI/urease)$_3$. Experimental conditions: 10 m*M* KCl + 0.1 m*M* HEPES buffer at a fixed V_g of −0.2 V and V_{ds} of 0.1 V.

the level of 1 m*M* urea in both, flow and static condition. At higher analyte concentrations, substrate inhibition of the enzyme may occur and the local alkaline pH may impair the functionality of the enzyme (Krajewska, 2009a, 2009b).

In another experiment, the effect of increasing numbers of polyelectrolyte–enzyme bilayers on the sensor response was studied. Therefore, assemblies with two and three bilayers, (PEI/urease)$_2$ and (PEI/urease)$_3$, respectively, were constructed and subjected to the same measurement

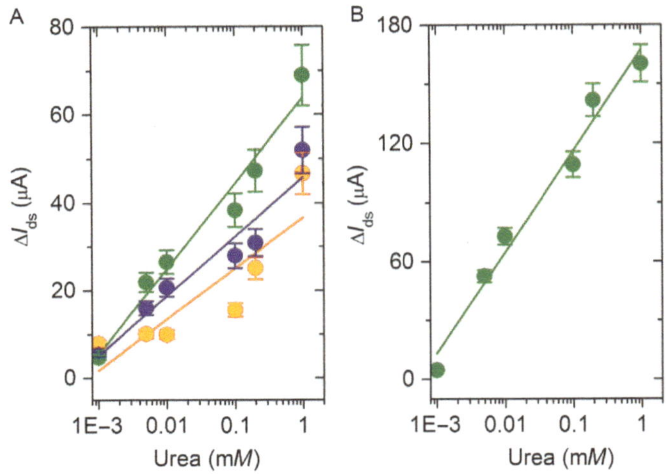

Fig. 8 (A) I_{ds} flow response as a function of the urea concentration of rGO FETs modified with (PEI/urease)$_1$ (*red*), (PEI/urease)$_2$ (*blue*), and (PEI/urease)$_3$ (*green*). (B) I_{ds} static response as a function of the urea concentration of a (PEI/urease)$_3$ modified rGO FET. Experimental conditions: 10 m*M* KCl + 0.1 m*M* HEPES buffer at a fixed V_g of −0.2 V and V_{ds} of 0.1 V.

(Fig. 7C). The ΔI_{ds} signal as a function of the logarithmic urea concentration was determined from the flow response (Fig. 8A) as well as the static response (Fig. 8B). Fig. 8A shows a linear pH dependence of sensors modified with one (red), two (blue), and three (green) bilayers. The sensitivity increased by 20% for two bilayers and by 68% for three bilayers compared to one PEI/urease bilayer. However, a significant signal enhancement was observed by analyzing the static response of the (PEI/urease)$_3$ assembly (Fig. 8B). This result shows the possibility of increasing the catalytic activity of the film by increasing the number of bilayers, thus augmenting the total enzyme loading in the assembly. Hence, higher changes in the local pH can be obtained upon hydrolysis of the analyte. At a urea concentration of 1 m*M*, the transistor modified with a (PEI/urease)$_3$ assembly showed a ΔI_{ds} of almost 70 μA under flow conditions. Taking into account the sensors pH sensitivity, this result is comparable with a pH change from 6 to 8.7, which is in good agreement with other studies concerning urease entrapped in polymeric matrices (Kazakova, Shabarchina, & Sukhorukov, 2011; Tsai & Doong, 2005).

The nanoconstruction of enzyme-containing multilayered films turns out to be especially attractive owing to the precise control over the loading of the enzyme. The resulting biosensors exhibit a LOD below 1 μ*M*, a fast response, and a good long-term stability, retaining almost 96% of the original response after 6 days of storage in buffer HEPES solution at 4°C, making them particularly interesting for online monitoring of urea (Sant et al., 2011).

4.2 Enzymatic Heavy Metal Sensor

Finally, taking advantage of the inhibition of urease activity by heavy metal ions, the transistor modified with a (PEI/urease)$_3$ assembly was applied for the quantification of Cu^{2+} ions. Krajewska et al. reported the strong binding of Cu^{2+} to cysteine and histidine residues crucial for urease activity, thus leading to the inhibition of the enzyme (Krajewska, 2008). Fig. 9 shows the sensor response I_{ds} as a function of the Cu^{2+} concentration in the presence of 0.2 mM urea. The increasing amount of Cu^{2+} in solution led to a clear reduction of the enzymatic conversion of urea by heavy metal inhibition of the enzyme, thus reducing the current output. A linear relationship between ΔI_{ds} and the logarithmic Cu^{2+} concentration with a LOD below 10 nM can be observed, whereas no enzymatic activity was found above 1000 nM Cu^{2+}. The transistors showed a faster response, as well as a lower LOD (more than one order of magnitude), as compared with previously reported heavy metal sensors based on urease (Tsai & Doong, 2005).

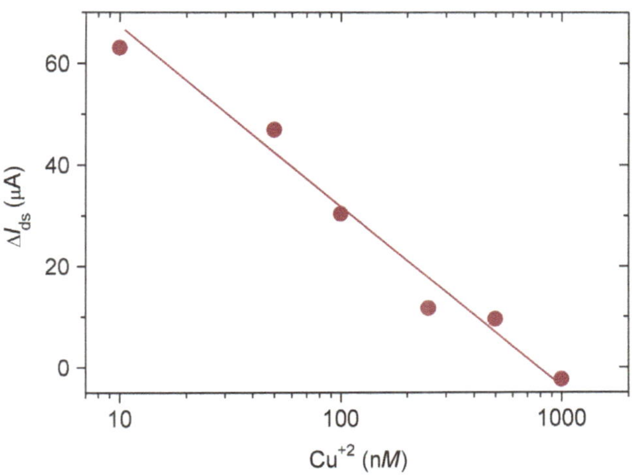

Fig. 9 I_{ds} of a (PEI/urease)$_3$ as a function of the Cu^{2+} concentration in the presence of 0.2 mM urea. Experimental conditions: in buffer of 10 mM KCl and 0.1 mM HEPES at a fixed V_g of −0.2 V and V_{ds} of 0.1 V.

5. SUMMARY AND CONCLUSIONS

This chapter presents a novel approach to construct highly sensitive biosensors based on the rational integration of enzymes—as recognition elements—on rGO-based FETs via LbL assembly. This assembly technique has neither affected the enzyme activity nor altered the electrical properties of rGO. The outstanding sensitivity of rGO-based FETs to small environmental changes at the sensor surface makes them excellent platforms for monitoring subtle pH changes derived from enzymatic substrate conversion close to the rGO surface. Owing to the architecture of microarray electrodes which is responsible for the high transconductance of the device, the pH sensitivity of the FETs was almost one order of magnitude higher than other graphene-based FETs. Thereby, a LOD of 1 μM and a linear range of 1–1000 μM could be reached for the detection of urea upon enzyme-catalyzed hydrolysis. By optimizing the enzyme loading as well as the arrangement of the assembly, highly sensitive, label-free biosensors could be developed for the real-time monitoring of urea. Taking advantage of the specific inhibition of enzymes by heavy metal ions, Cu^{2+} could be detected down to a concentration of 10 nM. The combination of pH-sensitive rGO-based FETs with highly selective enzymes embedded in a stabilizing polyelectrolyte assembly opens a wide range of possibilities for further expanding the field of biosensing.

ACKNOWLEDGMENTS

This work was supported by CONICET, ANPCyT (PICT-2010-2554, PICT-2013-0905), the Austrian Institute of Technology GmbH (AIT–CONICET Partner Group, Exp. 4947/11, Res. No. 3911, 28-12-2011), Universidad Nacional de La Plata (UNLP), the Austrian Federal Ministry for Transportation, Innovation, and Technology (GZ BMVIT-612.166/0001-III/I1/2010), by the FFG within the comet program, and from the governments of Lower and Upper Austria. E.P. acknowledges CONICET for a doctoral scholarship. O.A. is a staff researcher of CONICET. We thank Josef Breu for supplying graphene oxide flakes.

REFERENCES

Amine, A., Mohammadi, H., Bourais, I., & Palleschi, G. (2006). Enzyme inhibition-based biosensors for food safety and environmental monitoring. *Biosensors and Bioelectronics*, *21*(8), 1405–1423.

Ariga, K., Hill, J. P., & Ji, Q. (2007). Layer-by-layer assembly as a versatile bottom-up nanofabrication technique for exploratory research and realistic application. *Physical Chemistry Chemical Physics*, *9*, 2319–2340.

Ariga, K., Ji, Q., Mori, T., Naito, M., Yamauchi, Y., Abe, H., et al. (2013). Enzyme nanoarchitectonics: Organization and device application. *Chemical Society Reviews*, *42*(15), 6322.

Atta, N. F., Galal, A., & El-Ads, E. H. (2015). *Graphene—A platform for sensor and biosensor applications (p. Ch. 0)*. Rijeka: InTech.

Balasubramanian, K., & Kern, K. (2014). 25th anniversary article: Label-free electrical biodetection using carbon nanostructures. *Advanced Materials*, *26*(8), 1154–1175. https://doi.org/10.1002/adma.201304912.

Bao, Z., & Locklin, J. (2007). *Organic field-effect transistors*. CRC Press.

Bucur, C. B., Sui, Z., & Schlenoff, J. B. (2006). Ideal mixing in polyelectrolyte complexes and multilayers: Entropy driven assembly. *Journal of the American Chemical Society*, *128*(42), 13690–13691.

Carvounis, C. P., Nisar, S., & Guro-Razuman, S. (2002). Significance of the fractional excretion of urea in the differential diagnosis of acute renal failure. *Kidney International*, *62*(6), 2223–2229.

Choi, W., & Lee, J. (2012). Graphene synthesis and applications. In W. Choi & J. Lee (Eds.), *Nanomaterials and their applications*. CRC Press.

Cui, Y., Wei, Q., Park, H., & Lieber, C. M. (2001). Nanowire nanosensors for highly sensitive and selective detection of biological and chemical species. *Science*, *293*, 1289–1292. https://doi.org/10.1126/science.1062711.

De, S., Cramer, C., & Schönhoff, M. (2011). Humidity dependence of the ionic conductivity of polyelectrolyte complexes. *Macromolecules*, *44*, 8936–8943.

Decher, G., & Schlenoff, J. B. (2012). Layer-by-layer assembly (putting molecules to work). In G. Decher & J. B. Schlenoff (Eds.), *Multilayer thin films: Sequential assembly of nanocomposite materials* (2nd ed.): WILEY-VCH Verlag GmbH & Co.

Disawal, S., Qiu, J., Elmore, B. B., & Lvov, Y. M. (2003). Two-step sequential reaction catalyzed by layer-by-layer assembled urease and arginase multilayers. *Colloids and Surfaces B: Biointerfaces*, *32*(2), 145–156.

Du, X., Guo, H., Jin, Y., Jin, Q., & Zhao, J. (2015). Electrochemistry investigation on the graphene/electrolyte Interface. *Electroanalysis*, *27*, 2760–2765.

Fang, T., Konar, A., Xing, H., & Jena, D. (2007). Carrier statistics and quantum capacitance of graphene sheets and ribbons. *Applied Physics Letters*, *91*, 92109.

Forzani, E. S., Otero, M., Pérez, M. A., Teijelo, M. L., & Calvo, E. J. (2002). The structure of layer-by-layer self-assembled glucose oxidase and $Os(Bpy)_2ClPyCH_2NH-$ poly (allylamine) multilayers: Ellipsometric and quartz crystal microbalance studies. *Langmuir*, *18*, 4020–4029.

Fu, W., Jiang, L., van Geest, E. P., Lima, L. M. C., & Schneider, G. F. (2017). Sensing at the surface of graphene field-effect transistors. *Advanced Materials*, *29*, 1603610. https://doi.org/10.1002/adma.201603610.

Heller, I., Chatoor, S., Männik, J., Zevenbergen, M. A. G., Dekker, C., & Lemay, S. G. (2010). Influence of electrolyte composition on liquid-gated carbon nanotube and graphene transistors. *Journal of the American Chemical Society*, *132*, 17149–17156.

Hulanicki, A., Glab, S., & Ingman, F. (1991). Chemical sensors: Definitions and classification. *Pure and Applied Chemistry*, *63*(9), 1247–1250.

Iost, R. M., & Crespilho, F. N. (2012). Layer-by-layer self-assembly and electrochemistry: Applications in biosensing and bioelectronics. *Biosensors and Bioelectronics*, *31*(1), 1–10.

Karacaoğlu, S., Timur, S., & Telefoncu, A. (2003). Arginine selective biosensor based on arginase-urease immobilized in gelatin. *Artificial Cells, Blood Substitutes, and Immobilization Biotechnology*, *31*(3), 357–363.

Kazakova, L. I., Shabarchina, L. I., & Sukhorukov, G. B. (2011). Co-encapsulation of enzyme and sensitive dye as a tool for fabrication of microcapsule based sensor for urea measuring. *Physical Chemistry Chemical Physics*, *13*(23), 11110–11117.

Komaba, S., Fujino, Y., Matsuda, T., Osaka, T., & Satoh, I. (1998). Biological determination of Ag(I) ion and arginine by using the composite film of electroinactive polypyrrole and polyion complex. *Sensors and Actuators, B: Chemical*, *52*(1–2), 78–83.

Koncki, R., Wałcerz, I., Ruckruh, F., & Głab, S. (1996). Bienzymatic potentiometric electrodes for creatine and L-arginine determination. *Analytica Chimica Acta*, *333*(3), 215–222.

Krajewska, B. (2008). Mono- (Ag, Hg) and di- (Cu, Hg) valent metal ions effects on the activity of jack bean urease. Probing the modes of metal binding to the enzyme. *Journal of Enzyme Inhibition and Medicinal Chemistry*, *23*(4), 535–542.

Krajewska, B. (2009a). Ureases I. Functional, catalytic and kinetic properties: A review. *Journal of Molecular Catalysis B: Enzymatic*, *59*(1–3), 9–21.

Krajewska, B. (2009b). Ureases. II. Properties and their customizing by enzyme immobilizations: A review. *Journal of Molecular Catalysis B: Enzymatic*, *59*, 22–40.

Lakard, B., Magnin, D., Deschaume, O., Vanlancker, G., Glinel, K., Demoustier-Champagne, S., et al. (2011). Urea potentiometric enzymatic biosensor based on charged biopolymers and electrodeposited polyaniline. *Biosensors and Bioelectronics*, *26*(10), 4139–4145.

Larisika, M., Huang, J., Tok, A., Knoll, W., & Nowak, C. (2012). An improved synthesis route to graphene for molecular sensor applications. *Materials Chemistry and Physics*, *136*(2–3), 304–308.

Lee, C. S., Kyu Kim, S., & Kim, M. (2009). Ion-sensitive field-effect transistor for biological sensing. *Sensors*, *9*(9), 7111–7131.

Lisdat, F., Dronov, R., Möhwald, H., Scheller, F. W., & Kurth, D. G. (2009). Self-assembly of electro-active protein architectures on electrodes for the construction of biomimetic signal chains. *Chemical Communications (Cambridge, England)*, (3), 274–283.

Loh, K. P., Bao, Q., Ang, P. K., & Yang, J. (2010). The chemistry of graphene. *Journal of Materials Chemistry*, *20*(12), 2277.

Lvov, Y., & Caruso, F. (2001). Biocolloids with ordered urease multilayer shells as enzymatic reactors. *Analytical Chemistry*, *73*, 4212–4217.

Mello, L. D., & Kubota, L. T. (2002). Review of the use of biosensors as analytical tools in the food and drink industries. *Food Chemistry*, 77(2), 237–256.

Nehra, A., & Pal Singh, K. (2015). Current trends in nanomaterial embedded field effect transistor-based biosensor. *Biosensors and Bioelectronics*, *74*, 731–743. https://doi.org/10.1016/j.bios.2015.07.030.

Niyogi, S., Bekyarova, E., Itkis, M. E., Zhang, H., Shepperd, K., Hicks, J., et al. (2010). Spectroscopy of covalently functionalized graphene. *Nano Letters*, *10*(10), 4061–4066.

Ohno, Y., Maehashi, K., Yamashiro, Y., & Matsumoto, K. (2009). Electrolyte-gated graphene field-effect transistors for detecting pH and protein adsorption. *Nano Letters*, *9*(9), 3318–3322.

Piccinini, E., Alberti, S., Longo, G. S., Berninger, T., Breu, J., Dostalek, J., et al. (2018). Pushing the boundaries of interfacial sensitivity in graphene FET sensors: Polyelectrolyte multilayers strongly increase the Debye screening length. *The Journal of Physical Chemistry*, *122*, 10181–10188.

Piccinini, E., Bliem, C., Reiner-Rozman, C., Battaglini, F., Azzaroni, O., & Knoll, W. (2017). Enzyme-polyelectrolyte multilayer assemblies on reduced graphene oxide field-effect transistors for biosensing applications. *Biosensors and Bioelectronics*, *92*, 661–667.

Pumera, M. (2011). Graphene in biosensing. *Materials Today*, *14*(7–8), 308–315.

Pumera, M., Ambrosi, A., Bonanni, A., Chng, E. L. K., & Poh, H. L. (2010). Graphene for electrochemical sensing and biosensing. *TrAC—Trends in Analytical Chemistry*, *29*(9), 954–965.

Rajesh, Bisht, V., Takashima, W., & Kaneto, K. (2005). An amperometric urea biosensor based on covalent immobilization of urease onto an electrochemically prepared copolymer poly (N-3-aminopropyl pyrrole-co-pyrrole) film. *Biomaterials, 26*(17), 3683–3690.

Reiner-Rozman, C., Larisika, M., Nowak, C., & Knoll, W. (2015). Graphene-based liquid-gated field effect transistor for biosensing: Theory and experiments. *Biosensors and Bioelectronics, 70*, 21–27.

Rydzek, G., Ji, Q., Li, M., Schaaf, P., Hill, J. P., Boulmedais, F., et al. (2015). Electrochemical nanoarchitectonics and layer-by-layer assembly: From basics to future. *Nano Today, 10*(2), 138–167.

Sakr, O. S., & Borchard, G. (2013). Encapsulation of enzymes in layer-by-layer (LbL) structures: Latest advances and applications. *Biomacromolecules, 14*(7), 2117–2135.

Sant, W., Temple-Boyer, P., Chanie, E., Launay, J., & Martinez, A. (2011). On-line monitoring of urea using enzymatic field effect transistors. *Sensors and Actuators, B: Chemical, 160*(1), 59–64.

Scouten, W. H., Luong, J. H. T., & Stephen Brown, R. (1995). Enzyme or protein immobilization techniques for applications in biosensor design. *Trends in Biotechnology, 13*(5), 178–185.

Shao, Y., Wang, J., Wu, H., Liu, J., Aksay, I. A., & Lin, Y. (2010). Graphene based electrochemical sensors and biosensors: A review. *Electroanalysis, 22*(10), 1027–1036.

Sheldon, R. A., & Van Pelt, S. (2013). Enzyme immobilisation in biocatalysis: Why, what and how. *Chemical Society Reviews, 42*(42), 6223–6235.

Singh, V., Joung, D., Zhai, L., Das, S., Khondaker, S. I., & Seal, S. (2011). Graphene based materials: Past, present and future. *Progress in Materials Science, 56*(8), 1178–1271.

Singh, M., Verma, N., Garg, A. K., & Redhu, N. (2008). Urea biosensors. *Sensors and Actuators, B: Chemical, 134*(1), 345–351.

Sohn, I. Y., Kim, D. J., Jung, J. H., Yoon, O. J., Nguyen Thanh, T., Tran Quang, T., et al. (2013). PH sensing characteristics and biosensing application of solution-gated reduced graphene oxide field-effect transistors. *Biosensors and Bioelectronics, 45*(1), 70–76.

Soldatkin, A. P., Montoriol, J., Sant, W., Martelet, C., & Jaffrezic-Renault, N. (2002). Creatinine sensitive biosensor based on ISFETs and creatinine deiminase immobilised in BSA membrane. *Talanta, 58*(2), 351–357.

Stasyuk, N., Smutok, O., Gayda, G., & Gonchar, M. (2011). A new bi-enzyme potentiometric sensor for arginine analysis based on recombinant human arginase I and commercial urease. *Journal of Materials Science and Engineering, 1*, 819–827.

Stasyuk, N., Smutok, O., Gayda, G., Vus, B., Koval'chuk, Y., & Gonchar, M. (2012). Bi-enzyme l-arginine-selective amperometric biosensor based on ammonium-sensing polyaniline-modified electrode. *Biosensors and Bioelectronics, 37*(1), 46–52.

Stenberg, E., Persson, B., Roos, H., & Urbaniczky, C. (1991). Quantitative determination of surface concentration of protein with SPR using radiolabeled proteins. *Journal of Colloid and Interface Science, 143*, 513–526.

Sumner, J. B., & Hand, D. B. (1929). The isoelectric point of crystalline urease. *Journal of the American Chemical Society, 51*(4), 1255–1260.

Tang, Z., Wang, Y., Podsiadlo, P., & Kotov, N. A. (2006). Biomedical applications of layer-by-layer assembly: From biomimetics to tissue engineering. *Advanced Materials, 18*(24), 3203–3224.

Temple-Boyer, P., Benyahia, A., Sant, W., Pourciel-Gouzy, M. L., Launay, J., & Martinez, A. (2008). Modeling of urea-ENFETs for haemodialysis applications. *Sensors and Actuators, B: Chemical, 131*, 525–532.

Torsi, L., Magliulo, M., Manoli, K., & Palazzo, G. (2013). Organic field-effect transistor sensors: A tutorial review. *Chemical Society Reviews, 42*(22), 8612.

Tsai, H. C., & Doong, R. A. (2005). Simultaneous determination of pH, urea, acetylcholine and heavy metals using array-based enzymatic optical biosensor. *Biosensors and Bioelectronics*, *20*(9), 1796–1804.

Valle-Vega, P., Young, C. T., & Swaisgood, H. E. (1980). Arginase-urease electrode for determination of arginine and peanut maturity. *Journal of Food Science*, *45*, 1026–1030.

Wang, Y. Y., & Burke, P. J. (2013). A large-area and contamination-free graphene transistor for liquid-gated sensing applications. *Applied Physics Letters*, *103*(5), 052103.

Wang, Y. Y., & Burke, P. J. (2014). Polyelectrolyte multilayer electrostatic gating of graphene field-effect transistors. *Nano Research*, *7*, 1650–1658. https://doi.org/10.1007/s12274-014-0525-9.

Wang, Y., Li, Z., Wang, J., Li, J., & Lin, Y. (2011). Graphene and graphene oxide: Biofunctionalization and applications in biotechnology. *Trends in Biotechnology*, *29*(5), 205–212.

Wong, H.-S., & Akinwande, D. (2011). *Carbon nanotube and graphene device physics*. Cambridge University Press.

Xia, J., Chen, F., Li, J., & Tao, N. (2009). Measurement of the quantum capacitance of graphene. *Nature Nanotechnology*, *4*, 505–509.

Yan, F., Zhang, M., & Li, J. (2014). Solution-gated graphene transistors for chemical and biological sensors. *Advanced Healthcare Materials*, *3*, 313–331. https://doi.org/10.1002/adhm.201300221.

Zhang, M., Liao, C., Mak, C. H., You, P., Mak, C. L., & Yan, F. (2015). Highly sensitive glucose sensors based on enzyme-modified whole-graphene solution-gated transistors. *Scientific Reports*, *5*, 8311.

Zhang, A., & Lieber, C. M. (2016). Nano-bioelectronics. *Chemical Reviews*, *116*, 215–257. https://doi.org/10.1021/acs.chemrev.5b00608.

Zhao, H., Brown, P. H., & Schuck, P. (2011). On the distribution of protein refractive index increments. *Biophysical Journal*, *100*, 2309–2317. https://doi.org/10.1016/j.bpj.2011.03.004.

Zhu, Z., Garcia-Gancedo, L., Flewitt, A. J., Xie, H., Moussy, F., & Milne, W. I. (2012). A critical review of glucose biosensors based on carbon nanomaterials: Carbon nanotubes and graphene. *Sensors*, *12*(12), 5996–6022.

CHAPTER THREE

Stabilization of Laccase Through Immobilization on Functionalized GO-Derivatives

Alexandra V. Chatzikonstantinou*, Elena Gkantzou*, Dimitrios Gournis†, Michaela Patila*, Haralambos Stamatis*,[1]

*Department of Biological Applications and Technology, Laboratory of Biotechnology, University of Ioannina, Ioannina, Greece
†Department of Materials Science & Engineering, University of Ioannina, Ioannina, Greece
[1]Corresponding author e-mail address: hstamati@uoi.gr

Contents

Methods in Enzymology, Volume 609
ISSN 0076-6879
https://doi.org/10.1016/bs.mie.2018.05.014

Abstract

This chapter deals with the use of functionalized derivatives of graphene oxide as nanoscaffolds for the immobilization and stabilization of laccase from *Trametes versicolor*. Covalent and noncovalent immobilization approaches are described, while a novel method for the development of laccase-based multilayer nanoassemblies is also presented. Various biochemical, spectroscopic, and microscopic techniques were applied to characterize the nanobiocatalytic systems in respect to their microstructure and catalytic performance. Laccase-GO nanosystems were characterized with FTIR spectroscopy in order to confirm the functionalization of the nanomaterials, as well as to interpret the nanomaterial–enzyme interactions, while the multilayer structure of laccase-based multilayer nanoassemblies was confirmed by atomic force microscopy. The nanobiocatalytic systems presented here demonstrated exceptional stability and reusability compared with the free enzyme form, leading to robust biocatalytic systems appropriate for various applications of industrial interest.

ABBREVIATIONS

ABTS 2,2′-azino-bis(3-ethylbenzothiazoline-6-sulfonic acid)
AFM atomic force microscopy
ATR attenuated total reflection
BCA bicinchoninic acid
CD circular dichroism
CLEAs cross-linked enzyme aggregates
CNTs carbon nanotubes
DTA differential thermal analysis
EDC 1-ethyl-3-[3-dimethylaminopropyl] carbodiimide hydrochloride
f-GO functionalized graphene oxide
FTIR Fourier-transform infrared
GO graphene oxide
HEPES 4-(2-hydroxyethyl)-1-piperazineethanesulfonic acid
HRP horseradish peroxidase
NHS *N*-hydroxysuccinimide
SEM scanning electron microscopy
TG thermogravimetric analysis
TvL laccase from *Trametes versicolor*
UV–vis ultraviolet–visible
XRD X-ray diffraction

1. INTRODUCTION

1.1 Laccase as a Natural Biocatalyst

Laccases are multicopper oxidases (benzenediol: oxygen oxidoreductases; EC 1.10.3.2), typically extracellular monomeric glycoproteins, that can be found worldwide in nature, from bacteria to fungi and plants. Their origin determines their biological function; usually fungal laccases are involved in morphogenesis, stress defense, lignin degradation, and fungal plant-pathogen/host interactions. These enzymes own a special place among biotechnologically relevant enzymes, which is attributed to their wide catalytic versatility and remarkable process stability (Betancor, Johnson, & Luckarift, 2013; Mate & Alcalde, 2017). Laccases are able to oxidize a variety of compounds including phenols, polyphenols, nonphenolic substrates, and aromatic amines. Their low substrate specificity combined with their "eco-friendly" nature, since water is the only produced by-product, have made laccases highly proposed catalysts for diverse biotechnological applications. Hence, laccases have been applied both in environmental processes, such as treatment of wastewaters, delignification, degradation of industrial pollutants, and xenobiotics, and in the realm of biocatalysis, such as improvement of food properties, stabilization of beverages, production of antibiotics, as well as in organic synthesis (Betancor et al., 2013; Tzialla, Taha, Kalogeris, & Stamatis, 2009). Laccases from various fungi sources have also demonstrated improved catalytic performance and stabilization in nonconventional, ternary systems, broadening the extent of oxidation reactions they are able to catalyze (Tzialla et al., 2009).

The physiological role of laccase is the oxidation of various electron donors in the presence of chemical mediators that diffuse between the enzyme and the donor to facilitate the oxidation of large polymeric molecules. The enzyme consists of a multicopper active site. In particular, the active site involves a high-potential type-1 copper atom, which functions as the primary electron acceptor and shuttles electrons to a trinuclear copper cluster (type-2 and type-3), with the concomitant reduction of dioxygen to water, without the release of a H_2O_2 intermediate. These multicopper laccases are also called blue oxidases, because of the strong absorption band that the type-1 copper atom exhibits at around 600 nm, giving rise to a characteristic blue color (Betancor et al., 2013).

1.2 Biocatalytic Systems in Nanobiotechnology

Enzyme immobilization is commonly recognized as a technology to stabilize biocatalysts against chemical and environmental attacks. Especially during large-scale industrial applications, such obstacles may lead to high operational costs due to low recovery and reusability of the biocatalyst (de Jesús Rostro-Alanis et al., 2016). Nevertheless, once immobilized, enzymes demonstrate high stability through a favorable microenvironment for them to function. Namely, immobilization can lead to improved substrate accessibility, since aggregation of the hydrophilic protein particles is avoided and the catalytic properties of immobilized enzymes can be easily modulated, thus creating robust biocatalysts for the development of commercial biocatalytic processes (Hwang & Gu, 2013; Pavlidis, Patila, Bornscheuer, Gournis, & Stamatis, 2014). Furthermore, immobilization improves enzyme performance under extreme reaction conditions, such as the use of organic solvents and elevated temperatures, allowing applications in industrial synthesis, while a facile separation of the catalyst from the reaction products is also possible, a key advantage regarding reactions with dissolved enzymes (Hwang & Gu, 2013; Küchler, Yoshimoto, Luginbühl, Mavelli, & Walde, 2016).

There are extensive reports on laccase immobilization, in an effort to improve the enzyme's performance. Studies include carrier-based and carrier-free immobilized forms of laccase (CLEAs and Combi-CLEAs). When it comes to stability enhancement, the research focuses on carrier-based biocatalytic systems, in order to create a rigid biocatalyst that functions in harsh conditions and can easily be recycled (Ba & Vinoth Kumar, 2017).

The combined research in the fields of biotechnology and nanotechnology during the last decade has led to an upsurge of innovative biological nanosystems. Nanobiocatalysis constitutes one of the most thriving research fields nowadays, as new nanobiocatalytic systems are developed, based mainly on the immobilization of enzymes on nanostructured materials with tailor-made properties (Pavlidis, Patila, Bornscheuer, et al., 2014; Pavlidis, Patila, Polydera, Gournis, & Stamatis, 2014). Different modified nanoscale supports have been studied and found to reduce diffusion limitations, thus enhancing biocatalytic efficiency, as well as increasing enzyme loading given the superior surface area per mass unit (Pavlidis, Tsoufis, Enotiadis, Gournis, & Stamatis, 2010; Tzialla et al., 2010). Consequently, the functionalized nanocarriers serve as nanoscale-processing systems that provide a suitable microenvironment for the enzyme to react in the maximal efficiency. In this direction, nanomaterials are carefully designed to enable a long-term storage and recycling integrity for the immobilized biocatalyst, as well as large surface area and low mass transfer resistance (Hwang & Gu, 2013). Overall, elaborating the possibilities for biotechnological applications of immobilized enzymes requires an optimization of both the enzyme localization and the enzyme's storage and operational stability (Küchler et al., 2016).

A focal point to take into consideration is the fact that immobilization entails the interaction of the enzyme and the nanocarrier. Considering this, the optimal conditions for each nanobiocatalytic system have to be investigated. However, enzymes immobilized on nanomaterials are usually advantageous over native enzymes, showing wider temperature range, broader working pH, increased reusability and greater thermal, storage and operational stability. For example, nanoimmobilized laccase and lipase were shown to retain up to 90% of their activity in nonconventional media or in the presence of denaturing agents (Tzialla et al., 2010, 2009), while lipase was also shown to exhibit great storage stability while working as a nanobiocatalytic system (Pavlidis et al., 2010). Besides these, nanocarriers' properties, such as high surface-to-volume ratio, chemical, electrical, and optical properties, favor enzymes' performance by increasing enzyme loading and affecting their diffusion and catalytic activity (Chen, Zeng, Xu, Lai, & Tang, 2017).

Another aspect to consider pertains to the immobilization method. In particular, enzyme immobilization onto a support is possible through adsorption, covalent binding, entrapment, or encapsulation, with each method demonstrating different merits and demerits (Barrios-Estrada et al., 2018). Physical bonding (such as hydrophobic or van der Waals interactions) is generally too weak to keep the enzyme stable under industrial conditions. However, covalent binding has been extensively studied and is considered as a highly reliable method for achieving immobilization through specific binding sites as well as preventing enzyme leaching (de Jesús Rostro-Alanis et al., 2016; Hwang & Gu, 2013).

Concerning laccase immobilization, covalent binding has been widely studied during the last decade as it has been proven that the tertiary structure of the enzyme is stabilized, offering great resistance to denaturing agents. Providing the enzyme's rigidity, an optimal support should contain short spacer arms and a high density of reactive groups on its surface so as to be activated and react with nucleophilic groups on the protein (Fernández-Fernández, Sanromán, & Moldes, 2013).

1.3 Graphene-Based Nanomaterials and Their Application in Nanobiocatalysis

Graphene-based nanomaterials have rapidly become the most widely studied materials, as they apply to various research fields, from sensing to biomedicine. Their unique structural features and physical properties offer them an important advance over other materials for the design of novel biomaterials. Their mechanical, chemical, and electrochemical stability, along with their excellent electrical conductivity, make them ideal materials for the development of several types of biosensors for the detection of different kind of analytes and biological compounds (Baptista-Pires et al., 2014), and biofuel cells for power generation from biomass-derived energy carriers or biofuel substrates (Prasad, Chen, & Chen, 2014). Moreover, due to their surface functionalization and ultra-high surface area, GO-derivatives can be easily functionalized by small molecular dyes for bioimaging applications (Yang, Feng, Shi, & Liu, 2013). The unmodified graphene-moieties of GO-based nanomaterials are hydrophobic and governed by high π electron movement, which facilitates high drug loading through the formation of π–π stacking and hydrophobic interactions (Goenka, Sant, & Sant, 2014; Lin, Chen, & Huang, 2016). In addition, the large surface area and layered structure of GO-derivatives offer the capability for multivalent functionalization, and show high efficiency for biomacromolecules loading, providing biotechnologists with a great variety of nanoscaffolds for the development of robust nanobiocatalytic systems (Pavlidis, Patila, Bornscheuer, et al., 2014).

Isolated graphene is a layer of monoatomic thickness made by graphite type carbon. This nanomaterial serves as building block for fullerenes, single- and multiwalled carbon nanotubes (CNTs) and graphene oxide (GO). GO, an oxygen-rich derivative of graphene, is broadly used in immobilization studies, since it bears carboxyl, epoxy, and hydroxyl functional groups on its surface, facilitating its functionalization so as to serve as nanoscaffold for covalent immobilization of biocatalysts. These oxygen-containing groups are distributed randomly on the surface, making GO highly dispersible in water and other polar solvents. In summary, GO demonstrates such surface diversity that a variety of chemical modifications are feasible, including covalent attachment at the carboxylic groups, covalent attachment at the epoxy groups via ring opening reactions of amines and noncovalent functionalization through van der Waals and π–π interactions with polymers, surfactants, and other small molecules (Patila et al., 2013; Pavlidis, Patila, Bornscheuer, et al., 2014; Pavlidis, Patila, Polydera, et al., 2014; Pavlidis, Vorhaben, Gournis, et al., 2012; Pavlidis, Vorhaben, Tsoufis, et al., 2012).

The removal of the oxygen-containing groups of GO through chemical, thermal, or electrochemical reduction leads to the production of reduced graphene oxide (rGO). The reduction of GO is one of the most applicable techniques for large-scale production of graphene sheets, as the resulting rGO is very similar to graphene. Depending on the reduction method, rGO exhibits properties between those of GO and graphene. The surface charge and hydrophilicity of rGO is highly reduced, while its conductivity is restored and its optical absorbance is enhanced (Lin et al., 2016). The absence of oxygen-containing groups on rGO surface may affect interactions with different biomolecules, altering their catalytic characteristics. Our group has already demonstrated the importance of the presence of oxygen-containing groups on graphene oxide-based nanomaterials while they are used as matrices for the immobilization of cytochrome c (Patila, Pavlidis, et al., 2016). The results showed that when rGO-derivatives were used as carriers, the immobilization efficiency as well as the catalytic activity of cyt c was decreased, compared to GO-derivatives. This effect was ascribed to the development of stronger hydrophobic and π–π interactions between cyt c and rGO-derivatives which affected the protein conformation, and thus its catalytic behavior.

Graphene-based nanomaterials have been proven to introduce functionalities that enhance the fine-tuned actions of an immobilized enzyme, while increasing enzyme operational stability and catalytic activity, and improving protein electron transfer as well (Pavlidis, Patila, Bornscheuer, et al., 2014). However, it is critical to investigate the structural and functional changes that are caused by nanomaterial–enzyme interactions, in order to optimize a process. It has been demonstrated that different types of carbon-based nanomaterials (CNTs, fullerenes, graphene) have different effects on enzymatic structures and catalytic activities, depending on the type and orientation of enzymes, physical properties and chemical groups of nanomaterials and environmental conditions. More specifically, it is crucial to maintain the enzyme's active site unaffected by nanomaterials or other adjacent molecules. Hence, the physicochemical properties of each nanoscaffold together with the surrounding conditions of pH and temperature have to be carefully studied since there is direct correlation with the binding orientations, stability, and activity of the biocatalyst (Chen et al., 2017; Patila et al., 2013).

Several studies have been conducted concerning the effect of different graphene-based nanomaterials on activity and stability of biotechnologically relevant enzymes (Pavlidis, Patila, Bornscheuer, et al., 2014). Chen et al. found out that graphene and GO nanomaterials decrease the enzymatic stability of horseradish peroxidase (HRP), while GO in its reduced form enhances the enzyme's stability (Chen et al., 2017). They also concluded that all of these nanomaterials induce changes in the secondary structure of HRP. Another study concerning bio-nano interface reactions by Campbell et al. demonstrated that the different characteristics of the graphene-based nanomaterials studied, influence directly the catalytic behavior of three different enzymes (Campbell et al., 2014). The differences were attributed to the different immobilization methods used, while no significant conformational changes of the enzyme active site occurred upon immobilization. On the other hand, our team has shown that different graphene-based nanomaterials exert conformational changes on cytochrome *c*, resulting in a more rigid structure of the protein molecule (Patila et al., 2013; Patila, Pavlidis, et al., 2016). The catalytic activity of the biocatalyst might be lower but the stability and tolerance against denaturing agents are much greater than the free protein form. It has also been proven that functionalized multiwalled CNTs are able to optimize the catalytic performance of lipase enzyme (Pavlidis et al., 2010), while the catalytic behavior of different hydrolases was favorably affected through interactions with functionalized carbon-based nanomaterials (Pavlidis, Vorhaben, Gournis, et al., 2012).

An upcoming approach for nanobiocatalysts development is the immobilization of enzymes in multilayer systems through layer-by-layer deposition (Sakr & Borchard, 2013). This technique has been applied for the development of biotechnological devices and has potentials for developing artificial multienzyme complexes for in vitro cascade reactions. Patila et al. synthesized such multilayer nanoassemblies using graphene oxide and laccase, creating a nanobiocatalytic system with excellent biocatalytic behavior (Patila, Kouloumpis, Gournis, Rudolf, & Stamatis, 2016).

This chapter concerns laccase immobilization onto GO-derivatives, functionalized with different surface groups. Covalent and noncovalent immobilization techniques are presented, while biocatalytic and structural characterization studies are presented and discussed for the interpretation of the enzyme's stability and performance. A method for the synthesis of multilayer nanoassemblies is also introduced.

2. SYNTHESIS OF NANOMATERIALS

Graphene oxide (GO) is the *oxidized derivative* of *graphene*, which is obtained by the *oxidation of graphite* using *strong* oxidants (Lerf, He, Forster, & Klinowski, 1998; Yu, Zhang, Bulin, Li, & Xing, 2016). The presence of oxygen functional groups (hydroxyl, epoxy, and carboxyl groups) on the aromatic scaffold of GO allows these sheets to mediate ionic and nonionic interactions with a wide range of molecules (Kim et al., 2010). These groups are distributed through the basal planes and edges of the GO sheets and provide GO important swelling and intercalation properties. Due to the existence of plenty of oxygen-containing groups, GO can be further functionalized with amines or diamines and amino acids.

The attachment of the diamine molecules can be achieved through different mechanisms: (i) hydrogen-bonding interactions between the oxygen-containing functional groups of GO and the amine molecules, (ii) protonation of the amine by the weakly acidic sites of the GO layers, and (iii) chemical attachment of the amine to the GO surfaces via nucleophilic substitution reactions on the epoxy groups of GO (Kovtyukhova et al., 1999). The main insertion pathway, though, that has been proposed is the nucleophilic substitution reactions on the epoxy groups of GO (Bourlinos, Gournis, & Petridis, 2003).

The engineering of GO for grafting desirable functional groups onto its surface enables the design of functionalized nanomaterials with tailor-made properties and improved suitability for their application as nanoscaffolds for biomolecule immobilization (Pavlidis, Patila, Bornscheuer, et al., 2014). Biomolecules can be specifically bound on GO-derivatives, depending on the organic molecule used for the nanomaterial functionalization. The surface chemistry of the resulted nanomaterials significantly influences the interactions with biomolecules and thus affecting their conformation and biological function (Jin et al., 2012).

General safety: nitrile gloves, protective mask

Equipment:

1. Beakers
2. Spherical bottle
3. Magnetic stirrer
4. Pipette (3 mL)

Chemicals and materials

1. Graphite powder (purum, $\leq$0.2 mm; Fluka).
2. The following diamines and amino acids were employed for the surface modification of GO: $C_nH_{2n}(NH_2)_2$ ($n=2, 4, 6$), and $H_2N(CH_2)_nCOOH$ ($n=2, 4, 10$).
3. Sulfuric acid (95%–97%), nitric acid (65%), potassium chlorate (>98%), and absolute ethanol (100%) were purchased from Merck.
4. Distilled water.

2.1 Synthesis of GO

Graphene oxide aqueous dispersions are synthesized from graphite powder using a modified Staudenmaier's method (Enotiadis, Angjeli, Baldino, Nicotera, & Gournis, 2012; Staudenmaier, 1898).

1. 5 g of graphite powder is added to a solution consisting of 200 mL of H_2SO_4 and 100 mL of HNO_3, while cooling in a bath (0°C).
2. 100 g of powdered $KClO_3$ (100 g) is added slowly under continuous stirring and cooling.
3. The reaction is incubated for 18 h.
4. The oxidation product is mixed with distilled water until the pH value reaches 6.
5. The final product is dried at room temperature.

Tip 1: The addition of $KClO_3$ must be slowly and very carefully done, as an exothermic reaction takes place during this step.

Tip 2: During the production of GO via liquid exfoliation of graphite oxide by treatment of graphite with strong acids, the oxygen-containing functionalities added on its surface result in the swelling of the graphitic sheets. When the oxidation product is mixed with water (step 4), the appearance of negative electron charge due to the oxygen groups causes the formation of repulsive forces between the sheets, which leads to the exfoliation of graphite sheets for the production of graphene oxide (Enotiadis et al., 2012).

2.2 Modification of GO With Diamines and/or Amino Acids

1. 100 mg of GO are dispersed in 10 mL of water.
2. 300 mg of diamine or amino acid are dissolved in 10 mL of ethanol or water, respectively.
3. The amine/amino acid mixture is added to the GO dispersion.
4. The mixture is incubated under stirring for 20 h at room temperature.
5. The resulting solution is centrifuged at 4000 rpm and the precipitate containing the GO-derivatives is washed with 1:1 water/ethanol.
6. The samples are left to dry at room temperature and then collected and stored until used.

Tip 1: GO must be fully dispersed before the addition of the amine/amino acid solution.

Tip 2: Upon amine addition, the GO solid swells instantly, indicating the intercalation of the amine/amino acid in GO.

3. IMMOBILIZATION OF LACCASE ON GO-DERIVATIVES

The immobilization of laccase on GO-derivatives can be carried out via two different methods: physical adsorption and covalent binding. The nonspecific binding via physical adsorption is a simple process which enhances the immobilization efficiency without disrupting the surface of nanomaterials (Patila, Orfanakis, Polydera, Pavlidis, & Stamatis, 2017; Pavlidis, Patila, Bornscheuer, et al., 2014). Physical adsorption of proteins on GO-derivatives is governed by weak forces such as van der Waals forces, π–π stacking interactions, hydrophobic and electrostatic interactions (Gao & Kyratzis, 2008).

Covalent immobilization can be achieved in two different kinds of functionalized GO-derivatives; those with terminal carboxyl groups and those with terminal amine groups. In the first case, during the immobilization procedure, the f-GO-COOH derivatives first react with 1-ethyl-3-[3-dimethylaminopropyl] carbodiimide hydrochloride (EDC) to form an amine-reactive *O*-acylisourea intermediate which subsequently reacts with an amine group on the surface of the protein to produce a stable amine bond (Gao & Kyratzis, 2008). The addition of *N*-hydroxysuccinimide (NHS) stabilizes the intermediate by converting it to a semistable amine-reactive NHS ester, thus increasing the coupling efficiency (Sehgal & Vijay, 1994). Finally, the activated fGO reacts with the protein, whereas interprotein conjugation is eliminated because only the nanomaterial has been exposed to the cross-linker (Fig. 1).

In the case of immobilization on f-GO-NH_2 derivatives, the nanomaterials are firstly activated with the cross-linker glutaraldehyde. In a following step, the free terminal aldehyde groups are cross-linked to amine groups on the enzyme surface through the formation of a Shiff's base (Fig. 2).

General safety: nitrile gloves, protective mask

Equipment:

1. Falcon tubes (15 mL)
2. Micropipettes (0.1–1 mL)
3. Incubator with stirring
4. Ultrasonic bath
5. Centrifuge

Chemicals and materials

1. GO functionalized with terminal amine or carboxyl groups (fGO).
2. Laccase from *Trametes versicolor* (TvL) from Sigma-Aldrich Co., MO, USA. The activity of the enzyme was 13.6 U/mg (one unit corresponds to the amount of enzyme which converts 1 μmol of catechol per minute at pH 5.0 and 25°C).
3. 1-Ethyl-3-[3-dimethylaminopropyl] carbodiimide hydrochloride (EDC) from Sigma-Aldrich Co., MO, USA.
4. *N*-hydroxysuccinimide (NHS) from Sigma-Aldrich Co., MO, USA.

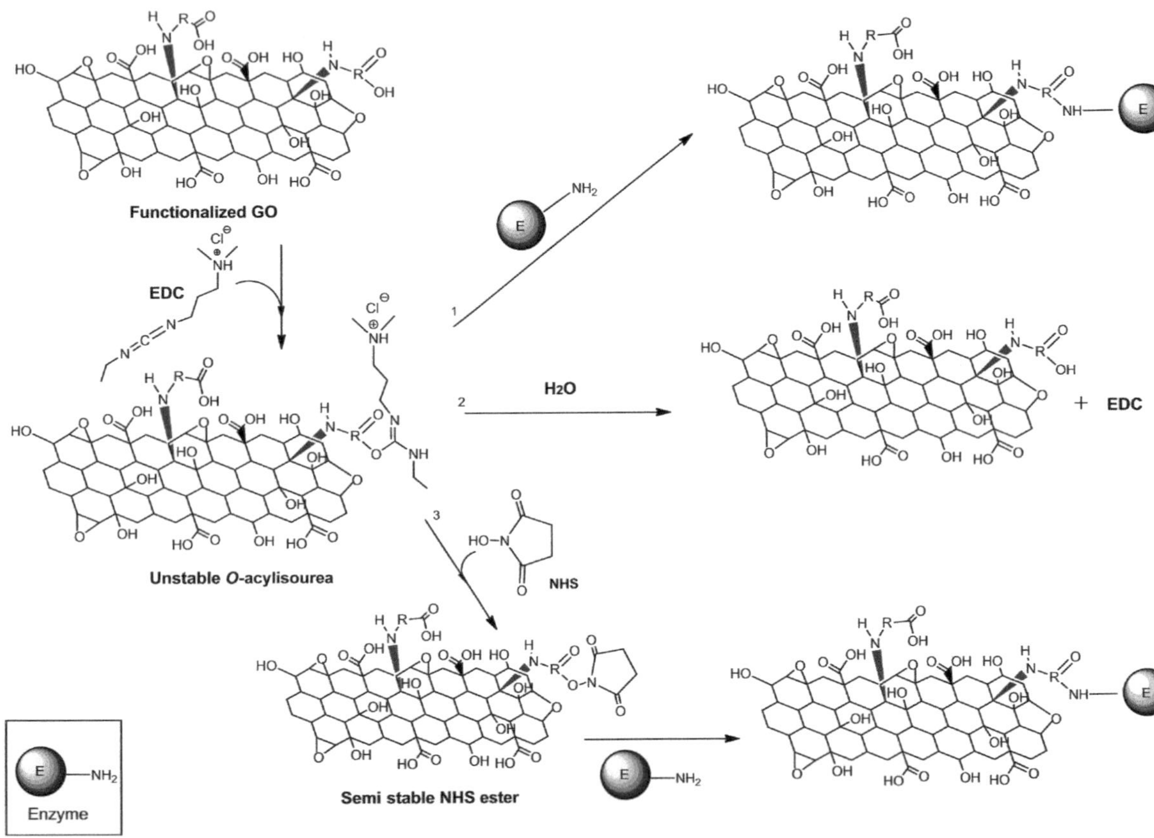

Fig. 1 Covalent immobilization of an enzyme on carboxyl-terminated GO-derivatives using carbodiimide chemistry.

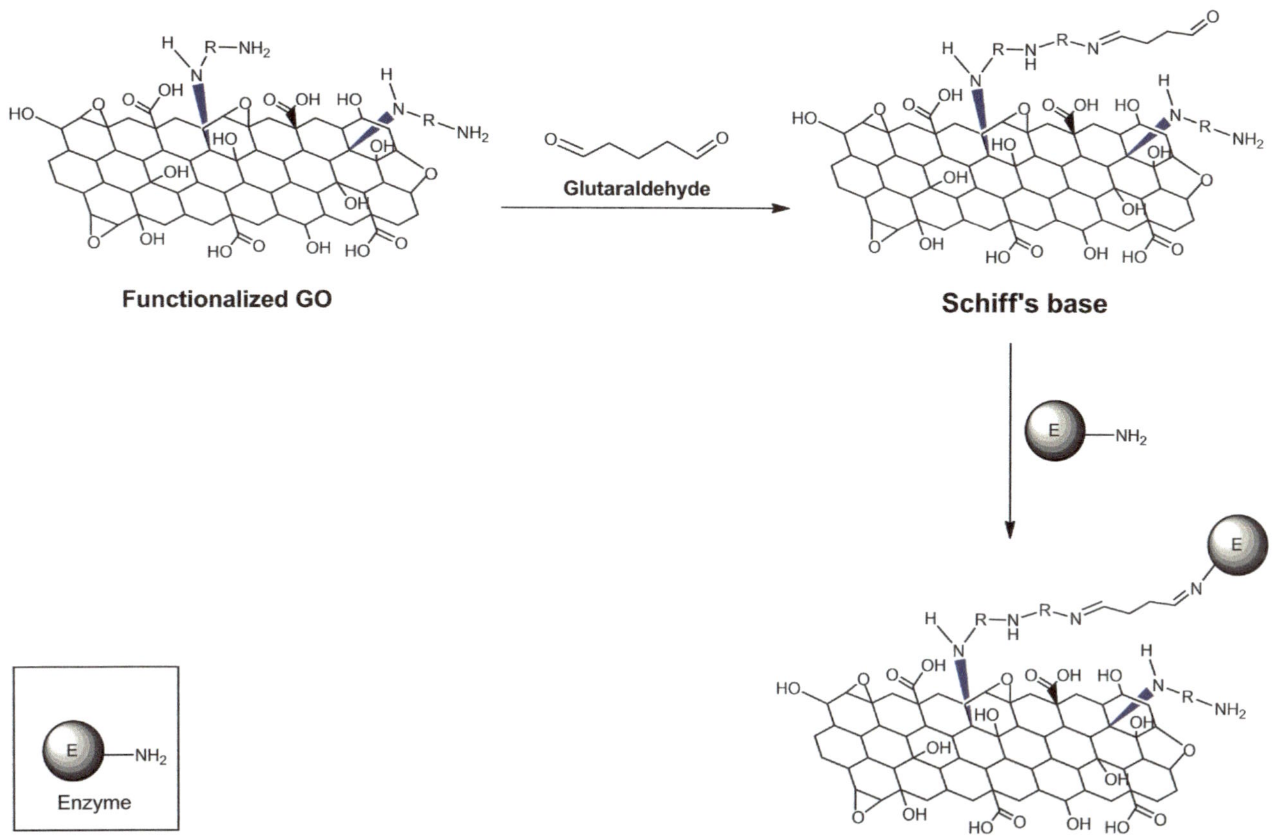

Fig. 2 Covalent immobilization of an enzyme on amine-terminated GO-derivatives using glutaraldehyde as cross-linker.

5. Glutardialdehyde solution 25% for electron microscopy from Merck KGaA, Darmstadt, Germany.
6. Tween-20 from Sigma-Aldrich Co., MO, USA.
7. Acetate buffer (0.1 *M*, pH 4.58).
8. HEPES (4-{2-hydroxyethyl}-1-piperazineethanesulfonic acid) buffer (0.1 *M*, pH 4.58).
9. HPLC water.

3.1 Noncovalent Immobilization of TvL on fGO

Duration

3 h.

Protocol

1. Weight 3 mg of fGO in a falcon tube.
2. Add 5 mL acetate buffer to the nanomaterials and leave at an ultrasonic bath for 30 min.
3. Suspend 3 mg of TvL in 1 mL acetate buffer.
4. Transfer the enzyme preparation (1 mL) to the fGO dispersion.
5. Incubate the reaction mixture for 1 h at 30°C.
6. Separate the immobilized TvL using centrifugation at 6000 rpm, followed by washing with acetate buffer, thrice.
7. Dry the immobilized TvL over silica gel.
8. Store the immobilized TvL at 4°C until used.

3.2 Covalent Immobilization of TvL on fGO With Terminal Amine Groups

Duration

5 h.

Protocol

1. Weight 3 mg of fGO in a falcon tube.
2. Add 9.13 mL acetate buffer and 0.110 mL Tween-20 to the nanomaterials and leave at an ultrasonic bath for 30 min.
3. Add 1.76 mL glutaraldehyde to the nanomaterial dispersion.
4. Incubate the mixture for 30 min at 30°C.
5. Remove the glutaraldehyde excess from the solution using centrifugation at 6000 rpm, followed by washing with acetate buffer, thrice.
6. Redisperse the activated nanomaterials in 5 mL acetate buffer.
7. Suspend 3 mg of TvL in 1 mL acetate buffer.
8. Transfer the enzyme preparation (1 mL) to the fGO dispersion.

9. Incubate the reaction mixture for 1 h at 30°C.
10. Separate the immobilized TvL using centrifugation at 6000 rpm, followed by washing with acetate buffer, thrice.
11. Dry the immobilized TvL over silica gel.
12. Store the immobilized TvL at 4°C until used.

3.3 Covalent Immobilization of TvL on fGO With Terminal Carboxyl Groups

Duration

5 h.

Protocol

1. Weight 3 mg of fGO in a falcon tube.
2. Add 5 mL HPLC water to the nanomaterials and leave at an ultrasonic bath for 30 min.
3. Add 1 mL HEPES buffer and 1.2 mL of a 10 mg mL^{-1} aqueous solution EDC to the nanomaterial dispersion.
4. Under fast stirring, add 2.3 mL of a 50 mg mL^{-1} NHS aqueous solution quickly. Incubate the mixture for 30 min at 30°C.
5. Remove the EDC/NHS excess from the solution using centrifugation at 6000 rpm, followed by washing with HEPES buffer, thrice.
6. Redisperse the activated nanomaterials in 5 mL HEPES buffer.
7. Suspend 3 mg of TvL in 1 mL HEPES buffer.
8. Transfer the enzyme preparation (1 mL) to the fGO dispersion.
9. Incubate the reaction mixture for 1 h at 30°C.
10. Separate the immobilized TvL using centrifugation at 6000 rpm, followed by washing with HEPES buffer, thrice.
11. Dry the immobilized TvL over silica gel.
12. Store the immobilized TvL at 4°C until used.

3.4 Synthesis of Multilayer Assemblies of TvL and fGO

The synthesis of multilayer nanoassemblies of enzymes and nanomaterials is a novel approach inspired form the Layer-by-Layer (LbL) technique. LbL deposition is a versatile method for producing thin-film multilayer films with controlled properties (Decher, 1997). The basic principle of the LbL method is the sequential deposition of polycations and polyanions through electrostatic interactions. The LbL technique is one beneficial technique for immobilizing bioactive molecules such as enzymes over or between the polyelectrolyte films (Schwinté et al., 2002).

Based on this technique, our group has developed multilayered nano-biocatalysts, which are formed through multipoint covalent immobilization of the enzyme onto the nanomaterials, instead of electrostatic interactions applied in classic LbL method (Patila, Kouloumpis, et al., 2016). The multi-point covalent immobilization of laccase on f-GO-NH_2 was achieved using glutaraldehyde as cross-linking agent between the enzyme and the amino groups located on the surface of the nanomaterials, as indicated in Fig. 3.

Duration

10h (depending on the number of TvL-fGO layers).

Protocol

1. Weight 3mg of fGO in a falcon tube.
2. Add 9.13mL acetate buffer and 0.110mL Tween-20 to the nanomaterials and leave at an ultrasonic bath for 30min.
3. Add 1.76mL glutaraldehyde to the nanomaterial dispersion. Incubate the mixture for 30min at 30°C.
4. Remove the glutaraldehyde excess from the solution using centrifugation at 6000rpm, followed by washing with acetate buffer, thrice.
5. Redisperse the activated nanomaterials in 2mL acetate buffer.
6. Suspend 3mg of TvL in 1mL acetate buffer.
7. Transfer the enzyme preparation (1mL) to the fGO dispersion.
8. Incubate the reaction mixture for 1h at 30°C (1st layer).
9. Add 2mL of fGO (3mg) previously activated with glutaraldehyde (steps 1–6) to the mixture.
10. Incubate the reaction mixture for 1h at 30°C (2nd layer).
11. Repeat the above procedure till the desired number of fGO-TvL layers is achieved.
12. Separate the final assemblies of fGO-TvL using centrifugation at 6000rpm.
13. Dry the fGO-TvL nanoassemblies over silica gel.
14. Store the fGO-TvL nanoassemblies at 4°C until used.

Tip 1: The nanomaterials have to be freshly prepared and fully dispersed in the buffer so that the fGO sheets exfoliate and enhance the surface area that will be used for high enzyme loading.

Tip 2: The surfactant Tween-20 is used for the coverage of the hydrophobic surface of the nanomaterials, in order to avoid the nonspecific interactions with the hydrophobic amino acids of the protein (Shim, Kam, Chen, Li, & Dai, 2002).

Tip 3: As glutaraldehyde is toxic, direct contact must be avoided. For this reason, nitrile gloves and mask are used and the activation procedure must take place in a fume hood.

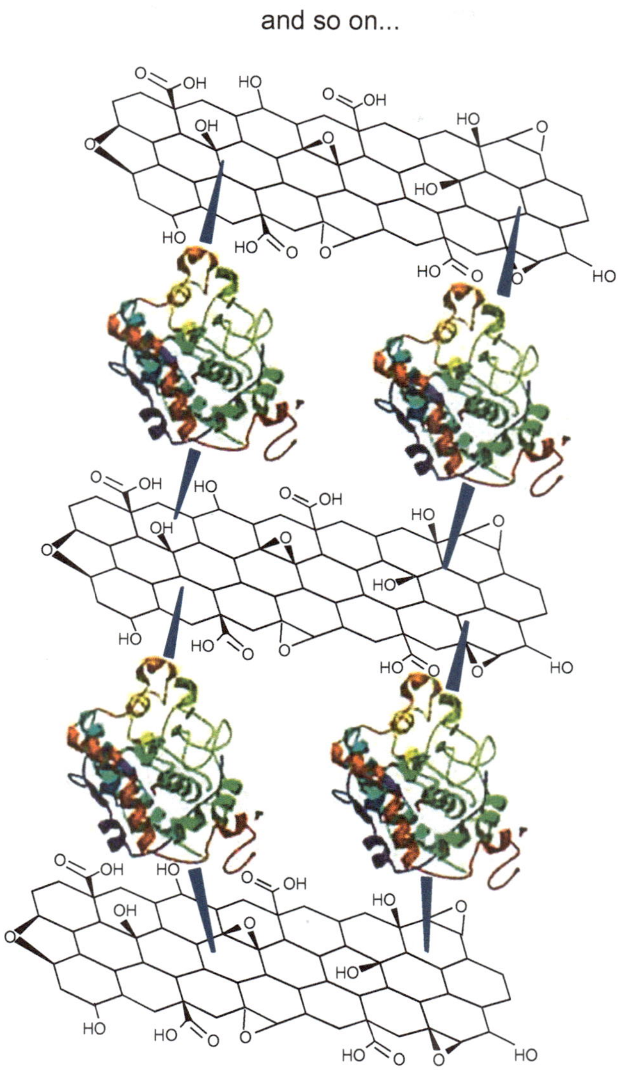

Fig. 3 Development of functionalized graphene oxide-laccase nanoassemblies. *Reprinted from Patila, M., Kouloumpis, A., Gournis, D., Rudolf, P., & Stamatis, H. (2016). Laccase-functionalized graphene oxide assemblies as efficient nanobiocatalysts for oxidation reactions.* Sensors (Switzerland), 16, *1–14. https://doi.org/10.3390/s16030287, licensed under CC BY 4.0.*

Tip 4: After the first centrifugation of the immobilized enzyme, 1 mL of the supernatant solution is collected and used for the determination of the immobilization yield. The rest of the supernatant is discarded.

Tip 5: In case of covalent immobilization using EDC/NHS chemistry, phosphate buffers should be avoided, as phosphate reacts with EDC and reduces its stability (Gilles, Hudson, & Borders, 1990). If phosphate buffer has to be used, a large excess of EDC can be used to compensate for the instability.

Tip 6: For the synthesis of enzyme–nanomaterial assemblies, start with activation of the nanomaterials for all steps, depending on how many layers are desired, in order to gain precious time.

4. CHARACTERIZATION OF THE IMMOBILIZED BIOCATALYSTS

A crucial step for the development of effective nanobiocatalytic systems through the immobilization of enzymes on GO-based nanomaterials is the characterization of the resulting biocatalysts. The immobilized enzymes are characterized in terms of their biochemical and functional properties, as well as of their morphological and structural characteristics. At first, the enzyme loading is estimated by determining the immobilization yield of the enzyme on the nanomaterial, using different protein quantification assays such as bicinchoninic acid (BCA) (Smith et al., 1985) or Bradford assay (Bradford, 1976). Moreover, the catalytic properties, i.e., the activity and stability, of the immobilized enzyme are investigated in comparison with the free enzyme form.

The resulting nanobiocatalytic systems can be further characterized in respect to their microstructure using different microscopic methods. atomic force microscopy (AFM), scanning electron microscopy (SEM), and transmission electron microscopy (TEM) are used to depict the successful attachment of the biomolecule on the surface of the nanomaterial. The size and width of each molecule can be calculated, confirming the immobilization of the biocatalyst on the nanocarrier (Orfanakis et al., 2018; Patila, Pavlidis, et al., 2016). The composition of the resulting immobilized system can be characterized by X-ray photoelectron spectroscopy (XPS). Deconvolution of C1s spectra of the sample results in an estimation of changes in the relative contribution of the carbon components of the system due to enzyme immobilization, while the appearance of N1s peaks confirms the formation of covalent bond between the enzyme and the nanomaterials (Patila, Pavlidis, et al., 2016). Thermogravimetric analysis (TG) and differential thermal analysis (DTA) can also be performed in order to confirm the successful immobilization of the enzyme onto the nanomaterials, by indicating the exothermal peaks due to decomposition and combustion of the sample (Patila, Pavlidis, et al., 2016). The exact protein content cannot be precisely calculated, however, an estimation of the enzyme content may be provided from the TG signal.

Immobilized enzymes can be further characterized in respect to conformational changes induced to the protein during the immobilization procedure. Different spectroscopic methods may provide information about alterations in the secondary or tertiary structure of an immobilized enzyme. For instance, circular dichroism (CD) spectroscopy provides insights into the secondary structure. Analysis of the far-UV spectra (200–260 nm) of an enzyme results in the distribution of the secondary elements of the protein, while the near-UV region (260–350 nm) provides information regarding the aromatic regions of the protein. Upon immobilization, changes in the secondary structure of the enzyme can be observed as changes in the CD spectra of the protein (Patila et al., 2013; Pavlidis, Vorhaben, Gournis, et al., 2012). Fluorescence and UV–visible (UV–vis) spectroscopy can also be applied for the determination of conformational changes in the microenvironment of the enzyme (Patila et al., 2013; Pavlidis, Vorhaben, Gournis, et al., 2012).

Fourier-transform infrared (FTIR) spectroscopy and attenuated total reflection (ATR) can be applied for both morphological and structural characterization of the immobilized enzyme. These techniques are used for the verification of the enzyme's presence on the nanomaterial carrier through the appearance of characteristic bonds in the IR spectra. Moreover, these methods can be further applied for the estimation of secondary elements in the enzyme's structure. The analysis of the Amide I band at approximately 1700–1600 cm^{-1}, which is attributed to the presence of the enzyme on the nanomaterial, makes it possible to obtain information on the effect of the immobilization on the secondary structure of the protein (Patila, Pavlidis, et al., 2016; Tzialla et al., 2010). The amide I band consists of several overlapping components that are assigned to different secondary structure elements (α-helix, β-sheets, β-turns, random coil) which can be identified by the second derivative spectra identification and fitting (Zhao, Du, Li, & Yang, 2012).

4.1 Determination of Immobilization Yield

Several parameters may affect the yield of an immobilization procedure as well as the catalytic activity of the immobilized biocatalyst. In general, the properties of a biocatalytic system are determined both by the protein structure and the surface chemistry of the support. Regarding noncovalent immobilization, the enzyme loading is mainly affected by the pH value and ionic strength of the solution, so it is necessary to take into consideration the enzyme's isoelectric point (p*I*). In the case of covalent immobilization, enzyme loading depends greatly on the linking molecules used and the functional groups of the supporting material (Mohamad, Marzuki, Buang, Huyop, & Wahab, 2015). Finally, there are parameters such as incubation time, reaction temperature, and enzyme/support ratio that should be considered as crucial factors affecting the immobilization yield (Chang, Shaw, Yang, Chang, & Shieh, 2008; Kuo et al., 2012).

General safety: nitrile gloves, protective mask

Equipment:

1. 96-Well plates
2. Eppendorf tubes (1.5 mL)
3. Micropipettes (2–20, 20–200, 100–1000 μL)
4. Pipette tips
5. Elisa Reader
6. UV–vis spectrophotometer

Chemicals and materials

1. BCA kit, Pierce™ BCA Protein Assay Kit, from ThermoFisher Scientific.
2. Bradford reagent (Coomassie Brilliant Blue G-250, from Sigma).
3. Milli-Q SP ultrapure water.

Methodology

A protein quantification assay such as bicinchoninic acid (BCA) (Smith et al., 1985) or Bradford assay (Bradford, 1976) may be used to determine the immobilization yield. Both of these assays are widely applied to measure the protein concentration in a solution. The amount of enzyme bound on the immobilization matrix can be estimated by calculating the difference between the amount of the protein in the supernatant after immobilization and the initial protein amount used.

The immobilization efficiency of covalently immobilized TvL onto various GO-derivatives is presented in Table 1. The immobilization yield strongly depends on the functionalization of GO. Nonfunctionalized GO exhibits the lowest capacity as immobilization carrier, while the immobilization efficiency is increased in the case of GO-derivatives, indicating the beneficial effect of the nanomaterial functionalization. The kind of diamine or amino acid used for the functionalization of GO, also seems to play a critical role for the successful grafting of the enzyme on the nanomaterials. As the alkyl chain of the organic molecule increases, the immobilization yield also increases. The use of longer alkyl chain is expected to increase

Table 1 Immobilization Yield (%) of TvL on GO-Derivatives, Calculated as the Ratio of Protein Immobilized on Nanomaterials to the Initial Protein Quantity Used

Nanomaterial Carrier	Immobilization Yield (%)
GO	43
GO-C_2-COOH	57
GO-C_{10}-COOH	67
GO-C_2-NH_2	60
GO-C_4-NH_2	67
GO-C_6-NH_2	84

the distance between the protein and the nanomaterials' surface and thus to reduce the undesired interactions upon immobilization (Patila, Pavlidis, et al., 2016), leading to the formation of stable covalent bonds.

4.1.1 BCA Assay

1. Add one part of reagent A from the kit (containing bicinchoninic acid and sodium hydroxide), in 50 parts of the reagent B (containing copper sulfate).
2. Add 0.06 mL of enzyme solution in 0.19 mL of the reagent mixture and mix gently.
3. Incubate the samples at 60°C for 30 min.
4. Allow the samples to cool at room temperature.
5. Measure the absorbance of the samples at 562 nm.
6. Quantification of observed absorptions is achieved by applying a standard albumin curve (1–20 μg mL^{-1} in the measuring solution). The immobilization yield was expressed as

$$\text{Immobilization yield } (\%) = 100 - \frac{a^* ABS_B + b}{a^* ABS_A + b} \times 100$$

where ABS_B is the absorbance at 562 nm of the enzyme solution before immobilization and ABS_A is the absorbance at 562 nm of the supernatant solution after the immobilization procedure. Parameters a and b are obtained from a standard albumin curve of the form $y = ax + b$, where y is the concentration of protein (μg mL^{-1}) and x is the absorbance at 562nm.

4.1.2 Bradford Assay

1. Add 0.2 mL (ultrapure water and suitable amount of the enzyme solution) in 0.8 mL of Bradford reagent and mix gently.
2. Incubate the samples at room temperature for 15 min.
3. Measure the absorbance of the samples at 595 nm.
4. Quantification of observed absorptions is achieved by applying a standard albumin curve (1–20 μg mL^{-1} in the measuring solution). The immobilization yield was expressed as

$$\text{Immobilization yield } (\%) = 100 - \frac{a^* ABS_B + b}{a^* ABS_A + b} \times 100$$

where ABS_B is the absorbance of the enzyme solution at 595 nm before immobilization and ABS_A is the absorbance at 595 nm of the supernatant solution after the immobilization procedure. Parameters a and b are obtained from a standard albumin curve of the form $y= ax + b$, where y is the concentration of protein (μg mL^{-1}) and x is the absorbance at 595nm.

Tip 1: Before performing the protein quantification assay, the supernatant solution must be free of any nanomaterial residues. An additional centrifugation is highly recommended.

Tip 2: The samples prepared for Bradford assay should be incubated at room temperature for at least 5 min. Absorbance will increase over time; samples should incubate at room temperature for no more than 1 h.

4.2 Determination of Free and Immobilized Laccase Activity

Determining the catalytic activity of an immobilized enzyme provides a perspective into the performance of the bound enzyme compared to the free enzyme form. In this way, a study can be conducted concerning the possible advantages or disadvantages of an immobilization procedure. The activity of an immobilized enzyme can mainly be affected by the substrate accessibility to the active site as well as the ease of diffusion of the reagents. The amount of the loaded enzyme is also of utmost importance (Zhang, Yuwen, & Peng, 2013).

General safety: nitrile gloves, protective mask

Equipment:

1. 96-Well plates
2. Micropipettes (2–20 and 20–200 μL)
3. Pipette tips
4. Elisa Reader
5. Ultrasonic bath

Chemicals and materials

1. Laccase from *Trametes versicolor* (TvL) from Sigma-Aldrich Co., MO, USA.
2. 2,2′-Azino-bis(3-ethylbenzothiazoline-6-sulfonic acid) (ABTS), from Sigma.
3. Sodium acetate buffer (0.1 *M*, pH 4.58).

Methodology

Laccase activity is determined by following the oxidation of ABTS at room temperature. ABTS is oxidized to the more stable state of a cation radical ($ABTS^{+}$). The enzyme activity is correlated to the concentration of the cation radical produced, which is responsible for the intense blue-green color observed in the assay solution, and can be determined by measuring the change in the absorbance at 415 nm (Wolfenden & Willson, 1982).

Protocol

1. Add a suitable amount of free or immobilized TvL (dispersed in acetate buffer) in 1 m*M* of ABTS in sodium acetate buffer (0.1 *M*, pH 4.58).
2. Measure the oxidation of ABTS by following the change in the absorbance at 415 nm.
3. Create a graph of kinetic data in terms of absorbance at 415 nm vs time and then use the initial slope of the graph to calculate the initial rates of the reaction.

Tip 1: In the present work, one activity Unit is defined as the amount of laccase that catalyzes the conversion of 1 μmol of ABTS per minute.

Tip 2: The immobilized enzyme has to be fully dispersed in the buffer. Use an ultrasonic bath for the dispersion, but only for a few seconds to avoid enzyme deactivation. The dispersion capacity mostly depends on the physicochemical properties of each material; therefore, the dispersion conditions may vary.

Tip 3: Use freshly prepared ABTS stock solution for every experimental batch.

4.3 Characterization of Nanobiocatalysts by FTIR Spectroscopy

FTIR characterization was conducted using a Shimadzu FT-IR 8400 spectrometer (Tokyo, Japan) equipped with a deuterated triglycine sulfate detector, with 2 cm^{-1} resolution and 1 cm^{-1} interval. The samples were measured as 2 wt% KBr pellets, and 32 scans were run and averaged to obtain a good signal-to-noise ratio. The changes occurring to the chemical functionalities on the surface of free TvL, GO-C_6-NH_2 and immobilized TvL on GO-C_6-NH_2 are presented in Fig. 4. GO-C_6-NH_2 displays most of the characteristic peaks of GO, such as a peak at 1626 cm^{-1} assigned to the C=O stretching vibrations of the —COOH groups, a weak band at 1394 cm^{-1} assigned to the O—H deformations of the C—OH groups and a strong band at 1060 cm^{-1} attributed to C—O stretching vibrations (Enotiadis et al., 2012). The disappearance of the band at 1230 cm^{-1}, which is assigned to asymmetric stretching of C—O—C bridges in epoxy groups, arises from the nucleophilic substitution reactions between the amine end groups of the organic molecules and the epoxy groups of the GO (Spyrou et al., 2015, 2014), confirming the successful functionalization of GO. Free TvL shows a —C—O stretching amide I band at 1634 cm^{-1}, a weak N—H bending amide III band at 1260 cm^{-1} and a weak peak at 1734 cm^{-1} assigned to the C=O stretching vibrations of the —COOH groups of the enzyme molecule. Immobilized TvL shows amide I and amide II bands at 1631 and 1550 cm^{-1}, respectively, confirming the successful immobilization of TvL onto GO-C_6-NH_2. The blue shift of the N—H stretching band at 3235 cm^{-1} suggests stronger amide bond in the TvL/GO-C_6-NH_2 conjugates than the TvL alone (Premaratne, Coats, & Krishnan, 2017). Finally, two extra bands at 2925 and 2856 cm^{-1}, due to asymmetric and symmetric stretching vibrations of —CH_2 groups, confirm the presence of the amine derivatives in the TvL/GO-C_6-NH_2.

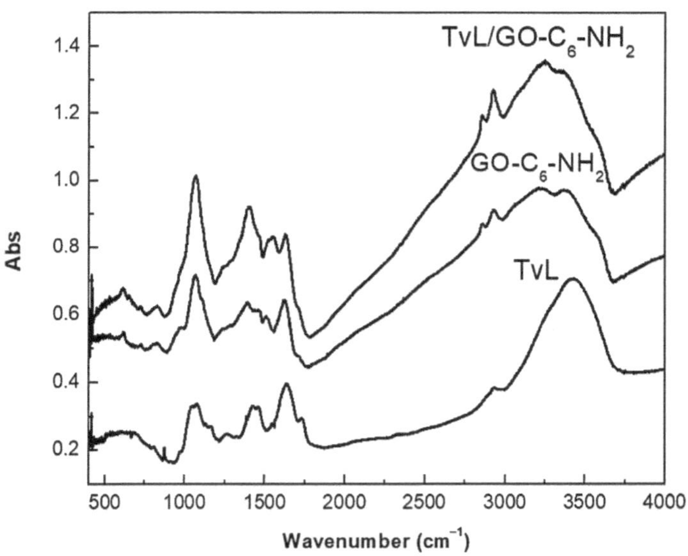

Fig. 4 FT-IR spectra of free TvL, GO-C_6-NH_2, and covalently immobilized TvL on GO-C_6-NH_2.

Tip 1: The samples must be well dispersed in KBr, in order to obtain a uniform pellet.

Tip 2: Deconvolution of the amide I band of TvL before and after immobilization, followed by curve fitting with a Gaussian function, yields several distinct underlying absorption bands which can be used to determine the secondary structural components of the protein (Patila, Pavlidis, et al., 2016; Tzialla et al., 2010).

Tip 3: Circular dichroism (CD) spectroscopy can also be applied for the study of secondary structure of the immobilized enzyme in a solution (Patila et al., 2013). Nevertheless, limitations may arise due to the high concentration of the biocatalyst needed, which is assigned at the same time to concentrated nanomaterial in the sample, which prevents the correct measurement of the sample.

4.4 Characterization of Laccase-Based Multilayer Nanoassemblies by AFM

AFM images were collected in tapping mode with a 3D Multimode Nanoscope, using Tap-300G silicon cantilevers (Ted Pella Inc., Redding, CA, USA) with a tip radius $<10\,nm$ and a force constant of $\approx 20–75\,N\,m^{-1}$. Samples were deposited onto silicon wafers (P/Bor, single side polished) from aqueous solutions by drop casting. The use of AFM has very important advantages, such as three-dimensional image design, great resolution in images as well as capability study of nonconductive materials. For these reasons, AFM is one of the most widespread techniques for morphological study and characterization of both nanostructured materials (Shao, Qian, Wu, Zhang, & Cai, 2013; Wang et al., 2014) and nanobiocatalysts (Pang, Li, & Zhang, 2015; Zhang et al., 2014).

AFM images of fGO-TvL multilayer (fGO-TvL-fGO-TvL-fGO-TvL-fGO) biocatalyst are shown in Fig. 5. The multilayer structure consists of four f-GO layers and is visible both in height (Fig. 5A) as well as in phase (Fig. 5B) images. This structure is further confirmed by the topographical height profile (Fig. 5C). Comparing the overall thickness of the fGO-TvL monolayer (found to be 12 nm) with the average thickness of multilayer (37 nm), it has been concluded that the multilayer structure consists of three layers of fGO-TvL ($3 \times 12 = 36\,nm$) plus one functionalized graphene oxide (f-GO) sheet (1 nm).

Tip 1: For AFM measurements, the samples must be diluted enough so as to avoid agglomeration.

Tip 2: The samples must be well dispersed before applied on the silicon wafer substrate for the AFM measurement.

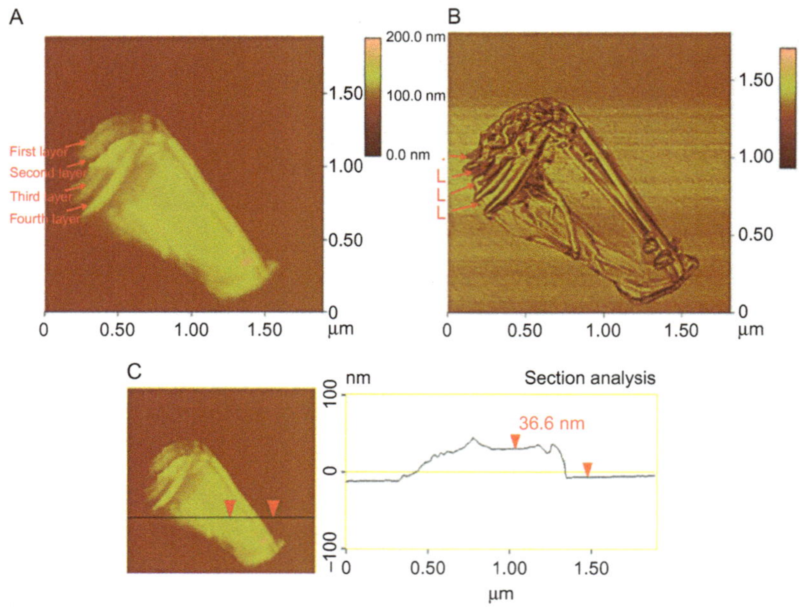

Fig. 5 (A) AFM height image, (B) phase image, and (C) and section analysis profile of fGO-TvL multilayer. *Reprinted from Patila, M., Kouloumpis, A., Gournis, D., Rudolf, P., & Stamatis, H. (2016). Laccase-functionalized graphene oxide assemblies as efficient nanobiocatalysts for oxidation reactions.* Sensors (Switzerland), 16, *1–14. https://doi.org/10.3390/s16030287, licensed under CC BY 4.0.*

5. STABILITY AND REUSABILITY OF NANOBIOCATALYSTS

5.1 Determination of the Stability of Immobilized TvL

Enzymes often show high decrease (even total loss) of catalytic activity when exposed to extreme conditions such as extreme values of pH, temperature, or organic solvents. From this point of view, enzyme immobilization has been widely applied as the most efficient technology to enhance enzyme recovery and stability against pH, thermal, and solvent exposure (Hermanova, Zarevucka, Bousa, Pumera, & Sofer, 2015). Furthermore, a crucial step for enzyme stabilization is the choice of the nanomaterials used as immobilization carriers. In our study, GO-derivatives with different alkyl chain length and terminal functional groups exhibited high potential in enzyme stabilization (Fig. 6). In all cases studied, the stability of covalently immobilized TvL on GO-derivatives is significantly higher than that of the

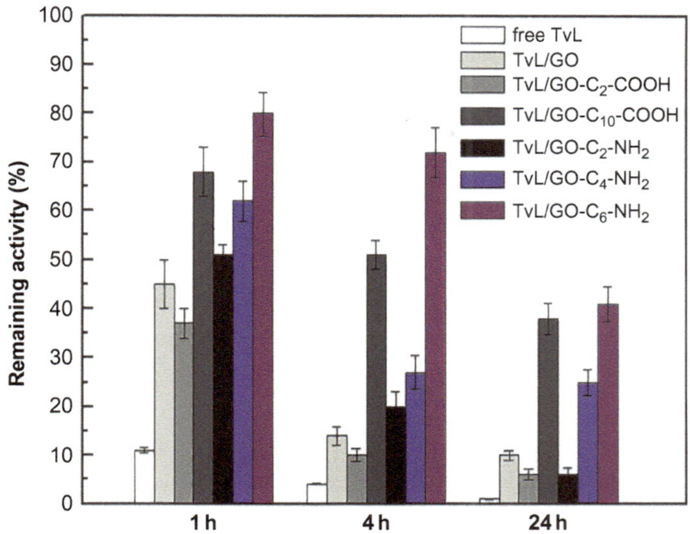

Fig. 6 Residual activity of free TvL and covalently immobilized TvL on GO-derivatives after 24h incubation in acetate buffer. As 100% is indicated the activity of TvL at $t = 0$ min.

free enzyme. For instance, after 24h of incubation at 60°C, the covalently immobilized TvL on GO-C_{10}-COOH and GO-C_6-NH_2 retain up to 38% and 41% of its initial activity, respectively, while free TvL almost deactivates, indicating that GO-derivatives enhance the thermal stability of the protein. This result also suggests that the stability of the immobilized enzyme is increased when the alkyl chain length of the fGO increases. Our group has previously proposed that the hydrophobic environment created by the increase of the alkyl chain of the nanomaterials protects in a higher extent the protein from denaturing conditions (Patila, Pavlidis, et al., 2016).

General safety: nitrile gloves, protective mask

Equipment:

1. 96-Well plates
2. Micropipettes (2–20, 20–200, 100–1000 μL)
3. Pipette tips
4. Eppendorf tubes (1.5 mL)
5. Thermoshaker
6. Elisa Reader
7. Ultrasonic bath

Chemicals and materials

1. Laccase from *Trametes versicolor* (TvL) from Sigma-Aldrich Co., MO, USA.
2. 2,2′-azino-bis(3-ethylbenzothiazoline-6-sulfonic acid) (ABTS), from Sigma.
3. Sodium acetate buffer (0.1 *M*, pH 4.58).
4. Immobilized TvL on GO-derivatives.

Methodology

The thermal stability of immobilized laccase at different temperatures is determined by calculating the remaining oxidation activity of laccase through the ABTS oxidation (Patila, Kouloumpis, et al., 2016).

Protocol

1. Prepare enzyme solutions with final concentration of 1 and 50 $\mu g\,mL^{-1}$ for free and immobilized TvL, respectively, in acetate buffer (0.1 *M*, pH 4.58).
2. Incubate the enzyme solutions at different temperatures, up to 24 h.
3. Take a specimen at predetermined time intervals to determine the remaining oxidation activity of laccase through the ABTS oxidation as described before.

Tip 1: Regarding the immobilized enzyme preparation, a well-dispersed mixture need to be achieved. Depending on the physicochemical properties of each material, appropriate ultrasonic activity is proposed.

Tip 2: The activity assay through ABTS oxidation has to be performed in room temperature, so cooling down of the samples before measurement is necessary.

Tip 3: The same protocol can be followed for the determination of enzyme stability at different pH values.

5.2 Determination of the Reusability of Immobilized TvL

Regarding industrial applications, multiple cycles of high yield processes are required so that stabilization of the biocatalyst through an immobilization protocol can be proved pivotal, especially when it comes to economic feasibility. Thereafter, each supporting material affects enzyme stabilization in a distinctive way, depending on the specific enzyme–support interactions (Mateo, Palomo, Fernandez-Lorente, Guisan, & Fernandez-Lafuente, 2007; Pavlidis, Vorhaben, Tsoufis, et al., 2012). The operational stability of a biocatalyst can be measured using different reactions and substrates, depending on the application of interest. Our group has already published a scientific research, in which enzyme–nanomaterial nanoassemblies were used for the continuous decolorization of pinacyanol chloride, a symmetric trimethinecyanine dye widely applied as an optical sensitizer in photographic processes (Patila, Kouloumpis, et al., 2016). fGO-TvL nanoassemblies, consisted of different number of layers of nanomaterial and enzyme, were investigated with respect to pinacyanol chloride decolorization in many catalytic cycles (Patila, Kouloumpis, et al., 2016). In all cases studied, the immobilized laccase was found to be able to decolorize the dye up to five reaction cycles (Fig. 7). After the completion of five reaction cycles the ability of fGO-TvL nanoassemblies to catalyze the decolorization of the dye was highly maintained (up to 94%), especially when multilayer fGO-TvL nanoassemblies were used as biocatalysts, indicating that the addition of extra fGO-TvL layers to the nanoassembly not only benefits the thermal stability and reaction efficiency, but also results in reusable nanobiocatalysts.

General safety: nitrile gloves, protective mask

Equipment:

1. Micropipettes (2–20, 20–200, 100–1000 μL)
2. Pipette tips
3. Eppendorf tubes (1.5 mL)
4. Thermoshaker

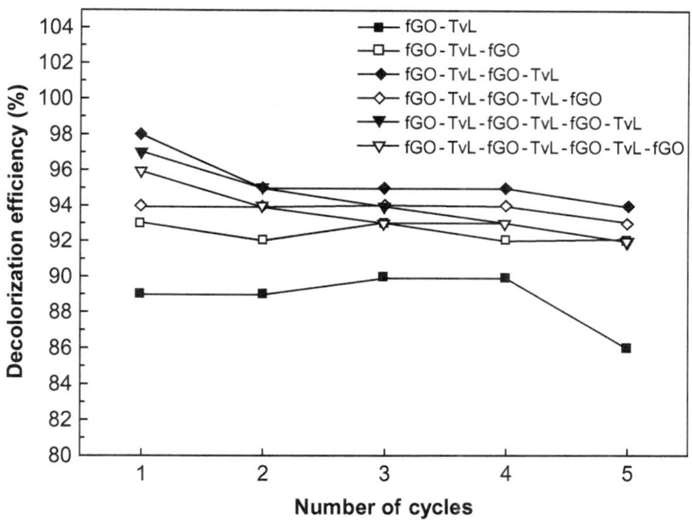

Fig. 7 Reusability of fGO-TvL nanoassemblies in the batch oxidation of pinacyanol chloride. *Reprinted from Patila, M., Kouloumpis, A., Gournis, D., Rudolf, P., & Stamatis, H. (2016). Laccase-functionalized graphene oxide assemblies as efficient nanobiocatalysts for oxidation reactions.* Sensors (Switzerland), 16, *1–14. https://doi.org/10.3390/s16030287, licensed under CC BY 4.0.*

5. UV–vis spectrophotometer
6. Ultrasonic bath
7. Centrifuge for Eppendorf tubes

Chemicals and materials

1. Pinacyanol chloride (1,11-diethyl-2,21-carbocyanine chloride), from Fluka (St. Louis, MO, USA).
2. Sodium acetate buffer (0.1 *M*, pH 4.58).
3. Immobilized TvL on GO-derivatives.

Methodology

The reusability of an enzyme can be determined by measuring the remaining activity of the biocatalyst after several operating catalytic cycles, with respect to the initial reaction rate.

Protocol

1. Add $100\,\mu g\,mL^{-1}$ pinacyanol chloride in 0.4 mL sodium acetate buffer (0.1 *M*, pH 4.58) and measure the initial absorbance of the dye at 603 nm.
2. Add 0.1 mL of immobilized TvL in the dye solution (to reach a final concentration of $500\,\mu g\,mL^{-1}$).
3. Incubate the reaction mixture at 30°C under stirring at 800 rpm for up to 5 h.
4. Separate the immobilized TvL, by centrifugation at 12,000 rpm, after each reaction cycle.
5. Monitor the remaining concentration of the pinacyanol chloride by measuring the absorbance at 603 nm. Use an extinction coefficient for pinacyanol chloride equal to $\varepsilon = 82{,}350\ M^{-1}\,cm^{-1}$ (Vazquez-Duhalt, Westlake, & Fedorak, 1995).
6. Wash the precipitate immobilized TvL three times with acetate buffer (0.1 *M*, pH 4.58).
7. Repeat the process until there is no enzymatic activity.
8. Estimate the decolorization yield after each reaction cycle through the following equation:

$$\text{Decolorization}\,(\%) = (C_i - C_t)/C_i \times 100$$

where C_i is the initial concentration and C_t is the concentration after incubation time.

Tip: During centrifugation at the end of each catalytic cycle, be sure that the immobilized enzyme is well precipitated, in order to maintain the same enzyme concentration in the reaction medium.

6. CONCLUSION

Herein, we present the use of functionalized graphene oxide derivatives, as nanoscaffolds for the effective immobilization and stabilization of laccase from *Trametes versicolor,* an enzyme with a wide range of applications. Nonspecific binding via physical adsorption is proposed as a simple process which enhances the immobilization efficiency, while covalent binding on functionalized GO-derivatives with terminal carboxyl or amine groups is also presented. It is interesting to note that nonfunctionalized GO exhibits the lowest capacity as an immobilization carrier, while the immobilization efficiency increases with the use of GO-derivatives, indicating the beneficial effect of nanomaterial functionalization on enzyme's performance. It is also worth mentioning that as the alkyl chain of the organic molecules grafted on GO increases, the immobilization yield follows this raise. This observation is attributed to the fact that the use of longer alkyl chain is expected to increase the distance between the protein and the nanomaterial's surface, thus reducing the undesired interactions upon immobilization.

Another immobilization approach is also presented, regarding multilayer nanostructured systems based on the LbL deposition as a versatile method to produce thin-film multilayers with controlled properties. In this approach, nanobiocatalysts were prepared through multipoint covalent attachment of the enzyme to functionalized GO sheets. These nanoassemblies were applied for the decolorization of pinacyanol chloride, demonstrating great catalytic capabilities of up to five reaction cycles. It was also concluded that the addition of extra fGO-TvL layers to the nanoassembly not only benefits the thermal stability and reaction efficiency, but also produces reusable nanobiocatalysts.

The nanobiocatalytic systems developed demonstrate significant enhancement of the biocatalyst's stability and reusability in comparison with the performance of the free enzyme form. Creating such efficient laccase-based nanobiocatalytic systems can contribute substantially to various industrial applications from degradation of pollutants to organic synthesis.

ACKNOWLEDGMENTS

This work was supported by the project "Synthetic Biology: from omics technologies to genomic engineering (OMIC-ENGINE)" (MIS 5002636) which is implemented under the action "Reinforcement of the Research and Innovation Infrastructure," funded by the Operational Programme "Competitiveness, Entrepreneurship and Innovation" (NSRF

2014–2020) and cofinanced by Greece and the European Union (European Regional Development Fund); by I.K.Y. Foundation scholarship which was funded by Operational Program "Human Resources Development, Education and Lifelong Learning," 2014–2020, with MIS 5000432, and cofinancing of the European Social Fund (ESF) and the Greek National Funds and the I.K.Y. Foundation scholarship which was funded by the Action of "Enhancement of post-doc researchers" from the funds of the Operational Program "Human Resources Development, Education and Life-Long Learning" with priority axis 6,8,9 and cofunded by the European Social Fund and the Greek National Fund (MIS 5001552).

The authors thank colleagues and collaborators named in joint publications for their valuable contributions.

REFERENCES

Ba, S., & Vinoth Kumar, V. (2017). Recent developments in the use of tyrosinase and laccase in environmental applications. *Critical Reviews in Biotechnology*, *37*, 819–832.

Baptista-Pires, L., Pérez-López, B., Mayorga-Martinez, C. C., Morales-Narváez, E., Domingo, N., Esplandiu, M. J., et al. (2014). Electrocatalytic tuning of biosensing response through electrostatic or hydrophobic enzyme-graphene oxide interactions. *Biosensors & Bioelectronics*, *61*, 655–662.

Barrios-Estrada, C., de Jesús Rostro-Alanis, M., Muñoz-Gutiérrez, B. D., Iqbal, H. M. N., Kannan, S., & Parra-Saldívar, R. (2018). Emergent contaminants: Endocrine disruptors and their laccase-assisted degradation—A review. *The Science of the Total Environment*, *612*, 1516–1531.

Betancor, L., Johnson, G. R., & Luckarift, H. R. (2013). Stabilized laccases as heterogeneous bioelectrocatalysts. *ChemCatChem*, *5*, 46–60.

Bourlinos, A. B., Gournis, D., & Petridis, D. (2003). Graphite oxide: Chemical reduction to graphite and surface modification with primary aliphatic amines and amino acids. *Langmuir*, *19*, 6050–6055.

Bradford, M. M. (1976). A rapid and sensitive method for quantitation of microgram quantities of protein utilizing the principle of protein-dye-binding. *Analytical Biochemistry*, *72*, 248–254.

Campbell, A. S., Dong, C., Meng, F., Hardinger, J., Perhinschi, G., Wu, N., et al. (2014). Enzyme catalytic efficiency: A function of bio-nano interface reactions. *ACS Applied Materials & Interfaces*, *6*, 5393–5403.

Chang, S. W., Shaw, J. F., Yang, K. H., Chang, S. F., & Shieh, C. J. (2008). Studies of optimum conditions for covalent immobilization of *Candida rugosa* lipase on poly(γ-glutamic acid) by RSM. *Bioresource Technology*, *99*, 2800–2805.

Chen, M., Zeng, G., Xu, P., Lai, C., & Tang, L. (2017). How do enzymes "meet" nanoparticles and nanomaterials? *Trends in Biochemical Sciences*, *42*, 914–930.

de Jesús Rostro-Alanis, M., Mancera-Andrade, E. I., Patiño, M. B. G., Arrieta-Baez, D., Cardenas, B., Martinez-Chapa, S. O., et al. (2016). Nanobiocatalysis: Nanostructured materials—A minireview. *Biocatalysis*, *2*, 1–24.

Decher, G. (1997). Fuzzy nanoassemblies: Toward layered polymeric multicomposites. *Science*, *277*, 1232–1237.

Enotiadis, A., Angjeli, K., Baldino, N., Nicotera, I., & Gournis, D. (2012). Graphene-based nafion nanocomposite membranes: Enhanced proton transport and water retention by novel organo-functionalized graphene oxide nanosheets. *Small*, *8*, 3338–3349.

Fernández-Fernández, M., Sanromán, M. Á., & Moldes, D. (2013). Recent developments and applications of immobilized laccase. *Biotechnology Advances*, *31*, 1808–1825.

Gao, Y., & Kyratzis, I. (2008). Covalent immobilization of proteins on carbon nanotubes using the cross-linker 1-ethyl-3-(3-dimethylaminopropyl)carbodiimide—A critical assessment. *Bioconjugate Chemistry*, *19*, 1945–1950.

Gilles, M. A., Hudson, A. Q., & Borders, C. L. (1990). Stability of water-soluble in aqueous solution. *Analytical Biochemistry*, *184*, 244–248.

Goenka, S., Sant, V., & Sant, S. (2014). Graphene-based nanomaterials for drug delivery and tissue engineering. *Journal of Controlled Release*, *173*, 75–88.

Hermanova, S., Zarevucka, M., Bousa, D., Pumera, M., & Sofer, Z. (2015). Graphene oxide immobilized enzymes show high thermal and solvent stability. *Nanoscale*, *7*, 5852–5858.

Hwang, E. T., & Gu, M. B. (2013). Enzyme stabilization by nano/microsized hybrid materials. *Engineering in Life Sciences*, *13*, 49–61.

Jin, L., Yang, K., Yao, K., Zhang, S., Tao, H., Lee, S., et al. (2012). Functionalized graphene oxide in enzyme engineering: A selective modulator for enzyme activity and thermostability functionalized graphene oxide in enzyme engineering. *ACS Nano*, *6*, 4864–4875.

Kim, J., Cote, L., Kim, F., Yuan, W., Shull, K., & Huang, J. (2010). Graphene oxide sheets at interfaces. *Journal of the American Chemical Society*, *132*, 8180–8186.

Kovtyukhova, N. I., Ollivier, P. J., Martin, B. R., Mallouk, T. E., Chizhik, S. A., Buzaneva, E. V., et al. (1999). Layer-by-layer assembly of ultrathin composite films from micron-sized graphite oxide sheets and polycations. *Chemistry of Materials*, *11*, 771–778.

Küchler, A., Yoshimoto, M., Luginbühl, S., Mavelli, F., & Walde, P. (2016). Enzymatic reactions in confined environments. *Nature Nanotechnology*, *11*, 409–420.

Kuo, C. H., Liu, Y. C., Chang, C. M. J., Chen, J. H., Chang, C., & Shieh, C. J. (2012). Optimum conditions for lipase immobilization on chitosan-coated Fe_3O_4 nanoparticles. *Carbohydrate Polymers*, *87*, 2538–2545.

Lerf, A., He, H., Forster, M., & Klinowski, J. (1998). Structure of graphite oxide revisited. *The Journal of Physical Chemistry. B*, *102*, 4477–4482.

Lin, J., Chen, X., & Huang, P. (2016). Graphene-based nanomaterials for bioimaging. *Advanced Drug Delivery Reviews*, *105*, 242–254.

Mate, D. M., & Alcalde, M. (2017). Laccase: A multi-purpose biocatalyst at the forefront of biotechnology. *Microbial Biotechnology*, *10*, 1457–1467.

Mateo, C., Palomo, J. M., Fernandez-Lorente, G., Guisan, J. M., & Fernandez-Lafuente, R. (2007). Improvement of enzyme activity, stability and selectivity via immobilization techniques. *Enzyme and Microbial Technology*, *40*, 1451–1463.

Mohamad, N. R., Marzuki, N. H. C., Buang, N. A., Huyop, F., & Wahab, R. A. (2015). An overview of technologies for immobilization of enzymes and surface analysis techniques for immobilized enzymes. *Biotechnology and Biotechnological Equipment*, *29*, 205–220.

Orfanakis, G., Patila, M., Catzikonstantinou, A. V., Lyra, K., Kouloumpis, A., Spyrou, K., et al. (2018). Hybrid nanomaterials of magnetic iron nanoparticles and graphene oxide as matrices for the immobilization of β-glucosidase: Synthesis, characterization, and biocatalytic properties. *Frontiers in Materials*, *5*, 11 pages.

Pang, R., Li, M., & Zhang, C. (2015). Degradation of phenolic compounds by laccase immobilized on carbon nanomaterials: Diffusional limitation investigation. *Talanta*, *131*, 38–45.

Patila, M., Kouloumpis, A., Gournis, D., Rudolf, P., & Stamatis, H. (2016). Laccase-functionalized graphene oxide assemblies as efficient nanobiocatalysts for oxidation reactions. *Sensors (Basel)*, *16*, 1–14.

Patila, M., Orfanakis, G., Polydera, A., Pavlidis, I., & Stamatis, H. (2017). Graphene-based nanobiocatalytic systems. In P. Grunwald (Ed.), *Biocatalysis and nanotechnology* (pp. 243–277). Pan Stanford Publishing Pte. Ltd. chapter 6.

Patila, M., Pavlidis, I. V., Diamanti, E. K., Katapodis, P., Gournis, D., & Stamatis, H. (2013). Enhancement of cytochrome c catalytic behaviour by affecting the heme environment using functionalized carbon-based nanomaterials. *Process Biochemistry*, *48*, 1010–1017.

Patila, M., Pavlidis, I. V., Kouloumpis, A., Dimos, K., Spyrou, K., Katapodis, P., et al. (2016). Graphene oxide derivatives with variable alkyl chain length and terminal functional groups as supports for stabilization of cytochrome c. *International Journal of Biological Macromolecules*, *84*, 227–235.

Pavlidis, I. V., Patila, M., Bornscheuer, U. T., Gournis, D., & Stamatis, H. (2014). Graphene-based nanobiocatalytic systems: Recent advances and future prospects. *Trends in Biotechnology*, *32*, 312–320.

Pavlidis, I. V., Patila, M., Polydera, A. C., Gournis, D., & Stamatis, H. (2014). Immobilization of enzymes and other biomolecules on graphene. In V. Georgakilas (Ed.), *Functionalization of graphene* (pp. 139–172). Wiley-VCH Verlag & Co. chapter 5.

Pavlidis, I. V., Tsoufis, T., Enotiadis, A., Gournis, D., & Stamatis, H. (2010). Functionalized multi-wall carbon nanotubes for lipase immobilization. *Advanced Engineering Materials*, *12*, 179–183.

Pavlidis, I. V., Vorhaben, T., Gournis, D., Papadopoulos, G. K., Bornscheuer, U. T., & Stamatis, H. (2012). Regulation of catalytic behaviour of hydrolases through interactions with functionalized carbon-based nanomaterials. *Journal of Nanoparticle Research*, *14*, 1–10.

Pavlidis, I. V., Vorhaben, T., Tsoufis, T., Rudolf, P., Bornscheuer, U. T., Gournis, D., et al. (2012). Development of effective nanobiocatalytic systems through the immobilization of hydrolases on functionalized carbon-based nanomaterials. *Bioresource Technology*, *115*, 164–171.

Prasad, K. P., Chen, Y., & Chen, P. (2014). Three-dimensional graphene-carbon nanotube hybrid for high-performance enzymatic biofuel cells. *ACS Applied Materials & Interfaces*, *6*, 3387–3393.

Premaratne, G., Coats, L., & Krishnan, S. (2017). NanoArmoring of enzymes by polymer-functionalized iron oxide nanoparticles. In *Methods in Enzymology* (pp. 225–257). Elsevier Inc. chapter 10.

Sakr, O. S., & Borchard, G. (2013). Encapsulation of enzymes in layer-by-layer (LbL) structures: Latest advances and applications. *Biomacromolecules*, *14*, 2117–2135.

Schwinté, P., Ball, V., Szalontai, B., Haikel, Y., Voegel, J. C., & Schaaf, P. (2002). Secondary structure of proteins adsorbed onto or embedded in polyelectrolyte multilayers. *Biomacromolecules*, *3*, 1135–1143.

Sehgal, D., & Vijay, I. (1994). A method for the high efficiency of water-soluble carbodiimide-mediated amidation. *Analytical Biochemistry*, *218*, 87–91.

Shao, Q., Qian, Y., Wu, P., Zhang, H., & Cai, C. (2013). Graphene oxide-induced conformation changes of glucose oxidase studied by infrared spectroscopy. *Colloids and Surfaces. B, Biointerfaces*, *109*, 115–120.

Shim, M., Kam, N. W. S., Chen, R., Li, Y., & Dai, H. (2002). Functionalisation of carbon nanotubes for biocompatability and biomolecular recognition. *Nano Letters*, *2*, 285–288.

Smith, P. K., Krohn, R. I., Hermanson, G. T., Mallia, A. K., Gartner, F. H., & Provenzano, M. D. (1985). Measurement of protein using bicinchoninic acid. *Analytical Biochemistry*, *150*, 76–85.

Spyrou, K., Calvaresi, M., Diamanti, E. K., Tsoufis, T., Gournis, D., Rudolf, P., et al. (2015). Graphite oxide and aromatic amines: Size matters. *Advanced Functional Materials*, *25*, 263–269.

Spyrou, K., Potsi, G., Diamanti, E. K., Ke, X., Serestatidou, E., Verginadis, I. I., et al. (2014). Towards novel multifunctional pillared nanostructures: Effective intercalation of adamantylamine in graphene oxide and smectite clays. *Advanced Functional Materials*, *24*, 5841–5850.

Staudenmaier, L. (1898). Verfahren zur darstellung der graphitslure. *Berichte der Deutschen Chemischen Gesellschaft*, *31*, 1481–1487.

Tzialla, A. A., Pavlidis, I. V., Felicissimo, M. P., Rudolf, P., Gournis, D., & Stamatis, H. (2010). Lipase immobilization on smectite nanoclays: Characterization and application to the epoxidation of α-pinene. *Bioresource Technology*, *101*, 1587–1594.

Tzialla, A. A., Taha, A. A., Kalogeris, E., & Stamatis, H. (2009). Improving the catalytic performance of fungal laccases in monoterpene-based reaction systems. *Biotechnology Letters*, *31*, 1451–1456.

Vazquez-Duhalt, R., Westlake, D. W. S., & Fedorak, P. M. (1995). Kinetics of chemically modified lignin peroxidase and enzymatic oxidation of aromatic nitrogen-containing compounds. *Applied Microbiology and Biotechnology*, *42*, 675–681.

Wang, Q., Cui, J., Li, G., Zhang, J., Li, D., Huang, F., et al. (2014). Laccase immobilized on a pan/adsorbents composite nanofibrous membrane for catechol treatment by a biocatalysis/adsorption process. *Molecules*, *19*, 3376–3388.

Wolfenden, B., & Willson, R. (1982). Radical-cations as reference chromogens in kinetic studies of one-electron transfer reactions: Pulse radiolysis studies of 2,2′-azinobis-ethyl benzthiazoline-6-sulphonate. *Journal of the Chemical Society, Perkin Transactions 2*, *2*, 805–812.

Yang, K., Feng, L., Shi, X., & Liu, Z. (2013). Nano-graphene in biomedicine: Theranostic applications. *Chemical Society Reviews*, *42*, 530–547.

Yu, H., Zhang, B., Bulin, C., Li, R., & Xing, R. (2016). High-efficient synthesis of graphene oxide based on improved hummers method. *Scientific Reports*, *6*, 1–7.

Zhang, G., Ma, J., Wang, J., Li, Y., Zhang, G., Zhang, F., et al. (2014). Lipase immobilized on graphene oxide as reusable biocatalyst. *Industrial and Engineering Chemistry Research*, *53*, 19878–19883.

Zhang, D., Yuwen, L., & Peng, L. (2013). Parameters affecting the performance of immobilized enzyme. *Journal of Chemistry—Hindawi Publishing Corporation*, *2013*, 1–7.

Zhao, H.-Z., Du, Q., Li, Z.-S., & Yang, Q.-Z. (2012). Mechanisms for the direct electron transfer of cytochrome c induced by multi-walled carbon nanotubes. *Sensors*, *12*, 10450–10462.

CHAPTER FOUR

Synthesis, Characterization, and Applications of Nanographene-Armored Enzymes

Maryam Khan, Qayyum Husain[1], Shamoon Asmat
Department of Biochemistry, Faculty of Life Sciences, Aligarh Muslim University, Aligarh, India
[1]Corresponding author: e-mail address: qayyumbiochem@gmail.com

Contents

Abstract

The unexpected discovery of graphene and especially the follow-up explosion of interest in its properties and applications marked the beginning of a new carbon era. Graphene-based nanostructured materials are highly useful because they show great promise in the field of biotechnology and biomedicine. Owing to their unique structural features, exceptional chemical, electrical, and mechanical properties, and their ability to

Methods in Enzymology, Volume 609
ISSN 0076-6879
https://doi.org/10.1016/bs.mie.2018.05.013

affect the microenvironment of biomolecules, graphene-armored nanomaterials are suitable for use in various applications, such as immobilization of enzymes, field-effect transistors, photovoltaic devices, and biosensors, which in turn is extremely vital to the development of biomedical instruments, clinical diagnosis, and disease treatment. In this chapter, we present our recently reported work to armor hydrolytic enzymes on graphene-based nanomaterial in order to develop novel scaffolds to build robust nanobiocatalysts. Synthesized graphene-Fe_3O_4 and polyaniline-coated silver graphene nanocomposite have been used to immobilize β-galactosidase and lipase, using noncovalent and covalent strategies, respectively. Herein, the methodologies of both techniques have been discussed in detail. Owing to the large surface area offered by the honeycomb like structure of graphene, very high amount of enzyme has been loaded on small amounts of the nanocomposite. The stability and reusability of the fabricated nanobiocatalysts have been compared with their free forms. Nanographene-armored enzymes demonstrated high catalytic stability and easy recovery from the reaction medium and can be applied in various biotechnological applications. Lastly, future prospects and possible challenges in this rapidly developing area have also been discussed.

ABBREVIATIONS

AChE	acetyl choline esterase
ADH	alcohol dehydrogenase
AFM	atomic force microscopy
AOx	alcohol oxidase
CNTs	carbon nanotubes
Cyt *c*	cytochrome *c*
DAOx	diamino oxidase
DLS	dynamic light scattering
EDS	energy dispersive X-ray spectroscopy
EDTA	ethylenediaminetetraacetic acid
FT-IR	Fourier-transform infrared spectroscopy
GO	graphene oxide
GOx	glucose oxidase
Gr	graphene
Gr@Fe_3O_4NC	graphene–iron oxide nanocomposite
Gr-NC	graphene nanocomposite
HRP	horseradish peroxidase
MNPs	magnetic nanoparticles
MP-11	microperoxidase-11
NC	nanocomposite
NM	nanomaterial
NPs	nanoparticles
ONPG	*o*-nitrophenyl β-D-galactopyranoside
PANI	polyaniline
PANI/Ag/GONC	polyaniline silver graphene oxide nanocomposite

p-NPP	*p*-nitrophenyl palmitate
PPO	polyphenol oxidase
RGO	reduced graphene oxide
Rh	hydrodynamic radius
SEM	scanning electron microcopy
TEM	transmission electron microscopy
Tyr	tyrosinase
VSM	vibrating sample magnetometer
XRD	X-ray diffraction

1. INTRODUCTION

Enzyme immobilization, a well-established and mature technology, leads to enhanced stability, facilitates reuse and easy separation from the reaction medium, thus allowing the modulation of catalytic properties of immobilized enzymes and creating robust biocatalysts suitable for their uses in various commercial applications (Al-Lolage, Meneghello, Ma, Ludwig, & Bartlett, 2017; Gomes et al., 2017; Husain, 2010; Husain & Qayyum, 2013; Husain & Ulber, 2011). The catalytic behavior of the enzyme is always affected by the nature of enzyme–support interactions and, thus, the availability of methods to immobilize enzymes with different materials is an important issue. Several approaches have been developed in the recent past for immobilization of enzymes (Ansari & Husain, 2012; Mateo, Palomo, Lorente, Guisan, & Lafuente, 2007; Sassolas, Blum, & Leca-Bouvier, 2012).

During the past 10 years, the synergistic interactions between nanotechnology and biotechnology have resulted in innovative functional biological nanosystems with potential applications in biotechnology, biosensing, and biomedical areas. Today, impressive achievements have been made at the cross-section of nanotechnology and biotechnology by employing modern nanomaterials (NM) to generate mechanical, optical, and electronic devices as well as biosensors (Husain, 2017a; Xiao, Bu, Yang, Zhang, & Xu, 2017; Yang et al., 2010). The development of effective nanobiocatalysts, in which enzymes are immobilized onto robust NM with tailor-made properties, is a typical example (Dutta, Biswas, & Saha, 2017; Husain, 2016, 2017b, 2017c).

Among the immobilization carriers used to date, nanostructured composite and hybrid materials, including metal oxide nanoparticles (NPs), nanofibers, and carbon-based NMs are now under focus of intense fundamental and applied research (Ansari & Husain, 2012; Ge, Yang, Zhu, Lu, & Liu, 2012; Husain, 2018a; Verma, Barrow, & Puri, 2013). One of the most advantageous features of NM is their potential for manipulating the environment of biomolecules and thus, of their biological function and stability (Husain, 2017d; Rana, Yeh, & Rotello, 2010). The unique properties of NM as immobilization support is due to their extremely small size, together with other desirable properties, such as conductivity, large surface area to volume ratio, high catalytic efficiency and surface reactivity, strong adsorption ability and magnetism, and these offer particularly exciting opportunities for the preparation of effective nanobiocatalysts for the development of unique enzyme applications (Husain, 2017e, 2018b; Johnson, Park, & Driscoll, 2011; Verma et al., 2013).

1.1 Synthesis of Graphene-Armored Materials

There are different synthetic routes to produce graphene (Gr) materials with distinct properties for various potential bioanalytical applications. Currently, the most common method to obtain Gr is through the chemical reduction of exfoliated graphene oxide (GO) (Alsharaeh, Othman, & Aldosari, 2014; Takahashi, Abiko, & Anzai, 2013; Wei et al., 2010; Yang, Feng, Hong, Cai, & Liu, 2013). This is called reduced GO (rGO) or chemically derived Gr. In addition to mechanical exfoliation, GO can also be chemically, thermally, or electrochemically engineered to manipulate its optoelectronic properties (Liu & Ma, 2014; Shi et al., 2015) and produce Gr-based materials (Allen, Tung, & Kaner, 2010; Chen, Tang, & Li, 2010), including single-layer Gr, few-layer Gr, GO, thermally reduced GO, and electrochemically reduced GO. In contrast to pristine Gr, chemical vapor deposition growth can be used to obtain large Gr films and to introduce heteroatoms like nitrogen, boron, etc. (Liu et al., 2014; Sun et al., 2010). The choice of peculiar Gr material should be made according to the targets to be detected and the biorecognition mechanism to be utilized.

1.2 Functionalization of Graphene-Armored Materials

Gr-based NMs can be engineered for grafting desirable functional groups (such as epoxide, carboxylic, and hydroxyl groups) onto their surface providing functionalized NM with tailor-made properties and improved suitability for their application as nanoscaffolds (Dai, 2013; Loh, Bao,

Ang, & Yang, 2010; Wang, Li, Wang, Li, & Lin, 2011). This in turn, offers the possibility for introduction of functionalities that enhance fine-tuned actions of the immobilized enzymes such as protection from protease cleavage (Wang, Li, et al., 2011), enhancement of transportation capability in living cells (Sun et al., 2008; Wang, Zhang, et al., 2011), switchable activities responding to external signals (Cho, Kim, & Choi, 2013), facilitation of electron transfer to the protein (Kuila et al., 2011; Putzbach & Ronkainen, 2013; Walcarius, Minteer, Wang, Lin, & Merkoci, 2013), and lastly incorporation of enzymes in microdevices and microchip bioreactors (Bao, Chen, Zhang, & Chen, 2011; Liang, Wang, Liu, Meng, & Qiu, 2013).

To obtain Gr-based nanoprobes for molecular diagnostics, it is necessary to functionalize Gr with special elements that bring the detection targets onto the Gr surface through partial interactions and sometimes for signal amplification or transduction (Xu, Shen, Li, & Ye, 2014; Yu et al., 2014; Zhou et al., 2014). Moreover, chemical doping (Jiang et al., 2014; Yang et al., 2013) and surface functionalization (Yu, Dong, Zhang, Wang, & Zhi, 2014) are used to obtain Gr with better surface properties and improved dispersion. Frequently, metal NPs (e.g., Au, Ag, Pt, and Pd) can be decorated through electrochemical deposition, electrospray, or in situ reduction (Singh et al., 2014; Yang, Asiri, Du, & Lin, 2014). The oxygenated groups on the surface of both Gr and GO facilitate the nucleation of metal NPs. Then, metal ions are reduced to give metal NPs (Song et al., 2014; Zheng, Li, Li, & Chen, 2014) particularly, metal NPs decorated on Gr are synthesized following a bottom-up approach. Based on different immobilization techniques, the NPs can mediate signal transduction or anchor biomolecular probes with high capacity in Gr-based biosensors (She et al., 2014; Ye et al., 2014). After different modes of functionalization, Gr can be used in various strategies of molecular diagnostics.

GO is easily functionalized, thanks to its abundant oxygen containing groups (Georgakilas, 2012). Since, the first isolation of freestanding Gr sheets, this two-dimensional (2D) carbon crystal has been highly anticipated to provide a new and unique chance for bioanalytical applications. In fact, Gr has already shown great potential in preparing various novel nanoprobes based on its great electrical properties (e.g., high electron transfer rate), optical properties (e.g., outstanding ability to quench fluorescence), and structural properties (e.g., extremely high surface to volume ratio). The large 2D aromatic surface of Gr may also be functionalized in order to enhance its bioanalytical properties (such as sensitivity, specificity, loading capacity, or biocompatibility), making it an ideal substrate for the synthesis and design of multicomponent nanoprobes (Pavlidis, Patila, Bornscheuer, Gounis, & Stamatis, 2014).

Various methods have been devised to functionalize Gr, including covalent methods, linker molecule mediating interaction, electrostatic interaction, and in situ synthesis of metal NPs. Chemical functional groups such as carboxylic (–COOH) and hydroxyl (–OH) groups can be covalently created on the Gr surface using strong acids and/or oxidants (Dai, 2013). In addition, microwave-assisted sulfonation has been used to create sulfonate (–SO_3) groups, while plasma (ammonia or nitrogen plasma) treatment has been used to create the amino (–NH_2) group on Gr (Choi et al., 2010). The grafting chemical moieties created on the Gr surface can serve as chemical handles for the conjugation of biological molecules (DNA, proteins, and carbohydrates) through covalent bonding (Wang, Li, et al., 2011). The properties, methods of functionalization, and various applications of Gr-based NMs have been summarized in Fig. 1.

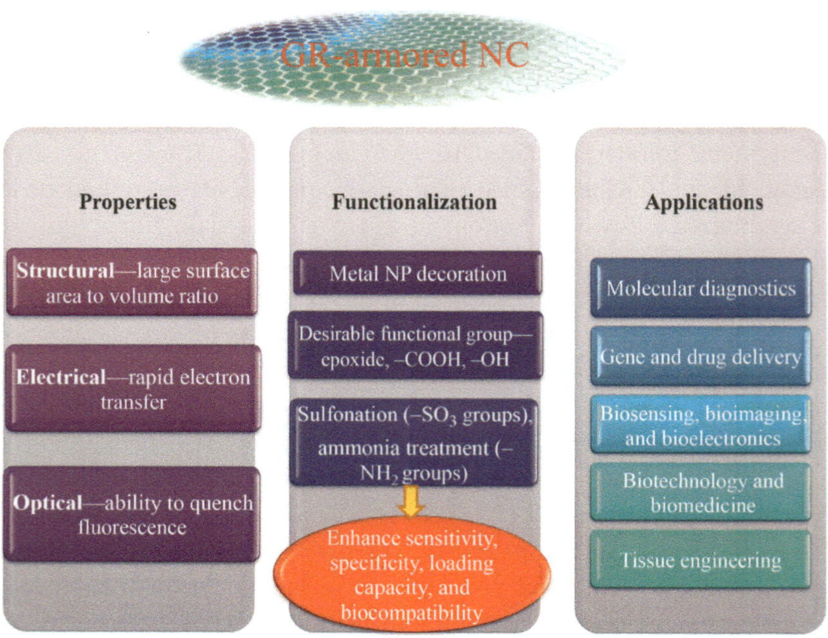

Fig. 1 Brief description of properties, ways of functionalization, and applications of Gr-armored NC.

Modification methods have also been developed to preserve the intrinsic properties of the original Gr materials. Gr can be viewed as the largest aromatic molecule. It can interact with some molecules by physical adsorption. For example, GO that is highly negatively charged, usually electrostatically adsorbs positively charged molecules. Additionally, van der Waals interaction or hydrophobic interaction may enhance the physical adsorption. Nevertheless, physical adsorption is not site specific. Thus, molecules (e.g., casein, BSA, and Tween-20) are often used to block the unwanted hydrophobic area in order to avoid nonspecific adhesion of nontarget molecules (Singh et al., 2014; Zhong et al., 2014). Graphene materials, particularly, GO and RGO, are useful materials for immobilizing NPs to exhibit better properties, such as higher conductivity, larger specific area, and remarkable thermal stability. By introducing metal NPs, metal oxides, or conductive polymers, the final Gr-based composite can achieve better catalytic and electrochemical activities for extending applications in biosensing. Several review articles have been published on Gr-based materials and their application in biosensing (Hernaez et al., 2017; Husain, 2017c; Yu et al., 2017; Zhu, Du, & Lin, 2017).

1.3 Advantages of Using Graphene-Based Materials in Health-Related Applications

Graphene and its derivatives play an ever-increasing role in the area of bioanalysis and biomedicine. In particular, they have been highly anticipated to provide new and unique opportunities for the improvement of novel nanoprobes for new biosensing methods and devices. Frequently, these new emerging nanoprobes have been applied in the areas related to human health such as diagnosis of viral diseases, nucleic acid mutations, and various cancers (Feng, Wu, & Qu, 2013; He, Zhang, Liu, Fang, & Zhang, 2014; Heldt et al., 2013). Till to date, the discovery of cancer diseases at an early stage is the most successful strategy to halt cancer progress by the detection of potential biomarkers. Thus, accurate, sensitive, and selective detection of nucleic acids, proteins, and small molecules is an urgent need due to their irreplaceable roles in human-related diseases (Zhu & Dong, 2014). GO has also been used as a photothermal agent for cancer treatment with encouraging therapeutic outcomes due to its high, intrinsic near-infrared (NIR) absorbance (Li, Yang, Ren, Qu, & Qu, 2012).

2. FACILE SYNTHESIS OF ENZYME–GRAPHENE-ARMORED NANOCOMPOSITES AS BIOCATALYSTS

This chapter deals with the fabrication of newly synthesized enzyme-Gr anchored NC for its possible applications in biocatalytic industry. The stable collaboration between nanosupports and biomolecules is of great importance to the area of nanobiotechnology. Based on the above-mentioned considerations, we chose *Aspergillus oryzae* β-galactosidase and *Aspergillus niger* lipase as model enzyme for the immobilization as they possess an intrinsic capacity to catalyze enantioselective hydrolysis, trans-esterification, esterification, and are extensively employed in the pharmaceutical, chemical, baking, food, textile, and biofuel industries, for the production of galactooligosaccharides, biochemical products, biosensors, paint, flavors, and biodiesel (Husain, 2016, 2017c). Nonetheless, despite its discernible intrinsic stability, the potential industrial applications of enzyme catalyzed reactions face frequent glitches, e.g., low stability and poor catalytic efficiency of enzyme, difficult retrieval of the enzyme from the reaction medium, lack of repetitive use, and the enzyme cannot be recycled. Thus, a fascinating approach to resolve the aforesaid snags is to immobilize the enzymes on different materials which includes NPs, NC, nanotubes, and layered double hydroxides. Additionally, the synthesis of numerous supports had led to the affordability of immobilized enzymes over a wider range of applications.

3. IMMOBILIZATION OF THE ENZYME ONTO GRAPHENE ANCHORED NANOCOMPOSITES

The discovery of Gr in 2004 added a new dimension to electrochemical biosensor research (Novoselov, Geim, & Morozov, 2004). Immediately after its isolation in 2004, this 2D material triggered scientific interest due to its extraordinary properties, which led to the Nobel Prize for Physics in 2010. The tremendous attention paid to Gr is based on its remarkable electrical, optical, and mechanical properties, such as large surface area, high electrical conductivity, excellent mechanical strength, and ease of functionalization (Geim & Novoselov, 2007). Despite its short history, and unique physicochemical properties, new Gr-based materials have made a profound impact in the fields of biotechnology and biomedicine, holding

Table 1 Various Gr-NC Bound Hydrolases and Their Potential Industrial Applications in Food Technology

Name of Enzyme	Name of NC	Application	References
Glucoamylase	GO/MNP-CC	Starch hydrolysis	Amirbandeh and Taheri-Kafrani (2016)
Amylase	MgNPs and GO	Application at very high temperatures	Dutta et al. (2017)
Cellulase	GO-MgNPs-	Enhanced hydrolysis of cellulosic compounds	Dutta, Biswas, and Saha (2016)
Cellulase	PEGylated GO	Enhanced straw hydrolysis	Xu et al. (2016)
β-Galactosidase	Gr@Fe_3O_4 NCs	Higher hydrolysis of lactose in batch processes	Khan, Husain, and Naqvi (2016)
Lipase	NH_2-GO and MWCNTs NC	Higher rate of esterification	Pavlidis et al. (2012)
Trypsin	GO-CO-NH-Fe_3O_4 NCs	Enhanced protein digestion efficiency and high throughput in proteomic study	Jiang et al. (2012)
Trypsin	GO-SiO_2 coated on PMMA	Highly efficient proteolysis in a microfluidic bioreactor	Bao, Zhang, and Chen (2013)
Trypsin	dGO nanosheets	High-throughput proteome identification	Jiang et al. (2014)

promising potential in applications such as in gene and drug delivery, bioremediation, bioimaging and biosensing, bioelectronics, tissue engineering and as antibacterial agents (Goenka, Sant, & Sant, 2014; Husain, 2017e; Krishna, Menard-Moyon, Verma, & Bianco, 2013). Table 1 summarizes various Gr-NC bound hydrolases and their potential industrial applications in food technology, while Table 2 depicts graphene nanocomposite (Gr-NC) bound enzymes and their application in the treatment of aromatic compounds from wastewater.

Table 2 Gr-NC Bound Enzymes and Their Environmental Application in the Treatment of Aromatic Compounds

Name of Enzymes	Type of NC	Compound Treated	References
Laccase	PAN/MMT/GO	Catechol	Wang et al. (2014)
Laccase	RGO/polymer GO nanoplatelets	RBBR	Ormategui, Veloso, Leal, Rodriguez-Couto, and Tomovska (2015)
Laccase	Carbon fullerene (C60), MWCNTs, *o*-MWCNTs, GO	Catechol and BPA	Pang, Li, and Zhang (2015)
Laccase	MGO	Decolorization of CV, MG, and BG	Chen et al. (2017)
Laccase	GO-$CuFe_2O_4$	Green synthesis of aryl sulfonyl benezenediols	Rouhani, Rostani, Salimi, and Pourshiani (2018)
Peroxidase	Co_3O_4-HRP/rGO/GCE	Chlorophenols	Chang et al. (2015)
Peroxidase	GO/Fe_3O_4 via EDC	Phenol, 95%	Chang, Huang, Ding, and Tang (2016)
Peroxidase	AuNPs/S-RGO	Bisphenol A	EzhiVilian et al. (2017)

The large surface area of Gr-armored NM creates an ideal immobilization support for various biomolecules including enzymes (Ge et al., 2012; Gokhale, Lu, & Lee, 2013; Lin et al., 2012; Pavlidis et al., 2012). The surface chemistry of these NM can influence their interactions with biomolecules and thus affect the adsorption as well as the conformational and biological function of conjugated proteins (Lin et al., 2012; Pavlidis et al., 2012). The large surface area and excellent electrical conductivity of Gr allows it to act as an "electron wire" between the redox center of an enzyme or protein and an electrode's surface. Rapid electron transfer facilitates accurate and selective detection of biomolecules with better sensitivity than conventional ones (Fig. 1). The addition of Gr leads to an increased sensing area and enhancement of the electron transfer rate (Zhang et al., 2016). Table 3 shows the list of various Gr-NC-based biosensors of oxidoreductive and hydrolytic enzymes, their analytes, linearity ranges of measurement and detection limits.

Table 3 Various Gr-NCs Based Biosensors of Oxidoreductive and Hydrolytic Enzymes, Their Analytes, Linearity Ranges of Measurement and Detection Limits

Name of Enzyme	Name NC	Analyte	Linearity Range	Detection Limit	References
ADH	AuNPs/PDA-RGO NC	Ethanol	5.0×10^{-8} to $4.2 \times 10^{-5} M$		Tian et al. (2013)
ChOx	Gr/PVP/PANI	Cholesterol	$50 \mu M$ to $10 mM$	$1 \mu M$	Ruecha, Rangkupan, Rodthongkum, and Chailapakul (2014)
Cyt *c*	GO-AuNPs-MWCNTS	Rebaudioside	$0.001–0.05 mM$	$0.264 \mu M$	Bathinapatla, Kanchi, Singh, Sabela, and Bisetty (2016)
DAOx	Gr–Pt NPs	Histamine	$0.1–300 \mu M$	$2.54 \times 10^{-8} M$	Apetrei and Apetrei (2016)
Laccase	RGO/Rh NCs GCE	17 β-estradiol	$0.9–11 pM$	$0.54 pM$	Povedano et al. (2017)
MP 11	Phospholipid-Gr-AuNPs	H_2O_2	2.0×10^{-5} to $2.8 \times 10^{-4} M$	$2.6 \times 10^{-6} M$	Wang, Liu, Ren, Wang, and Wang (2015)
HRP	AuPd-RGO-NW-PWE	H_2S	AuPd-RGO-NW-PWE		Yang et al. (2017)
PPO	PEDOT-RGO-Fe_2O_3-NCs GCE	Catechol	4×10^{-8} to $6.20 \times 10^{-5} M$	$7 \times 10^{-9} M$	Sethuraman, Mthuraja, Anandha, and Manisankar (2016)
Tyr	RGO-IrO_2NCs-SPE	Captopril	—	$0.019 \mu M$	Kurbanoglu, Rivas, Ozkan, and Merkoçi (2017)

Continued

Table 3 Various Gr-NCs Based Biosensors of Oxidoreductive and Hydrolytic Enzymes, Their Analytes, Linearity Ranges of Measurement and Detection Limits—cont'd

Name of Enzyme	Name NC	Analyte	Linearity Range	Detection Limit	References
Tyr	Gr-Au-CS-NC-SPCE	Phenolics	0.05–15 μ*M*	0.016 μ*M*	Fartas, Abdullah, Yusof, Sulaiman, and Saiman (2017)
GOx	MWCNT/GO hybrid biocomposite	Glucose	0.05–23.2 m*M*	28 μ*M*	Palanisamy, Karuppiah, and Chen (2014)
GOx	RGO/Ag NC electrode	Glucose	0.5–12.5 m*M*	0.16 m*M*	Palanisamy, Cheemalapati, and Chen (2014)
GOx	GO/lucigenin NC	Glucose	1.0×10^{-6} to 5.0×10^{-3} g mL^{-1}	9.9×10^{-7} g mL^{-1}	Gao, Zhang, and Cui (2014)
GOx	Gr-CNT-ZnO NC	Glucose	10 μ*M* to 6.5 m*M*	4.5 (±0.08) μ*M*	Hwa and Subramani (2014)
GOx	CSGr MNPs	Glucose	26 m*M*	16 μ*M*	Zhang, Chen, Ruo, Zhong, and Wu (2015)
GOx	PEI RGO-HAuNPs	Con A	1.0–20 ng mL^{-1}	0.31 ng mL^{-1}	Zhang et al. (2015)
GOx	Gr-CNTs	Glucose	2–8 m*M*		Terse-Thakoor, Komori, Ramnani, Lee, and Mulchandani (2015)
GOx	RGO nanosheets	Glucose	0.1–15.5 m*M*	5 μ*M*	Halder, Zhang, and Chi (2017)

GOx	RGO-Fe_3O_4 NC	Glucose	0.05–1 m*M*	0.1 μ*M*	Pakapongpan and Poo-Arporn (2017)
GOx	MRGO/PtAuNPs modified hybrid electrode	H_2O_2	0.5–8 m*M*		Hossain and Park (2017)
	CS-PDDA-MRGO/Pt AuNPs modified hybrid electrode	Glucose	0.01–8 m*M*	1 μ*M*	
AChE	Au-PPy-RGO NC-$(NH_4)_2SiF_6$	Paraoxon-ethyl	1.0 n*M* to 5 μ*M*	0.5 n*M*	Yang, Wang, Cao, Zhang, and Zhang (2014)
AChE	Gr-Au NCs GCE	Chlorpyrifos	0.05–150 $\mu g L^{-1}$	0.05 $\mu g L^{-1}$	Zhai, Guo, Sun, Zheng, and Wang (2014)
AChE	Cd0.5Zn0.5S-RGO NC	Acetylcholine	0.001–1 $\mu g mL^{-1}$	0.3 $ng mL^{-1}$	Liu, Zhang, Yan, Zhang, and Liu (2014)
AChE	Fe_2O_3NPs/RGO/PEDOT-FTO electrode	Acetylcholine	4.0 n*M* to 800 μ*M*	4.0 n*M*	Chauhan, Chawla, Pundir, and Jain (2017)
Urease	Sulfonated Gr/PANI NC	Urea	0.12–12.3 m*M*	0.050 m*M*	Das and Yoon (2015)

Modified from Husain, Q. (2017e). Biosensor applications of graphene–nanocomposites bound oxidoreductive and hydrolytic enzymes. *Analytical Methods, 9*, 6734–6746.

Lately, polymer–Gr composite supports have gained a lot of attention due to their characteristic properties such as low cytotoxicity, large surface area, lightweight, unique conjugated structure, better water dispersibility. Their functionalization for their applications in biomedicine, bioenergy, and nanobiotechnology is attractive. GO possess a 2D, single-layered sp^2 hybrid structure with adequate surface groups, offering an exceptional double sided, easily accessible substrate for multivalent functionalization and effective loading of molecules from small organic ones to biomacromolecules. Considering the binding between GO and biomacromolecules, such as enzymes, GO has been reported in numerous studies, typically for biosensing, as a support for immobilization of several enzymes, including horseradish peroxidase (HRP), glucose oxidase (GOx), amylase, β-galactosidase, and hemoglobin (Das, Talat, Srivastava, & Kayastha, 2018; Khan et al., 2016; Mansouri, Babadi, Bagheri, & Abd Hamid, 2017; Vineh, Saboury, Poostchi, Rashidi, & Parivar, 2018; Zhan, Tan, Wang, & Hou, 2018). Different oxidoreductive and hydrolytic enzymes immobilized on Gr-NC along with their industrial applications have been illustrated in Fig. 2.

In this chapter, we portray how the hydrolytic enzymes are amassed onto GO sheets to form graphene-armored nanobiocatalysts to further stabilize

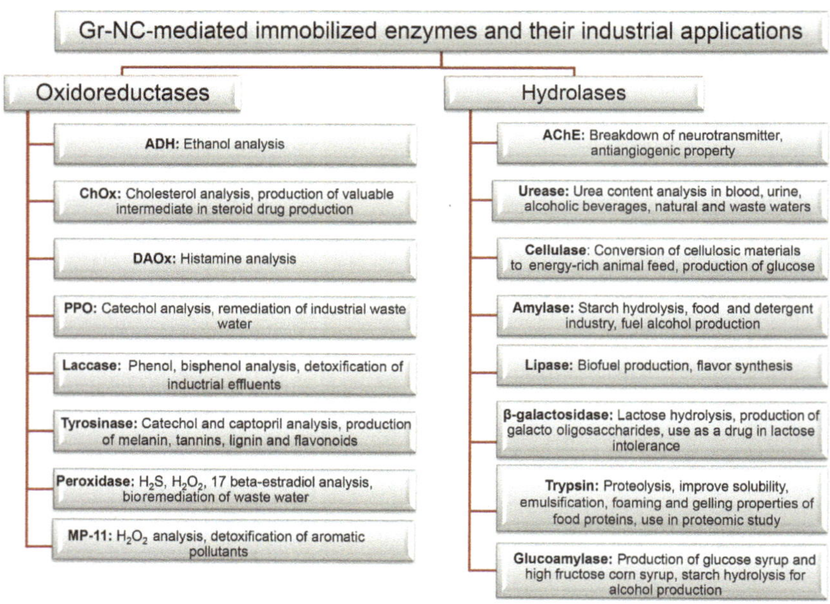

Fig. 2 Immobilization of enzymes onto Gr-armored NC and their industrial applications.

both β-galactosidase and lipase against denaturing environments while retaining their activities. Gr-iron oxide NC (Gr@Fe_3O_4 NCs) was synthesized using coprecipitation method, while the polyaniline-coated silver-functionalized GO anchored NC (PANI/Ag/GO NC) was prepared through in situ polymerization. β-Galactosidase was immobilized on Gr@Fe_3O_4NC employing physical adsorption; on the other hand, lipase was immobilized on PANI/Ag/GONC utilizing covalent method, protecting the enclosed enzyme while contributing fair access to the active site. The details of the characterization and fabrication are described in the subsequent sections. We state that the amalgamation of covalently immobilized enzyme with the polyaniline-coated silver-functionalized GO anchored NC boost the catalytic and operational stability of the biocatalyst even under severe conditions where native, undefended enzyme denatures. The outcomes displayed here completely support our claim, though simple and effective immobilization techniques and tools demand further exploration.

3.1 Graphene–Iron Oxide NC

3.1.1 Synthesis of Gr@Fe_3O_4 NC

General safety: Safety glasses, latex, or nitrile gloves, lab coat

Chemicals and reagents

1. Graphite—Sigma-Aldrich Chemicals Co., USA
2. $FeCl_2$—Sigma-Aldrich Chemicals Co., USA
3. $FeCl_3$—Sigma-Aldrich Chemicals Co., USA
4. NaOH—Merck & Co., USA

Equipment

1. Beaker (50–500 mL)
2. Stir/hot plate
3. Analytical balance
4. pH meter
5. Microtubes (1.5 mL)—Eppendorf, India
6. Magnet (6000G)
7. Magnetic stirrer with bar
8. Hot air oven

Methodology

GO was synthesized via modified Hummers method and was subsequently exfoliated by ultrasonication to attain an aqueous dispersion of

GO. Gr@Fe_3O_4 NC was synthesized by coprecipitating prehydrolyzed ferric and ferrous salts in the presence of GO (Zubir et al., 2014).

1. Prepare an aqueous solution (100 mL) containing $FeCl_3 \cdot 6H_2O$ (4.0 m*M*) and $FeCl_2 \cdot 4H_2O$ (2.0 m*M*) with an initial pH of 1.48.
2. Add GO solution (50 mL, 0.55 mg mL^{-1}) into the earlier mixture only when pH is adjusted to pH 4 via addition of NaOH (1.0 *M*) to prevent stacking of GO sheets at low pH.
3. Stir the solution for 30 min in order to obtain a stable and homogeneous mixture.
4. Now add NaOH (1.0 *M*) into mixture and during this process maintain the pH of solution at about 10.
5. Again stir the mixture constantly for 30 min at room temperature.
6. Separate the black precipitate, i.e., Gr@Fe_3O_4 NC from the reaction using a permanent magnet.
7. Once the magnetic nanoparticles (MNPs) are settled, wash them several times with deionized water and ethanol to remove any unbound and extra reagents.
8. Dry the precipitate well in a vacuum oven at 60°C for 48 h.
9. Also prepare pure Fe_3O_4 NPs via an analogous method without the addition of GO solution.
10. Store NC and NPs as a dried powder or as a suspension in deionized water or buffer until further use.

3.1.2 Characterization of Gr@Fe_3O_4 NC

General safety: Safety glasses, latex, or nitrile gloves, lab coat

Chemicals and reagents

1. Potassium bromide (AR Grade)—Sisco Research Laboratories (SRL), India
2. Sodium acetate (AR Grade) buffer (pH 4.5, 0.1 *M*)—SRL, India
3. Fe_3O_4 NPs and Gr@Fe_3O_4 NC (diluted solutions)

Equipment

1. Carbon-coated copper grids—300 mesh
2. JEOL-2100 transmission electron microscope, Japan
3. Sputter coater—JEOL JFC 1600, Japan
4. JSM-6510 LV scanning electron microscope, Japan
5. X-ray diffractometer (Rigaku, Japan-Miniflex-II)
6. BRUKER RFS 27: Stand alone FT-Raman Spectrometer
7. FT-IR Perkin Elmer instrument, USA (Spectrum-II, wave no. 4000-400)
8. Vibrating sample magnetometer (VSM, ADE EV5)

Methodology

The sizes and morphologies of Gr@Fe_3O_4 NC, Fe_3O_4 NPs and adsorbed β-galactosidase were examined by transmission electron and scanning electron microscopy (TEM and SEM), respectively.

1. Prepare small amounts of samples from Fe_3O_4 NPs, Gr@Fe_3O_4 NCs, and bound NC for TEM analysis by drop coating diluted solutions on carbon-coated copper grids at normal atmospheric conditions.
2. Analyze obtained samples under TEM at 200 kV.
3. Mount small samples from Gr@Fe_3O_4 NC, Fe_3O_4 NPs, and air dry immobilized enzyme on carbon tape on a copper stub, followed by 60 nm of gold deposition using a sputter coater.
4. Examine the samples under SEM at 15 kV. The crystal structure of the NPs and the NCs have been interpreted by X-ray diffraction (XRD) pattern obtained by X-ray diffractometer using a monochromatized X-ray beam with nickel-altered Cu Kα radiation. Keep scan speed of four steps per second with an angle range of 20–80 degrees.
5. The Raman spectra of Fe_3O_4 NPs and Gr@Fe_3O_4NC were recorded on a "Stand alone FT-Raman Spectrometer" using laser source Nd: YAG1064 nm and a resolution of 2 cm^{-1}. Fourier-transform infrared spectroscopy (FT-IR) spectra of the Fe_3O_4 NPs, Gr@Fe_3O_4 NC, and air dried immobilized enzyme were recorded using FT-IR Perkin Elmer Instruments, USA. Calibrate using polystyrene film and use KBr pellet method to run the samples. Acquire background spectra of clean atmosphere, prior to each sample measurement using identical acquisition parameters.
6. To assess the magnetic properties of Fe_3O_4 NPs as well as Gr@Fe_3O_4 NC wth a vibrating sample magnetometer (VSM) by measuring the applied field dependence of magnetization between −15,000 and 15,000 H (G) at room temperature. Lastly, the enzyme concentration was measured with the help of a UV–vis spectrophotometer.

The design of nanobiocatalyst by attaching β-galactosidase to Gr@Fe_3O_4 NC for scalable and reusable bio- and electrocatalytic applications is mentioned in the next section.

3.1.3 Immobilization of β-Galactosidase on Gr@Fe_3O_4 NC

General safety: Safety glasses, latex, or nitrile gloves, lab coat

Chemicals

1. Sodium acetate (AR grade) buffer (pH 4.5, 0.1 *M*)—SRL, India
2. Gr@Fe_3O_4NC (500 mg)
3. β-Galactosidase (*A. oryzae*)—Sigma-Aldrich Chemicals Co., USA

Materials and equipment
1. Falcons (15 mL)—Tarson, India
2. Microtubes (1.5 mL)—Eppendorf, India
3. Shaking incubator—Biogen. India
4. Micropipettes (20.0–100.0 μL, 0.1–1.0 mL)
5. Ultrasonicator—Life Care
6. Magnet (6000G)
7. UV–vis spectrophotometer (Shimadzu, UV-1800)
8. Centrifuge-REMI cooling centrifuge

Methodology
1. Independently adsorb free β-galactosidase on varying concentrations (ranging from 2 to 18 mg) of Gr@Fe_3O_4 NC by continuously shaking the mixture in sodium acetate buffer, pH 4.5 at 25°C.
2. Vary the enzyme concentration from 1 to 5 $mg mL^{-1}$, keeping the NC concentration (10 mg) fixed.
3. Also increase the immobilization time from 1 to 10 h.
4. Calculate the yield in all cases and take the highest yielding parameters to progress further.
5. Suspend and sonicate Gr@Fe_3O_4 NC (500 mg) in 0.1 *M* sodium acetate buffer, pH 4.5.
6. Load β-galactosidase (49.12 U) in the colloidal suspension.
7. Gently stir the mixture for 5 h and then either magnetically separate or centrifuge at 3000 × *g* for 10 min at 4°C.
8. The precipitate obtained is washed thrice with assay buffer and preserved at 4°C for further use.
9. Store the supernatant and the washes at 4°C for the calculation of immobilization yield.
10. Determination of immobilization yield.

The immobilization efficiency was determined in terms of immobilization yield as follows:

$$\text{Immobilization yield}(\%) = A - B/A \times 100\%$$

where A is the initially added activity of β-galactosidase/lipase in the immobilization system and B is the residual enzyme in the supernatant and washing solution after the immobilization process.

3.1.4 Stability Studies of Immobilized β-Galactosidase

Chemicals and reagents
1. β-Galactosidase—Sigma-Aldrich Chemicals Co., USA
2. Glycine–HCl (pH 2.0 and 3.0)—SRL, India

3. Sodium acetate buffer, pH (4.0, 4.5, and 5.0)—SRL, India
4. Sodium phosphate buffer, pH (6.0 and 7.0)—SRL, India
5. Tris–HCl buffer, pH (8.0 and 9.0)—SRL, India
6. Crushed ice
7. Galactose (1.0%–5.0%, w/v)—SRL, India
8. *o*-Nitrophenyl β-D-galactopyranoside, ONPG (0.1–1.0 m*M*)—SRL, India

Equipment

1. Water bath (30–110°C)—Scientech Instruments, Delhi
2. Ice maker and crusher—Labman
3. Ice box
4. Microtubes (1.5 mL)—Eppendorf, India
5. UV–vis spectrophotometer (Shimadzu, UV-1800)

Methodology

1. pH–activity profiles (Ansari & Husain, 2010)
   - **(a)** Prepare buffers of glycine–HCl (pH 2.0 and 3.0), sodium acetate (pH 4.0, 4.5, and 5.0), sodium phosphate (pH 6.0 and 7.0), and Tris–HCl (pH 8.0 and 9.0).
   - **(b)** Set the molarity of each buffer as 0.1 *M*.
   - **(c)** Incubate the free and immobilized enzymes in the buffers of different pHs.
   - **(d)** Consider the activity at pH 4.5 as control (100%) for the calculation of residual percent activity.
2. Temperature–activity profiles (Ansari & Husain, 2010)
   - **(a)** Incubate free and immobilized β-galactosidase (0.08 U) at different temperatures starting from 20 to 80°C in 0.1 *M* sodium acetate buffer of pH 4.5 for 15 min.
   - **(b)** Consider the enzyme activity at 50°C as control (100%) for the calculation of residual percent activity for free and immobilized enzyme.
3. Thermal denaturation (Ansari & Husain, 2010)
   - **(a)** Incubate free and immobilized β-galactosidase at 60°C in 0.1 *M* sodium acetate buffer, pH 4.5 for varying times.
   - **(b)** Take out aliquots of each preparation (20 mL) in triplicates at intervals of 15 min and chill them quickly in crushed ice for 5 min.
   - **(c)** Bring the enzyme to room temperature and perform assays as described earlier.
   - **(d)** Consider the activity of enzyme without incubation at 60°C as control (100%) for the calculation of residual percent activity.
4. Effect of galactose (Ansari & Husain, 2010)
   - **(a)** Incubate free and immobilized enzyme preparations (0.08 U) with various concentrations of galactose (1.0%–5.0%, w/v).

   - **(b)** Evaluate the activity of both enzymes in 0.1 *M* sodium acetate buffer, pH 4.6 at 37°C by performing enzyme assay.
   - **(c)** Consider the activity of enzyme without added galactose as control (100%) for the calculation of remaining percent activity.
5. Reusability of immobilized enzyme (Husain, Ansari, Alam, & Azam, 2011)
   - **(a)** Take immobilized enzyme preparation (0.08 U) in triplicates for assaying the activity of the enzyme after repeated use.
   - **(b)** Perform the reaction in 1.5 mL microtubes for 10 min at 37°C.
   - **(c)** Immediately after incubation, centrifuge the mixture at 3000 × *g* for another 10 min at 4°C.
   - **(d)** Decant the supernatant and add 0.1 *M* sodium carbonate into it, in order to stop the reaction.
   - **(e)** After each assay, wash the immobilized enzyme with 0.1 *M* sodium acetate buffer, pH 4.5, by centrifugation at 3000 × *g* for 10 min.
   - **(f)** Store the obtained pellet in assay buffer at 4°C and repeat this process for a period of 10 days.
   - **(g)** Consider the activity determined on the first day as control (100%) for the calculation of residual percent activity after repeated use.
6. Storage stability of free and immobilized β-galactosidase (Ansari & Husain, 2010)
   - **(a)** Store free and adsorbed β-galactosidase at 4°C in 0.1 *M* sodium acetate buffer of pH 4.5 for 2 months.
   - **(b)** Take aliquots from each preparation (50 μL) in triplicates at a gap of 5 days and then analyze them for remaining enzyme activity.
   - **(c)** The activity determined on the first day was considered as control (100%) for the calculation of residual percent activity.

3.1.5 Genotoxicity Assessment of Gr@Fe_3O_4 NCs

Reagents and chemicals

1. Histopaque-1077—Sigma-Aldrich Chemicals Co., USA.
2. RPMI-1640 medium—Sigma-Aldrich Chemicals Co., USA.
3. Low melting agarose—Sigma-Aldrich Chemicals Co., USA.
4. pBR^{322} DNA—Thermo Fischer Scientific, USA.
5. Ethylenediaminetetraacetic acid (EDTA)—SRL, India.
6. Tris—SRL, India.
7. Bromophenol Blue—Merck & Co., USA.
8. Glycerol—Merck & Co., USA.

9. Triton-X—SRL, India.
10. Ethidium bromide—Sigma-Aldrich Chemicals Co., USA.

Equipment

1. Beakers (500–1000 mL)
2. Incubator
3. Microtubes (1.5 mL) with stand—Eppendorf, India
4. Electrophoretic gel apparatus
5. UV-transilluminator
6. Fully frosted microscopic slides
7. Cover slips
8. Flat tray
9. Refrigerator
10. Ice packs

An image analysis system (Comet 5.5, Kinetic Imaging, Liverpool, UK) attached to an Olympus fluorescent microscope (CX 41, Olympus Optical Co., Tokyo, Japan) and a COHU 4910-integrated CC camera (equipped with 510–560 nm excitation and 590 nm barrier filters, COHU, San Diego, CA, USA).

Methodology

1. Plasmid nicking assay (Khan, Husain, & Ansari, 2013)
   - **(a)** Prepare a reaction mixture (30 mL) containing 10 m*M* Tris–HCl buffer, pH 7.5, pBR^{322} DNA (0.5 mg) with free enzyme (0.08 U), Gr@Fe_3O_4 NC (40 mg), and immobilized enzyme (10 mL) and incubate for 2 h at 37°C.
   - **(b)** Add 10 mL of solution containing 40 m*M* EDTA, 0.05% (w/v) bromophenol.
   - **(c)** Blue (tracking dye) and 50% (v/v) glycerol after incubation.
   - **(d)** Perform electrophoresis in submarine 1.0% (w/v) agarose gel.
   - **(e)** View and photograph ethidium bromide stained gel under a UV-transilluminator.
2. Comet assay (Khan et al., 2016)
   - **(a)** Use fully frosted microscopic slides precoated with 1.0% (w/v) normal melting agarose (PBS buffer lacking Ca^{2+} and Mg^{2+} ions).
   - **(b)** Mix diluted lymphocytes (10,000 cells) with 80 mL of 1.0% (w/v) low melting point agarose to form a cell suspension which is pipetted over the first layer and covered immediately by a cover slip.
   - **(c)** Put the slides on a flat tray and maintain on ice for 10 min to solidify agarose.

(d) Remove cover slips and layer 0.5% (w/v) low melting point agarose (80 mL). Allow to solidify on ice for 5 min.
(e) Remove the cover slips and within 20 min, treat cells with free enzyme (0.08 U), free NC (200 mg), and NC-adsorbed β-galactosidase (50 mL) for 1 h at 4°C.
(f) Perform all reactions on the slides in triplicates.
(g) Immerse the slides in cold lysis buffer containing 2.5 *M* NaCl, 100 m*M* EDTA, and 10 m*M* Tris–HCl, pH 10 for 1 h at 4°C and add 1% Triton-X 100 (just prior to use).
(h) After lysis, allow DNA to unwind for 30 min in an alkaline electrophoretic solution consisting of 300 m*M* NaOH and 1.0 m*M* EDTA, pH > 13.
(i) Run electrophoresis at 4°C in field strength of $0.7\,V\,cm^{-1}$ and 300 mA current.
(j) Neutralize slides with cold 0.4 *M* Tris buffer, pH 7.5, and subsequently stain with 75 mL ethidium bromide ($20\,mg\,mL^{-1}$) and cover with cover slips.
(k) Lastly, place the slides in a humidified chamber to prevent drying of gel and analyze the same day.
(l) Score comets at 100 × magnification. Analyze images from 50 cells (25 from each replicate slide).
(m) Tail length in millimeter (migration of DNA from the nucleus) is the parameter taken to assess DNA damage in human lymphocytes and it is automatically generated by the Comet 5.5 image analysis system.

3.2 Synthesis and Characterization of Gr@Fe_3O_4 NC

The synthetic methods for the preparation of Gr@Fe_3O_4 NC usually require tedious steps and toxic reagents such as hydrazine (Chandra et al., 2010) and $NaBH_4$ (Shin et al., 2009). Hence, there is an urgent need to develop a safe and efficient method to prepare Gr@Fe_3O_4 NCs. In this study, we have prepared Gr@Fe_3O_4 NCs by coprecipitation method and by employing eco-friendly reagents. Since pure Gr sheets have rare functional moieties on their surfaces, graphite oxide has appeared as a suitable substitute for Gr on account of having a large number of chemically reactive sites, such as epoxy and hydroxyl groups on the basal plane and the carboxylic groups at the surface (Fig. 3), and due to the presence of π–π conjugated system, which oxidizes Fe^{2+} ions for the nucleation and growth of Fe_3O_4 NPs (Yang et al., 2016).

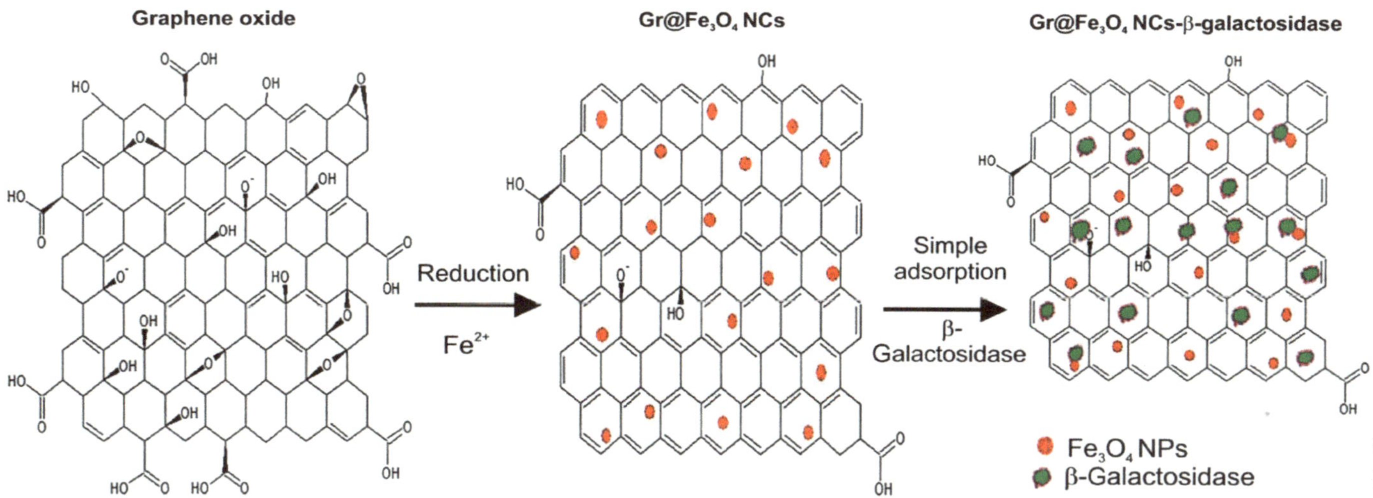

Fig. 3 Gr@Fe_3O_4 NC synthesis and immobilization of β-galactosidase on Gr@Fe_3O_4 NC. *Reproduced from Khan, M., Husain, Q., Naqvi, A. H. (2016). Graphene based magnetic nanocomposites as versatile carriers for high yield immobilization and stabilization of β-galactosidase.* RSC Advances, 6, *53493–53503.*

Fig. 3 depicts the procedure for the preparation of Gr@Fe_3O_4 NC and immobilization of β-galactosidase on Gr@Fe_3O_4 NC. Graphite oxide was reduced by the addition of Fe^{2+} ions which are attached to the surface of graphite oxide via electrostatic interactions and was oxidized to Fe^{3+} ions by the O^{2-} groups present on the surface of graphite oxide. NaOH was added in order to prevent restacking of graphite oxide sheets by increasing the pH of the medium. Consequently, the Fe^{2+} ions coprecipitated with the acquired Fe^{3+} ions into Fe_3O_4, leading to the formation of Gr@Fe_3O_4 NC (Yang et al., 2016). The enzyme was simply adsorbed on the surface of prepared NCs via hydrogen bonding and van der Waals interactions between the surfaces of protein and Gr@Fe_3O_4 NC (Khan, Qayyum, Alam, & Husain, 2011).

Free and β-galactosidase bound Gr@Fe_3O_4 NC were characterized by TEM, SEM, and FT-IR spectroscopy. XRD and VSM were carried out to monitor the crystallinity and confirm the magnetic character of the NPs and the NC, respectively. The surface morphologies and particle sizes of pure Fe_3O_4 NPs, Gr@Fe_3O_4 NC, and NC bound β-galactosidase were determined by using SEM and TEM. The incorporation of Fe_3O_4 NPs onto Gr sheets has been determined by TEM and SEM images (Fig. 4A–D). As seen in the representative TEM images of Gr@Fe_3O_4 NC. The surface of Gr was densely covered by cylindrical-shaped Fe_3O_4 NPs with an average size of about 25 nm (Fig. 4C and D). The distribution of Fe_3O_4 NPs on Gr sheet was evidently even, and no aggregation or large vacancy on Gr has been noticed. Fig. 4B and F illustrates the selection area electron diffraction (SAED) pattern of NC and Fe_3O_4 NPs, respectively. The observed NCs were found to be polycrystalline in nature with a slight visibility of circular rings while the Fe_3O_4 NPs were appeared crystalline in nature. He and Gao (2010) have also reported identical results. Fig. 4E and F demonstrates TEM and SEM images of Fe_3O_4 NPs. Roughly spherical Fe_3O_4 NPs are observed with an average particle size of 55 nm. The space resolved lattice fringes as shown in Fig. 4H agreed well with the lattice spacing planes of Fe_3O_4 NPs. In an earlier study also, similar findings have been described (Zubir et al., 2014).

The binding of β-galactosidase to the matrix has also been shown by SEM and TEM analyses. Fig. 5A and B clearly demonstrated the adsorption of protein onto functionalized Gr-NC. Rigid binding of enzyme to the support can be evidently illustrated by SEM analysis at 50, 10, and 5 mm

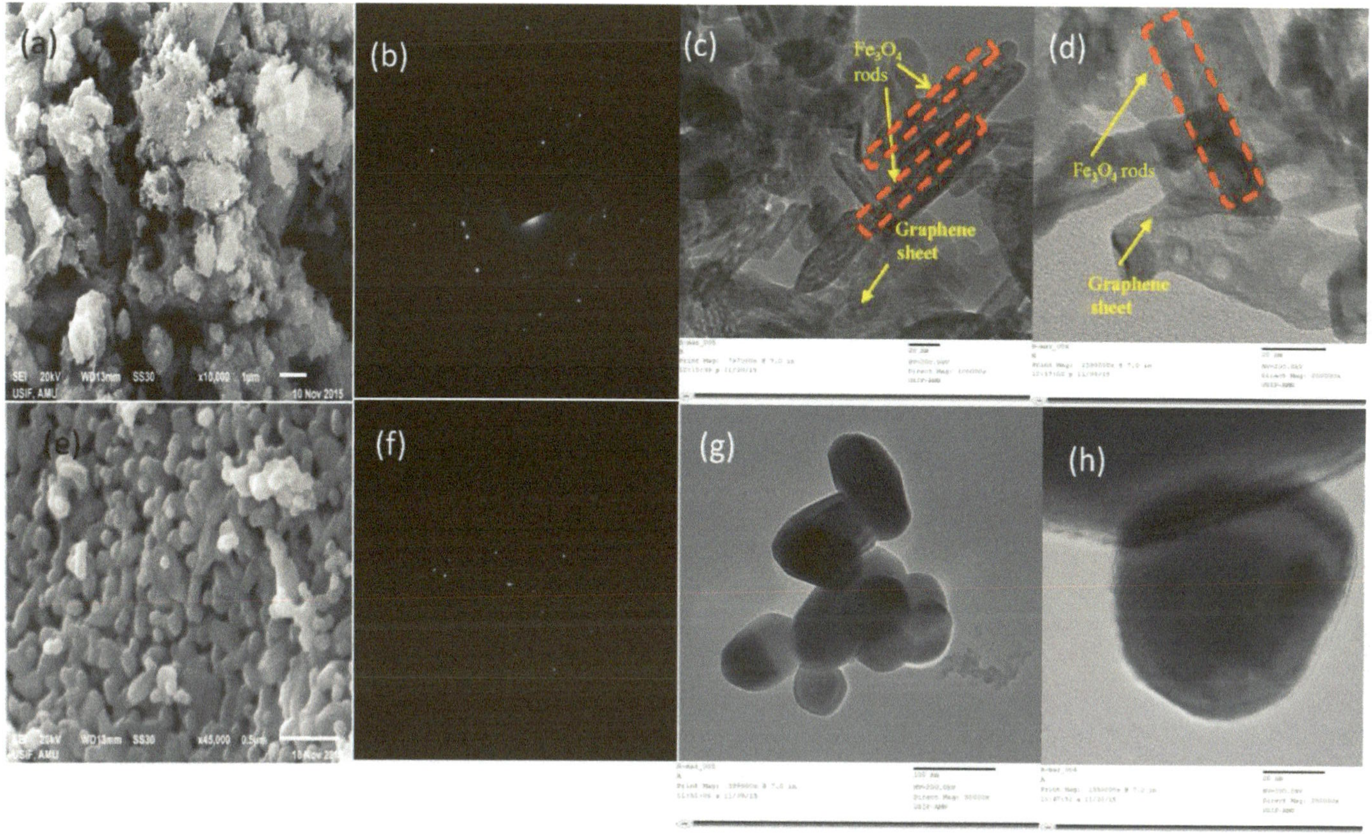

Fig. 4 (A) SEM image of Gr@Fe_3O_4 NC, (B) SAED pattern of Gr@Fe_3O_4 NC, (C and D) TEM images of Gr@Fe_3O_4 NC, (E) SEM image of Fe_3O_4 NPs, (F) SAED pattern of Fe_3O_4 NPs, (G) TEM image of Fe_3O_4 NPs, and (H) lattice fringes of a single Fe_3O_4 NPs. *Reproduced from Khan, M., Husain, Q., Naqvi, A. H. (2016). Graphene based magnetic nanocomposites as versatile carriers for high yield immobilization and stabilization of β-galactosidase.* RSC Advances, 6, *53493–53503.*

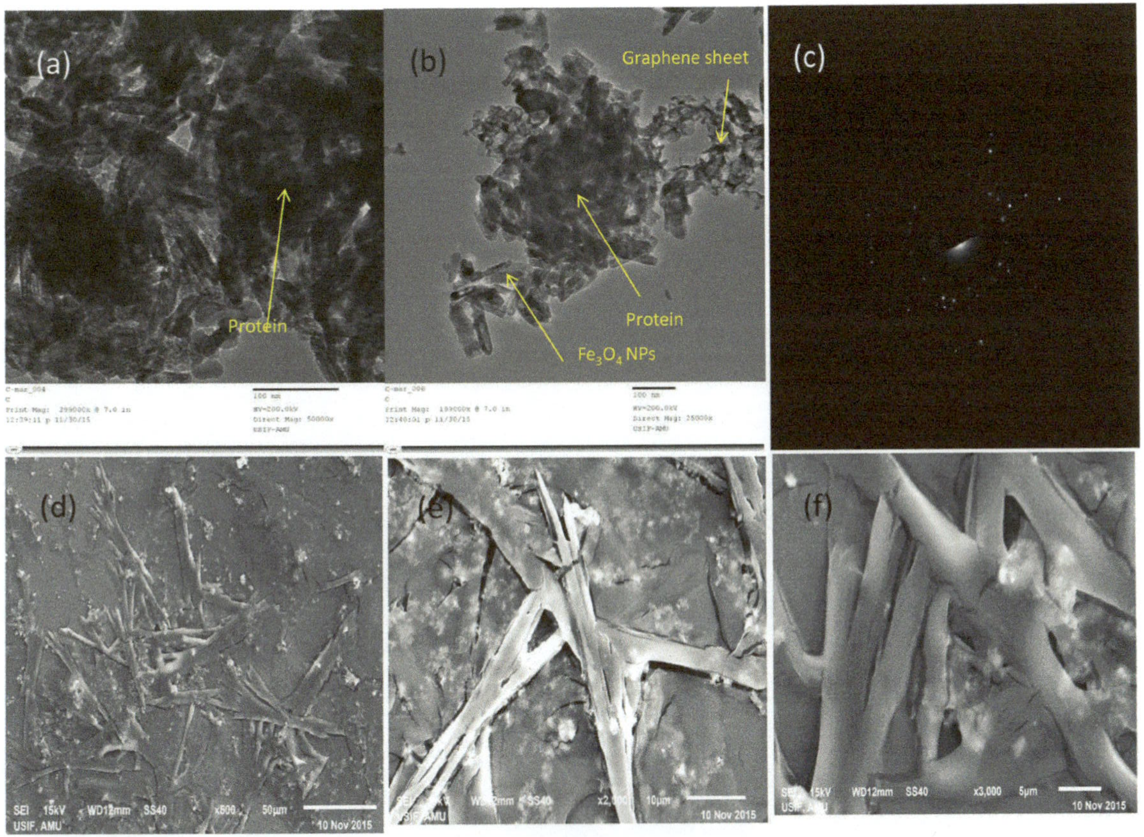

Fig. 5 TEM and SEM images of enzyme bound Gr@Fe_3O_4 NC (A and B) TEM images, (C) SAED pattern, and (D–F) SEM images at 50, 10, and 5 μm. *Reproduced from Khan, M., Husain, Q., Naqvi, A. H. (2016). Graphene based magnetic nanocomposites as versatile carriers for high yield immobilization and stabilization of β-galactosidase.* RSC Advances, 6, *53493–53503.*

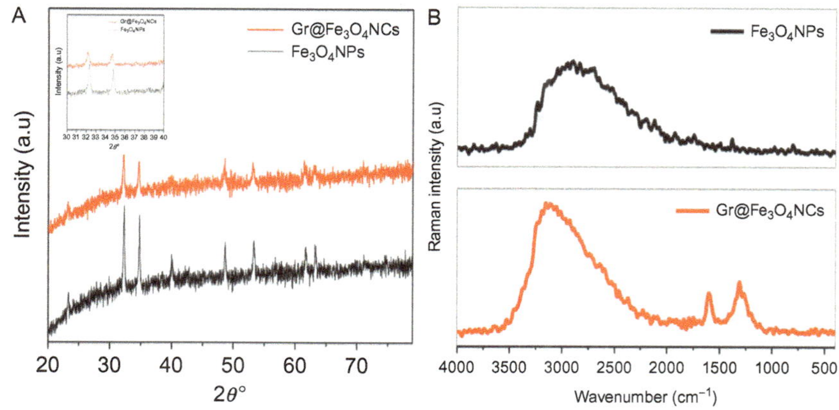

Fig. 6 (A) XRD pattern of Gr@Fe_3O_4 NC and Fe_3O_4 NPs and (B) FT-Raman spectra of Fe_3O_4 NPs and Gr@ Fe_3O_4 NC. *Reproduced from Khan, M., Husain, Q., Naqvi, A. H. (2016). Graphene based magnetic nanocomposites as versatile carriers for high yield immobilization and stabilization of β-galactosidase.* RSC Advances, 6, *53493–5350.*

(Fig. 5D–F), respectively. Moreover, the immobilization was also responsible for alteration in the characteristic SAED pattern. Fig. 5C shows distinguishable concentric rings which indicates that the Gr@Fe_3O_4 NC changes into highly crystalline form after binding with β-galactosidase.

Fig. 6A demonstrates XRD patterns for both Fe_3O_4 NPs and Gr@Fe_3O_4 NC. The crystallinity in both samples was corroborated by the presence of peaks at $2\theta = 32.31$, 34.65, 39.92, 48.71, and 53.26 degrees for Fe_3O_4 NPs and 32.49, 34.99, 40.14, 48.88, and 53.57 degrees for Gr@Fe_3O_4 NC, respectively. These peaks were found consistent with the standard XRD data in JCDPS (file no. 19-0629). No diffraction peak of other impurities is detected, demonstrating that both prepared sample are pure and crystalline in nature which was in accordance with the results reported earlier (Yang et al., 2016). Raman spectroscopic studies were also performed in order to assess the incorporation of Fe_3O_4 NPs on the Gr sheets. Fig. 6B illustrates the Raman spectra of Fe_3O_4 NPs and Gr@Fe_3O_4 NC. The Fe_3O_4 NPs spectrum showed a major peak at 2907 cm^{-1} and a small peak at 1382 cm^{-1}, corresponding to C—H and CH_3 bonding, respectively. The Raman spectrum of Gr@Fe_3O_4 NC revealed two prominent peaks at 1602 and 1310 cm^{-1} which demonstrated the presence of well-documented D (defects/disorder-induced modes) and G bands (in-plane stretching tangential modes) in Gr sheets. The intensity ratio of D and G bands (ID/IG ratio) for GO was normally more than 1.0, as it was evidently recognized (Eigler, Dotzer, & Hirsch, 2012), but on reduction of graphite oxide to Gr in the presence of Fe^{2+} ions, the intensity ID/IG ratio was decreased to 0.82. Similar findings have also been previously mentioned by Narayanan et al. (2012) who synthesized and characterized rGO–Fe_3O_4 multifunctional freestanding membranes. These bands were not observed in Fe_3O_4 NPs. The peak at 1602 cm^{-1} further indicated the presence of aromatic ring chain vibrations (C=C) and showed attachment of Fe_3O_4 NPs to Gr sheets.

Fe_3O_4 NPs as well as the Gr@Fe_3O_4 NC were exposed to external magnetic field sweeping from −15,000 to 15,000 H (g) at room temperature by the use of VSM to assess their magnetic properties. Fig. 7A represents the magnetic hysteresis (M–H) loops of Gr@Fe_3O_4 NC and it exhibited an

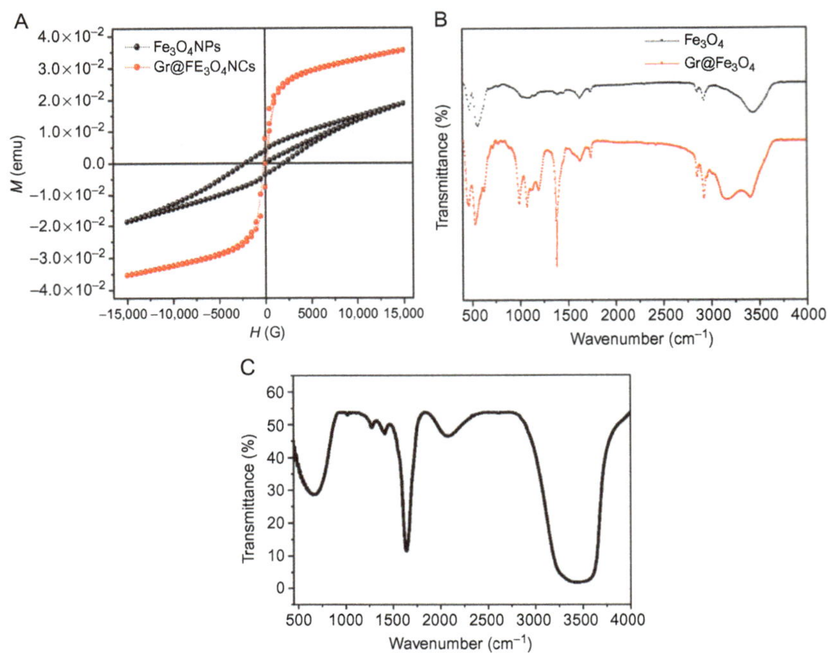

Fig. 7 (A) Magnetic hysteresis loops of Fe_3O_4 NPs and Gr@Fe_3O_4 NC measured at room temperature, (B) FT-IR spectra of Fe_3O_4 NPs and Gr@Fe_3O_4 NC, and (C) Enzyme bound Gr@Fe_3O_4 NC. *Reproduced from Khan, M., Husain, Q., Naqvi, A. H. (2016). Graphene based magnetic nanocomposites as versatile carriers for high yield immobilization and stabilization of β-galactosidase.* RSC Advances, 6, *53493–53503.*

S-shaped curve with negligible retentivity and coercivity indicating no remaining magnetization upon removal of external magnetic field (He & Gao, 2010). The NC, hence, exhibited superparamagnetic behavior while the naked Fe_3O_4 NPs showed paramagnetic behavior. The saturation magnetization (M_s) for Gr@Fe_3O_4 NC was noticed nearly two folds higher than that of the naked ones, suggesting that the particles are highly superparamagnetic, which is a characteristic of a ferromagnetic material. Fig. 7B demonstrates the characteristic FT-IR spectra of Fe_3O_4 NPs and of free as well as bound Gr@Fe_3O_4 NC. In the first two samples, the peaks at 568 and 534 cm^{-1}, respectively, correspond to the Fe–O vibrations related to the magnetite phase (Mahdevi et al., 2013). The peak at 1631 cm^{-1} in the absorption spectra of Gr@Fe_3O_4 NC is the characteristic peak for C=C demonstrating the presence of Gr. The peaks observed at 2925 and 2854 cm^{-1} in Fe_3O_4 NPs and Gr@Fe_3O_4 NC, respectively, were due to C—H stretching. The FT-IR bands corresponding to OH stretching mode were shifted to lower wave number in FT-IR spectra of Gr@Fe_3O_4 NC, 3407 cm^{-1} than the pure Fe_3O_4 NPs, 3435 cm^{-1} due to the interaction taking place between the NPs and the Gr sheets (Baby, Aravind, Arockiadoss, Rakhi, & Ramaprabhu, 2010). It demonstrated the role of O—H group during interaction of NPs and Gr. Fig. 7C depicts the absorption spectrum of the immobilized enzyme. The β-galactosidase peaks from 3320 to 3547 cm^{-1} confirmed enzyme binding with the support (Ansari & Husain, 2011a). Furthermore, the peak at 1636 cm^{-1} corresponding to N—H bend also ascertained the presence of amide band I.

3.3 Binding Studies of Immobilized β-Galactosidase

Although the preparation of carrier is quite important for enzyme immobilization and its performance, the selection and optimization of immobilization conditions are equally critical for immobilized enzyme to express its maximum activity. Therefore, three parameters affecting the immobilization of β-galactosidase, i.e., enzyme concentration, Gr@Fe_3O_4 NC concentration, and immobilization time have also been standardized in this study. After optimizing all conditions, maximum immobilization yield was obtained at 1 mg mL^{-1} of enzyme, 10 mg of Gr@Fe_3O_4 NC, and 5 h of immobilization time. Table 4 demonstrates the calculation of immobilization yield of β-galactosidase on Gr@Fe_3O_4 NC in terms of enzyme units (U). Consequently, the immobilization strategy achieved 90% yield after optimizing all the earlier factors.

Table 4 Immobilization of β-Galactosidase on Gr@Fe_3O_4 NCs

Enzyme: NCs (mg)	Enzyme Activity Loaded, *X* (U)	Enzyme Activity in Washes, *Y* (U)	Enzyme Activity Bound on Gr@Fe_3O_4 NCs		Immobilization Yield (%), *B*/*A*
			Theoretical $(X - Y) = A$	Actual/ Immobilized = *B*	
10:100	49.12	0.075	49.04	43.84	89.7

Each value represents the mean for three independent experiments performed in duplicates, with average standard deviations, ≤5%.

Reproduced from Khan, M., Husain, Q., Naqvi, A. H. (2016). Gr based magnetic nanocomposites as versatile carriers for high yield immobilization and stabilization of β-galactosidase. *RSC Advances, 6,* 53493–53503.

3.4 Stability Studies of Immobilized β-Galactosidase

3.4.1 *Effect of pH and Temperature on the Activity of Free and Immobilized β-Galactosidase*

The catalytic activity of enzyme depends on its conformational structure. Even a minor change in the native conformation of the protein may result in the loss of its catalytic activity (Husain et al., 2011). Fig. 8A illustrates the pH–activity profiles for free and Gr@Fe_3O_4 NC bound β-galactosidase. The free and immobilized enzyme exhibited identical pH–optima at pH 4.5. However, the immobilized enzyme showed a significant broadening in pH–activity profile and retained higher fractions of catalytic activity at both acidic and basic sides of pH optimum, compared to that of the native enzyme. Gr@Fe_3O_4 NC bound β-galactosidase retained 42% activity at pH 8.0 whereas free enzyme lost nearly 98% of its original activity under similar experimental conditions. The Gr@Fe_3O_4 NC after binding to the β-galactosidase protected its confirmation at the active site, thus, it retained remarkably higher activity when compared to that of the free form (Husain et al., 2011).

Fig. 8B demonstrates the temperature–activity profiles for the free and NC bound enzyme. The temperature optima for both free and immobilized β-galactosidase were found to be same at 50°C. The immobilized enzyme maintained greater fraction of catalytic activity at higher temperatures than the free β-galactosidase. Gr@Fe_3O_4 NC bound β-galactosidase preserved 71% of its initial activity at 70°C while its free form retained only 31% of its activity under identical exposure. Husain et al. (2011) in an earlier study also noticed similar findings. Soluble and ZnO-adsorbed β-galactosidase showed temperature optima at 50°C whereas ZnO–NP bound β-galactosidase exhibited broadening in temperature optima from 50 to 60°C. ZnO–NP bound β-galactosidase exhibited 74% activity at

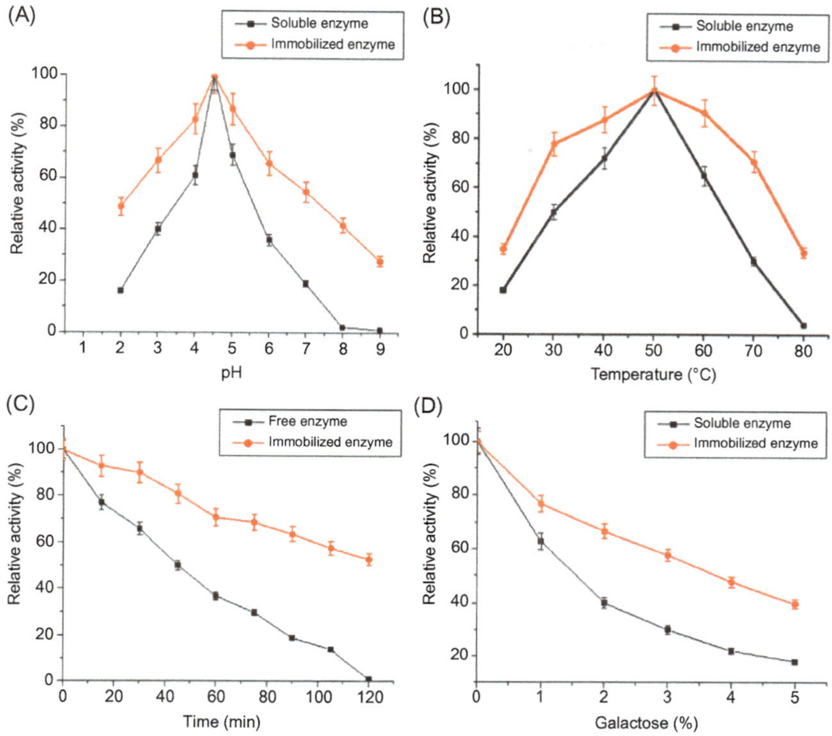

Fig. 8 (A) The pH–activity profiles for free and immobilized β-galactosidase. (B) Temperature–activity profiles for the free and immobilized β-galactosidase. (C) Thermal denaturation of the free and immobilized β-galactosidase. (D) Effect of galactose on the activity of the free and immobilized β-galactosidase. The activity of the enzyme in the absence of galactose was considered as control (100%) for the calculation of remaining activity at varying galactose concentrations. Data represent mean $\pm$ SD of three individual experiments performed in duplicates, *p*-value ≤ 0.05 for each figure. *Reproduced from Khan, M., Husain, Q., Naqvi, A. H. (2016). Graphene based magnetic nanocomposites as versatile carriers for high yield immobilization and stabilization of β-galactosidase.* RSC Advances, *6, 53493–53503.*

70°C while native ZnO-adsorbed and soluble enzyme retained only 58% and 31% activity under similar experimental conditions, respectively.

Gr@Fe_3O_4 NCs bound enzyme retained 53% of its initial activity, while the free enzyme lost almost its all activity under similar conditions. Husain et al. (2011) also demonstrated higher heat stability retained by the ZnO bound β-galactosidase compared to its free form under similar heat exposure. The loss in enzyme activity at higher temperatures might be due to denaturation of enzyme molecules. However, the immobilization matrix provided rigidity to the enzyme structure and thus, it made the immobilized enzyme highly stable against inactivation mediated by heat treatment (Singh et al., 2014).

3.4.2 Effect of Galactose

Some earlier workers have reported that β-galactosidase can be competitively inhibited by its own product, galactose (Haider & Husain, 2007; Husain, 2010; Park & Oh, 2010). Therefore, an attempt has been made to evaluate the effect of galactose (1.0%–5.0%, w/v) on the activity of free and NC bound β-galactosidase (Fig. 8D). The findings showed that immobilized β-galactosidase appeared significantly more resistant to galactose-mediated inhibition, when compared to the soluble enzyme free form at varying concentrations. The immobilized β-galactosidase maintained 40% of its initial activity in the presence of 5% galactose, while the free enzyme retained only 18% of its activity under similar experimental conditions.

3.5 Reusability and Storage Stability

Reusability is one of the most important parameters to investigate the significance of immobilized enzyme in various sectors on large scale. Fig. 9A illustrates the residual activity of Gr@Fe_3O_4 NC bound β-galactosidase after repeated uses. The immobilized enzyme retained 80% of its activity after its ninth repeated use. The high operational stability of the bound enzyme can effectively reduce operational cost in practical applications. Leaching of the enzyme from the support was responsible for activity loss which is because of the weakening of the binding forces between matrix and enzyme during their reuses (Singh et al., 2014). The free and immobilized β-galactosidase preparations are stored at 4°C, and activity measurements are carried out at the gap of 5-day intervals for 60 days (Table 5). The Gr@Fe_3O_4 NC bound β-galactosidase exhibited markedly high stability than its free form. The immobilized enzyme retained 79% of its initial activity during storage, while the free β-galactosidase lost more than 50% of its activity under identical storage conditions. Several earlier investigators have also found an improvement in the stability of enzymes on very long storage. Improved storage stability of the immobilized enzyme has been reported by several earlier workers (Ansari & Husain, 2011b; Bayraktar, Serilmiz, Karakas, Celem, & Onal, 2011; Singh & Kayastha, 2012).

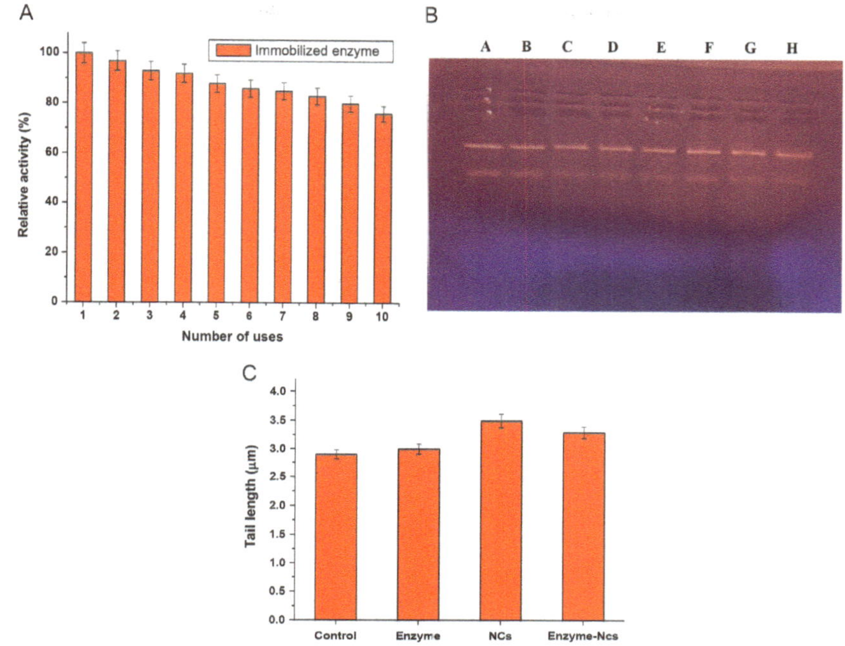

Fig. 9 (A) Reusability assay of immobilized β-galactosidase. The reusability of adsorbed β-galactosidase was monitored for 10 successive days. NC bound enzyme was taken in triplicates, and the activity determined on the first day was taken as the control (100%). p-Value ≤ 0.05 was considered statistically significant, (B) Plasmid nicking assay: Lanes A and B: pBR^{322} DNA alone, Lanes C and D: pBR^{322} DNA + enzyme (0.08 U), Lanes E and F: pBR^{322} DNA + Gr@Fe_3O_4 NC (40 μg), and Lanes G and H: pBR^{322} DNA + Gr@Fe_3O_4 NC bound β-galactosidase. (C) DNA damage in human lymphocytes by the enzyme alone (0.08 U), Gr@Fe_3O_4 NC (200 μg), and Gr@Fe_3O_4 NC bound enzyme. Data represent mean ± SD of three individual experiments performed in triplicates, p-value ≤ 0.05. *Reproduced from Khan, M., Husain, Q., Naqvi, A. H. (2016). Graphene based magnetic nanocomposites as versatile carriers for high yield immobilization and stabilization of β-galactosidase.* RSC Advances, 6, *53493–53503.*

3.6 Toxicity Assessment

It is essential to investigate the genotoxic effect of Gr@Fe_3O_4 NC in order to use it as an enzyme immobilization support for use in food technology. The plasmid nicking assay is one of the important parameters, which is used to evaluate the effect of various genotoxic chemicals on the integrity of DNA. In this study, pBR^{322} DNA was exposed to the enzyme alone, the Gr@Fe_3O_4 NC and NC bound β-galactosidase. Fig. 9B showed no damaging effect on the plasmid DNA, since the band intensity of the treated

Table 5 Storage Stability of Soluble and NC Bound β-Galactosidase Preparations When Kept at 4°C in 0.1 *M* Sodium Acetate Buffer, pH 4.5 for Over 60 Days

	Remaining Activity (%)	
Number of Days	**Free β-Galactosidase**	**Immobilized β-Galactosidase**
Control	100	100
5	94 ± 3.95	99 ± 4.98
10	87 ± 3.80	97 ± 4.85
15	84 ± 3.72	94 ± 4.70
20	76 ± 3.65	93 ± 4.65
25	72 ± 3.56	92 ± 4.62
30	67 ± 3.43	89 ± 4.45
35	63 ± 3.31	87 ± 4.35
40	62 ± 3.28	85 ± 4.27
45	60 ± 3.20	84 ± 4.19
50	56 ± 3.15	83 ± 4.05
55	52 ± 3.09	81 ± 3.96
60	49 ± 3.02	79 ± 3.88

The activity on first day was taken as control (100%) for the determination of remaining percent activity. p-Value ≤ 0.05 was considered statistically significant.
Reproduced from Khan, M., Husain, Q., Naqvi, A. H. (2016). Graphene based magnetic nanocomposites as versatile carriers for high yield immobilization and stabilization of β-galactosidase. *RSC Advances, 6*, 53493–53503.

sample was recorded similar to the control in all the cases. Ansari and Husain (2011a) have earlier also reported identical findings.

Single cell gel electrophoresis, commonly known as comet assay, is a very sensitive technique to assess DNA strand breaks in individual cells and has been recommended by various workers as a valid indicator for the examination of genotoxicity (Tice et al., 2000; Vnenchak & Rokosz, 1997). Human lymphocytic DNA of human lymphocytes was treated in vitro by free enzyme, the free Gr@Fe_3O_4 NC, and NC bound β-galactosidase, and the tail length, a parameter of DNA damage was plotted against concentration of different samples. Tail lengths obtained for lymphocytic DNA exposed to free enzyme, free Gr@Fe_3O_4 NC, and NC bound the β-galactosidase were found to be around 3.5 and 3.3 mm approximately, as compared to control (3.0 mm), it evidently demonstrates that the support used in this study for immobilization of β-galactosidase is not toxic and highly biocompatible in nature (Fig. 9C). The toxicity assessment for any immobilized β-galactosidase system has been rarely reported. However, it is equally important to assess the toxicity of the support, because it can be applied at large scale in the food industry to give orally to the lactose intolerant people worldwide (Ansari & Husain, 2011a).

3.7 Polyaniline-Coated Silver-Functionalized Graphene NC

3.7.1 Synthesis of PANI/Ag/GONC

General safety: Safety glasses, Latex, or nitrile gloves, lab coat

Reagents and chemicals

Silver nitrate—Sigma-Aldrich Chemicals Co., USA
Graphite flakes—Sigma-Aldrich Chemicals Co., USA

Equipment

1. Beaker (50–500 mL)
2. Stir/hot plate
3. Analytical balance
4. pH meter
5. Microtubes (1.5 mL)—Eppendorf (India)
6. Magnet (6000G)
7. Magnetic stirrer with bar

Methodology

The PANI/Ag/GONC was synthesized through in situ polymerization procedure. To be precise, following steps were taken in order to successfully synthesize the above-mentioned NC. Formulation of Gr hybrid support involved the preparation of GO solution from graphite. Preparation of GO is outlined later:

1. Add potassium persulfate (0.1 *M*) into 10% aniline solution under constant stirring conditions at very low speed in the molar ratio of 1:1 for 3 h.
2. Now add ammonium persulfate solution dropwise to the aniline solution in 1 *M* HCl and keep for 45 min under constant stirring condition. A dark green colored precipitate is obtained which indicates successful formation of PANI.
3. Prepare 2 m*M* of silver nitrate in Tris–HCl buffer, pH 8.5 and add this slowly drop by drop to the as prepared PANI solution. Gently stir for 2 h in an ice bath.
4. Synthesize GO via oxidative treatment of natural graphite flake utilizing a modified "Hummers method" following ultrasonication treatment for 1 h.

5. Add GO to solution comprising PANI/Ag and actively stir in a water bath containing ice for 2h and keep the obtained solution in a refrigerator overnight at 4°C.
6. Wash the obtained prepared product many times with distilled water and ethanol.
7. Finally, air dry in the hot incubator at 60°C for 20h and preserve it for further use.

3.7.2 Characterization of PANI/Ag/GONC

General safety: Safety glasses, latex, or nitrile gloves, lab coat

Chemicals and reagents

1. Potassium bromide
2. Sodium acetate buffer (pH 4.5, 0.1 *M*)
3. ANL@PANI/Ag/GONC (diluted solutions)

Equipment

1. UV/vis JASCO Spectrophotometer, Japan
2. Carbon-coated copper grids—300 mesh
3. Sputter coater—JEOL JFC 1600, Japan
4. Perkin Elmer FT-IR spectrometer, Spectrum Two, CT, USA
5. Perkin Elmer Pyris Thermogravimetric Analyzer, USA
6. JEOL-2100 100/120kV Transmission Electron Microscope, JEOL, Tokyo, Japan
7. JSM 6510LV Scanning Electron Microscope, JEOL, Tokyo, Japan
8. Oxford Instruments INCAx sight EDAX spectrometer
9. Atomic Force Microscopy (AFM) Probe (RTESP, Veeco model Icon with controller and Nanoscope 5 integrated system)
10. DynaPro-TC-04 Dynamic Light Scattering (DLS) equipment (Protein Solutions, Wyatt Technology, Santa Barbara, CA) equipped with a temperature-controlled microsampler

3.7.2.1 Fourier-Transform Infrared Spectroscopy

FT-IR analysis was carried out in order to determine the interactions of enzyme with NC.

Methodology

1. Use Perkin Elmer FT-IR Spectrometer, Spectrum Two (CT, USA) for analyzing the interactions of the functional groups between the enzyme and NC.
2. Dilute all with spectroscopic grade potassium bromide to fix the samples in a clearer sheet for FT-IR measurements.

3. Record the spectra in a wavenumber range of 4000–400 cm^{-1} for the samples.
4. Take the average value of 12 scans for the samples.

3.7.2.2 Transmission Electron Microscopy

TEM analysis is basically done to understand the size and morphology of the synthesized NC as well as the nanobiocatalyst.

Methodology

1. Prepare aqueous samples of ANL and ANL@PANI/Ag/GONC making the concentration of about 0.1 m*M*.
2. Ultrasonicate the samples to make the mixture homogenously distributed.
3. Using a micropipette drop-cast 10 μL of the sample on the copper grid.
4. Now using filter paper blot the excess solution.
5. Dry the aliquots and record images using TEM instrument at accelerating voltage of 200 keV.

3.7.2.3 Scanning Electron Microscopy and Energy Dispersive X-Ray Spectroscopy (EDS)

SEM studies were performed to analyze the surface morphology of the synthesized NC before and after conjugation with lipase while EDS was employed to analyze the elemental composition of synthesized NC.

Methodology

1. Prepare aqueous samples of ANL and PANI/Ag/GONC bound ANL and ultrasonicate them for even distribution.
2. Using a micropipette drop-cast 10 μL of the sample on a glass cover slip and blot excess solution using filter paper.
3. Use JSM SEM instrument to record images.
4. Using INCAx sight EDAX spectrometer, elemental composition of NC was analyzed.

3.7.2.4 AFM

The tapping mode in AFM was employed to visualize roughness of the NC surface before and after immobilization of lipase.

Methodology

1. Dilute and ultrasonicate all the samples to around 0.1 m*M* concentrations.
2. Load 0.2 mL solution on the cover glass slit using a micropipette.

3. Use RTESP, Veeco model microscope and set resonance frequency and amplitude to 340 Hz and 13 mV, respectively.
4. Carry out microscopic scans on variable surface locations of the samples and record the AFM images.

3.7.2.5 DLS Measurements

DLS studies were carried for size distribution analysis of synthesized NC before and after conjugation with the enzyme.

Methodology

1. Dilute the samples to 0.01 m*M* equivalent enzyme concentration and ultrasonicate them for uniform distribution of the particles.
2. Use DynaPro-TC-04 DLS instrument and set the wavelength to 830 nm for executing size distribution analysis of the samples.
3. Record the DLS measurement outcomes at 830 nm wavelength.
4. For each experiment, take around 20 measurements and analyze the mean hydrodynamic radius (Rh) and polydispersity index using Dynamics 6.10.0.10 software.

3.7.3 Immobilization of Lipase on PANI/Ag/GONC (Zhu & Sun, 2012)

General safety: Safety glasses, latex, or nitrile gloves, lab coat

Chemicals

1. Tris buffer (pH 7.0, 20 m*M*) and sodium phosphate buffer (pH 7.4, 25 m*M*)
2. PANI/Ag/GONC (500 mg)
3. Lipase (*A. niger*)

Materials and equipment

1. Falcons (15 mL)—Tarson, India
2. Microtubes (1.5 mL)—Eppendorf, India
3. Shaking incubator—Biogen, India

Methodology

The above-mentioned GO anchored nanosupport was used to make hybrid enzyme–NC biocatalyst. The typical protocol for the synthesis of biocatalyst is as follows:

1. For analyzing optimum conditions aiming to maximum yield immobilization of enzyme onto the GO anchored NC, during immobilization of native *A. niger* lipase (ANL), vary concentrations of PANI/Ag/GO NC (5–20 mg) and concentrations of crosslinker for different time intervals.
2. Dissolve the earlier formulations in 50 m*M* phosphate buffer, pH 7 at 30°C.

3. Ultrasonicate GO-based NCs for 30 min.
4. Activate the amino groups of above-mentioned support with specific glutaraldehyde (crosslinker) concentration for different time periods from 2 to 12 h.
5. Wash the NC several times with double distilled water to eliminate unbound glutaraldehyde.
6. Dry in hot air incubator at 60°C.
7. Now, dissolve free ANL (10 mg) in 100 mL sodium phosphate buffer, pH 7.0 and add this slowly to a certain amount of PANI/Ag/GONC.
8. Stir the reaction mixture (180 rpm) on a magnetic stirrer for a specified time and isolate the immobilized preparation of Gr-armored enzyme nanobiocatalyst via centrifugation.
9. Remove the uncrosslinked lipase by washing till no protein is spotted and perform activity assay in order to detect the amount of uncrosslinked ANL.
10. Store the immobilized biocatalyst in assay buffer at 4°C for further experiments.

3.7.4 Activity and Stability Studies of ANL@PANI/Ag/GONC

Reagents and chemicals

1. *p*-Nitrophenyl palmitate (*p*-NPP)—Sigma-Aldrich Chemicals Co., USA
2. Sodium phosphate buffer (20 m*M*, pH 7.4)—SRL, India
3. Triton X-100—SRL, India
4. Tris–HCl buffer (25 m*M*, pH 7.0)—SRL, India
5. Acetone—SRL, India
6. Acetonitrile—SRL, India
7. Isopropanol—SRL, India
8. Butanol—SRL, India
9. Nitroaniline—SRL, India
10. Toluene—SRL, India
11. Chlorobenzene—SRL, India

Equipment

1. Water bath (30–110°C)—Scientech Instruments, India
2. Ice maker and crusher—Labman, India
3. Ice box
4. Microtubes (1.5 mL)—Eppendorf, India
5. UV–vis spectrophotometer—Shimadzu, UV-1800, Japan)

Methodology

1. Use UV Spectrophotometer (JASCO) to determine catalytic activity of the samples using hydrolysis assay.
2. Prepare the samples of soluble and immobilized enzyme of equal enzyme concentration, i.e., 1 mg mL^{-1}.
3. Prepare stock solution of lipase (1 mg mL^{-1}) in 50 m*M* phosphate buffer, pH 7.0.
4. Prepare substrate solution with *p*-NPP (150 m*M* in isopropanol), 150 mL Triton X-100, and 67.5 mL 25 m*M* Tris–HCl buffer, pH 7.0. The reaction mixture consisted of 2 mL substrate and 0.1 mL lipase solution.
5. Incubate free and immobilized lipase independently (6.2 U), in appropriate buffers of varying pH; sodium acetate (pH 4.0–5.0), sodium phosphate (6.0 and 7.0), and Tris–HCl (pH 8.0, 9.0, and 10.0) and perform hydrolysis assay to determine the catalytic activity of the enzyme.
6. To evaluate the optimum temperature of the enzyme, incubate equal amounts of free and immobilized lipase at various temperatures (25–55°C) in 50 m*M* sodium phosphate buffer, pH 7.0 for 10 min and measure the *p*-NPP hydrolytic activity by both enzyme preparations.
7. To perform hydrolysis assay, incubate the reaction mixture at 37°C for 15 min in a water bath under continuous stirring condition, and add 0.25 mL 0.1 *M* Na_2CO_3 in order to terminate the reaction.
8. To analyze the thermal stability of both the free and immobilized lipase preparations, incubate both enzyme preparations simultaneously in 50 m*M* sodium phosphate buffer, pH 7 at 60°C for 2 h. Withdraw specific aliquots from both enzyme preparations at definite time intervals and chill them quickly in crushed ice for 5 min. The samples are then brought to room temperature until enzyme assay is performed.
9. For storage stability analysis, incubate the samples at 4 and 30°C in 2 mL of 50 m*M* sodium phosphate buffer, for a period of 30 days.
10. For solvent stability studies, incubate free ANL, and ANL@PANI/Ag/GONC (6.2 U) in 100 μL (33%, v/v) of various solvents such as nitroaniline, toluene, acetone, isopropanol, acetonitrile, butanol, and chlorobenzene at 30°C for 1 h. Also, incubate the obtained nanobiocatalyst in 0.05 *M* sodium phosphate buffer, pH 7.0 for the reference point in determining the effect of organic solvents.

3.7.5 Reusability Assay

1. Perform reusability assay of synthesized nanobiocatalyst-ANL@PANI/Ag/GONC (6.2 U) in triplicates in 50 m*M* sodium phosphate buffer, pH 7.0 for 11 successive cycles.
2. After each repetitive use, wash ANL@PANI/Ag/GONC with 50 m*M* sodium phosphate buffer, pH 7.0 and remove the pellet by centrifugation at 5000 × *g* for 10 min and store in assay buffer at 4°C.
3. The activity obtained on the first day was considered as control (100%), for the calculation of remaining percent activity after each repetitive use.

3.8 Characterization of Free and ANL Bound PANI/Ag/GONC

Fig. 10 illustrates SEM and TEM microphotographs of PANI/Ag/GONC and ANL@PANI/Ag/GONC. The morphology of the NCs before and

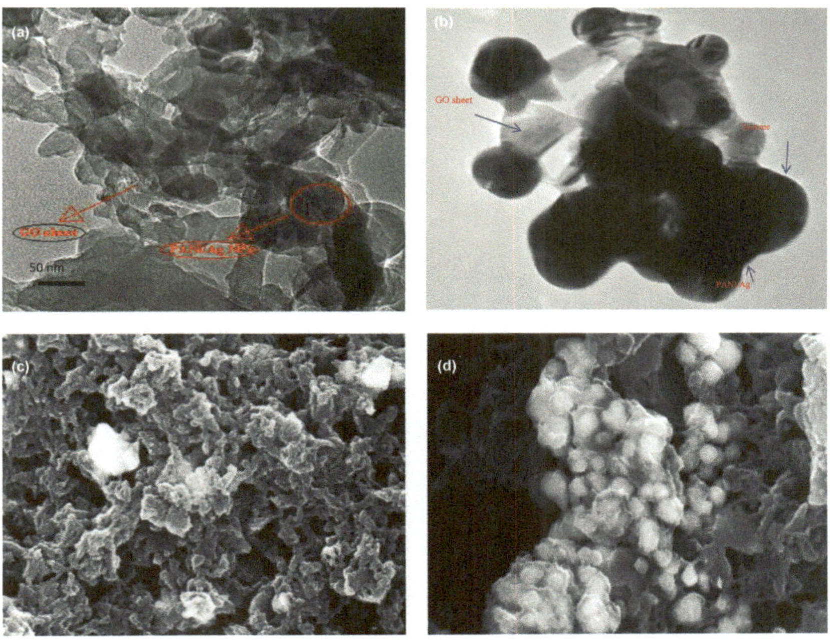

Fig. 10 Transmission electron micrographs of (A) PANI/Ag/GONC and (B) ANL@PANI/Ag/GONC. Scanning electron micrographs of (C) PANI/Ag/GONC and (D) ANL@PANI/Ag/GONC. *Reproduced from Asmat, S., Husain, Q., & Azam, A. (2017). Lipase immobilization on facile synthesized polyaniline-coated silver-functionalized graphene oxide nanocomposites as novel biocatalysts: stability and activity insights.* RSC Advances, *7(9), 5019–5029.*

after binding to the enzyme was affirmed and monitored by these techniques. The results showed that the NC are spherical in shape having a mean particle size of 50 nm and validate the capability of GO nanosheets to attach with PANI-coated Ag-NPs on their surface. The NC remained spherical in shape even after conjugation with ANL, however, the average diameter hiked from 50 to 72 nm, it signified that certain aggregation due to overlapping of some bigger particles with the smaller ones. The results showed the binding of considerable amounts of enzyme on the NC surface and demonstrated that the armoring of lipase with PANI/Ag/GONC did not alter the morphology of the NPs. Our findings are inconsistent with earlier reported results where the investigators have used Au/NP@Fe_3O_4 NPs as an immobilization support (Cao, Wen, Svec, Tan, & Lv, 2016).

To further affirm the morphology of the NC in these micrographs, AFM was performed to analyze the peak to peak valley distance in microphotographs as a pointer for the roughness in the surface of NC before and after immobilization of enzyme. As evident from Fig. 11, the NC surface before enzyme immobilization showed smaller peak-to-valley distance and was comparatively smooth, while after conjugation of enzyme with the NC was conjugated, there was a noticeable surge in peak-to-valley distance which demonstrated high enzyme loading. FT-IR was employed to

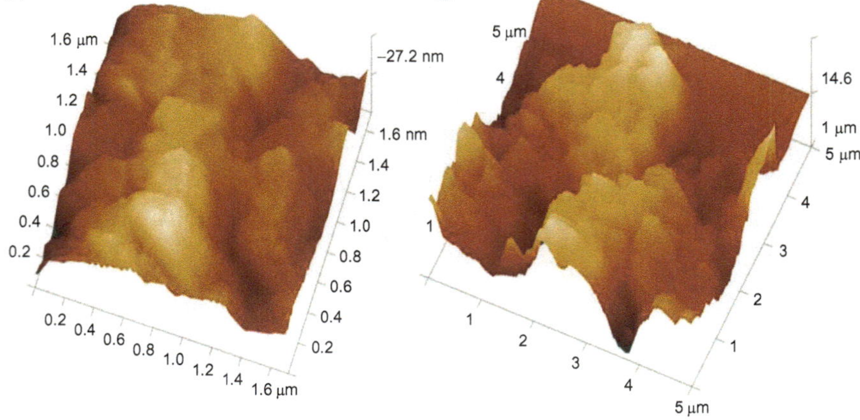

Fig. 11 Atomic force micrographs of (A) PANI/Ag/GONC and (B) ANL@PANI/Ag/GONC. Tapping mode of AFM is performed using commercial etched silicon tips as AFM probes with distinctive resonance frequency of c. 340 Hz and 13 mV amplitude (RTESP, Veeco, Japan). *Reproduced from Asmat, S., Husain, Q., & Azam, A. (2017). Lipase immobilization on facile synthesized polyaniline-coated silver-functionalized graphene oxide nanocomposites as novel biocatalysts: stability and activity insights.* RSC Advances, *7(9), 5019–5029.*

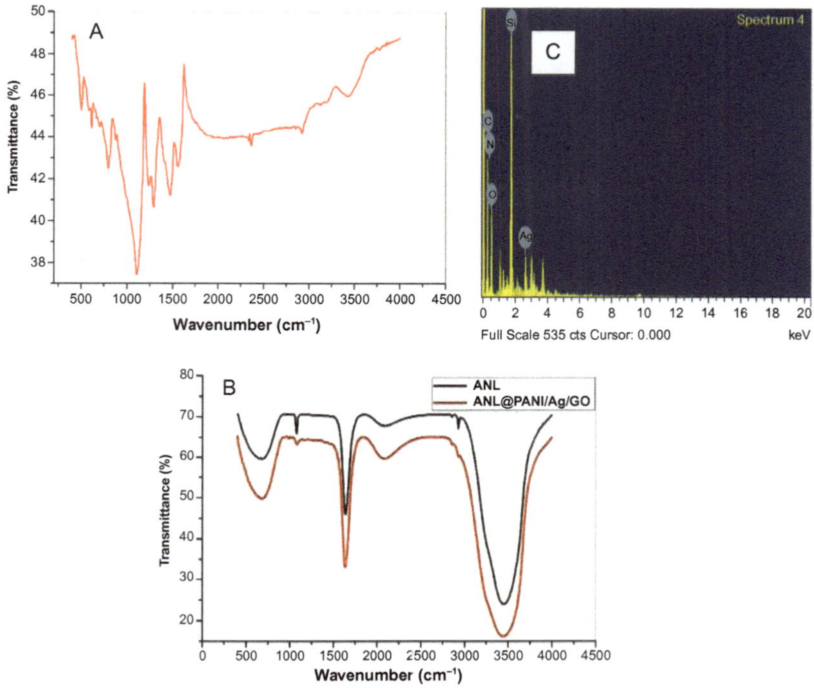

Fig. 12 FT-IR analysis is performed to monitor the linkages present between lipase and PANI/Ag/GONC. FT-IR spectra of (A) PANI/Ag/GONC, (B) ANL and ANL@PANI/Ag/GONC, these measurements were carried out on Perkin Elmer FT-IR spectrometer, Spectrum Two (CT, USA), and (C) EDAX analysis. *Reproduced from Asmat, S., Husain, Q., & Azam, A. (2017). Lipase immobilization on facile synthesized polyaniline-coated silver-functionalized graphene oxide nanocomposites as novel biocatalysts: stability and activity insights.* RSC Advances, *7(9), 5019–5029.*

examine the presence of characteristic functional groups of NC before and after immobilization and the changes occurring in them. Fig. 12 represents the spectra of synthesized NC, free ANL, and ANL@PANI/Ag/GONC. The FT-IR spectra of NC displays peak in the wavenumber range 1085, 1440, 1580, and 3425 cm^{-1} which indicate C—O bonds, C—N stretching vibration of slightly reduced Ag–GO, C=C bonds and oxygenated functional groups –OH, respectively (Chen, Feng, & Li, 2012). The FT-IR spectrum of ANL revealed dominant signals in the wavenumber range 3420–3150 cm^{-1} (–OH stretching vibrations and –NH stretching vibrations), at 2924 cm^{-1} (C—H stretching vibrations), 1652 cm^{-1} (N—H bending vibrations) whose intensity was considerably increased after binding with PANI/Ag/GONC, clearly signifying the successful immobilization of the enzyme onto NC (Hou, Qi, & Zhu, 2015). EDAX mapping examines

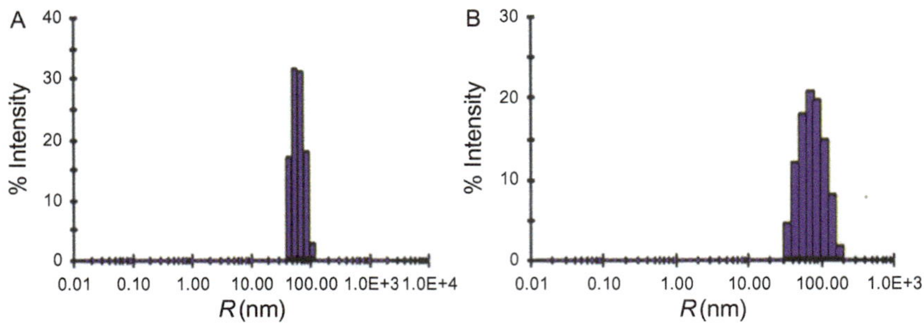

Fig. 13 DLS measurements showing hydrodynamic size distribution of (A) PANI/Ag/GONC and (B) ANL@PANI/Ag/GONC. *Reproduced from Asmat, S., Husain, Q., & Azam, A. (2017). Lipase immobilization on facile synthesized polyaniline-coated silver-functionalized graphene oxide nanocomposites as novel biocatalysts: stability and activity insights.* RSC Advances, *7(9), 5019–5029.*

the elemental distribution and structure of the NC (Rezaei, Akhavan, Hashemi, & Shamsara, 2016). The outcome can be seen in Fig. 12 in which the characteristic peaks of C, Ag, O, N_2, and Si are visible, with Gr designated through C peaks. The NC was visualized on the silica cover slip and thus the corresponding peak for Si was visible.

DLS is a technique which is used to analyze the size distribution profile of particles in solutions or suspension which provides us information regarding the conjugation of enzyme with NM (Venditti et al., 2015). Herein, it was employed to inspect the size and polydispersity of PANI/Ag/GONC and ANL@PANI/Ag/GONC. As shown in Fig. 13, the hydrodynamic radii (Rh) and percent polydispersity of free NCs were 62 nm and 24.5%, respectively. These values increased remarkably after binding with ANL. The Rh and percent polydispersity values were 100 nm and 32.6%, after immobilization, respectively, and this might be possibly due to aggregation of bigger particles with smaller ones.

3.9 Immobilization Efficiency of ANL@PANI/Ag/GONC

Glutaraldehyde has been used as a crosslinker to immobilize ANL on PANI/Ag/GONC, which was achieved by covalent attachment as displayed in the schematic representation of the formulation of ANL@PANI/Ag/GONC (Fig. 14). The immobilized lipase formulation showed an effectiveness factor (η) of 0.94. The η value of an immobilized enzyme preparation represents the ratio of actual to theoretical activity of the immobilized enzyme preparation (Matto, Naqash, & Husain, 2008). Table 6 displays the immobilization efficiency of bound lipase.

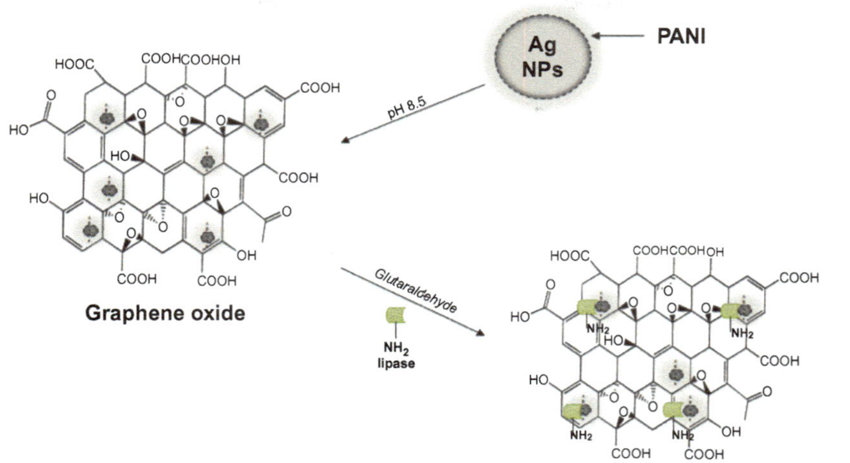

Fig. 14 Schematic illustration of preparation of ANL@PANI/Ag/GONC. *Reproduced from Asmat, S., Husain, Q., & Azam, A. (2017). Lipase immobilization on facile synthesized polyaniline-coated silver-functionalized graphene oxide nanocomposites as novel biocatalysts: stability and activity insights.* RSC Advances, *7(9), 5019–5029.*

Table 6 Immobilization of Lipase on PANI/Ag/GONC

		Enzyme Activity Bound to PANI/Ag/GONC			
Enzyme Loaded (X) (U)	**Enzyme Activity in Washes (Y) (U)**	**Theoretical ($X-Y$) = (A)**	**Actual (B)**	**Effectiveness Factor (η) = (B/A)**	**Immobilization Yield (%) ($B/A\times 100$)**
7	0.4	6.6	6.2	0.94	94

Each value demonstrates the mean for three independent experiments performed in triplicates.
Reproduced from Asmat, S., Husain, Q., & Azam, A. (2017). Lipase immobilization on facile synthesized polyaniline-coated silver-functionalized graphene oxide nanocomposites as novel biocatalysts: stability and activity insights. *RSC Advances*, 7(9), 5019–5029.

3.10 Stability Studies of ANL@PANI/Ag/GONC

Fig. 15A demonstrates the biocatalytic activity of free and PANI/Ag/GONC bound ANL at various temperatures and in buffers of various pH values. The free ANL showed its temperature optima at 40°C, whereas the fabricated nanobiocatalyst exhibited its temperature optima at 45°C. The immobilized ANL retained over 90% of its initial activity from 35 to 50°C demonstrating that ANL@PANI/Ag/GONC was more resistant

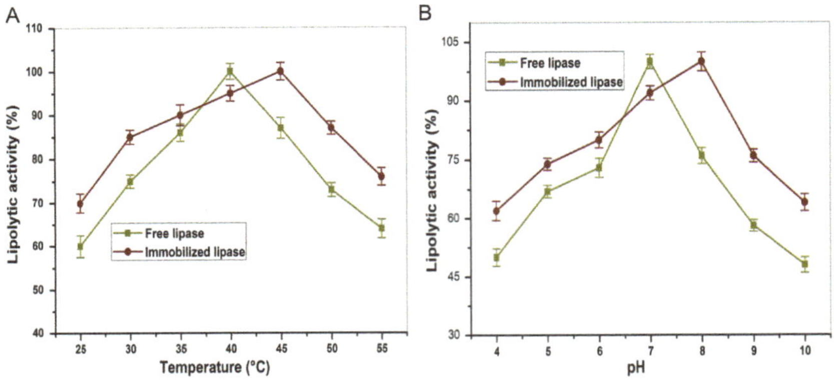

Fig. 15 (A) Temperature–activity profiles for free and immobilized lipase. The activity obtained at 40°C was taken as reference (100%) for the calculation of remaining percent activity. (B) Optimal pH–activity profiles for free and immobilized lipase. The activity at pH 7 was taken as control (100%) for the determination of remaining percent activity. The *symbols* show (–■–) free and (–●–) immobilized lipase. *Reproduced from Asmat, S., Husain, Q., & Azam, A. (2017). Lipase immobilization on facile synthesized polyaniline-coated silver-functionalized graphene oxide nanocomposites as novel biocatalysts: stability and activity insights.* RSC Advances, *7(9), 5019–5029.*

against heat as compared to free ANL (Hou et al., 2015). ANL and ANL@PANI/Ag/GONC exhibited their maximum activity at pH 7.0 and 8.0, respectively. However, ANL@PANI/Ag/GONC showed an increased resistance in a pH range from 7.0 to 9.0. ANL@PANI/Ag/GO retained higher fractions of catalytic activity in alkaline region as compared to free ANL. It might be due to electrostatic interactions and hydrogen bonding between ANL and the nanosupport (Suplatov et al., 2014).

Fig. 16A depicts the storage stability for free and PANI/Ag/GONC bound ANL, where both preparations have been incubated at temperatures 4 and 30°C. The results showed that the constructed nanobiocatalyst retained significantly higher activity when stored at both temperatures. ANL@PANI/Ag/GONC maintained 88% of its original activity after 30 days of storage at 4°C, whereas free enzyme showed about 65% of the initial activity under similar storage conditions. The decrease in the catalytic activity could be due to conformational changes in ANL, which can be improved upon immobilization as it provides rigidity to the enzyme (Yang et al., 2016).

Fig. 16B presents the thermal stability analysis for free and immobilized ANL. The free enzyme lost 51% of its activity after 2h incubation at 60°C, while fabricated nanobioconjugate retained 85% activity under similar

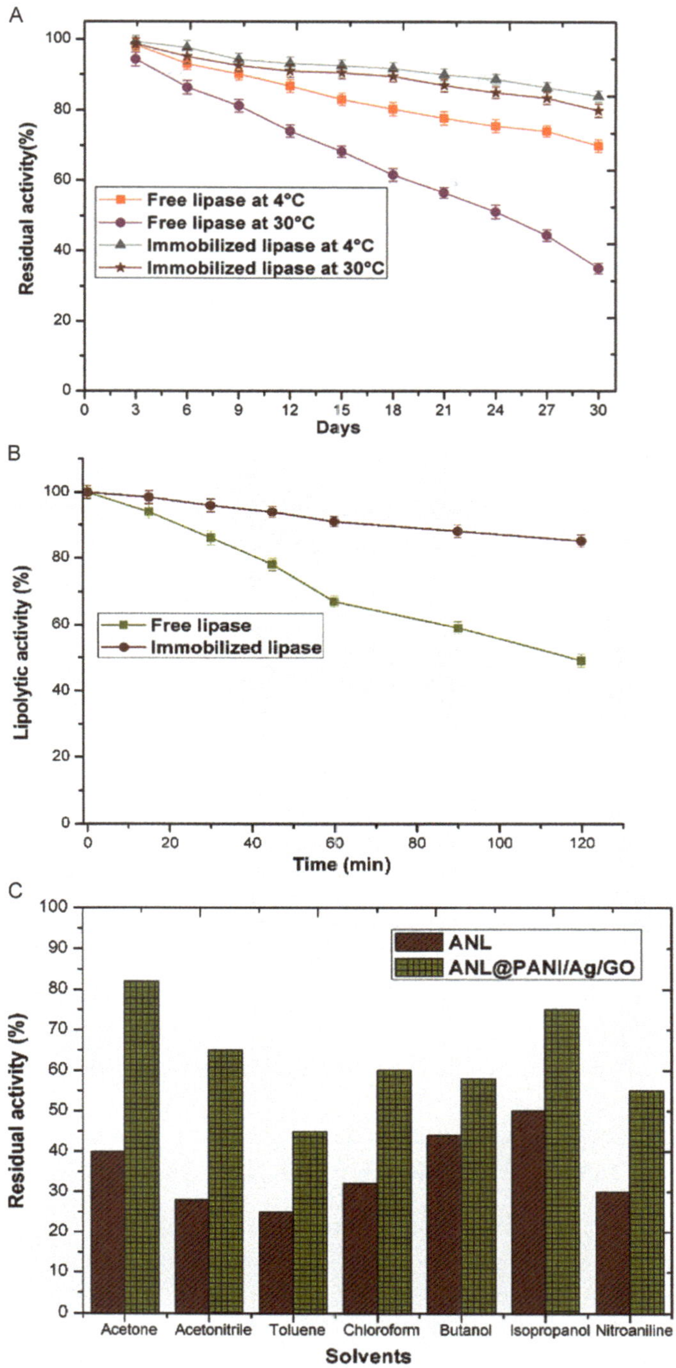

Fig. 16 Stability of free and PANI/Ag/GONC bound ANL. (A) Storage stability at 4°C and 30°C; (B) thermal denaturation; and (C) solvent stability. *Reproduced from Asmat, S., Husain, Q., & Azam, A. (2017). Lipase immobilization on facile synthesized polyaniline-coated silver-functionalized graphene oxide nanocomposites as novel biocatalysts: stability and activity insights.* RSC Advances, *7(9), 5019–5029.*

conditions. ANL@PANI/Ag/GONC showed better solvent tolerance as compared to its free enzyme in the presence of seven investigated solvents (Fig. 16C). Isopropanol exhibited lowest toxicity to free ANL, while acetone displayed lowest toxicity to ANL@PANI/Ag/GONC as its residual activity was high (81.5%). Likewise, 1 h incubation with toluene caused nearly 75% loss of activity in free ANL, but ANL@PANI/Ag/GONC maintained 55% of its original activity during similar exposure. We have recently reported similar enhancement in enzyme activity by solvent exposure (Asmat, Husain, & Khan, 2018). The prepared lipase@Ppy-MA/TiO_2-NC preparation displayed remarkable improved activity against solvent tolerance (150% and 125% activity recovery in acetone and isopropanol) as compared to its free counterpart. Usually, upon contact with organic solvents or ionic liquids, stripping of water from the protein surface is seen, in addition to solvent penetration into the enzyme, causing the protein to unfold and in turn, deactivation of the enzyme results (Romero et al., 2012). However, it has been observed that after the immobilization of lipase on PANI/Ag/GONC, organic solvents do not simply strip off the aqueous layer from the protein surface, thereby sustaining catalytic conformation and inner rigidity of the enzyme.

3.11 Reusability Analysis of ANL@PANI/Ag/GONC

Enzyme recovery provides a number of cost advantages that are often an essential prerequisite for establishing an economically viable enzyme catalyzed process. ANL@PANI/Ag/GONC maintained nearly 85% activity after its 11th repeated use (Fig. 17). There was a slight loss in its catalytic activity that might be during centrifugation after each repeated use. The PANI/Ag/GONC bound lipase retained significantly high catalytic activity even after its 11th repeated use, thereby enhancing its industrial applicability.

4. SUMMARY AND FUTURE OUTLOOK

Gr-based NM have rapidly become the most widely studied carbon-based materials owing to their unique structural features and exceptionally high chemical, electrical, and mechanical properties, as well as their capability for multivalent functionalization and efficient loading of biomacromolecules. The remarkable progress in synthesis and surface engineering of Gr-NM has opened new avenues exploring their use as nanoscaffolds for the development of nanobiocatalytic systems. These novel

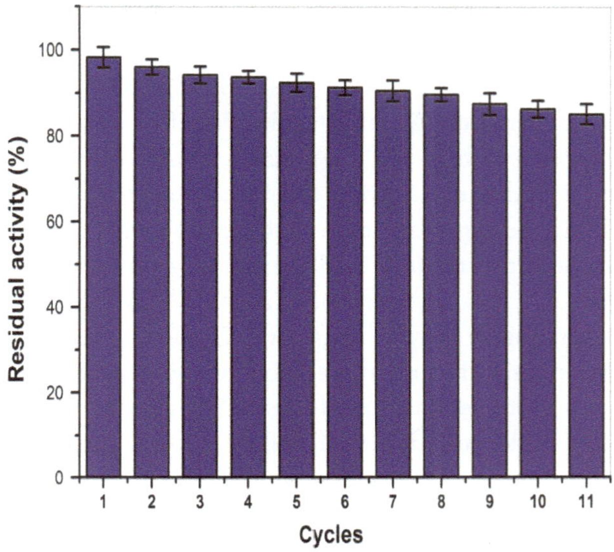

Fig. 17 Reusability assay of ANL@PANI/Ag/GONC. The enzyme activity calculated on the first day was taken as the reference (100%) for the determination of the remaining percentage activity after each reuse. *Reproduced from Asmat, S., Husain, Q., & Azam, A. (2017). Lipase immobilization on facile synthesized polyaniline-coated silver-functionalized graphene oxide nanocomposites as novel biocatalysts: stability and activity insights.* RSC Advances, *7(9), 5019–5029.*

bioconjugates differ from traditional immobilized enzymes in terms of catalytic efficiency, operational stability, and application potential.

In this chapter, an effort has been made to describe techniques employed in the synthesis and characterization of Gr-NC. The Gr-NC has been further used to immobilize industrially important hydrolytic enzymes; β-galactosidase and lipase. Two different approaches for the production of Gr-armored proteins have been discussed here. The first approach involved the physical adsorption of β-galactosidase on Gr@Fe_3O_4 NC, while the other made the use of glutaraldehyde, a crosslinking agent to covalently bind lipase with ANL@PANI/Ag/GONC.

In the first study, efficient immobilization of *A. oryzae* β-galactosidase onto functionalized Gr sheets (Gr@Fe_3O_4 NC) has been investigated. This support was selected because it provided high stability to enzyme and it is also nontoxic in nature. The immobilized β-galactosidase exhibited remarkably high storage stability and was found less inhibitory to its own product galactose as compared to its free form. Furthermore, this fabricated biocatalyst may be simply separated from the reaction mixture by magnetization or low speed centrifugation, and exhibited excellent reusability over several successive repeated uses. Hence, such robust nanobiocatalyst can serve as a powerful biorecognition probe for biosensor applications. One can detect the lactose concentrations in milk easily and efficiently by applying the biosensor armored with the help of such type of material. Utility of this method can be further extended in the development of therapeutic agents for targeted drug delivery for lactose intolerant patients.

In the second investigation of this chapter, a polymer modified silver incorporated GO NC (ANL@PANI/Ag/GONC) has been fabricated and used to immobilize enzyme lipase from *A. niger*. Lipase armored to this GO NC appeared significantly more stable to various kinds of denaturants, particularly it was noticed that PANI/Ag/GONC bound enzyme exhibited high resistance to denaturation mediated by various organic solvents. The immobilized enzyme can be applied for catalysis of some novel reactions such as acylation, esterification, transesterification, and interesterification.

The methods employed in the earlier studies have advantages as well as drawbacks. Enzyme leaching, which is a major problem associated with noncovalent immobilization techniques can be significantly reduced by crosslinking the enzyme with glutaraldehyde, which subsequently provides multipoint attachment of enzyme with the support. As can be inferred from the earlier investigations, in the second approach involving covalent method of binding, immobilized lipase retained 94% activity as compared to physically adsorbed β-galactosidase which was able to maintain 89% activity. Similarly, when the reusability was assessed, the immobilized lipase preparation exhibited 85% enzyme activity even after its 11 reuses, whereas the immobilized β-galactosidase was able to retain about 80% of the activity after its ninth cycle of repetitive use. Although the progress in the field is very promising, some critical points still need to be addressed in order to expand the applications of Gr-NC as immobilization matrices in nanobiotechnology. Further investigation is required to gain a deeper understanding of the effects of Gr-NC on the structure and functions of enzymes and other proteins. For instance, studying the effect of the type of functionalization as well as the structure and physicochemical characteristics of engineered Gr-NM on the immobilization efficiency and orientation of proteins will provide further understanding of the interactions of NM with protein molecules, which could lead to optimization of nanobiocatalytic systems.

Extensive characterization of enzymes and armored nanosupport would allow a better understanding of the role of binding moieties involved in the immobilization process and this would further help in scaling up these nanobiocatalysts in a cost effective and an efficient manner. For example, circular dichroism and 3D spectroscopic studies can be conducted to reveal the conformational changes in the immobilized protein with regard to protein-structure and function, which would broaden our knowledge on such fascinating nanobiocatalytic materials. Furthermore, an added focus would be to develop efficient Gr-NC with high biocompatibility, reduced toxicity, and negligible environmental effect, is essential for creating assemblies and functional devices that will expand the applications of Gr-NC in the field of biocatalytic transformations, enzyme engineering, biofuel and energy production, enzyme-based biosensing, and bioassays. Systematic comparisons of the performance of different kinds of Gr-NC is needed to establish a principle for the application of Gr-related NM in biosensors and other fields. Compared with carbon nanotubes (CNTs), limited research has been carried out on Gr/conducting polymer NC and Gr/carbon paste electrodes. Finally, the mass production of single-layer Gr is a major challenge that, if overcome, could benefit. In any case, the future of Gr is indeed bright.

ACKNOWLEDGMENTS

USIF, AMU, is duly acknowledged for providing the facilities of TEM and SEM. VSM and FT-Raman technique were performed in SAIF, IIT, Madras, while AFM was conducted at IIT Kanpur instrumentation facility. Centre of Excellence in Materials Science (Nanomaterials), Department of Applied Physics, Z.H. College of Engineering & Technology, AMU is duly acknowledged for providing TGA.

FUNDING

M.K. gratefully acknowledges the financial assistance in terms of "SRF" by "UGC-MANF, New Delhi." We are also grateful for funding support by UGC in terms of "University Non-NET fellowship" to one of our coauthors, S.M.

REFERENCES

Allen, M. J., Tung, V. C., & Kaner, R. B. (2010). Honeycomb carbon: A review of graphene. *Chemical Reviews*, *110*, 132–145.

Al-Lolage, F. A., Meneghello, M., Ma, S., Ludwig, R., & Bartlett, P. N. (2017). A flexible method for the stable, covalent immobilization of enzymes at electrode surfaces. *ChemElectroChem*, *4*, 1528–1534.

Alsharaeh, E. H., Othman, A. A., & Aldosari, M. A. (2014). Microwave irradiation effect on the dispersion and thermal stability of RGO nanosheets within a polystyrene matrix. *Materials (Basel, Switzerland)*, *7*, 5212–5224.

Amirbandeh, M., & Taheri-Kafrani, A. (2016). Immobilization of glucoamylase on triazine-functionalized Fe_3O_4/graphene oxide nanocomposite: Improved stability and reusability. *International Journal of Biological Macromolecules*, *93*, 1183–1191.

Ansari, S. A., & Husain, Q. (2010). Lactose hydrolysis by β galactosidase immobilized on concanavalin A-cellulose in batch and continuous mode. *Journal of Molecular Catalysis B: Enzymatic, 63*, 68–74.

Ansari, S. A., & Husain, Q. (2011a). Immobilization of *Aspergillus oryzae* 456 ß-galactosidase on concanavalin A-layered calcium alginate-cellulose beads and its application in lactose hydrolysis in continuous spiral bed reactors. *Polish Journal of Chemical Technology, 13*, 15–20.

Ansari, S. A., & Husain, Q. (2011b). Immobilization of *Kluyveromyces lactis* β galactosidase on concanavalin A layered aluminium oxide nanoparticles—Its future aspects in biosensor applications. *Journal of Molecular Catalysis B: Enzymatic, 70*, 119–126.

Ansari, S. A., & Husain, Q. (2012). Potential applications of enzymes immobilized on/in nano materials: A review. *Biotechnology Advances, 30*, 512–523.

Apetrei, I. M., & Apetrei, C. (2016). Amperometric biosensor based on diamine oxidase/platinum nanoparticles/graphene/chitosan modified screen-printed carbon electrode for histamine detection. *Sensors, 16*, 422–429.

Asmat, S., Husain, Q., & Khan, M. S. (2018). A polypyrrole–methyl anthranilate functionalized worm-like titanium dioxide nanocomposite as an innovative tool for immobilization of lipase: Preparation, activity, stability and molecular docking investigations. *New Journal of Chemistry, 42*, 91–102.

Baby, T. T., Aravind, S. S. J., Arockiadoss, T., Rakhi, R. B., & Ramaprabhu, S. (2010). Metal decorated graphene nanosheets as immobilization matrix for amprometric biosensor. *Sensors and Actuators B: Chemical, 145*, 71–77.

Bao, H., Chen, Q., Zhang, L., & Chen, G. (2011). Immobilization of trypsin in the layer-by-layer coating of graphene oxide and chitosan on in-channel glass fiber for microfluidic proteolysis. *Analyst, 136*, 5190–5196.

Bao, H., Zhang, L., & Chen, G. (2013). Immobilization of trypsin via graphene oxide-silica composite for efficient microchip proteolysis. *Journal of Chromatography, 1310*, 74–81.

Bathinapatla, A., Kanchi, S., Singh, P., Sabela, M. I., & Bisetty, K. (2016). An ultrasensitive performance enhanced novel cytochrome c biosensor for the detection of rebaudioside A. *Biosensors & Bioelectronics, 77*, 116–123.

Bayraktar, H., Serilmiz, M., Karakas, T. E., Celem, B., & Onal, S. (2011). Immobilization and stabilization of α-galactosidase on Sepabeads EC-EA and EC-HA. *International Journal of Biological Macromolecules, 49*, 855–860.

Cao, Y., Wen, L., Svec, F., Tan, T., & Lv, Y. (2016). Magnetic AuNP@ Fe_3O_4 nanoparticles as reusable carriers for reversible enzyme immobilization. *Chemical Engineering Journal, 286*, 272–281.

Chandra, V., Park, J., Chun, Y., Lee, J. W., Hwang, I. C., & Kim, K. S. (2010). Water dispersible magnetite-reduced graphene oxide composites for arsenic removal. *ACS Nano, 4*, 3979–3986.

Chang, Q., Huang, J., Ding, Y., & Tang, H. (2016). Catalytic oxidation of phenol and 2,4-dichlorophenol by using horseradish peroxidase immobilized on graphene oxide/Fe_3O_4. *Molecules, 21*, 1044–1051.

Chang, Q., Jiang, G., Tang, H., Li, N., Huang, J., & Wu, L. (2015). Enzymatic removal of chlorophenols using horseradish peroxidase immobilized on superparamagnetic Fe_3O_4/graphene oxide nanocomposites. *Chinese Journal of Catalysis, 36*, 961–968.

Chauhan, N., Chawla, S., Pundir, C. S., & Jain, U. (2017). An electrochemical sensor for detection of neurotransmitter-acetylcholine using metal nanoparticles, 2D material and conducting polymer modified electrode. *Biosensors & Bioelectronics, 89*, 377–383.

Chen, D., Feng, H., & Li, J. (2012). Graphene oxide: Preparation, functionalization, and electrochemical applications. *Chemical Reviews, 112*, 6027–6053.

Chen, J., Leng, J., Yang, X., Liao, L., Liu, L., & Xiao, A. (2017). Enhanced performance of magnetic graphene oxide-immobilized laccase and its application for the decolorization of dyes. *Molecules*, *22*, 221–230.

Chen, D., Tang, L., & Li, J. (2010). Graphene-based materials in electrochemistry. *Chemical Society Reviews*, *39*, 3157–3180.

Cho, Y., Kim, H., & Choi, Y. (2013). A graphene oxide-photosensitizer complex as an enzyme-activatable theranostic agent. *Chemical Communications*, *49*, 1202–1204.

Choi, B. G., Park, H., Yang, M. H., Jung, Y. M., Lee, S. Y., Hong, W. H., et al. (2010). Microwave-assisted synthesis of highly water-soluble graphene towards electrical DNA sensor. *Nanoscale*, *2*, 2692–2697.

Dai, L. (2013). Functionalization of graphene for efficient energy conversion and storage. *Accounts of Chemical Research*, *46*, 31–42.

Das, R., Talat, M., Srivastava, O. N., & Kayastha, A. M. (2018). Covalent immobilization of peanut β-amylase for producing industrial nano-biocatalysts: A comparative study of kinetics, stability and reusability of the immobilized enzyme. *Food Chemistry*, *245*, 488–499.

Das, G., & Yoon, H. H. (2015). Amperometric urea biosensors based on sulfonated graphene/polyaniline nanocomposite. *International Journal of Nanomedicine*, *10*, 55–66.

Dutta, N., Biswas, S., & Saha, M. K. (2016). Biophysical characterization and activity analysis of nano-magnesium supplemented cellulase obtained from a psychrobacterium following graphene oxide immobilization. *Enzyme and Microbial Technology*, *95*, 248–258.

Dutta, A., Biswas, S., & Saha, M. K. (2017). Nano-magnesium aided activity enhancement and biophysical characterization of a psychrophilic α-amylase immobilized on graphene oxide nanosupport. *Journal of Bioscience and Bioengineering*, *124*, 15–22.

Eigler, S. M., Dotzer, C., & Hirsch, A. (2012). Formation and decomposition of CO_2 intercalated graphene oxide. *Carbon*, *50*, 3666–3673.

EzhiVilian, E. T., Giribabu, K., Choe, S. R., Muruganantham, R., Lee, H., Roh, C., et al. (2017). A spick-and-span approach to the immobilization of horseradish peroxidase on Au nanospheres incorporated with a methionine/graphene biomatrix for the determination of endocrine disruptor bisphenol A. *Sensors and Actuators B: Chemical*, *251*, 804–812.

Fartas, F. M., Abdullah, J., Yusof, N. A., Sulaiman, Y., & Saiman, M. I. (2017). Biosensor based on tyrosinase immobilized on graphene-decorated gold nanoparticle/chitosan for phenolic detection in aqueous. *Sensors*, *17*, 1132–1139.

Feng, L. Y., Wu, L., & Qu, X. G. (2013). New horizons for diagnostics and therapeutic applications of graphene and graphene oxide. *Advanced Materials*, *25*, 168–186.

Gao, L., Zhang, H., & Cui, H. (2014). A general strategy to prepare homogeneous and reagentless GO/lucigenin & enzyme biosensors for detection of small biomolecules. *Biosensors & Bioelectronics*, *57*, 65–70.

Ge, J., Yang, C., Zhu, J., Lu, D., & Liu, Z. (2012). Nanobiocatalysis in organic media: Opportunities for enzymes in nanostructures. *Topics in Catalysis*, *55*, 1070–1080.

Geim, A. K., & Novoselov, K. S. (2007). The rise of graphene. *Nature Materials*, *6*, 183–191.

Georgakilas, V. (2012). Functionalization of graphene: Covalent and non-covalent approaches, derivatives and applications. *Chemical Reviews*, *112*, 6156–6214.

Goenka, S., Sant, V., & Sant, S. (2014). Graphene-based nanomaterials for drug delivery and tissue engineering. *Journal of Controlled Release*, *173*, 75–88.

Gokhale, A. A., Lu, J., & Lee, I. (2013). Immobilization of cellulase on magnetoresponsive graphene nano-supports. *Journal of Molecular Catalysis B: Enzymatic*, *90*, 76–86.

Gomes, A. A., Telli, E. P., Miletti, L. C., Skoronski, E., Ghislandi, M. G., Da Silva, G. F., et al. (2017). Improved enzymatic performance of graphene-immobilized β-glucosidase A in the presence of glucose-6-phosphate. *Biotechnology and Applied Biochemistry*, *15*, 1569–1574.

Haider, T., & Husain, Q. (2007). Calcium alginate entrapped preparations of *Aspergillus oryzae* β galactosidase: Its stability and applications in the hydrolysis of lactose. *International Journal of Biological Macromolecules*, *41*, 72–80.
Halder, A., Zhang, M., & Chi, Q. (2017). Electroactive and biocompatible functionalization of graphene for the development of biosensing platforms. *Biosensors & Bioelectronics*, *87*, 764–771.
He, H., & Gao, C. (2010). Supraparamagnetic, conductive, and processable multifunctional graphene nanosheets coated with high-density Fe_3O_4 nanoparticles. *ACS Applied Materials & Interfaces*, *2*, 3201–3210.
He, L. H., Zhang, Y. C., Liu, S. L., Fang, S. M., & Zhang, Z. H. (2014). A nanocomposite consisting of plasma-polymerized propargylamine and graphene for use in DNA sensing. *Microchimica Acta*, *181*, 1981–1989.
Heldt, C. L., Sieloff, A. K., Merillat, J. P., Minerick, A. R., King, J. A., Perger, W. F., et al. (2013). Stacked graphene nanoplatelet paper sensor for protein detection. *Sensors and Actuators B: Chemical*, *181*, 92–98.
Hernaez, M., Zammareno, C. R., Melendi-Espina, S., Bird, L. R., Mayes, A. G., & Arregui, F. J. (2017). Optical fibre sensors using graphene-based materials: A review. *Sensors*, *17*, 155.
Hossain, M. F., & Park, J. Y. (2017). Fabrication of sensitive enzymatic biosensor based on multi-layered reduced graphene oxide added PtAu nanoparticles-modified hybrid electrode. *PLoS One*, *12*, 173553.
Hou, C., Qi, Z., & Zhu, H. (2015). Preparation of core–shell magnetic polydopamine/alginate biocomposite for *Candida rugosa* lipase immobilization. *Colloids and Surfaces B: Biointerfaces*, *128*, 544–551.
Husain, Q. (2010). β-Galactosidases and their potential applications: A review. *Critical Reviews in Biotechnology*, *30*, 41–62.
Husain, Q. (2016). Magnetic nanoparticles as a tool for the immobilization/stabilization of hydrolases and their applications: An overview. *Biointerface Research in Applied Chemistry*, *6*, 1585–1606.
Husain, Q. (2017a). Nanomaterials as novel support for the immobilization of amylolytic enzymes and their application: A review. *Biocatalysis*, *3*, 31–53.
Husain, Q. (2017b). Nanomaterials immobilized cellulolytic enzymes and their industrial applications: A literature review. *JSM Biochemistry and Molecular Biology*, *4*, 1029–1045.
Husain, Q. (2017c). Nanosupport bound lipases, their stability and applications. *Biointerface Research in Applied Chemistry*, 7, 2194–2216.
Husain, Q. (2017d). High yield immobilization and stabilization of oxidoreductases using magnetic nanosupports and their potential applications: An update. *Current Catalysis*, *6*, 168–187.
Husain, Q. (2017e). Biosensor applications of graphene-nanocomposites bound oxidoreductive and hydrolytic enzymes. *Analytical Methods*, *9*, 6734–6746.
Husain, Q. (2018a). Nanocarriers immobilized proteases and their industrial applications: An overview. *Journal of Nanoscience and Nanotechnology*, *18*, 486–499.
Husain, Q. (2018b). Carbon nanotubes mediated immobilized glucose oxidase biosensors as an effective and sensitive analytical tool. *Biointerface Research in Applied Chemistry*, *8*, 3060–3074.
Husain, Q., Ansari, S. A., Alam, F., & Azam, A. (2011). Immobilization of *Aspergillus oryzae* β galactosidase on zinc oxide nanoparticles via simple adsorption mechanism. *International Journal of Biological Macromolecules*, *49*, 37–43.
Husain, Q., & Qayyum, S. (2013). Biological and enzymatic treatment of bisphenol A and other endocrine disrupting compounds: A review. *Critical Reviews in Biotechnology*, *33*, 260–292.

Husain, Q., & Ulber, R. (2011). Immobilized peroxidase as a valuable tool in the remediation of aromatic pollutants and xenobiotic compounds: A review. *Critical Reviews in Environmental Science and Technology*, *41*, 770–804.
Hwa, K. Y., & Subramani, B. (2014). Synthesis of zinc oxide nanoparticles on graphene–carbon nanotube hybrid for glucose biosensor applications. *Biosensors & Bioelectronics*, *62*, 127–133.
Jiang, D., Liu, Q., Wang, K., Qian, J., Dong, X., Yang, Z., et al. (2014). Enhanced non-enzymatic glucose sensing based on copper nanoparticles decorated nitrogen-doped graphene. *Biosensors & Bioelectronics*, *54*, 273–278.
Jiang, B., Yang, K., Zhang, L., Liang, Z., Peng, X., & Zhang, Y. (2014). Dendrimer-grafted graphene oxide nanosheets as novel support for trypsin immobilization to achieve fast on-plate digestion of proteins. *Talanta*, *122*, 278–284.
Jiang, B., Yang, K., Zhao, Q., Wu, Q., Liang, Z., Zhang, L., et al. (2012). Hydrophilic immobilized trypsin reactor with magnetic graphene oxide as support for high efficient proteome digestion. *Journal of Chromatography A*, *1254*, 8–13.
Johnson, P. A., Park, H. J., & Driscoll, A. J. (2011). Enzyme nanoparticle fabrication: Magnetic nanoparticle synthesis and enzyme immobilization. *Methods in Molecular Biology*, *679*, 183–191.
Khan, M. J., Husain, Q., & Ansari, S. A. (2013). Polyaniline-assisted silver nanoparticles: A novel support for the immobilization of α-amylase. *Applied Microbiology and Biotechnology*, *97*, 1513–1522.
Khan, M., Husain, Q., & Naqvi, A. H. (2016). Graphene based magnetic nanocomposites as versatile carriers for high yield immobilization and stabilization of β-galactosidase. *RSC Advances*, *6*, 53493–53503.
Khan, M. J., Qayyum, S., Alam, F., & Husain, Q. (2011). Effect of tin oxide nanoparticle binding on the structure and activity of α-amylase from *Bacillus amyloliquefaciens*. *Nanotechnology*, *22*, 455708.
Krishna, K. V., Menard-Moyon, C., Verma, S., & Bianco, A. (2013). Graphene-based nanomaterials for nanobiotechnology and biomedical applications. *Nanomedicine*, *8*, 1669–1688.
Kuila, T., Bose, S., Khanra, P., Mishra, A. K., Kim, N. H., & Lee, J. H. (2011). Recent advances in graphene-based biosensors. *Biosensors & Bioelectronics*, *26*, 4637–4678.
Kurbanoglu, S., Rivas, L., Ozkan, S. A., & Merkoçi, A. (2017). Electrochemically reduced graphene and iridium oxide nanoparticles for inhibition-based angiotensin-converting enzyme inhibitor detection. *Biosensors & Bioelectronics*, *88*, 122–129.
Li, M., Yang, X., Ren, J., Qu, K., & Qu, X. (2012). Using graphene oxide high near-infrared absorbance for photothermal treatment of Alzheimer's disease. *Advanced Materials*, *24*, 1722–1728.
Liang, R. P., Wang, X. N., Liu, C. M., Meng, X. Y., & Qiu, J. D. (2013). Construction of graphene oxide magnetic nanocomposites-based on-chip enzymatic microreactor for ultrasensitive pesticide detection. *Journal of Chromatography A*, *1315*, 28–35.
Lin, Y. C., Lu, C. C., Yeh, C. H., Jin, C., Suenaga, K., & Chiu, P. W. (2012). Graphene annealing: How clean can it be? *Nano Letters*, *12*, 414–419.
Liu, N., & Ma, Z. F. (2014). Au-ionic liquid functionalized reduced graphene oxide immunosensing platform for simultaneous electrochemical detection of multiple analytes. *Biosensors & Bioelectronics*, *51*, 184–190.
Liu, Y. P., Yuan, L., Yang, M., Zheng, Y., Li, L. J., Gao, L., et al. (2014). Giant enhancement in vertical conductivity of stacked CVD graphene sheets by self-assembled molecular layers. *Nature Communications*, *20*, 5461.
Liu, X., Zhang, J., Yan, R., Zhang, Q., & Liu, X. (2014). Preparation of graphene nanoplatelet-titanate nanotube composite and its advantages over the two single components as biosensor immobilization materials. *Biosensors & Bioelectronics*, *51*, 76–81.

Loh, K. P., Bao, Q., Ang, P. K., & Yang, J. (2010). The chemistry of graphene. *Journal of Materials Chemistry*, *20*, 2277–2289.

Mahdevi, M., Ahmad, M. B., Haron, M. J., Namvar, F., Nadi, B., Rahman, M. Z. A., et al. (2013). Synthesis, surface modification and characterization of biocompatible magnetic iron oxide nanoparticles for biomedical applications. *Molecules*, *18*, 7533–7548.

Mansouri, N., Babadi, A. A., Bagheri, S., & Abd Hamid, S. B. (2017). Immobilization of glucose oxidase on 3D graphene thin film: Novel glucose bioanalytical sensing platform. *International Journal of Hydrogen Energy*, *42*, 1337–1343.

Mateo, C., Palomo, J. M., Lorente, G. F., Guisan, J. M., & Lafuente, R. F. (2007). Improvement of enzyme activity, stability and selectivity via immobilization techniques. *Enzyme Microbial Technology*, *40*, 1451–1463.

Matto, M., Naqash, S., & Husain, Q. (2008). An economical and simple bioaffinity support for the immobilization and stabilization of tomato (Lycopersicon esculentum) peroxidase. *Acta Chimica Slovenica*, *55*, 671–676.

Narayanan, T. N., Liu, Z., Lakshmy, P., Gao, P., Nagoka, Y., Kumar, D. S., et al. (2012). Synthesis of reduced graphene oxide–Fe_3O_4 multifunctional freestanding membranes and their temperature dependent electronic transport properties. *Carbon*, *50*, 1338–1345.

Novoselov, K. S., Geim, A. K., & Morozov, S. V. (2004). Electric field effect in atomically thin carbon films. *Science*, *306*, 666–669.

Ormategui, N., Veloso, A., Leal, G. P., Rodriguez-Couto, S., & Tomovska, R. (2015). Design of stable and powerful nanobiocatalysts, based on enzyme laccase immobilized on self-assembled 3D graphene/polymer composite hydrogels. *ACS Applied Materials & Interfaces*, *7*, 14104–14112.

Pakapongpan, S., & Poo-Arporn, R. P. (2017). Self-assembly of glucose oxidase on reduced graphene oxide-magnetic nanoparticles nanocomposite-based direct electrochemistry for reagentless glucose biosensor. *Materials Science & Engineering C: Materials for Biological Applications*, *76*, 398–405.

Palanisamy, S., Cheemalapati, S., & Chen, S. M. (2014). Amperometric glucose biosensor based on glucose oxidase dispersed in multiwalled carbon nanotubes/graphene oxide hybrid biocomposite. *Materials Science and Engineering C: Materials for Biology Applications*, *34*, 207–213.

Palanisamy, S., Karuppiah, C., & Chen, S. M. (2014). Direct electrochemistry and electrocatalysis of glucose oxidase immobilized on reduced graphene oxide and silver nanoparticles nanocomposite modified electrode. *Colloids and Surfaces B: Biointerfaces*, *114*, 164–169.

Pang, R., Li, M., & Zhang, C. (2015). Degradation of phenolic compounds by laccase immobilized on carbon nanomaterials: Diffusional limitation investigation. *Talanta*, *131*, 38–45.

Park, A., & Oh, D. (2010). Effects of galactose and glucose on the hydrolysis reaction of a thermostable β-galactosidase from *Caldicellulosiruptor saccharolyticus*. *Applied Microbiology and Biotechnology*, *85*, 1427–1435.

Pavlidis, I. V., Patila, M., Bornscheuer, U. T., Gounis, D., & Stamatis, H. (2014). Graphene-based nanobiocatalytic systems: Recent advances and future prospects. *Trends in Biotechnology*, *32*, 6–21.

Pavlidis, I. V., Vorhaben, T., Tsoufis, T., Rudolf, P., Bornscheuer, U. T., Gournis, D., et al. (2012). Development of effective nanobiocatalytic systems through the immobilization of hydrolases on functionalized carbon-based nanomaterials. *Bioresource Technology*, *115*, 164–171.

Povedano, E., Cincotto, F. H., Parrado, C., Diez, P., Sanchez, A., Canevari, T. C., et al. (2017). Decoration of reduced graphene oxide with rhodium nanoparticles for the design of a sensitive electrochemical enzyme biosensor for 17β-estradiol. *Biosensors and Bioelectronics*, *89*, 343–351.

Putzbach, W., & Ronkainen, N. J. (2013). Immobilization techniques in the fabrication of nanomaterial-based electrochemical biosensors: A review. *Sensors*, *13*, 4811–4840.

Rana, S., Yeh, Y. C., & Rotello, V. M. (2010). Engineering the nanoparticle–protein interface: Applications and possibilities. *Current Opinion in Chemical Biology*, *14*, 828–834.

Rezaei, A., Akhavan, O., Hashemi, E., & Shamsara, M. (2016). Ugi four-component assembly process: An efficient approach for one-pot multifunctionalization of nanographene oxide in water and its application in lipase immobilization. *Chemistry of Materials*, *28*, 3004–3016.

Romero, C. M., Pera, L. M., Loto, F., Vallejos, C., Castro, G., & Baigori, M. D. (2012). Purification of an organic solvent-tolerant lipase from *Aspergillus niger* MYA 135 and its application in ester synthesis. *Biocatalysis and Agricultural Biotechnology*, *1*, 25–31.

Rouhani, S., Rostani, A., Salimi, A., & Pourshiani, O. (2018). Graphene oxide/CuFe2O4 nanocomposite as a novel scaffold for the immobilization of laccase and its application as a recyclable nanobiocatalyst for the green synthesis of arylsulfonyl benzenediols. *Biochemical Engineering Journal*, *133*, 1–11.

Ruecha, N., Rangkupan, R., Rodthongkum, N., & Chailapakul, O. (2014). Novel paper-based cholesterol biosensor using graphene/polyvinylpyrrolidone/polyaniline nanocomposite. *Biosensors and Bioelectronics*, *52*, 13–19.

Sassolas, A., Blum, L. J., & Leca-Bouvier, B. D. (2012). Immobilization strategies to develop enzymatic biosensors. *Biotechnology Advances*, *30*, 489–511.

Sethuraman, V., Mthuraja, P., Anandha, R. J., & Manisankar, P. (2016). A highly sensitive electrochemical biosensor for catechol using conducting polymer reduced graphene oxide-metal oxide enzyme modified electrode. *Biosensors & Bioelectronics*, *84*, 112–129.

She, G. W., Huang, X., Jin, L. L., Qi, X. P., Mu, L. X., & Shi, W. S. (2014). Electrochemical sensors: SnO_2 nanoparticle-coated ZnO nanotube arrays for high-performance electrochemical sensors. *Small*, *10*, 4685–4692.

Shi, X. F., Xia, X. Y., Cui, G. W., Deng, N., Zhao, Y. Q., Zhuo, L. H., et al. (2015). Multiple exciton generation application of PbS quantum dots in ZnO@PbS/graphene oxide for enhanced photocatalytic activity. *Applied Catalysis B: Environmental*, *163*, 123–128.

Shin, H. J., Kim, K. K., Benayad, A., Yoon, S. M., Park, H. K., Jung, I. S., et al. (2009). Efficient reduction of graphite oxide by sodium borohydride and its effect on electrical conductance. *Advanced Functional Materials*, *19*, 1987–1992.

Singh, N., & Kayastha, A. M. (2012). Cicer α-galactosidase immobilization onto chitosan and Amberlite MB-150: Optimization, characterization, and its applications. *Carbohydrate Research*, *358*, 61–66.

Singh, N., Srivastava, G., Talat, M., Raghubhanshi, H., Srivastava, O. N., & Kayastha, A. M. (2014). Cicer α-galactosidase immobilization onto functionalized graphene nanosheets using response surface method and its applications. *Food Chemistry*, *142*, 430–438.

Song, J., Xu, L., Xing, R. Q., Li, Q. L., Zhou, C. Y., Liu, D. L., et al. (2014). Synthesis of Au/graphene oxide composites for selective and sensitive electrochemical detection of ascorbic acid. *Scientific Reports*, *4*, 7515. https://doi.org/10.1038/srep07515.

Sun, X., Liu, z., Welsher, k., Robinson, J. T., Goodwin, A., Zaric, S., et al. (2008). Nano-graphene oxide for cellular imaging and drug delivery. *Nano Research*, *1*, 203–212.

Sun, Z. Z., Yan, Z., Yao, J., Beitler, E., Zhu, Y., & Tour, J. M. (2010). Growth of graphene from solid carbon sources. *Nature*, *468*, 549–552.

Suplatov, D., Panin, N., Kirilin, E., Shcherbakova, T., Kudryavtsev, P., & Svedas, V. (2014). Computational design of a pH stable enzyme: Understanding molecular mechanism of penicillin acylase's adaptation to alkaline conditions. *PLoS One*, *9*, e100643.

Takahashi, S., Abiko, N., & Anzai, J. (2013). Redox response of reduced graphene oxide-modified glassy carbon electrodes to hydrogen peroxide and hydrazine. *Materials (Basel, Switzerland)*, *6*, 1840–1850.

Terse-Thakoor, T., Komori, K., Ramnani, P., Lee, I., & Mulchandani, A. (2015). Electrochemically functionalized seamless three-dimensional graphene–carbon nanotube hybrid for direct electron transfer of glucose oxidase and bioelectrocatalysis. *Langmuir*, *31*, 13054–13061.

Tian, J., Deng, S. Y., Li, D. L., Shan, D., He, W., Zhang, X. J., et al. (2013). Bioinspired polydopamine as the scaffold for the active AuNPs anchoring and the chemical simultaneously reduced graphene oxide: Characterization and the enhanced biosensing application. *Biosensors & Bioelectronics*, *49*, 466–471.

Tice, R. R., Agurell, E., Anderson, D., Burlinson, B., Hartmann, A., Kobayashi, H., et al. (2000). Single cell gel/comet assay: Guidelines for in vitro and in vivo genetic toxicology testing. *Environmental and Molecular Mutagenesis*, *35*, 206–221.

Venditti, I., Palocci, C., Chronopoulou, L., Fratoddi, I., Fontana, L., Diociaiuti, M., et al. (2015). *Candida rugosa* lipase immobilization on hydrophilic charged gold nanoparticles as promising biocatalysts: Activity and stability investigations. *Colloids and Surfaces B: Biointerfaces*, *131*, 93–101.

Verma, M. L., Barrow, C. J., & Puri, M. (2013). Nanobiotechnology as a novel paradigm for enzyme immobilization and stabilization with potential applications in biodiesel production. *Applied Microbiology and Biotechnology*, *97*, 23–39.

Vineh, M. B., Saboury, A. A., Poostchi, A. A., Rashidi, A. M., & Parivar, K. (2018). Stability and activity improvement of horseradish peroxidase by covalent immobilization on functionalized reduced graphene oxide and biodegradation of high phenol concentration. *International Journal of Biological Macromolecules*, *106*, 1314–1322.

Vnenchak, K. M., & Rokosz, K. (1997). The comet assay for detection of potential genotoxicity of polluted water. *Folia Biologica*, *45*, 153–239.

Walcarius, A., Minteer, S. D., Wang, J., Lin, Y., & Merkoci, A. (2013). Nanomaterials for bio-functionalized electrodes: Recent trends. *Journal of Materials Chemistry B*, *1*, 4878–4908.

Wang, Q., Cui, J., Li, G., Zhang, J., Li, D., Huang, F., et al. (2014). Laccase immobilized on a PAN/adsorbents composite nanofibrous membrane for catechol treatment by a biocatalysis/adsorption process. *Molecules*, *19*, 3376–3388.

Wang, Y., Li, Z., Wang, J., Li, J., & Lin, Y. (2011). Graphene and graphene oxide: Biofunctionalization and applications in biotechnology. *Trends in Biotechnology*, *29*, 205–212.

Wang, T., Liu, J., Ren, J., Wang, J., & Wang, E. (2015). Mimetic biomembrane-AuNPs-graphene hybrid as matrix for enzyme immobilization and bioelectrocatalysis study. *Talanta*, *143*, 438–441.

Wang, H., Zhang, Q., Chu, X., Chen, T., Ge, J., & Yu, R. (2011). Graphene oxide-peptide conjugate as an intracellular protease sensor for caspase-3 activation imaging in live cells. *Angewandte Chemie (International ed. in English)*, *50*, 7065–7069.

Wei, Z. Q., Wang, D. B., Kim, S., Kim, S. Y., Hu, Y. K., Yakes, M. K., et al. (2010). Nanoscale tunable reduction of graphene oxide for graphene electronics. *Science*, *11*, 1373–1376.

Xiao, P., Bu, F., Yang, G., Zhang, Y., & Xu, Y. (2017). Integration of graphene, nano sulfur, and conducting polymer into compact, flexible lithium–sulfur battery cathodes with ultrahigh volumetric capacity and superior cycling stability for foldable devices. *Advanced Materials*, *29*. https://doi.org/10.1002/adma.201703324.

Xu, X. W., Shen, J. F., Li, N., & Ye, M. X. (2014). Facile synthesis of reduced graphene oxide/$CoWO_4$ nanocomposites with enhanced electrochemical performances for supercapacitors. *Electrochimica Acta*, *150*, 23–34.

Xu, J., Sheng, Z., Wang, X., Liu, X., Xia, J., Xiong, P., et al. (2016). Enhancement in ionic liquid tolerance of cellulase immobilized on PEGylated graphene oxide nanosheets: Application in saccharification of lignocellulose. *Bioresource Technology, 200*, 1060–1074.

Yang, Y., Asiri, A. M., Du, D., & Lin, Y. (2014). Acetylcholinesterase biosensor based on a gold nanoparticle-polypyrrole-reduced graphene oxide nanocomposite modified electrode for the amperometric detection of organophosphorus pesticides. *Analyst, 139*, 3055–3060.

Yang, K., Feng, L. Z., Hong, H., Cai, W. B., & Liu, Z. (2013). Preparation and functionalization of graphene nanocomposites for biomedical applications. *Nature Protocols, 8*, 2392–2403.

Yang, W. R., Ratinac, K. R., Ringer, S. P., Thordarson, P., Gooding, J. J., & Braet, F. (2010). Carbon nanomaterials in biosensors: Should you use nanotubes or graphene? *Angewandte Chemie International Edition, 49*, 2114–2138.

Yang, F. C., Wang, J., Cao, Y., Zhang, L., & Zhang, X. (2014). A highly sensitive ascorbic acid sensor based on carbon-supported CoPd nanoparticles. *Sensors and Actuators B: Chemical, 205*, 20–25.

Yang, D., Wang, X., Shi, J., Wang, X., Zhang, S., Han, P., et al. (2016). In situ synthesized rGO–Fe_3O_4 nanocomposites as enzyme immobilization support for achieving high activity recovery and easy recycling. *Biochemical Engineering Journal, 105*, 273–280.

Yang, H., Zhang, Y., Li, L., Sun, G., Zhang, L., Ge, S., et al. (2017). Real-time and in situ enzyme inhibition assay for the flux of hydrogen sulfide based on 3D interwoven AuPd-reduced graphene oxide network. *Biosensors & Bioelectronics, 87*, 53–58.

Ye, X. L., Du, Y. L., Duan, K. Y., Lu, D. B., Wang, C. M., & Shi, X. Z. (2014). Fabrication of nano-ZnS coated PEDOT-reduced graphene oxide hybrids modified glassy carbon-rotating disk electrode and its application for simultaneous determination of adenine, guanine, and thymine. *Sensors and Actuators B: Chemical, 203*, 271–281.

Yu, Y., Chen, Z., He, S., Zhang, B., Li, X., & Yao, M. (2014). Direct electron transfer of glucose oxidase and biosensing for glucose based on PDDA-capped gold nanoparticle modified graphene/multi-walled carbon nanotubes electrode. *Biosensors & Bioelectronics, 52*, 147–152.

Yu, X., Cheng, H., Zhang, M., Zhao, Y., Qu, L., & Shi, G. (2017). Graphene-based smart materials. *Nature Reviews Materials, 2*, 17046.

Yu, K. L. Q., Dong, K. T., Zhang, Y. P., Wang, Y. Q., & Zhi, Q. Q. (2014). Tuned n/n or n/p heterojunctions for reduced graphene oxide and titania nanosheets and their electrochemical properties. *Materials Chemistry and Physics, 148*, 803–809.

Zhai, C., Guo, Y., Sun, X., Zheng, Y., & Wang, X. (2014). An acetylcholinesterase biosensor based on graphene-gold nanocomposite and calcined layered double hydroxide. *Enzyme and Microbial Technology, 58–59*, 8–13.

Zhan, T. R., Tan, Z. W., Wang, X. J., & Hou, W. (2018). Hemoglobin immobilized in g-C3N4 nanoparticle decorated 3D graphene-LDH network: Direct electrochemistry and electrocatalysis to trichloroacetic acid. *Sensors and Actuators B: Chemical, 255*, 149–158.

Zhang, J., Chen, S., Ruo, Y., Zhong, X., & Wu, X. (2015). An ultrasensitive electrochemiluminescent biosensor for the detection of concanavalin a based on poly(ethylenimine) reduced graphene oxide and hollow gold nanoparticles. *Analytical and Bioanalytical Chemistry, 407*, 447–453.

Zhang, W., Li, X., Zou, R., Wu, H., Shi, H., Yu, S., et al. (2015). Multifunctional glucose biosensors from Fe_3O_4 nanoparticles modified chitosan/graphene nanocomposites. *Scientific Reports, 5*, 11129–11138.

Zhang, Y., Shen, J., Li, H., Wang, L., Cao, D., Feng, X., et al. (2016). Recent progress on graphene-based electrochemical biosensors. *The Chemical Record, 16*, 273–294.

Zheng, Y. L., Li, P., Li, H. B., & Chen, S. B. (2014). Controllable growth of cobalt oxide nanoparticles on reduced graphene oxide and its application for highly sensitive glucose sensor. *International Journal of Electrochemical Science*, *9*, 7369–7381.

Zhong, Z. Y., Li, M. X., Qing, Y., Dai, N., Guan, W., Liang, W., et al. (2014). Signal-on electrochemical immunoassay for APE1 using ionic liquid doped Au nanoparticle/graphene as a nanocarrier and alkaline phosphatase as enhancer. *Analyst*, *139*, 6563–6568.

Zhou, H. H., Han, G. Y., Fu, D. Y., Chang, Y. Z., Xiao, Y. M., & Zhai, H. J. (2014). Petal-shaped poly (3, 4-ethylenedioxythiophene)/sodium dodecyl sulfate-graphene oxide intercalation composites for high-performance electrochemical energy storage. *Journal of Power Sources*, *272*, 203–210.

Zhu, C. Z., & Dong, S. J. (2014). Energetic graphene-based electrochemical analytical devices in nucleic acid, protein and cancer diagnostics and detection. *Electroanalysis*, *26*, 14–29.

Zhu, C. Z., Du, D., & Lin, Y. H. (2017). Graphene-like 2D nanomaterial-based bio-interfaces for biosensing applications. *Biosensors and Bioelectronics*, *89*, 43–55.

Zhu, J., & Sun, G. (2012). Lipase immobilization on glutaraldehyde-activated nanofibrous membranes for improved enzyme stabilities and activities. *Reactive and Functional Polymers*, *72*, 839–845.

Zubir, N. A., Yacou, C., Motuzas, J., Zhang, X., Joao, C., & Costa, D. (2014). Structural and functional investigation of graphene oxide–Fe_3O_4 nanocomposites for the heterogeneous Fenton-like reaction. *Scientific Reports*, *4*, 4594.

FURTHER READING

Kim, J., Bayram, J., Park, C., Cheng, H., Dimitrakopoulos, C. W., Ott, C., et al. (2014). Principle of direct van der Waals epitaxy of single-crystalline films on epitaxial graphene. *Nature Communications*, *5*, 4836–4844.

Vigneshvar, S., Sudhakumari, C. C., Senthilkumaran, B., & Prakash, H. (2016). Recent advances in biosensor technology for potential applications—An overview. *Frontiers in Bioengineering and Biotechnology*, *4*, 3389.

CHAPTER FIVE

PAMAM Dendrimer Modified Reduced Graphene Oxide Postfunctionalized by Horseradish Peroxidase for Biosensing H_2O_2

Sheela Berchmans[1], Manju Venkatesan, Chirnajeevi Srinivasa Rao Vusa, Palaniappan Arumugam
CSIR-Central Electrochemical Research Institute, Karaikudi, Tamil Nadu, India
[1]Corresponding author: e-mail addresses: sheelab@cecri.res.in; sheelaberchmans@yahoo.com

Contents

Methods in Enzymology, Volume 609
ISSN 0076-6879
https://doi.org/10.1016/bs.mie.2018.05.012

Abstract

In this chapter, we describe the tethering of horseradish peroxidase (HRP) to reduced graphene oxide (RGO) for sensing H_2O_2 in serum. To accomplish this, RGO was synthesized through a green route by reducing graphene oxide (GO) prepared by Hummers method with carrot extract. The RGO was then covalently functionalized by electrochemical amination using fourth generation, amine-terminated PAMAM dendrimers. Subsequently, HRP was postfunctionalized through glutaraldehyde linkage. The synthesized RGO and the functionalization steps were well characterized by spectroscopic, microscopic, and electrochemical techniques. The application of HRP tethered RGO was demonstrated for H_2O_2 sensing in blood serum. This work provides scope for extending this functionalization strategy for other carbonaceous materials as well.

ABBREVIATIONS

ALD atomic layer deposition
ATRP atomic transfer radical polymerization
Ct-RGO carrot reduced graphene oxide
DAN 1,5-diaminonaphthalene
EGO exfoliated graphite oxide
GO graphene oxide
GR graphene
HPC-Py pyrene-containing hydroxypropyl cellulose
NP 1-nitropyrene
PTCDA perylene-3,4,9,10-tetracarboxylic dianhydride
RGO reduced graphene oxide
SCMC sodium carboxymethylcellulose
SLS sodium lignosulfonate

1. INTRODUCTION

Graphene (GR), the 2D material with a honeycomb structure, is a single-atom-thick material with sp^2-hybridized carbons and possesses exquisite electronic and mechanical properties which are superior to many of the existing materials. Many applications have been envisaged for this wonder material at the laboratory level which ranges from swift radiofrequency/logic devices (Weiss et al., 2012), reinforced composites with thermal and electrical conductivities (Stankovich et al., 2006), chemical/biological sensors (Cohen-Tanugi & Grossman, 2012; Shao et al., 2010), desalination membranes (Cohen-Tanugi & Grossman, 2012; O'Hern et al., 2014), photocatalysts (Zhang, Zhang, & Xu, 2012), transparent electrodes for liquid crystal displays (Pang, Hernandez, Feng, & Müllen, 2011) to solar cells (Liu et al., 2008; Wang, Zhi, & Müllen, 2008). Large-scale preparation of graphene often leads to a compromise of the properties of the pristine carbon. However, the large-scale processability of the material is limited by its solubility in very few solvents. The possibility of reaggregation leads to graphite-like agglomerates resulting in the shortfall of its properties. The exceptionally high electron mobility could not be put to practical use due to the lack of band gap in the materials. Further, the reactivity of graphene is very low as the extensive delocalization of π-bonding systems due to the strong interaction of all p_z orbitals, which makes the 2D structure more inert. Though the strong hydrophobic nature of the 2D material is suitable for electronic applications, it is not appropriate for biosensor applications. These challenges can be overcome by suitable functionalization of graphene which can be carried out through covalent or noncovalent routes (Georgakilas et al., 2012; Mali, Greenwood, Adisoejoso, Phillipson, & De Feyter, 2015).

Noncovalent method involves the physisorption of organic molecules which is a mild procedure and it would not involve compromising the properties of the pristine carbon. Covalent functionalization would lead to chemisorption of reagents which would result in robust functionalization. Covalent functionalization will also lead to the conversion of sp^2 to sp^3 carbons and would change the properties of graphene to a large extent. The covalent modification can improve their solubility, and molecular recognition moieties could be introduced, which makes it a good candidate for sensing applications. The covalent modification can also be used to introduce band gaps into graphene for applications in electronics. Subsequently, research focus also grew around graphene oxide (GO) whose structure is made up of oxygen linked to the carbon skeleton of graphene. Being hydrophilic, GO can be dispersed in water and consists of aromatic sp^2 and aliphatic sp^3 carbons making it amenable for further functionalization. It can be easily reduced to graphene-like material (Vusa, Berchmans, & Alwarappan, 2014; Wang, Zhou, Zhang, Boey, & Zhang, 2009; Zhu et al., 2010) and is referred to as reduced graphene oxide (RGO). The RGO exhibits interesting, physical, chemical, and morphological properties though not equivalent to graphene, and it has become a better choice for large-scale processability of graphene-like materials. Also, we can add tailored functionalities for various applications to RGO. The functionalization of GO and RGO can be carried out by noncovalent and covalent interactions for many useful applications.

2. FUNCTIONALIZATION OF GRAPHENE (OR) RGO BY NONCOVALENT INTERACTIONS

Noncovalent intermolecular interactions involving π-system of graphene are critical to the stabilization of proteins, enzyme–drug complexes, DNA–protein complexes, organic supramolecules, and functional nanomaterials (Burley & Petsko, 1985; Hong et al., 2001; Lee et al., 2009; Meyer, Castellano, & Diederich, 2003; Singh, Lee, Hwang, & Kim, 2007). The van der Waals and ionic interactions between graphene and molecules are made use of during noncovalent functionalization. There are many reports which describe the noncovalent functionalization of graphene with a pyrene moiety. Xu et al. prepared aqueous dispersions of GO functionalized with 1-pyrene butyrate in the presence of a base which was subsequently reduced with hydrazine resulting in graphene material with conductivity seven times greater than that of the GO precursor (Xu, Bai, Lu, Li, & Shi, 2008). Noncovalent modification of graphene with 1-pyrenebutanoic acid succinimidyl ester is shown to retain the optical adsorption of pristine graphene in the visible region and it is shown to improve the power conversion efficiency by 1.71%. This is higher compared to the value of 0.71% obtained for pristine carbon. Hence, graphene being more flexible and less expensive will be able to replace indium-doped tin oxide for optical applications (Wang, Chen, Zhong, Zhu, & Loh, 2009). A similar functionalization was carried out by Kodali et al. (2011) and functionalization of graphene by perylene carboxylic acid laminated onto polydimethylsiloxane membranes is found to exhibit unique optical and molecular sensing properties while preserving the conducting property of pristine graphene (Kodali et al., 2011).

Chemical reduction of exfoliated graphite oxide (EGO) by sodium lignosulfonate (SLS), sodium carboxymethylcellulose (SCMC), and pyrene-containing hydroxypropyl cellulose (HPC-Py) (Yang, Pan, Huang, & Li, 2010) has led to the synthesis of graphene nanoplatelets and the product was well characterized by several spectroscopic techniques. A solar cell fabricated using graphene (single- and double-layer graphene sheets) functionalized with pyrene and perylenediimide as electron donor and acceptor molecules (Su et al., 2009) showed increased conductivity. Noncovalent functionalization of graphene with 1,5-diaminonaphthalene (DAN) and 1-nitropyrene (NP) (Cheng, Shiue, Tsai, Wang, & Chen, 2011) as electron-donating and electron-withdrawing molecules has been carried out, and the functionalized graphene has been well characterized by various techniques. Graphene could be functionalized by carboxylate-terminated perylene molecules (Wang, Tabakman, & Dai, 2008) and functionalized graphene was used to grow metal oxide by atomic layer deposition (ALD) method, well characterized by several techniques.

In a subsequent work, the room temperature molecular resolution of a self-assembled perylene-3,4,9,10-tetracarboxylic dianhydride (PTCDA) monolayer on epitaxial graphene using scanning tunneling microscopy was reported (Wang & Hersam, 2009). The noncovalent functionalization of graphene nanoplatelets with single-stranded DNA has been shown starting with GO precursor which was then reduced by hydrazine (Liang, Wu, Feng, & Müllen, 2009). Subsequently, gold nanoparticles-coated DNA functionalized onto GO (Liu, Li, Li, Li, & Deng, 2010) and GO/DNA composite hydrogels (Xu, Wu, Sun, Bai, & Shi, 2010) were reported. Further many molecules like porphyrins (Geng & Jung, 2010; Tu, Lei, Zhang, & Ju, 2010; Zhang, Tang, Lei, Dong, & Ju, 2011), polystyrene (Wang, Li, et al., 2009), nafion (Ansari, Kelarakis, Estevez, & Giannelis, 2010; Choi, Park, et al., 2010; Wu, Zhao, Hu, & Chen, 2011), Congo red (Ansari et al., 2010), Methylene green (Liu, Gao, et al., 2009), poly(3,4-ethylenedioxythiophene) (PEDOT) embedded with fluoroalkyl phosphate-based ionic liquid (Saxena, Deepa, Joshi, Bhandari, & Srivastava, 2011), 4-bromobenzene diazonium tetrafluoroborate (Lim, Lee, Shin, Shin, & Choi, 2010), chitosan (Fang, Long, Zhao, Wang, & Chen, 2010), PEDOT:PSS (Jo et al., 2011), coronene carboxylate acceptor molecule (Ghosh, Rao, George, & Rao, 2010), thionine (Chen, Zhai, Lu, Zhang, & Zheng, 2011), poly(*N*-isopropyl acrylamide) (Liu, Yang, et al., 2010), poly(2,5-bis(3-sulfonatopropoxy)-1,4-ethynyl-phenylene-alt-1,4-ethynylphenylene) sodium salt (PPE-SO_3^-) (Liu, Lerner, & Yan, 2010; Liu, Li, et al., 2010; Liu, Yang, et al., 2010), hemin–graphene hybrid (Guo et al., 2011), polyethylene sorbitan laurate (Park et al., 2010), Pluronic copolymer (Zu & Han, 2009), SWNTs (Ghosh, Rao, Voggu, & George, 2010), amine-terminated polymers (Choi, Han, et al., 2010), coil–rod–coil conjugated triblock copolymer (PEG-OPE) (Qi et al., 2010), ionic liquid polymer, etc. (Kim, Lee, Kim, & Suh, 2010) have been used to functionalize graphene by noncovalent approaches.

3. COVALENT FUNCTIONALIZATION OF GRAPHENE, RGO

The covalent functionalization of graphene leads to robust bond formation and it leads to the disturbance of the sp^2 carbon-based structure leading to a decrease in charge carrier mobility. However, functionalization improves the solubility of graphene for making composites with other materials. The existing protocols that have been developed for the functionalization of other carbon allotropes like fullerenes and carbon nanotubes (CNTs) proved to work well, in the case of graphene too. The organic intermediates such as free radicals, carbenes, nitrenes, and arynes can easily form covalent adducts with graphene through free-radical addition (Park & Yan, 2013). Furthermore, graphene can take the role of a Diels–Alder substrate by acting as either the diene or the dienophile depending on the reaction conditions (Sarkar, Bekyarova, Niyogi, & Haddon, 2011). Atomic covalent functionalization at the basal plane of graphene surface which involves hydrogenation, halogenation, and oxygenation has also been demonstrated (Johns & Hersam, 2013).

In the case of GO, the functionalization is made through the oxygen groups. Functionalized graphene was prepared via chemical unzipping of CNTs with nitrophenyls through the diazonium salt reaction. The product, upon heating, produced a free radical and attacked the sp^2 carbon atoms of graphene. This reaction was shown to decrease the conductivity of graphene in a controlled manner due to disruption in the sp^2/sp^3 ratio (Kosynkin et al., 2009; Sinitskii et al., 2010). A similar approach leads to the introduction of the band gap in the material in a controlled fashion (Niyogi et al., 2010). The hydroxylated aryl groups grafted covalently on graphene by the diazonium addition reaction act as initiators for the polymerization of styrene via the atomic transfer radical polymerization (ATRP) method, leading to the grafting of polymer chains to the graphene surface (Fang, Wang, Lu, Yang, & Nutt, 2009). Another free-radical addition method used for the functionalization includes the reaction of benzoyl peroxide with graphene sheets, which was produced by the isolation of graphene mechanically and subsequently deposited on a silicon substrate (Liu, Ryu, et al., 2009).

1,3-Dipolar cycloadditions to graphene have been shown with the formation of azomethine ylide by the condensation of 3,4-hydroxybenzaldehyde and sarcosine where it was found that the introduction of hydroxyl groups improved the solubility of the material in polar solvents such as ethanol and DMF (Georgakilas et al., 2010). Quintana et al. employed 1,3-dipolar cycloaddition to functionalize graphene sheets by making use of paraformaldehyde precursors and a specifically designed NH_2-terminated α-amino acid that can selectively bind to gold nanorods (Quintana et al., 2010). Several other reports describe the functionalization of graphenes with azides, nitrenes, and arynes through which various functional groups and polymeric chains could be introduced into graphene sheets (Choi, Kim, Kim, Lee, & Kim, 2009; He & Gao, 2010; Liu, Lerner, et al., 2010; Strom, Dillon, Hamilton, & Barron, 2010; Vadukumpully, Gupta, Zhang, Xu, & Valiyaveettil, 2011; Zhong et al., 2010).

GO has been the preferred choice as the starting material to synthesize graphene derivatives through the covalent attachment of organic groups on its surface. Abundant literature is available for the functionalization of GO with chromophores, polymers, and other organic molecules. In principle, reduction of GO should be able to get rid of the oxygen groups leading to the conversion of sp^3 to sp^2 carbon structure, which is more aromatic and defect free. However, experiments show that a certain number of oxygen groups and defects remain even after the reduction of GO. The reduction of GO has been explored through various approaches starting from conventional reduction with reducing agents like hydrazine, sodium borohydride, bio-based reducing agents, bacterial reduction, thermal reduction, and so on (Dubin et al., 2010; Schniepp et al., 2006; Wang, Chen, et al., 2009, Wang, Li, et al., 2009, Wang, Zhou, et al., 2009; Zhu et al., 2010).

4. GRAPHENE IN BIOSENSING

Biosensors are analytical devices which are made up of a unique union of biological recognition units like enzymes, microorganisms, antibodies or tissues, and a suitable transducer. Electrochemical biosensors are highly advantageous and preferred due to their high sensitivity, selectivity, high signal-to-noise ratio, relative simplicity, and rapid response time (Ronkainen, Halsall, & Heineman, 2010). Carbon is considered as the most suited electrode material owing to its chemical inertness, low cost, and wide potential window within which it can operate. The GO is often favored for biosensor applications for its increased hydrophilicity. However, its low conductivity is disadvantageous which can be circumvented by the elimination of its oxygen functionalities by appropriate reducing agents. This gives rise to RGO with the desired properties such as large surface area, high conductivity, and ease of modification.

Graphene-related materials can act as electrochemical nanoprobes if one could monitor the distinct electrochemical signal that arises during the electrochemical reduction of the oxygen-containing groups. This concept has been demonstrated in the case of thrombin sensing (Bonanni, Chua, Zhao, Sofer, & Pumera, 2012; Loo, Bonanni, & Pumera, 2013; Wang & Dai, 2015). The electrocatalytic ability of graphene-related materials which are further enhanced by the decoration of atoms like potassium (K), nitrogen (N), and sulfur (S) has been shown to be good sensing materials (Guo et al., 2013; Li et al., 2012; Ma, Wang, Wang, & Xia, 2015; Shang et al., 2008; Wang, Shao, Matson, Li, & Lin, 2010; Yang, Feng, Wang, & Müllen, 2011). Further GR-like materials can be functionalized with redox probes (Bai, Chai, Pu, & Yuan, 2014; Bai et al., 2012), recognition components (Wang et al., 2013), or target molecules (Wang et al., 2013) for improved analytical performance.

Direct electron transfer of redox enzymes like glucose oxidase is promoted in GR-based biosensors (Baptista-Pires et al., 2014; Liang et al., 2013; Zhang, Liao, Chu, Liu, & Zhang, 2014). Also, the assembly of glucose oxidase on GR, utilized in online microdialysis systems, has been used to monitor the extracellular concentration of glucose in vivo (Gu et al., 2012). Coimmobilization of a bienzyme (cholesterol oxidase and catalase) in the presence of GR was investigated to detect cholesterol (Gholivand & Khodadadian, 2014). Besides redox enzymes, other proteins have also been assembled on GR structures (Huang, Yao, Liang, & Qiu, 2013; Song et al., 2014). Also, GR-based electrochemical immunosensing systems have also been reported (Haque et al., 2012; Park, Singh, Kim, & Yang, 2014; Srivastava et al., 2013). Some of the interesting examples would be electrochemical immunosensing of okadaic acid (Eissa & Zourob, 2012) and vascular endothelial growth factor receptor 2 (VEGFR2) (Wei et al., 2014). Anchoring of the DNA probe on GR is highly facilitated by the presence of various groups and the π-conjugated regions, and the anchoring can be carried out by either noncovalent or covalent methods for electrochemical biosensor applications (Bai et al., 2012; Benvidi, Rajabzadeh, Mazloum-Ardakani, Heidari, & Mulchandani, 2014; Dubuisson, Yang, & Loh, 2011; Liang, Chen, Guo, & Li, 2011; Luo et al., 2013; Yang et al., 2013).

In this chapter, we discuss a green and facile strategy to reduce GO using carrot extract and the characterization of the resulting RGO. The mixture of antioxidants present in carrot extract plays a vital role in reducing GO. Also, we discuss functionalization of RGO using poly(amidoamine) dendrimers

(PAMAM dendrimers) at room temperature by an electrochemical approach involving covalent linkage between RGO and the dendrimer. Further, the PAMAM-modified RGO was tethered with HRP enzyme by postfunctionalization strategy. The high density of amine functional groups allowed high loadings of the enzymes, and its applicability toward sensing H_2O_2 is also discussed.

5. SYNTHESIS OF CARROT EXTRACT REDUCED GRAPHENE OXIDE

The carrot is a rich source of potential antioxidants especially carotenoids and their content ranges from 60 to 120 mg/100 g, but some varieties contain up to 300 mg/100 g. But, 80% is made up of mainly β-carotene (Fikselova, Silhar, Marecek, & Francakova, 2008). Carotenoids such as lycopene, lutein, β-carotene, and 3-epilutein are found to be potential antioxidants in many biological processes. Carotenoids react with many reducible species such as peroxy radicals, hydroxyl radicals, singlet oxygen, hydrogen peroxide, superoxide, and organic hydrogen peroxides in biological systems. The mixture of carotenoids can enhance the antioxidant property by their synergistic effects (Burton & Ingold, 1984; Carail & Caris-Veyrat, 2006; Heber & Lu, 2002). Hence, carrot extract contains a mixture of these carotenoids, and it can be an effective reducing mixture toward many reducible species.

5.1 Equipment

Regular vegetable cutter, mixer, glass flasks, magnetic stirrer with a temperature controller, magnetic stir bar, and filtration setup.

5.2 Materials

Graphite powder (<20 μm), Analytical Grade $KMnO_4$, $NaNO_3$, H_2SO_4, HCl, NaOH, H_2O_2, and fresh carrots.

5.3 Preparation Method

5.3.1 Preparation of Graphene Oxide by Hummers Method

a. 10 g of graphite was weighed and transferred to a round-bottom flask.

b. 5 g of sodium nitrite was weighed and added to the above round-bottom flask and the flask was kept in an ice bath to reduce the temperature below 10°C.

c. The 250 mL of concentrated sulfuric acid was added drop by drop to the above-cooled mixture under stirred conditions. During above conditions, the temperature of the reaction mixture was found to increase. Then, the reaction mixture is continuously stirred in an ice bath until the temperature is reduced to 20°C.

d. Then, 30 g of $KMnO_4$ was weighed and added for >1 h and stirred vigorously. During this addition, the temperature was increased to 35°C and the same temperature is maintained for 30 min.

e. Then, 450 mL of distilled water is added in aliquots during which the temperature is increased to 70°C.

f. After 20 min 1.4 L of hot water and 25 mL of 30% aqueous solution of H_2O_2 are added to the reaction mixture. The reaction mixture turns from yellow to dark brown.

g. The reaction suspension is then repeatedly washed with dilute HCl and water.

h. The resulting product is dried at 50°C under vacuum for 72 h, which results in the formation of dark brown solid of GO.

5.3.2 Preparation of Carrot Extract

a. The carrots were procured from the local market and cleaned in cold water.

b. The skin of the cleaned carrot is peeled off and chopped into pieces.

c. The chopped carrot slices were weighed (20 g) and mixed with 100 mL of deionized water and boiled at 100°C for 50 min.

d. Then the solution is filtered through Whatman 40 filter paper. The filtrate thus prepared, probably a mixture of antioxidants like β-carotene, lutein, and lycopene, acts as the reducing agent for the reduction of GO.

5.3.3 Synthesis of Reduced Graphene Oxide (Ct-RGO) by a Green Approach

a. The above preparation of GO is dispersed in deionized water ($1\,mg\,mL^{-1}$) and sonicated for 1 h.

b. 100 mL of the carrot extract is added to the flask.

c. Then 200 mL of 1 *M* NaOH is added.

d. About 20 mL of the GO suspension is added to a flask containing a mixture of 1 *M* NaOH and 100 mL of the carrot extract.

e. The above mixture is stirred for 1 h at 90°C.

f. The color of the reaction mixture slowly turns dark black within 1 h, and the reaction is allowed to continue for another 1 h for complete reduction of GO.

g. Then, the reaction mixture is cooled to room temperature.

h. The cooled mixture is repeatedly washed with deionized water to remove soluble impurities

i. The resulting product is dried under vacuum.

The schematic of the synthesis of graphene from GO using carrot extract is shown in Fig. 1.

5.4 Characterization of the Reduced Graphene Oxide Prepared by the Green Approach

The synthesized graphene material is characterized by Raman and X-ray diffraction, scanning electron microscopy (SEM) and transmission electron microscopy (TEM).

5.4.1 Equipment

Raman spectrometer with 514 nm Ar laser source, X-ray diffraction instrument that uses Ni-filtered Cu Kα radiation ($\lambda = 1.54$ Å), SEM with an accelerating voltage in the range 0.3–30 kV, and TEM with an accelerating voltage of 220 kV.

5.4.2 Materials

Microscopic cover glass, Cu grid, and a conducting substrate.

5.4.3 Procedure

The Ct-RGO suspension is dropcasted onto the microscopic cover glass for Raman and XRD analysis. The Ct-RGO suspension dropcasted on Cu grid is used for TEM analysis and the Ct-RGO suspension dropcasted on a conducting substrate is used for SEM analysis.

5.4.4 Characterization

The appearance of characteristic D and G bands at 1336 and 1593 cm^{-1} and the appearance of a sharp 2D peak at 2635 cm^{-1} in the Raman spectrum of synthesized GO and Ct-RGO, as shown in Fig. 2A and B, provide evidence for the formation of few layers of graphene. The appearance of the 2D band and an increase in D/G ratio from 0.979 to 1.198 indicate the reduction of GO by carrot extract which manifests a mixture of antioxidants as the potential candidates to reduce GO. During this reduction process, the mixture of

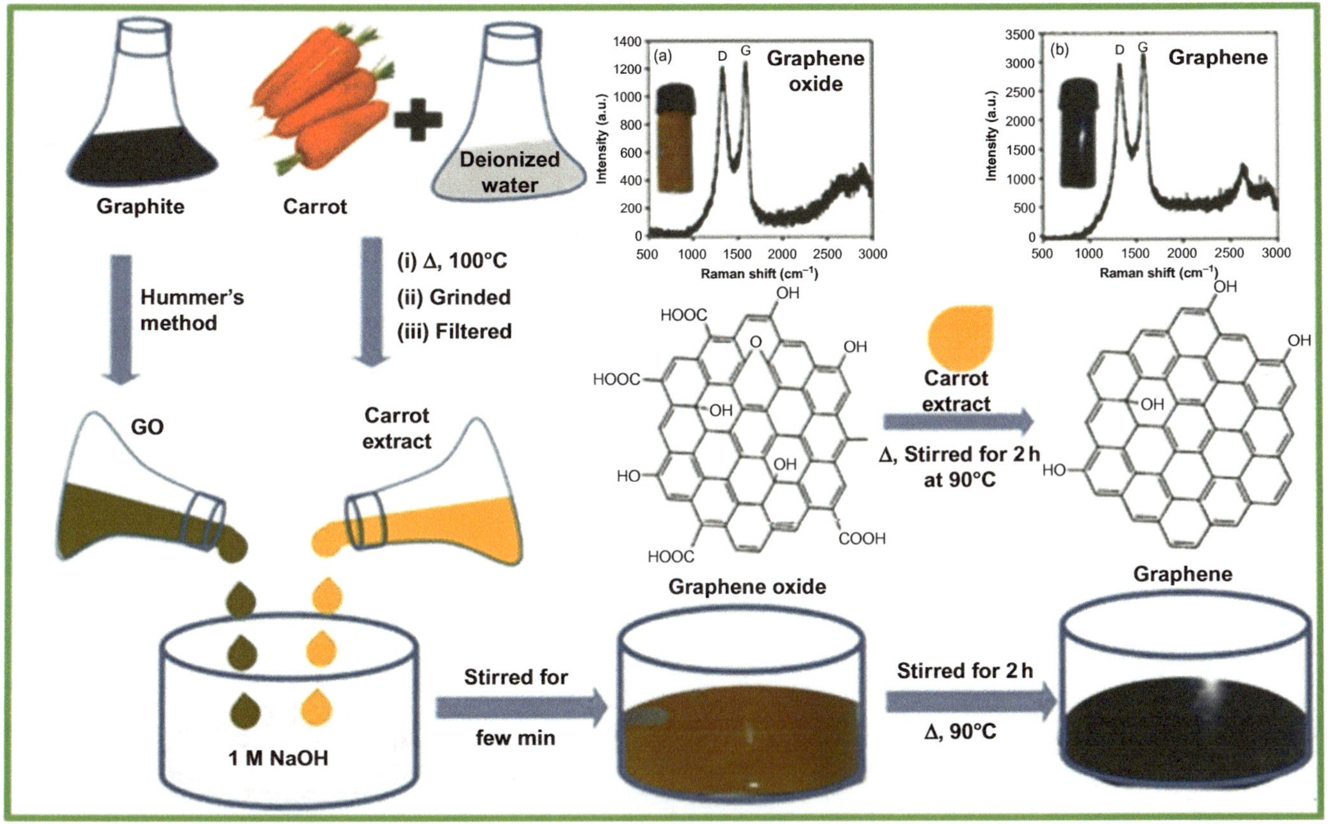

Fig. 1 The schematic of the graphene synthesis from GO by using carrot extract as the reducing agent.

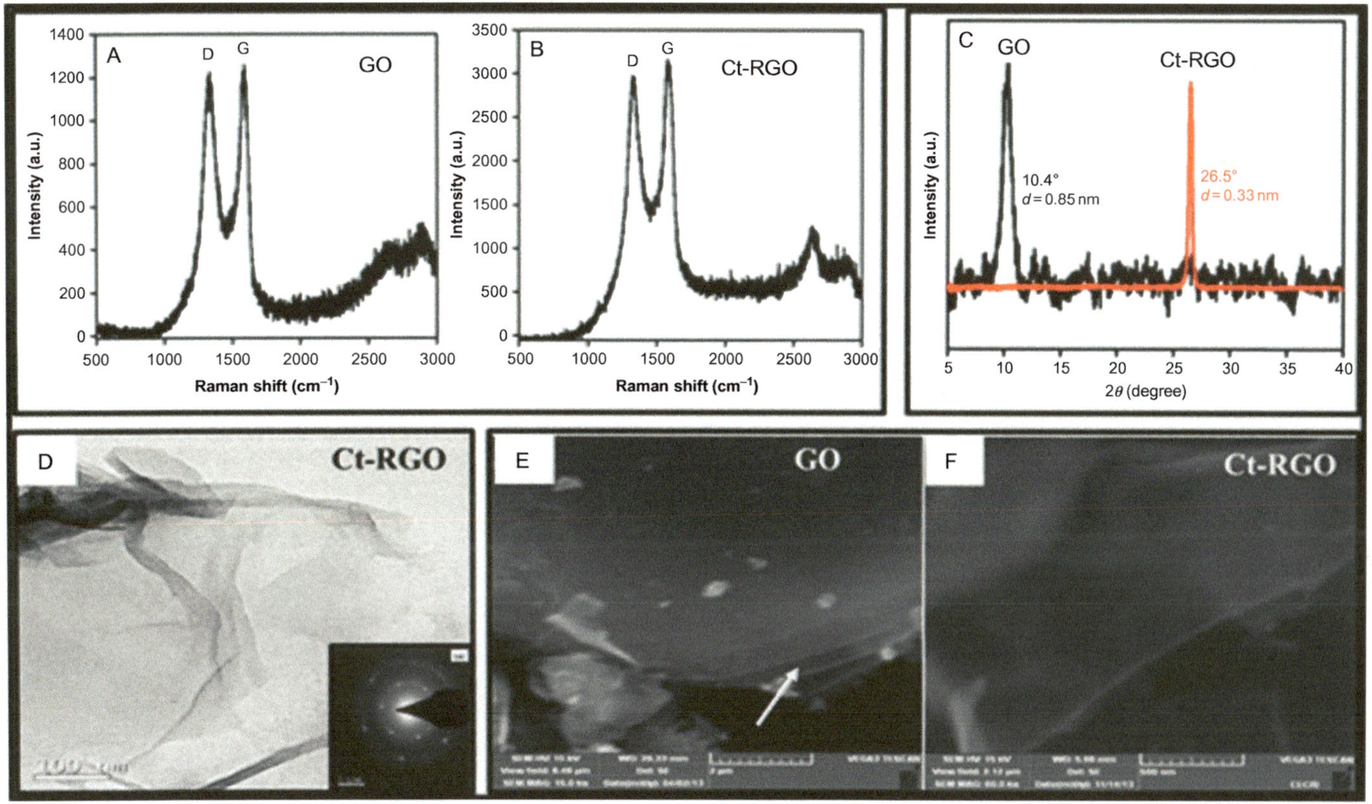

Fig. 2 (A) Raman spectra of graphene oxide and (B) Ct-RGO. (C) XRD spectra of GO and Ct-RGO. (D) TEM analysis of Ct-RGO (*inset* shows the SAED pattern of Ct-RGO) and SEM analysis of (E) GO and (F) Ct-RGO. *Reproduced from Vusa, C. S. R., Berchmans, S., & Alwarappan, S. (2014). Facile and green synthesis of graphene.* RSC Advances, *4(43), 22470–22475 with permission from the Royal Society of Chemistry.*

carotenoid oxidation products can be formed (Carail & Caris-Veyrat, 2006). Among them, the epoxide and apo- or seco-carotenoids are formed mainly. Hydrolysis of epoxide results in diol formation and oxygen anions are formed due to the dissociation of protons from the diol group. The oxygen anions attack the epoxide and hydroxyl groups of GO by S_N2 pathway and result in RGO. The carotenoids are oxidized to aldehydes (Zaid, Chong, Teo, Ng, & Chong, 2015).

The XRD spectrum of GO and Ct-RGO shows the peak at 10.4 degrees with *d*-spacing of 0.85 nm and 26.5 degrees with *d*-spacing of 0.33 nm, respectively, in Fig. 2C. The decrease in *d*-spacing indicates the successful reduction of oxygen functional groups of GO. As shown in Fig. 2D, the TEM analysis shows typical transparent, silky, and wrinkled morphology of synthesized Ct-RGO with the typical sixfold symmetry of SAED. The SEM images of GO also show stacked layer morphology of GO layers, whereas Ct-RGO shows separated layers of Ct-RGO which confirms the separation of graphene layers (Fig. 2E and F).

6. FUNCTIONALIZATION OF GRAPHENE BY PAMAM DENDRIMERS VIA ELECTROCHEMICAL GRAFTING FOR ENZYMATIC SENSING APPLICATION

Electrochemically assisted functionalization procedures are highly advantageous owing to their experimental simplicity, the shorter time scale of the reaction, mild experimental conditions, and minimum requirement of basic electrochemical equipment. With the use of commercially available, inexpensive reagents a great variety of monolayers with tailored functionalities could be introduced on carbon materials. Further, the density of the modified layer could be controlled from a closely packed monolayer to a sparse layer. By electrochemical oxidation of primary and secondary amines and by the electrochemical oxidation of alcohols and aryl acetates, functionalization of graphene can be carried out. Further graphene can be functionalized by electrochemical reduction of aryl diazonium cations (Downard, 2000). We have already reported the electrochemical functionalization of graphene using various amines containing different terminal functional groups such as –COOH, $-NH_2$, –OH, SH, and $-SO_3H$. The resulting monolayer-functionalized graphene can be used for postfunctionalizing of enzymes, DNA, antigen, antibodies, and so on

(Vusa, Venkatesan, Aneesh, Berchmans, & Arumugam, 2017). Also, we have functionalized the graphene surface with nitrogen and its functional groups using ammonium carbamate as a nitrogen source (Manju, Rao Vusa, Arumugam, & Berchmans, 2017). In order to modify the graphene material with high loadings of the enzyme by covalent attachment, we grafted amine-terminated dendrimers to the graphene surface which can subsequently be linked to the enzyme through glutaraldehyde linkage (Rao Vusa, Manju, Berchmans, & Arumugam, 2016). Also, the nanometer-sized 3D dendrimers can provide microenvironment, while the presence of high-density amine functional groups may allow high loading of enzymes, antigen, antibodies, etc. The anodic oxidation of amines leads to the formation of amine cation radical which then forms a C—N bond between graphene and the modifying reagent.

PAMAM dendrimers are a versatile class of molecules which possess a spherical shape (3D structure), high density of surface active groups, good structural homogeneity, and good biocompatibility. They find applications in various fields such as biochemistry, sensors, catalysis, and information storage applications. It can be demonstrated that the fourth-generation amine-terminated PAMAM dendrimer molecules could be electrochemically grafted to carrot extract RGO and well characterized by techniques such as cyclic voltammetry and XPS. The PAMAM dendrimer-grafted graphene surfaces are then postfunctionalized by the enzyme HRP for H_2O_2 detection owing to its importance in diseases like cancer, Alzheimer, Parkinson, diabetes, mitochondrial infraction, and atherosclerosis, and industries such as pharmaceutical, chemical, food, and paper industries. The applicability of the graphene-based sensing platform could be shown in serum analysis where good recovery of H_2O_2 was observed. The schematic representation of graphene functionalization, enzyme immobilization, and its enzymatic sensing application is shown in Fig. 3.

6.1 Equipment

Emery sheets for electrode polishing, electrochemical equipment, electrochemical cell made of glass, glassy carbon electrode (GCE) ($\phi = 3$ mm) as working electrode, Pt foil as Counter electrode, Normal Calomel Electrode Hg/Hg_2Cl_2 (1 *M* KCl) as reference electrode in aqueous solution and silver wire as pseudo-reference in nonaqueous solutions are used in this work.

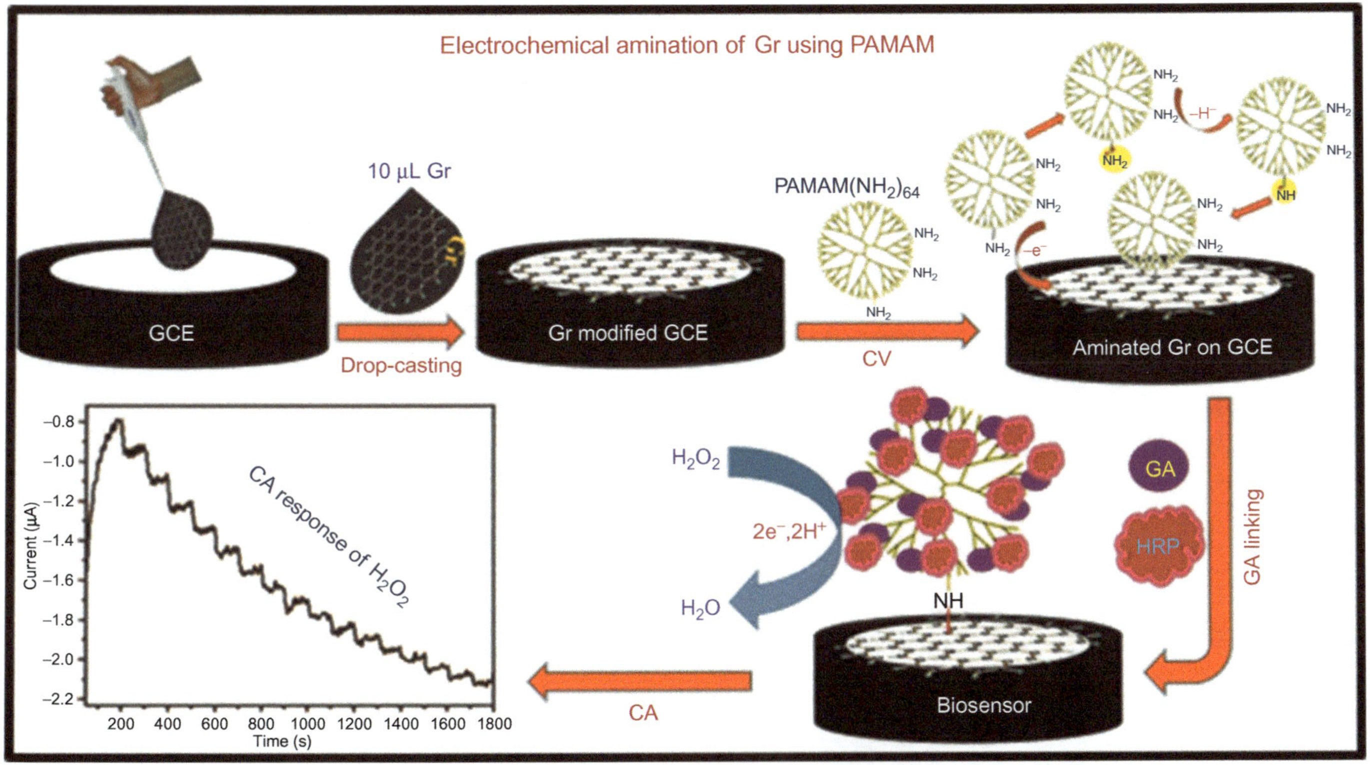

Fig. 3 Schematic diagram describing the electrochemical grafting of PAMAM on graphene followed by postfunctionalization of HRP enzyme for H_2O_2 sensing.

6.2 Materials

Graphene, fourth-generation poly(amidoamine) dendrimer (10 wt% in methanol), $LiClO_4$, KH_2PO_4, sodium hydroxide, acetonitrile, dimethylformamide (DMF), and hydrogen peroxide (H_2O_2).

6.3 Methods

6.3.1 Electrode Preparation

Prior to surface modification, GCE ($\phi = 3$ mm) is polished using 0.05 μm α-Al_2O_3 and emery sheets followed by sonication in alcohol and double-distilled water. Then, the cleaned GCE is dried in N_2 stream.

6.3.2 Modification of GCE by Ct-RGO

5 mL of Ct-RGO suspension (1 mg mL^{-1} in DMF) is dropcasted on GCE and allowed to dry at room temperature for 3 h and this electrode is designated as GCE/Ct-RGO.

6.3.3 Electrochemical Grafting of PAMAM on Ct-RGO

Electrochemical grafting of PAMAM on Ct-RGO is done by scanning the GCE/Ct-RGO between the potentials +0.0 and +1.0 V vs NCE at the scan rate of 20 mV s^{-1} for five cycles, in an aqueous solution of 0.1 *M* $LiClO_4$ containing 20 μ*M* PAMAM. The resulted PAMAM tethered surface is denoted as GCE/Ct-RGO–PAMAM. A GCE without Ct-RGO is also functionalized using the same conditions as a control and denoted as GCE–PAMAM.

6.3.4 Following the Process of Electrochemical Grafting by Cyclic Voltammetry

Cyclic voltammograms are recorded during the electrochemical grafting of PAMAM, shown in Fig. 4. As shown in Fig. 4, the irreversible oxidation peaks at +0.750 V and +0.770 V indicate one-electron oxidation of amino group to its cation radical and the formation of C—N bond between PAMAM and carbon surface by the loss of a proton. The significant reduction of peak current after five cycles indicates that the formation of an irreversibly modified thin film of dendrimers on GCE and GCE/Ct-RGO. The molecular density of PAMAM on graphene estimated from the charge consumed (first scan) during electrochemical grafting of surfaces by one-electron anodic oxidation is found to be 37.51×10^{15} molecules per cm^2.

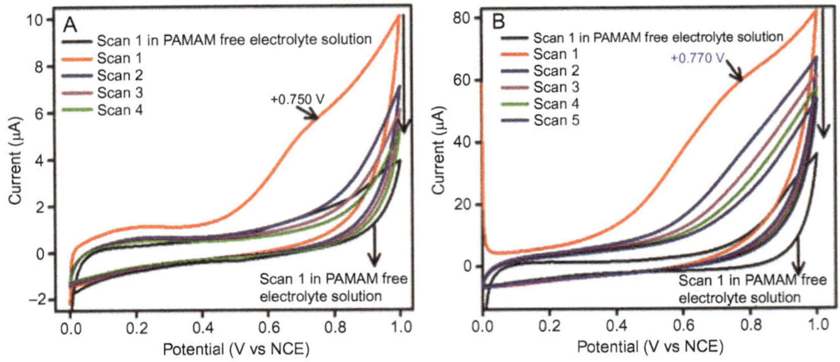

Fig. 4 Cyclic voltammograms obtained during electrochemical grafting of PAMAM on (A) GCE and (B) GCE–Ct-RGO using 20 μM PAMAM-NH_2 dendrimer in 0.1 M $LiClO_4$ at the scan rate of 20 $mV s^{-1}$. *Reproduced from Rao Vusa, C. S., Manju, V., Berchmans, S., & Arumugam, P. (2016). Electrochemical amination of graphene using nanosized PAMAM dendrimers for sensing applications.* RSC Advances, *6(40), 33409–33418 with permission from the Royal Society of Chemistry.*

7. POSTFUNCTIONALIZATION OF GCE/CT-RGO–PAMAM BY THE HORSERADISH PEROXIDASE ENZYME

7.1 Equipment

Microcentrifuge tubes, micropipettes and tips, refrigerator, and microcentrifuge tube stand.

7.2 Materials

Horseradish peroxidase (HRP, type I, essentially salt-free, lyophilized powder, 50–150 U per mg solid), glutaraldehyde (GA, 25% aqueous solution), and PBS buffer pH 7.

7.3 Phosphate Buffer Preparation (PBS, pH 7)

a. The 0.1 M phosphate buffer is prepared using KH_2PO_4 and NaOH.
b. The 6.8 g of KH_2PO_4 is added to standard measuring flask and dissolved using 100 mL water and 1.2 g of NaOH is dissolved using 100 mL water.
c. The prepared NaOH solution is added to the KH_2PO_4 solution and the volume is made up to 500 mL.

7.4 Enzyme Immobilization

a. The GCE/Ct-RGO–PAMAM and GCE–PAMAM are activated with 1% aqueous glutaraldehyde solution for 3 h.

b. The activated electrodes are washed with water and incubated in PBS (pH 7) containing HRP (5 mg mL^{-1}) at 4°C for 12 h.
c. After 12 h, the electrodes are removed from the HRP solution, and washed with PBS repeatedly to remove the physically adsorbed HRP.
d. The as-prepared HRP-modified working electrode is stored at 4°C when not in use.

8. APPLICATION OF HRP-MODIFIED GCE/CT-RGO–PAMAM–GA–HRP FOR H_2O_2 SENSING

8.1 Equipment

Electrochemical equipment, electrochemical cell made of glass, GCE/Ct-RGO–PAMAM–GA–HRP electrode as working electrode, Pt foil as the counter electrode, and the normal Calomel electrode (NCE) Hg/Hg_2Cl_2 (1 *M* KCl) as a reference electrode.

8.2 Materials

H_2O_2 (30%, w/v) and phosphate buffer (pH 7).

8.3 Method

Amperometry is performed to evaluate the performance of the electrode for H_2O_2 sensing at −0.2 V vs NCE in phosphate buffer pH 7. Once a steady-state current is obtained, the different concentrations of H_2O_2 are injected into the electrochemical cell. Fig. 5 shows the amperometric response obtained for HRP-modified PAMAM grafted Ct-RGO, along with the responses obtained for the control experiments carried out with the electrodes GCE, GCE–PAMAM, GCE–PAMAM–GA–HRP at −0.2 V in N_2-saturated phosphate buffer solution, during successive additions of 25 μ*M* of H_2O_2.

From Fig. 5, we observed that the reduction of H_2O_2 is observed at −0.2 V only in the case of electrodes GCE–PAMAM–GA–HRP and GCE/Ct-RGO–PAMAM–GA–HRP. Also, we found that the sensitivity and linear range of GCE/Ct-RGO–PAMAM–GA–HRP are found to be 28.96 mA mM^{-1} cm^{-2} and 50–800 m*M*, while GCE–PAMAM–GA–HRP platform showed a linear range of 200 μ*M* to 1.1 m*M* with a sensitivity of 5.68 mA mM^{-1} cm^{-2}. The corresponding calibration plots are shown in Fig. 5C and D. The sensitivity of the GCE/Ct-RGO–PAMAM–GA–HRP is five times higher than GCE–PAMAM–GA–HRP for H_2O_2 reduction which is attributed to the high density of amino groups.

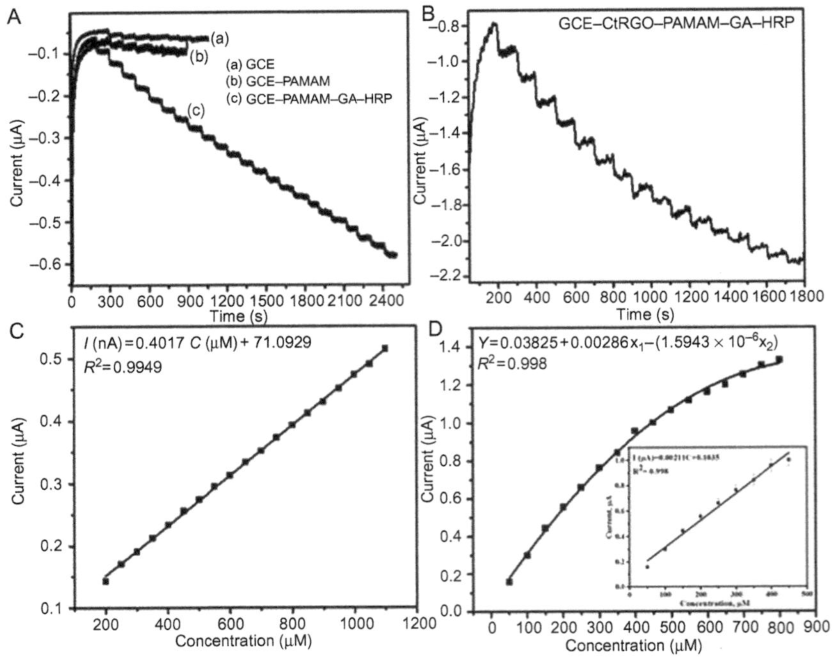

Fig. 5 The *i–t* curves of GCE, GCE–PAMAM, GCE–PAMAM–HRP (A) and GCE/Ct-RGO–PAMAM–HRP (B) at −0.2 V during the successive addition of H_2O_2 in PBS (pH 7.2). The calibration plots for GCE–PAMAM–GA–HRP (C) and GCE/Ct-RGO–PAMAM–GA–HRP (D). *Inset* in (D) shows a linear range of GCE/Ct-RGO–PAMAM–GA–HRP for H_2O_2 reduction. *Reproduced from Rao Vusa, C. S., Manju, V., Berchmans, S., & Arumugam, P. (2016). Electrochemical amination of graphene using nanosized PAMAM dendrimers for sensing applications.* RSC Advances, *6(40), 33409–33418 with permission from the Royal Society of Chemistry.*

9. APPLICATION OF HRP-MODIFIED GCE/CT-RGO–PAMAM–GA–HRP FOR H_2O_2 SENSING IN SERUM

9.1 Equipment

Electrochemical equipment, electrochemical cell made of glass, GCE/Ct-RGO–PAMAM–GA–HRP electrode as working electrode, Pt foil as the counter electrode, Hg/Hg_2Cl_2 (1 N KCl), the normal calomel electrode (NCE) as a reference electrode.

9.2 Materials

H_2O_2-spiked serum, H_2O_2, phosphate buffer pH 7.

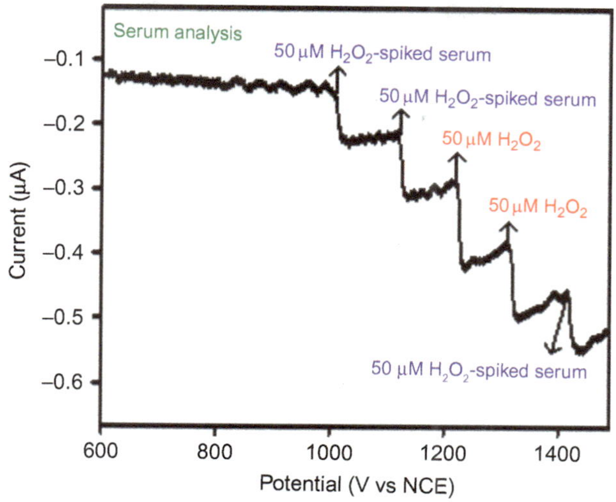

Fig. 6 Amperometric response of the GCE–Ct-RGO–PAMAM–GA–HRP electrode for additions of H_2O_2-spiked serum and H_2O_2. *Reproduced from Rao Vusa, C. S., Manju, V., Berchmans, S., & Arumugam, P. (2016). Electrochemical amination of graphene using nanosized PAMAM dendrimers for sensing applications.* RSC Advances, *6(40), 33409–33418 with permission from the Royal Society of Chemistry.*

9.3 Method

To validate the applicability of HRP-functionalized graphene-modified electrode for the analysis of H_2O_2 in real samples, recovery studies are carried out using H_2O_2-spiked serum samples by amperometry, as shown in Fig. 6. The % mean recovery is found to be 96.8% showing the efficacy of the enzyme-functionalized graphene-modified electrode.

10. CONCLUSIONS

From this work, it is evident that the graphene surface could be functionalized by amines through electrochemical grafting. The process involves the anodic oxidation of amines which yield amine cation radical which subsequently forms the C—N bond with graphene. Herein, it is shown that amine-terminated fourth-generation PAMAM dendrimers could be covalently tethered to the graphene surface followed by further functionalization with the enzyme HRP for H_2O_2 biosensing. The resulting enzymatic sensor platform exhibits good stability and affinity toward analyte along with high sensitivity which is due to robust nature of grafted PAMAM layer and high loading of enzymes as the fourth-generation PAMAM dendrimers possess 64 amine functional groups in its structure. Hence, by grafting few PAMAM molecules, the number of enzymes can be immobilized in the 3D nanometer-sized microenvironment. This method can be easily adapted to other carbonaceous materials for a variety of applications.

ACKNOWLEDGMENTS

The authors acknowledge Indian Council of Medical Research (ICMR), University Grants Commission (UGC), and Council of scientific and Industrial research (CSIR) for senior research fellowship.

REFERENCES

Ansari, S., Kelarakis, A., Estevez, L., & Giannelis, E. P. (2010). Oriented arrays of graphene in a polymer matrix by in situ reduction of graphite oxide nanosheets. *Small*, *6*(2), 205–209.

Bai, L., Chai, Y., Pu, X., & Yuan, R. (2014). A signal-on electrochemical aptasensor for ultrasensitive detection of endotoxin using three-way DNA junction-aided enzymatic recycling and graphene nanohybrid for amplification. *Nanoscale*, *6*(5), 2902–2908.

Bai, L., Yuan, R., Chai, Y., Yuan, Y., Wang, Y., & Xie, S. (2012). Direct electrochemistry and electrocatalysis of a glucose oxidase-functionalized bioconjugate as a trace label for ultrasensitive detection of thrombin. *Chemical Communications*, *48*(89), 10972–10974.

Bai, L., Yuan, R., Chai, Y., Zhuo, Y., Yuan, Y., & Wang, Y. (2012). Simultaneous electrochemical detection of multiple analytes based on dual signal amplification of single-walled carbon nanotubes and multi-labeled graphene sheets. *Biomaterials*, *33*(4), 1090–1096.

Baptista-Pires, L., Pérez-López, B., Mayorga-Martinez, C. C., Morales-Narváez, E., Domingo, N., Esplandiu, M. J., et al. (2014). Electrocatalytic tuning of biosensing response through electrostatic or hydrophobic enzyme–graphene oxide interactions. *Biosensors and Bioelectronics*, *61*, 655–662.

Benvidi, A., Rajabzadeh, N., Mazloum-Ardakani, M., Heidari, M. M., & Mulchandani, A. (2014). Simple and label-free electrochemical impedance Amelogenin gene hybridization biosensing based on reduced graphene oxide. *Biosensors and Bioelectronics*, *58*, 145–152.

Bonanni, A., Chua, C. K., Zhao, G., Sofer, Z., & Pumera, M. (2012). Inherently electroactive graphene oxide nanoplatelets as labels for single nucleotide polymorphism detection. *ACS Nano*, *6*(10), 8546–8551.

Burley, S. K., & Petsko, G. A. (1985). Aromatic-aromatic interaction: A mechanism of protein structure stabilization. *Science*, *229*(4708), 23.

Burton, G. W., & Ingold, K. U. (1984). Beta-carotene: An unusual type of lipid antioxidant. *Science*, *224*(4649), 569–573.

Carail, M., & Caris-Veyrat, C. (2006). Carotenoid oxidation products: From villain to saviour? *Pure and Applied Chemistry*, *78*, 1493.

Chen, C., Zhai, W., Lu, D., Zhang, H., & Zheng, W. (2011). A facile method to prepare stable noncovalent functionalized graphene solution by using thionine. *Materials Research Bulletin*, *46*(4), 583–587.

Cheng, H.-C., Shiue, R.-J., Tsai, C.-C., Wang, W.-H., & Chen, Y.-T. (2011). High-quality graphene p–n junctions via resist-free fabrication and solution-based noncovalent functionalization. *ACS Nano*, *5*(3), 2051–2059.

Choi, E.-Y., Han, T. H., Hong, J., Kim, J. E., Lee, S. H., Kim, H. W., et al. (2010). Noncovalent functionalization of graphene with end-functional polymers. *Journal of Materials Chemistry*, *20*(10), 1907–1912.

Choi, J., Kim, K.-J., Kim, B., Lee, H., & Kim, S. (2009). Covalent functionalization of epitaxial graphene by azidotrimethylsilane. *The Journal of Physical Chemistry C*, *113*(22), 9433–9435.

Choi, B. G., Park, H., Park, T. J., Yang, M. H., Kim, J. S., Jang, S.-Y., et al. (2010). Solution chemistry of self-assembled graphene nanohybrids for high-performance flexible biosensors. *ACS Nano*, *4*(5), 2910–2918.

Cohen-Tanugi, D., & Grossman, J. C. (2012). Water desalination across nanoporous graphene. *Nano Letters*, *12*(7), 3602–3608.

Downard, A. J. (2000). Electrochemically assisted covalent modification of carbon electrodes. *Electroanalysis*, *12*(14), 1085–1096.

Dubin, S., Gilje, S., Wang, K., Tung, V. C., Cha, K., Hall, A. S., et al. (2010). A one-step, solvothermal reduction method for producing reduced graphene oxide dispersions in organic solvents. *ACS Nano*, *4*(7), 3845–3852.

Dubuisson, E., Yang, Z., & Loh, K. P. (2011). Optimizing label-free DNA electrical detection on graphene platform. *Analytical Chemistry*, *83*(7), 2452–2460.

Eissa, S., & Zourob, M. (2012). A graphene-based electrochemical competitive immunosensor for the sensitive detection of okadaic acid in shellfish. *Nanoscale*, *4*(23), 7593–7599.

Fang, M., Long, J., Zhao, W., Wang, L., & Chen, G. (2010). pH-responsive chitosan-mediated graphene dispersions. *Langmuir*, *26*(22), 16771–16774.

Fang, M., Wang, K., Lu, H., Yang, Y., & Nutt, S. (2009). Covalent polymer functionalization of graphene nanosheets and mechanical properties of composites. *Journal of Materials Chemistry*, *19*(38), 7098–7105.

Fikselova, M., & Slovenska Polnohospodarska Univ., Nitra (Slovak Republic). Katedra Skladovania a Spracovania Rastlinnych Produktov, S. Silhar, Vyskumny Ustav Potravinarsky, Modra (Slovak Republic). Biocentrum, J. Marecek, Slovenska Polnohospodarska Univ., Nitra (Slovak Republic). Katedra Skladovania a Spracovania Rastlinnych Produktov and H. Francakova, Slovenska Polnohospodarska Univ., Nitra (Slovak Republic). Katedra Skladovania a Spracovania Rastlinnych Produktov. (2008). Extraction of carrot (*Daucus carota* L.) carotenes under different conditions. *Czech Journal of Food Sciences*, *26*, 268–274.

Geng, J., & Jung, H.-T. (2010). Porphyrin functionalized graphene sheets in aqueous suspensions: From the preparation of graphene sheets to highly conductive graphene films. *The Journal of Physical Chemistry C*, *114*(18), 8227–8234.

Georgakilas, V., Bourlinos, A. B., Zboril, R., Steriotis, T. A., Dallas, P., Stubos, A. K., et al. (2010). Organic functionalisation of graphenes. *Chemical Communications*, *46*(10), 1766–1768.

Georgakilas, V., Otyepka, M., Bourlinos, A. B., Chandra, V., Kim, N., Kemp, K. C., et al. (2012). Functionalization of graphene: Covalent and non-covalent approaches, derivatives and applications. *Chemical Reviews*, *112*(11), 6156–6214.

Gholivand, M. B., & Khodadadian, M. (2014). Amperometric cholesterol biosensor based on the direct electrochemistry of cholesterol oxidase and catalase on a graphene/ionic liquid-modified glassy carbon electrode. *Biosensors and Bioelectronics*, *53*, 472–478.

Ghosh, A., Rao, K. V., George, S. J., & Rao, C. N. R. (2010). Noncovalent functionalization, exfoliation, and solubilization of graphene in water by employing a fluorescent coronene carboxylate. *Chemistry—A European Journal*, *16*(9), 2700–2704.

Ghosh, A., Rao, K. V., Voggu, R., & George, S. J. (2010). Non-covalent functionalization, solubilization of graphene and single-walled carbon nanotubes with aromatic donor and acceptor molecules. *Chemical Physics Letters*, *488*(4), 198–201.

Gu, H., Yu, Y., Liu, X., Ni, B., Zhou, T., & Shi, G. (2012). Layer-by-layer self-assembly of functionalized graphene nanoplates for glucose sensing in vivo integrated with on-line microdialysis system. *Biosensors and Bioelectronics*, *32*(1), 118–126.

Guo, Y., Deng, L., Li, J., Guo, S., Wang, E., & Dong, S. (2011). Hemin–graphene hybrid nanosheets with intrinsic peroxidase-like activity for label-free colorimetric detection of single-nucleotide polymorphism. *ACS Nano*, *5*(2), 1282–1290.

Guo, P., Xiao, F., Liu, Q., Liu, H., Guo, Y., Gong, J. R., et al. (2013). One-pot microbial method to synthesize dual-doped graphene and its use as high-performance electrocatalyst. *Scientific Reports*, *3*, 3499.

Haque, A.-M. J., Park, H., Sung, D., Jon, S., Choi, S.-Y., & Kim, K. (2012). An electrochemically reduced graphene oxide-based electrochemical immunosensing platform for ultrasensitive antigen detection. *Analytical Chemistry*, *84*(4), 1871–1878.

He, H., & Gao, C. (2010). General approach to individually dispersed, highly soluble, and conductive graphene nanosheets functionalized by nitrene chemistry. *Chemistry of Materials*, *22*(17), 5054–5064.

Heber, D., & Lu, Q.-Y. (2002). Overview of mechanisms of action of lycopene. *Experimental Biology and Medicine*, *227*(10), 920–923.

Hong, B. H., Lee, J. Y., Lee, C.-W., Kim, J. C., Bae, S. C., & Kim, K. S. (2001). Self-assembled arrays of organic nanotubes with infinitely long one-dimensional H-bond chains. *Journal of the American Chemical Society*, *123*(43), 10748–10749.

Huang, C.-F., Yao, G.-H., Liang, R.-P., & Qiu, J.-D. (2013). Graphene oxide and dextran capped gold nanoparticles based surface plasmon resonance sensor for sensitive detection of concanavalin A. *Biosensors and Bioelectronics*, *50*, 305–310.

Jo, K., Lee, T., Choi, H. J., Park, J. H., Lee, D. J., Lee, D. W., et al. (2011). Stable aqueous dispersion of reduced graphene nanosheets via non-covalent functionalization with conducting polymers and application in transparent electrodes. *Langmuir*, *27*(5), 2014–2018.

Johns, J. E., & Hersam, M. C. (2013). Atomic covalent functionalization of graphene. *Accounts of Chemical Research*, *46*(1), 77–86.

Kim, T., Lee, H., Kim, J., & Suh, K. S. (2010). Synthesis of phase transferable graphene sheets using ionic liquid polymers. *ACS Nano*, *4*(3), 1612–1618.

Kodali, V. K., Scrimgeour, J., Kim, S., Hankinson, J. H., Carroll, K. M., de Heer, W. A., et al. (2011). Nonperturbative chemical modification of graphene for protein micropatterning. *Langmuir*, *27*(3), 863–865.

Kosynkin, D. V., Higginbotham, A. L., Sinitskii, A., Lomeda, J. R., Dimiev, A., Price, B. K., et al. (2009). Longitudinal unzipping of carbon nanotubes to form graphene nanoribbons. *Nature*, *458*, 872.

Lee, J. Y., Hong, B. H., Kim, W. Y., Min, S. K., Kim, Y., Jouravlev, M. V., et al. (2009). Near-field focusing and magnification through self-assembled nanoscale spherical lenses. *Nature*, *460*, 498.

Li, X.-R., Kong, F.-Y., Liu, J., Liang, T.-M., Xu, J.-J., & Chen, H.-Y. (2012). Synthesis of potassium-modified graphene and its application in nitrite-selective sensing. *Advanced Functional Materials*, *22*(9), 1981–1988.

Liang, J., Chen, Z., Guo, L., & Li, L. (2011). Electrochemical sensing of l-histidine based on structure-switching DNAzymes and gold nanoparticle-graphene nanosheet composites. *Chemical Communications*, *47*(19), 5476–5478.

Liang, B., Fang, L., Yang, G., Hu, Y., Guo, X., & Ye, X. (2013). Direct electron transfer glucose biosensor based on glucose oxidase self-assembled on electrochemically reduced carboxyl graphene. *Biosensors and Bioelectronics*, *43*, 131–136.

Liang, Y., Wu, D., Feng, X., & Müllen, K. (2009). Dispersion of graphene sheets in organic solvent supported by ionic interactions. *Advanced Materials*, *21*(17), 1679–1683.

Lim, H., Lee, J. S., Shin, H.-J., Shin, H. S., & Choi, H. C. (2010). Spatially resolved spontaneous reactivity of diazonium salt on edge and basal plane of graphene without surfactant and its doping effect. *Langmuir*, *26*(14), 12278–12284.

Liu, H., Gao, J., Xue, M., Zhu, N., Zhang, M., & Cao, T. (2009). Processing of graphene for electrochemical application: Noncovalently functionalize graphene sheets with water-soluble electroactive methylene green. *Langmuir, 25*(20), 12006–12010.

Liu, L.-H., Lerner, M. M., & Yan, M. (2010). Derivitization of pristine graphene with well-defined chemical functionalities. *Nano Letters, 10*(9), 3754–3756.

Liu, J., Li, Y., Li, Y., Li, J., & Deng, Z. (2010). Noncovalent DNA decorations of graphene oxide and reduced graphene oxide toward water-soluble metal-carbon hybrid nanostructures via self-assembly. *Journal of Materials Chemistry, 20*(5), 900–906.

Liu, Z., Liu, Q., Huang, Y., Ma, Y., Yin, S., Zhang, X., et al. (2008). Organic photovoltaic devices based on a novel acceptor material: Graphene. *Advanced Materials, 20*(20), 3924–3930.

Liu, H., Ryu, S., Chen, Z., Steigerwald, M. L., Nuckolls, C., & Brus, L. E. (2009). Photochemical reactivity of graphene. *Journal of the American Chemical Society, 131*(47), 17099–17101.

Liu, J., Yang, W., Tao, L., Li, D., Boyer, C., & Davis, T. P. (2010). Thermosensitive graphene nanocomposites formed using pyrene-terminal polymers made by RAFT polymerization. *Journal of Polymer Science Part A: Polymer Chemistry, 48*(2), 425–433.

Loo, A. H., Bonanni, A., & Pumera, M. (2013). Thrombin aptasensing with inherently electroactive graphene oxide nanoplatelets as labels. *Nanoscale, 5*(11), 4758–4762.

Luo, L.-Q., Zhang, Z., Ding, Y.-P., Deng, D.-M., Zhu, X.-L., & Wang, Z.-X. (2013). Label-free electrochemical impedance genosensor based on 1-aminopyrene/graphene hybrids. *Nanoscale, 5*(13), 5833–5840.

Ma, F.-X., Wang, J., Wang, F.-B., & Xia, X.-H. (2015). The room temperature electrochemical synthesis of N-doped graphene and its electrocatalytic activity for oxygen reduction. *Chemical Communications, 51*(7), 1198–1201.

Mali, K. S., Greenwood, J., Adisoejoso, J., Phillipson, R., & De Feyter, S. (2015). Nanostructuring graphene for controlled and reproducible functionalization. *Nanoscale*, 7(5), 1566–1585.

Manju, V., Rao Vusa, C. S., Arumugam, P., & Berchmans, S. (2017). A facile and versatile electrochemical tuning of graphene for oxygen reduction reaction in acidic, neutral and alkali media. *ChemistrySelect, 2*(27), 8541–8552.

Meyer, E. A., Castellano, R. K., & Diederich, F. (2003). Interactions with aromatic rings in chemical and biological recognition. *Angewandte Chemie International Edition, 42*(11), 1210–1250.

Niyogi, S., Bekyarova, E., Itkis, M. E., Zhang, H., Shepperd, K., Hicks, J., et al. (2010). Spectroscopy of covalently functionalized graphene. *Nano Letters, 10*(10), 4061–4066.

O'Hern, S. C., Boutilier, M. S. H., Idrobo, J.-C., Song, Y., Kong, J., Laoui, T., et al. (2014). Selective ionic transport through tunable subnanometer pores in single-layer graphene membranes. *Nano Letters, 14*(3), 1234–1241.

Pang, S., Hernandez, Y., Feng, X., & Müllen, K. (2011). Graphene as transparent electrode material for organic electronics. *Advanced Materials, 23*(25), 2779–2795.

Park, S., Mohanty, N., Suk, J. W., Nagaraja, A., An, J., Piner, R. D., et al. (2010). Biocompatible, robust free-standing paper composed of a TWEEN/graphene composite. *Advanced Materials, 22*(15), 1736–1740.

Park, S., Singh, A., Kim, S., & Yang, H. (2014). Electroreduction-based electrochemical-enzymatic redox cycling for the detection of cancer antigen 15-3 using graphene oxide-modified indium–tin oxide electrodes. *Analytical Chemistry, 86*(3), 1560–1566.

Park, J., & Yan, M. (2013). Covalent functionalization of graphene with reactive intermediates. *Accounts of Chemical Research, 46*(1), 181–189.

Qi, X., Pu, K.-Y., Li, H., Zhou, X., Wu, S., Fan, Q.-L., et al. (2010). Amphiphilic graphene composites. *Angewandte Chemie International Edition, 49*(49), 9426–9429.

Quintana, M., Spyrou, K., Grzelczak, M., Browne, W. R., Rudolf, P., & Prato, M. (2010). Functionalization of graphene via 1,3-dipolar cycloaddition. *ACS Nano*, *4*(6), 3527–3533.

Rao Vusa, C. S., Manju, V., Berchmans, S., & Arumugam, P. (2016). Electrochemical amination of graphene using nanosized PAMAM dendrimers for sensing applications. *RSC Advances*, *6*(40), 33409–33418.

Ronkainen, N. J., Halsall, H. B., & Heineman, W. R. (2010). Electrochemical biosensors. *Chemical Society Reviews*, *39*(5), 1747–1763.

Sarkar, S., Bekyarova, E., Niyogi, S., & Haddon, R. C. (2011). Diels–Alder chemistry of graphite and graphene: Graphene as diene and dienophile. *Journal of the American Chemical Society*, *133*(10), 3324–3327.

Saxena, A. P., Deepa, M., Joshi, A. G., Bhandari, S., & Srivastava, A. K. (2011). Poly(3,4-ethylenedioxythiophene)-ionic liquid functionalized graphene/reduced graphene oxide nanostructures: Improved conduction and electrochromism. *ACS Applied Materials & Interfaces*, *3*(4), 1115–1126.

Schniepp, H. C., Li, J.-L., McAllister, M. J., Sai, H., Herrera-Alonso, M., Adamson, D. H., et al. (2006). Functionalized single graphene sheets derived from splitting graphite oxide. *The Journal of Physical Chemistry B*, *110*(17), 8535–8539.

Shang, N. G., Papakonstantinou, P., McMullan, M., Chu, M., Stamboulis, A., Potenza, A., et al. (2008). Catalyst-free efficient growth, orientation and biosensing properties of multilayer graphene nanoflake films with sharp edge planes. *Advanced Functional Materials*, *18*(21), 3506–3514.

Shao, Y., Wang, J., Wu, H., Liu, J., Aksay, I. A., & Lin, Y. (2010). Graphene based electrochemical sensors and biosensors: A review. *Electroanalysis*, *22*(10), 1027–1036.

Singh, N. J., Lee, H. M., Hwang, I.-C., & Kim, K. S. (2007). Designing ionophores and molecular nanotubes based on molecular recognition. *Supramolecular Chemistry*, *19*(4–5), 321–332.

Sinitskii, A., Dimiev, A., Corley, D. A., Fursina, A. A., Kosynkin, D. V., & Tour, J. M. (2010). Kinetics of diazonium functionalization of chemically converted graphene nanoribbons. *ACS Nano*, *4*(4), 1949–1954.

Song, Y., Liu, H., Tan, H., Xu, F., Jia, J., Zhang, L., et al. (2014). pH-switchable electrochemical sensing platform based on chitosan-reduced graphene oxide/concanavalin A layer for assay of glucose and urea. *Analytical Chemistry*, *86*(4), 1980–1987.

Srivastava, S., Kumar, V., Ali, M. A., Solanki, P. R., Srivastava, A., Sumana, G., et al. (2013). Electrophoretically deposited reduced graphene oxide platform for food toxin detection. *Nanoscale*, *5*(7), 3043–3051.

Stankovich, S., Dikin, D. A., Dommett, G. H. B., Kohlhaas, K. M., Zimney, E. J., Stach, E. A., et al. (2006). Graphene-based composite materials. *Nature*, *442*, 282.

Strom, T. A., Dillon, E. P., Hamilton, C. E., & Barron, A. R. (2010). Nitrene addition to exfoliated graphene: A one-step route to highly functionalized graphene. *Chemical Communications*, *46*(23), 4097–4099.

Su, Q., Pang, S., Alijani, V., Li, C., Feng, X., & Müllen, K. (2009). Composites of graphene with large aromatic molecules. *Advanced Materials*, *21*(31), 3191–3195.

Tu, W., Lei, J., Zhang, S., & Ju, H. (2010). Characterization, direct electrochemistry, and amperometric biosensing of graphene by noncovalent functionalization with picket-fence porphyrin. *Chemistry—A European Journal*, *16*(35), 10771–10777.

Vadukumpully, S., Gupta, J., Zhang, Y., Xu, G. Q., & Valiyaveettil, S. (2011). Functionalization of surfactant wrapped graphene nanosheets with alkylazides for enhanced dispersibility. *Nanoscale*, *3*(1), 303–308.

Vusa, C. S. R., Berchmans, S., & Alwarappan, S. (2014). Facile and green synthesis of graphene. *RSC Advances*, *4*(43), 22470–22475.

Vusa, C. S. R., Venkatesan, M., Aneesh, K., Berchmans, S., & Arumugam, P. (2017). Tactical tuning of the surface and interfacial properties of graphene: A versatile and rational electrochemical approach. *Scientific Reports*, 7, 8354.
Wang, Y., Chen, X., Zhong, Y., Zhu, F., & Loh, K. P. (2009). Large area, continuous, few-layered graphene as anodes in organic photovoltaic devices. *Applied Physics Letters*, *95*(6), 063302.
Wang, Z., & Dai, Z. (2015). Carbon nanomaterial-based electrochemical biosensors: An overview. *Nanoscale*, 7(15), 6420–6431.
Wang, Q. H., & Hersam, M. C. (2009). Room-temperature molecular-resolution characterization of self-assembled organic monolayers on epitaxial graphene. *Nature Chemistry*, *1*, 206.
Wang, D.-W., Li, F., Zhao, J., Ren, W., Chen, Z.-G., Tan, J., et al. (2009). Fabrication of graphene/polyaniline composite paper via in situ anodic electropolymerization for high-performance flexible electrode. *ACS Nano*, *3*(7), 1745–1752.
Wang, Y., Shao, Y., Matson, D. W., Li, J., & Lin, Y. (2010). Nitrogen-doped graphene and its application in electrochemical biosensing. *ACS Nano*, *4*(4), 1790–1798.
Wang, X., Tabakman, S. M., & Dai, H. (2008). Atomic layer deposition of metal oxides on pristine and functionalized graphene. *Journal of the American Chemical Society*, *130*(26), 8152–8153.
Wang, H., Zhang, Y., Li, H., Du, B., Ma, H., Wu, D., et al. (2013). A silver–palladium alloy nanoparticle-based electrochemical biosensor for simultaneous detection of ractopamine, clenbuterol and salbutamol. *Biosensors and Bioelectronics*, *49*, 14–19.
Wang, X., Zhi, L., & Müllen, K. (2008). Transparent, conductive graphene electrodes for dye-sensitized solar cells. *Nano Letters*, *8*(1), 323–327.
Wang, Z., Zhou, X., Zhang, J., Boey, F., & Zhang, H. (2009). Direct electrochemical reduction of single-layer graphene oxide and subsequent functionalization with glucose oxidase. *The Journal of Physical Chemistry C*, *113*(32), 14071–14075.
Wei, T., Tu, W., Zhao, B., Lan, Y., Bao, J., & Dai, Z. (2014). Electrochemical monitoring of an important biomarker and target protein: VEGFR2 in cell lysates. *Scientific Reports*, *4*, 3982.
Weiss, N. O., Zhou, H., Liao, L., Liu, Y., Jiang, S., Huang, Y., et al. (2012). Graphene: An emerging electronic material. *Advanced Materials*, *24*(43), 5782–5825.
Wu, H., Zhao, W., Hu, H., & Chen, G. (2011). One-step in situball milling synthesis of polymer-functionalized graphene nanocomposites. *Journal of Materials Chemistry*, *21*(24), 8626–8632.
Xu, Y., Bai, H., Lu, G., Li, C., & Shi, G. (2008). Flexible graphene films via the filtration of water-soluble noncovalent functionalized graphene sheets. *Journal of the American Chemical Society*, *130*(18), 5856–5857.
Xu, Y., Wu, Q., Sun, Y., Bai, H., & Shi, G. (2010). Three-dimensional self-assembly of graphene oxide and DNA into multifunctional hydrogels. *ACS Nano*, *4*(12), 7358–7362.
Yang, S., Feng, X., Wang, X., & Müllen, K. (2011). Graphene-based carbon nitride nanosheets as efficient metal-free electrocatalysts for oxygen reduction reactions. *Angewandte Chemie International Edition*, *50*(23), 5339–5343.
Yang, T., Li, Q., Li, X., Wang, X., Du, M., & Jiao, K. (2013). Freely switchable impedimetric detection of target gene sequence based on synergistic effect of ERGNO/PANI nanocomposites. *Biosensors and Bioelectronics*, *42*, 415–418.
Yang, Q., Pan, X., Huang, F., & Li, K. (2010). Fabrication of high-concentration and stable aqueous suspensions of graphene nanosheets by noncovalent functionalization with lignin and cellulose derivatives. *The Journal of Physical Chemistry C*, *114*(9), 3811–3816.

Zaid, R. M., Chong, F. C., Teo, E. Y. L., Ng, E.-P., & Chong, K. F. (2015). Reduction of graphene oxide nanosheets by natural beta carotene and its potential use as supercapacitor electrode. *Arabian Journal of Chemistry*, *8*(4), 560–569.

Zhang, X., Liao, Q., Chu, M., Liu, S., & Zhang, Y. (2014). Structure effect on graphene-modified enzyme electrode glucose sensors. *Biosensors and Bioelectronics*, *52*, 281–287.

Zhang, S., Tang, S., Lei, J., Dong, H., & Ju, H. (2011). Functionalization of graphene nanoribbons with porphyrin for electrocatalysis and amperometric biosensing. *Journal of Electroanalytical Chemistry*, *656*(1), 285–288.

Zhang, N., Zhang, Y., & Xu, Y.-J. (2012). Recent progress on graphene-based photocatalysts: Current status and future perspectives. *Nanoscale*, *4*(19), 5792–5813.

Zhong, X., Jin, J., Li, S., Niu, Z., Hu, W., Li, R., et al. (2010). Aryne cycloaddition: Highly efficient chemical modification of graphene. *Chemical Communications*, *46*(39), 7340–7342.

Zhu, Y., Murali, S., Stoller, M. D., Velamakanni, A., Piner, R. D., & Ruoff, R. S. (2010). Microwave assisted exfoliation and reduction of graphite oxide for ultracapacitors. *Carbon*, *48*(7), 2118–2122.

Zu, S.-Z., & Han, B.-H. (2009). Aqueous dispersion of graphene sheets stabilized by pluronic copolymers: Formation of supramolecular hydrogel. *The Journal of Physical Chemistry C*, *113*(31), 13651–13657.

CHAPTER SIX

Preparation, Characterization, and Application of Enzyme Nanoparticles

Neelam Yadav*, Jagriti Narang†,1, Anil Kumar Chhillar*, Chandra S. Pundir‡,1

*Centre for Biotechnology, Maharshi Dayanand University, Rohtak, Haryana, India
†Amity Institute of Nanotechnology, Amity University, Noida, Uttar Pradesh, India
‡Department of Biochemistry, Maharshi Dayanand University, Rohtak, Haryana, India
[1]Corresponding authors: e-mail address: jags_biotech@yahoo.co.in; pundircs@rediffmail.com

Contents

Methods in Enzymology, Volume 609
ISSN 0076-6879
https://doi.org/10.1016/bs.mie.2018.07.001

Abstract

Enzymes are fundamental biocatalysts, which regulate various metabolic reactions. They exhibit high substrate specificity, sensitivity, and exceptional catalytic activity under ideal conditions and, hence, have been used as industrial catalysts. Enzyme nanoparticles have attracted the scientific community as they can be used for environment protection, biochemical engineering, and biomedicine. It is necessary to understand the nature of enzyme nanoparticles interactions with their analyte. Various types of enzyme/protein nanoparticles have been immobilized onto a matrix (electrode or membrane) for the fabrication of biosensors. Among the various nanoparticles and nanomaterials, organic nanoparticles have received more attention due to their fascinating properties. Therefore, the future research should be focused to develop advanced biosensors based on enzyme nanoparticles that could be used for early diagnosis and management of chronic diseases. This chapter explains various enzyme nanoparticles-based biosensors, such as GOX, HRP, uricase, cholesterol oxidase, hemoglobin, and their biomedical applications.

1. INTRODUCTION

1.1 Introduction to Enzyme Kinetics

Enzymes are biological machines, which increase the rate of chemical reactions approximately by as much as 10^8 times by decreasing the activation energy (E_a). E_a is defined as the minimum amount of energy required for a substrate to reach its transition state. Metabolic reactions require small quantities of enzymes. Enzymes (E) bind with a specific substrate (S) to form E–S conjugate which is transformed into product (P); during product formation, enzymes undergo distinct forms of catalysis. The activity of an enzyme depends upon its characteristic parameters such as optimum pH, temperature, incubation time, and appropriate value of the Michaelis–Menten constant (K_m). K_m is defined as the concentration of substrate required for achieving one-half the maximum velocity (V_{max}). Generally, enzymes show a hyperbolic relationship between their activity and substrate concentration and follow Michaelis–Menten kinetics during the reaction given by Eq. (1):

$$v = \frac{V_{max}[S]}{K_m + [S]} \tag{1}$$

In the above equation v is the initial velocity of the reaction, V_{max} is the maximum velocity, [S] is the substrate concentration, and K_m is the Michaelis–Menten constant. The activity of an enzyme also depends on activators, which increase the rate of the reaction, or inhibitors, which decrease the rate of the reaction. Enzymes often show exceptional substrate selectivity, which means that they bind to specific substrate/analyte but ignore all the others. This is a limitation for most enzymes because each substrate might require a different enzyme to catalyze the corresponding reaction.

1.2 What Are Nanoparticles?

Nanotechnology is a branch of multidisciplinary sciences including life sciences, material science, and information technology. It is implemented at the atomic, molecular, and supramolecular levels on nanoscale, i.e., 1–100 nm. This technology is being used to understand, create, and use material structures, tools, and systems with vital novel properties and applications (Silva, 2004). Hence, it is a promising field of science, which is capable of resolving problems and issues that are difficult to solve in engineering and medical sciences (Nikalje, 2015). Every biological and artificial system indicates association at the nanoscale, such as nanocrystals, nanotubes, or nanobiomotors. Nanoparticles (NPs) are often made by combining molecules into objects, arranged in a functional order by considering various length scales, or dismantle bulk material into particles of nanoscale. The subunits of the particles are held together by means of weak molecular forces including van der Waal forces, hydrogen bonds, electrostatic dipoles, fluidics, and diverse surface forces (Roco, 2003). Various nanostructures such as nanotubes, nanofibers, nanorods, and NPs have been scrutinized to investigate the characteristic features and their applications. The NPs may be organic, inorganic, or a mixed phase; however, organic NPs including protein/enzyme have attracted attention in the past decades for the fabrication of improved biosensors (Pundir, 2015; Shukla, Deshpande, Shukla, & Tiwari, 2012; Shukla, Mishra, Mamba, & Omotayo, 2014).

1.3 Enzyme Nanoparticles

Covalent linking was used to make enzyme nanoparticles (ENPs) of size <100 nm (Deshapriya et al., 2015; Stromer & Kumar, 2016), but enzyme particles formed by desolvation are often in the range of 100–200 nm (Pundir, 2015). These ENPs exhibit several exceptional characteristic properties such as optical, electronic, electric, thermal, chemical, mechanical, and catalytic while supporting a large surface area, compared to the bulk material but less than those of the free enzymes. These properties of ENPs improve the activity of enzyme-based biosensors. Direct immobilization of proteins or enzymes onto NPs may cause denaturation which could result in enzyme inactivation. This problem has been resolved by preparing ENPs and self-cross-linked before their attachment. ENPs-based biosensors have enhanced the analytical parameters such as detection limit and current response (Sharma, Shrivastav, Gupta, & Srivastava, 2011). NPs of horseradish peroxidase (HRP) were synthesized very first time (Liu, Lin, Ostatna, & Wang, 2005). Subsequently, NPs of glucose oxidase (GOD) (Sharma et al., 2011), cholesterol oxidase (ChOx) (Chawla, Rawal, Sonia, & Pundir, 2013), uricase (Chauhan, Kumar, & Pundir, 2014), and hemoglobin (Hb) (Yadav, Chhillar, & Pundir, 2018)-based amperometric biosensors have been developed.

These ENPs differ from single enzyme nanoparticles (SENs), which are single enzyme encapsulated with nanoscale organic/inorganic macromolecules that improve the catalytic activity and thermal stability. Trypsin SENs were used to improve the enzyme stability at increased temperature, while chymotrypsin SENs have been used to improve cellulose degradation (Blanchette, Lacayo, Fischer, Hawang, & Thelen, 2012; Khosravi, Vossoughi, Sharokhian, & Alemzadeh, 2012; Kim & Grate, 2003).

2. EXPERIMENTAL

2.1 Chemicals and Reagents

(i) Enzyme (1–2 mg)
(ii) Ethanol, desolvating agent (4 mL)
(iii) 2.5% glutaraldehyde (1.8 mL)
(iv) Cysteine or cysteamine dihydrochloride (0.12 g)
(v) Analyte to detect

2.2 Equipment

(i) Support material [electrode (Au, pencil graphite, screen printed, etc.), fluorine tin oxide nitrocellulose (NC) membrane]
(ii) Weighing balance
(iii) Magnetic stirrer
(iv) Magnetic beads
(v) Centrifuge
(vi) Potentiostat
(vii) Water bath
(viii) Ultraviolet (UV)-spectrophotometer

2.3 Methods to Prepare ENPs

NPs of soluble proteins have been prepared by their aggregation by following methods:

(i) Emulsification in plant oil (Scheffel, Rhodes, Natrajan, & Wagner, 1972)
(ii) Desolvation by ethanol or natural salts and the aggregated protein is cross-linked by glutaraldehyde (Coester, Langer, & Von, 2000)
(iii) Coacervation via anhydrous ethanol, glutaraldehyde, and ethanolamine (Sailaja, Amareshwar, & Chakravarty, 2011)
(iv) Cross-linking in water and oil emulsion at high pressure (Ezpeleta et al., 1999)

ENPs were prepared by desolvation by using ethanol as desolvating agent followed by cross-linking with glutaraldehyde (Liu et al., 2005).

2.4 Preparation of ENPs by Desolvation

Various steps of desolvation method for ENP preparation are described below.

2.4.1 Desolvation

The lyophilized form of an enzyme (1–2 mg mL^{-1}) was dissolved in reaction buffer. Ethanol was added as a desolvating agent, dropwise at a rate of 0.1–0.5 mL min^{-1} in 1:2 ratio, under constant stirring at a speed of 500 rpm at 4°C. Addition of the desolvating agent facilitated the formation of enzyme/protein aggregates to form NPs in the range of 1–100 nm.

2.4.2 Cross-linking

After the formation of enzyme aggregates they are cross-linked by the addition of 2.5%–8.0% glutaraldehyde (1.8 mL) under constant stirring at 500 rpm at 4°C in an ice bath for 24 h. Addition of glutaraldehyde resulted in intermolecular cross-linking within ENPs by forming Schiff base linkages.

2.4.3 Functionalization

The ENPs were further functionalized by adding 0.12 g cysteamine dihydrochloride or cysteine under constant stirring, which introduced $-NH_2$ or –SH groups on the ENPs. These functions enhance the attachment of the ENPs on the surface of the electrode.

2.4.4 Purification

The cross-linked, surface functionalized ENPs were separated by centrifuging the above preparation at 14,000 rpm at 4°C for 10 min. The supernatant was removed and the pellet dispersed into the buffer by ultrasonication for 5 min. Finally, the ENPs were separated by centrifugation at 12,000 rpm, $15{,}000 \times g$ at 4°C for 10 min. Schematic representation of the formation of ChOx NPs was shown in Fig. 1.

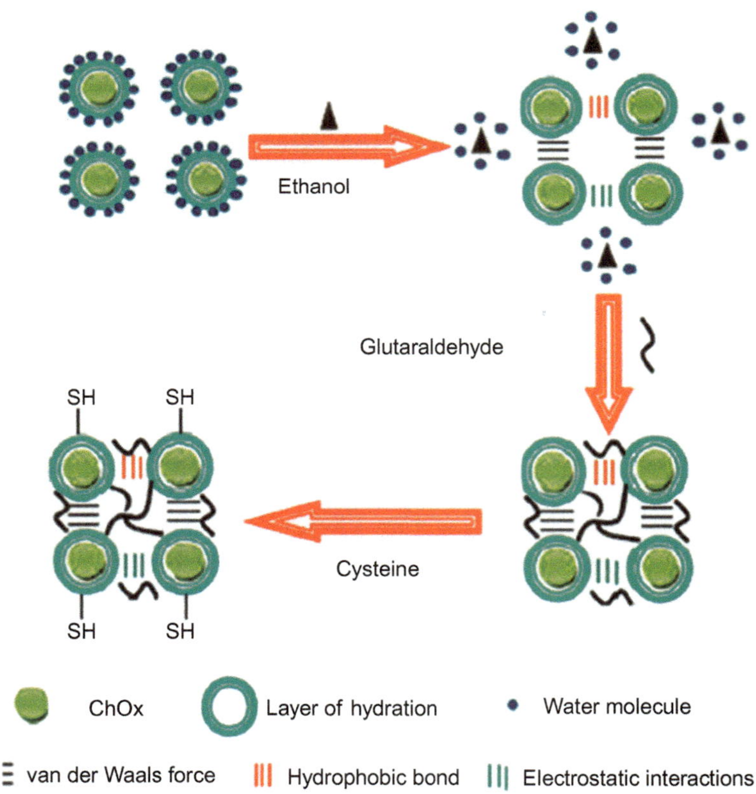

Fig. 1 Schematic representation of formation of cholesterol oxidase nanoparticles (CHOxNPs) by the desolvation method and introduction of –SH group by cysteine on the surface of ChOxNPs (Chawla et al., 2013).

2.5 Characterization of ENPs

ENPs have characteristic features, which were determined by various techniques as described below.

2.5.1 Transmission Electron Microscopy

The shape and size of ENPs were characterized by transmission electron microscopy (TEM) images. The spherical shape and the diameter of ENPs aggregates range from 100 to 200 nm (Fig. 2). The diameter of free enzyme molecules is in the range of 2–18 nm. A comparison of TEM size of free enzymes and ENPs is shown in Table 1.

2.5.2 Colorimetric Methods

ENPs exhibited characteristic colors. For instance, ENPs of HRP were white in color. The ENP aggregates acquired the color due to a combination of light scattering and absorption. The particles are too large, on the order of the wavelengths of visible light.

2.5.3 UV Absorption Spectra

Different enzymes and their ENPs showed absorbance at their characteristic wavelengths. For instance, native Hb exhibited a Soret absorbance peak

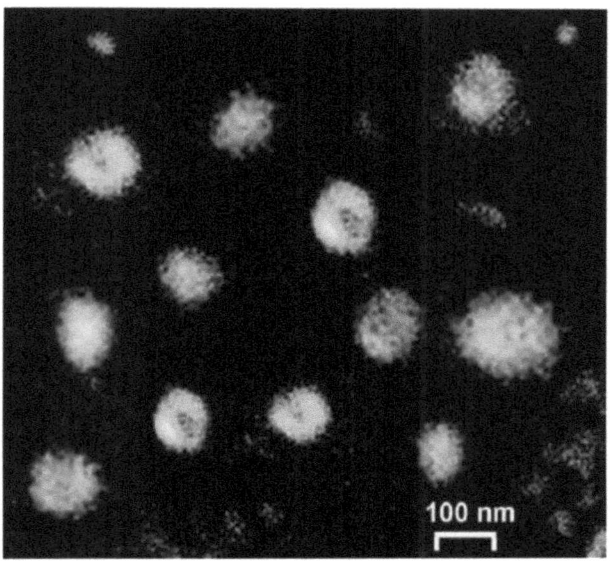

Fig. 2 Transmission electron microscopic (TEM) images illustrating configuration, size, and color of horseradish peroxidase nanoparticles (Liu et al., 2005).

Table 1 A Comparison of Average Diameter of Enzyme Nanoparticles (ENPs) and Free Enzyme by TEM

ENPs	Diameter of ENPs (nm)	Diameter of Free Enzyme	Ref.
Horseradish peroxidase	100	14.2	Liu et al. (2005)
Glucose oxidase	117	2–8	Kundu, Yadav, and Pundir (2012)
Cholesterol oxidase	100–200	–	Chawla et al. (2013)
Uricase	100	14–18	Chauhan et al. (2014)
Hb	97	—	Yadav et al. (2018)
ChE	35–40	—	Aggarwal, Malik, Prashant, Jaiwal, and Pundir (2016)
ChOx	56.97		Aggarwal et al. (2016)
Lipase	83–218	—	Pundir and Aggarwal (2017)
GK	144–329	—	Pundir and Aggarwal (2017)
GPO	22–68	—	Pundir and Aggarwal (2017)

at 410 nm, while hemoglobin nanoparticles (HbNPs) showed no Soret band but had an absorbance peak at 280 nm which is due to the protein absorption. The loss of Soret band could be due to the large size of HbNPs (Yadav et al., 2018). Fig. 3 shows the comparative UV spectra of native Hb and HbNPs.

2.5.4 FTIR Spectra

The FTIR spectra of native Hb and HbNPs are compared. The native Hb showed the characteristic peaks at 1062, 1317, 1553, 1679, and 3497 cm^{-1}, while HbNPs had peaks at 1090, 1305, 1561, 1701, and 3488 cm^{-1}. These characteristic peaks showed the presence of various functional groups of native Hb and HbNPs. The band at 1062 cm^{-1} indicates the presence of amine groups. In addition, there was a band located at 3497 cm^{-1}, which could be assigned to the O–H stretching, indicating the presence of hydroxyl groups. The FTIR spectra of the Hb and HbNPs exhibited a key shift at carbonyl, hydroxyl, and amino groups of the protein, which may be due to ethanol, glutaraldehyde, and cysteamine dihydrochloride treatment during the synthesis of HbNPs (Fig. 4).

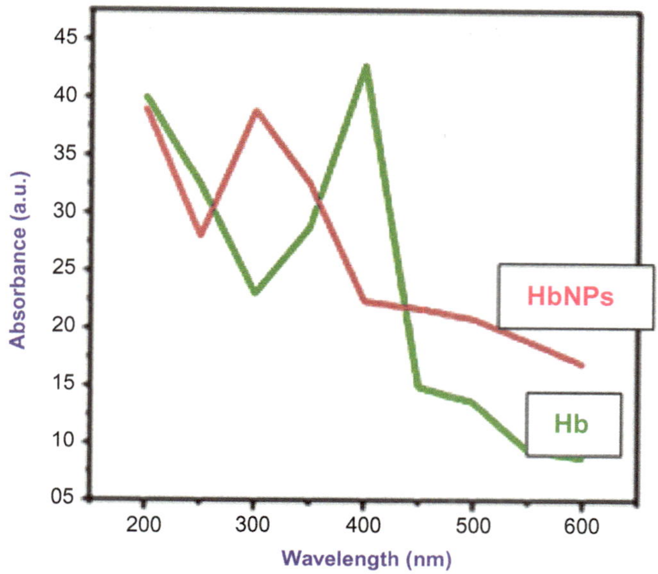

Fig. 3 UV spectra of native Hb and HbNPs (Yadav et al., 2018).

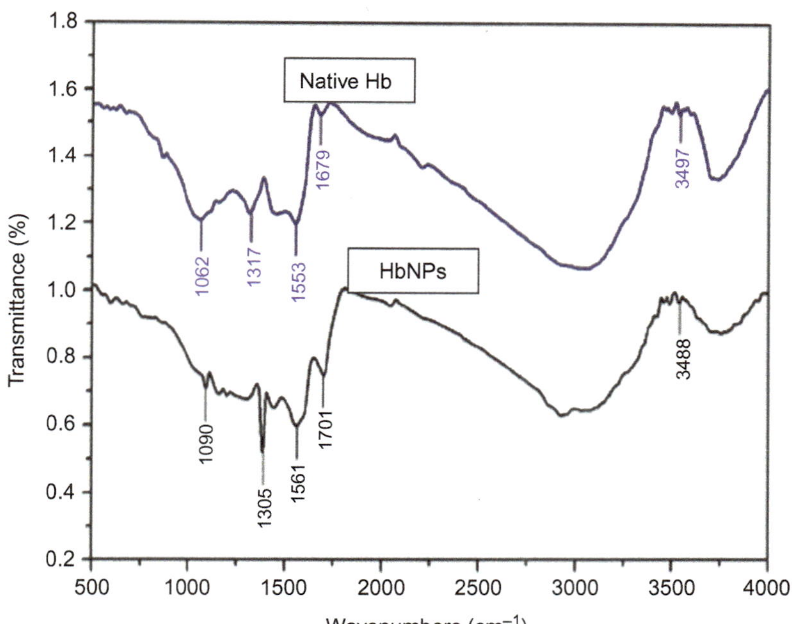

Fig. 4 FTIR spectra of monomeric native Hb and HbNPs (Yadav et al., 2018).

3. IMMOBILIZATION OF ENPs

The attachment of ENPs on to an inorganic/organic matrix is to be done in such a way that the attached ENPs do not lose their function and the resulting biocatalyst is usable multiple times. ENPs are immobilized onto support matrix via adsorption, entrapment, microencapsulation, covalent binding, and metal binding and aggregated by bifunctional reagents. For instance, aggregates of NPs of HRP (Liu et al., 2005), Hb (Yadav et al., 2018), lactate oxidase (Dagar & Pundir, 2017), cholesterol oxidase (ChOx) (Aggarwal et al., 2016), uricase (Chauhan et al., 2014), and many more have been immobilized covalently onto Au electrode. Besides, NPs of GOD isolated from *Aspergillus niger* have been immobilized on Pt electrode and NC membranes.

3.1 Characterization of ENPs Immobilized Onto Matrix

Confinement of ENPs onto their support material was confirmed by scanning electron microscopy (SEM). SEM images of bare electrode or membrane showed the uniform surfaces, whereas electrodes immobilized with ENPs showed globular, beaded configurations and clusters of aggregated ENPs on the entire surface (Fig. 5). This confirms the immobilization of ENP onto support material (electrode/membrane) (Yadav et al., 2018).

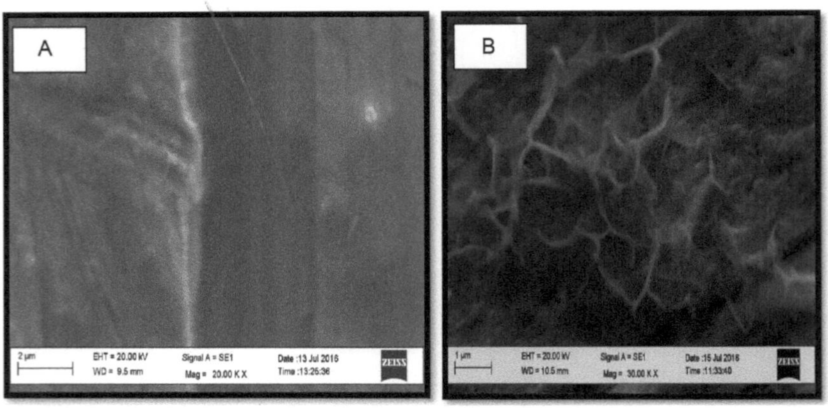

Fig. 5 (A) SEM images of bare Au electrode and (B) HbNPs immobilized Au electrode. *From Yadav et al. (2018).*

3.2 Kinetic Properties of ENPs

For immobilized ENPs, one needs to determine optimum pH, temperature, the working range, reusability, storage stability, K_m, and response time. The kinetic parameters were studied after covalent immobilization either onto metal wire like Au/Pt or on NC membranes. A comparison of various parameters is given in Table 2.

Table 2 K_m Value of Enzyme Nanoparticles (ENPs)-Based Biosensors

ENPs-Based Biosensor	K_m Value (m*M*)	Ref.
HbNPs/AuE	0.1 ± 0.001	Narwal, Yadav, Thakur, and Pundir (2017)
Uricase NPs/AuE	0.058	Chauhan et al. (2014)

3.2.1 Optimum pH

ENPs showed maximum activity at a particular pH, and ENPs attached to supports exhibited some change in their optimum pH in comparison to a monomeric enzyme. For instance, elevated pH with a difference of 0.4 for HRP (Liu et al., 2005), pH with a difference of 0.5 for GOD (Kundu et al., 2012), ChOx (Chawla et al., 2013), uricase (Chauhan et al., 2014), and HbNPs (Fig. 6) (Yadav et al., 2018). This alteration in optimum pH was because of the modified configuration of ENPs after their immobilization.

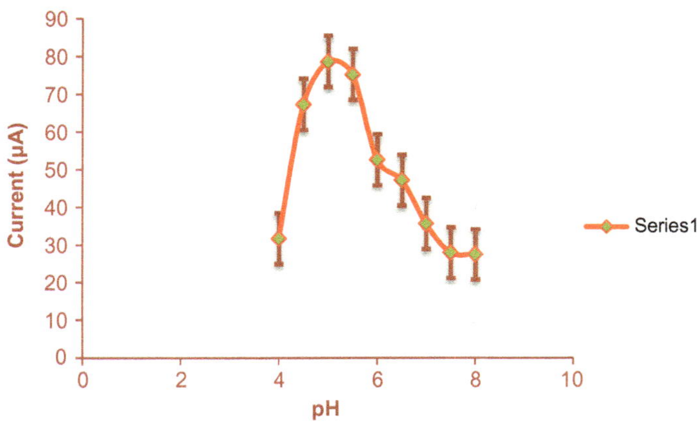

Fig. 6 Effect of pH on current response of HbNPs/AuE in 0.1 *M* sodium acetate buffer. *From Yadav et al. (2018).*

3.2.2 Optimum Temperature and Thermostability

At the optimum temperature, the attached ENPs showed significant activity improvements when compared to the free enzyme. The optimum temperature for ChENPs + ChOxNPs/AuE was 40°C (Aggarwal et al., 2016), and for lipase NPs + GKNPs + GPONPs/AuE, it was 35°C (Pundir & Aggarwal, 2017). ENPs exhibited higher stability in their immobilized form as compared to the free enzyme. For example, the thermostability of GOD NPs was greater as compared to free GOD because of inter- and intramolecular covalent bonding that inhibited the structural alterations of the enzyme at elevated temperatures (Fig. 7).

3.2.3 Response Time

ENP immobilization had shorter response times. For example, response time of HbNPs/AuE was 2 s (Yadav et al., 2018), while response time of Hb/cMWCNT-Fe_3O_4NP/CHIT/Au electrode was 8 s (Batra, Lata, Sharma, & Pundir, 2012), ChENPs + ChOxNPs/AuE was 5 s (Aggarwal et al., 2016), lipase NPs/GKNPs/GPONPs/AuE was 5 s (Pundir & Aggarwal, 2017), while lipase-based biosensor showed response time 30 min (Chen et al., 2014), uricase NPs/AuE was 7 s (Chauhan et al., 2014), whereas response time of

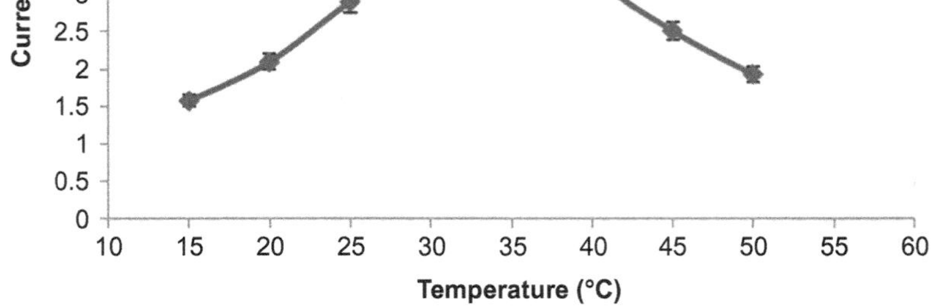

Fig. 7 Effect of temperature on current response of lipase NPs, GKNPs, and GPONPs/AuE. *From Pundir, C. S., & Aggarwal, V. (2017). Amperometric triglyceride bionanosensor based on nanoparticles of lipase, glycerol kinase, glycerol-3-phosphate oxidase.* Analytical Biochemistry, 517, *56–63.*

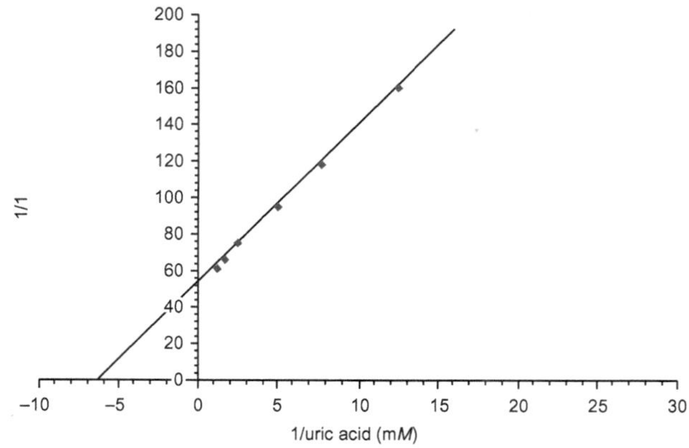

Fig. 8 Lineweaver–Burk plot of the effect of uric acid concentration on response of uric acid biosensor based on uricase NPs-modified Au electrode. K_m was calculated using the following equation: $1/v = K_m/V_{max}.1/[S] + 1/V_{max}$, where v is the biosensor response in current (mA), and S is the uric acid concentration in m*M*. Slope of LB plot $= K_m/V_{max}$; intercept of the plot $= 1/V_{max}$. *From Chauhan, N., Kumar, A., & Pundir, C. S. (2014). Construction of an uricase nanoparticles modified Au electrode for amperometric determination of uric acid.* Applied Biochemistry and Biotechnology, 174, *1683–1694.*

uricase biosensor was 330 s (Cete, Yaşar, & Arslan, 2006) and ChOxNPs/AuE was 8 s (Chawla et al., 2013), while monomeric ChOx-based biosensor exhibited response time of 30 s (Guo, Chen, Li, Nie, & Yao, 2004).

3.2.4 K_m Values

Immobilization also decreased or retained the K_m values of ENPs as compared to those of the corresponding free enzymes. The K_m values of uricase NPs immobilized onto Au electrode were calculated from the Lineweaver–Burk plot (Fig. 8), which was similar to that of the free enzyme (Chauhan et al., 2014).

3.2.5 Working Range

Working range is the concentration range over which the sensor is responsive. Limit of detection (LOD) is the lowest concentration detected by the biosensor calculated by using the formula: $LOD = 3.3 \times SD/S$, where SD is the standard deviation of the response, and S is the slope of the calibration curve (Valipour, 2015). The working range of immobilized ENPs is narrow as compared to that of the free enzyme. The working ranges of various biosensors are collected in Table 3.

Table 3 Working Range of Various Enzyme Nanoparticles (ENPs)-Based Biosensors

ENPs-Based Biosensor	Working Range	Ref.
HbNPs/AuE	0.1–100 n*M*	Yadav et al. (2018)
HbNPs/AuE	1.0–1200 μ*M*	Narwal et al. (2017)
ChENPs + ChOxNPs/AuE	10–700 mg/dL	Aggarwal et al. (2016)
Lipase NPs + GKNPs + GPO/AuE	10–100 mg/dL	Pundir and Aggarwal (2017)
ChOx NPs/AuE	12.5–700 mg/dL	Chawla et al. (2013)
Uricase NPs/AuE	0.005–0.8 m*M*	Chauhan et al. (2014)

4. STABILITY AND REUSABILITY

4.1 Storage Stability

At lower temperatures, immobilized ENPs showed better storage stability than the free enzyme and these data are collected in Table 4. This is usually the case for most immobilized enzymes.

4.2 Reusability

ENPs immobilized on different matrices avoid the frequent recharging with ENPs, which, in turn, reduces the cost and labor. The reusability of immobilized ENPs was better than free enzymes. Immobilized ENPs can be reused 100–200 times, and hence, ENPs were economic and have been employed in diverse applications.

4.3 Correlation of Measurements With a Standard Method

The performance of the biosensor was correlated by a colorimetric method. The concentration determined by the biosensor was validated by plotting the concentration obtained by the colorimetric method on the x-axis and the concentration obtained by biosensing method on the y-axis (Narwal et al., 2017). The values obtained were analyzed by regression and the correlation coefficient has been calculated by Eq. (2):

$$\text{The correlation coefficient } (r) = \frac{n\left(\sum xy\right) - \left(\sum x\right)\left(\sum y\right)}{\sqrt{\left[n\sum x^2 - \left(\sum x\right)^2\right]\left[n\sum y^2 - \left(\sum y\right)^2\right]}} \tag{2}$$

where

x values obtained by HPLC

y values obtained by biosensing

Table 4 Comparison of Kinetic Properties of Immobilized Enzyme Nanoparticles With Those of Free Enzymes

	Glucose Oxidase From *A. niger*		Cholesterol Oxidase from Microorganism		Uricase From *Candida* sp.		Hemoglobin	
Property	**Free**	**Immobilized NPs**	**Free**	**Immobilized NPs**	**Free**	**Immobilized NPs**	**Free**	**Immobilized NPs**
Support used	—	NC membrane	—	Au wire	—	Au wire	Au wire	Au wire
Method of immobilization	—	Covalent	—	Covalent	—	Covalent	Covalent	Covalent
Type of electrode	—	DO metric	—	Amperometric	—	Amperometric	Amperometric	Amperometric
Optimum pH	5.5	6.0	5.5	6.0	8.0	8.5	5.0	
Optimum temperature	3.5	35–45	37	35	30	40	30	20
Incubation time (s)	120	60	300	8	300	7	8	2
K_m (m*M*)	6.2	4.17	—	ND	0.056	0.058	—	—
Linearity (m*M*)	Up to 13.8	0.056–1.39	Up to 6.4	0.69–38.88	Up to 1.5	0.005–0.8	3–90 n*M*	0.1 n*M*–100 m*M*
Limit of detection (m*M*)	0.138	0.55	0.01	0.04	160	0.005	0.02 n*M*	0.1 n*M*
Half-life at 4°C in days	180	180	60	90	60	210	60	120
No. of reuses	—	150	—	180	—	200	—	
Ref.	Kalia, Goyal, and Pundir (1999)	Kundu et al. (2012)	Varma and Nene (2003)	Chawla et al. (2013)	Freitas, Spencer, Vasao, and Abrahao-Neto (2009)	Chauhan et al. (2014)	Batra et al. (2012)	Yadav et al. (2018)

ND, not detected.

5. APPLICATIONS OF ENPs

Biosensors are the quantitative analytical devices, which consist of a biological recognition entity such as enzyme, antibody, phage, aptamer, or single-stranded DNA coupled with appropriate physicochemical, optical, thermometric, piezoelectric, and magnetic transducers (Elif & Mustafa, 2015). In other words, biosensors are chemical sensors in which the biological recognition element uses a biochemical mechanism interfacing the optoelectronic system (Cammann, 1977; Turner, Karube, & Wilson, 1987) (Fig. 9).

There are of various types of biosensors. Among them electrochemical biosensors have been used widely for the analysis of processed foods and have given more attention in various fields, due to their small size, cost-effectiveness, fast response time, possibility of achieving low detection limits, and good functioning (Wang et al., 2014). Various ENP-based biosensors are described below.

5.1 Biosensor Based on Immobilization of Nanoparticles of Cholesterol Esterase and Cholesterol Oxidase

An amperometric biosensor based on covalent immobilization of cholesterol esterase nanoparticles (ChENPs) and cholesterol oxidase (ChOxNPs) onto Au electrode for the determination of cholesterol in serum (Fig. 10) (Aggarwal et al., 2016). The biosensor showed an optimum response within 5 s at pH 5.5, 40°C at 0.25 V vs Ag/AgCl. The linear range of the biosensor was 10–700 mg dL^{-1}. This biosensor was evaluated in terms of various analytical parameters such as analytical recovery of added cholesterol concentration at 100 and 200 mg dL^{-1} were 95% and 96%, with coefficients of variation as <2% and <3%, respectively. The storage stability of working electrode was 60 days at 4°C.

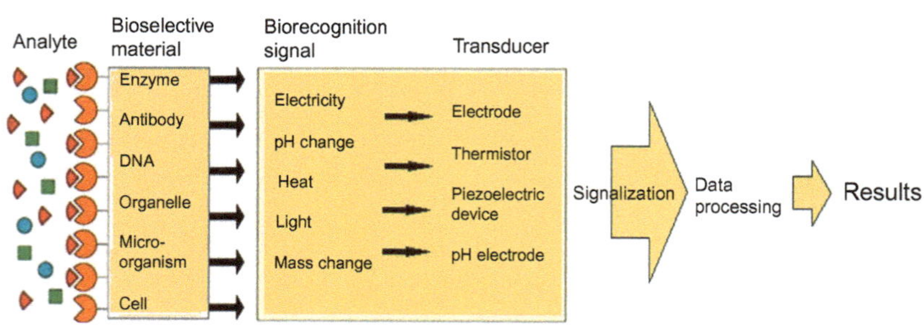

Fig. 9 Principle and working of a biosensor (https://doi.org/10.5772/54127).

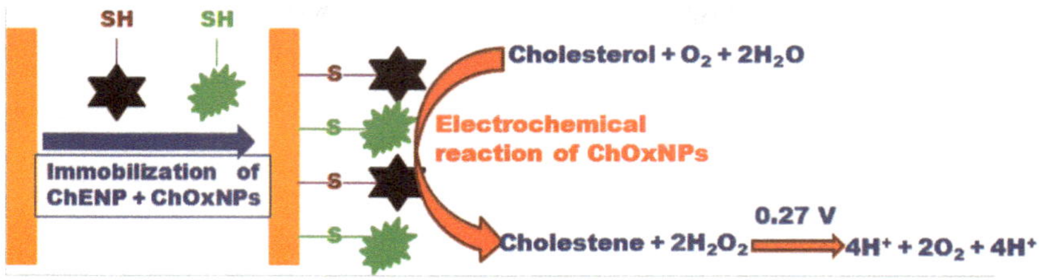

Fig. 10 Schematic representation of immobilization and working of ChENPs- and ChOxNPs-based cholesterol biosensor. *From Aggarwal, V., Malik, J., Prashant, A., Jaiwal, P. K., & Pundir, C. S. (2016). Amperometric determination of serum total cholesterol with nanoparticles of cholesterol esterase and cholesterol oxidase.* Analytical Biochemistry, 500, *6–11.*

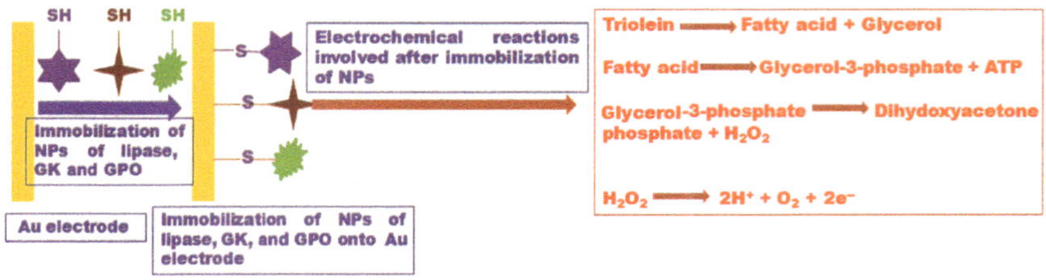

Fig. 11 Schematic representation of attachment of functionalized lipase, glycerol kinase (GK), and glycerol-3-phosphate oxidase (GPO) NPs and their electrochemical reaction onto Au electrode (Pundir & Aggarwal, 2017).

5.2 Biosensor Based on Immobilization of Nanoparticles of Lipase, Glycerol Kinase, and Glycerol-3-Phosphate Oxidase

The porcine pancreas lipase, glycerol kinase (GK) from *Cellulomonas* sp., and glycerol-3-phosphate oxidase (GPO) from *Aerococcus viridans* were prepared by the above desolvation method (Pundir & Aggarwal, 2017) (Fig. 11). These NPs were characterized by TEM and FTIR and immobilized onto Au electrode. The performance of the biosensor was maximum at pH 6.5 at 35°C. The working potential of biosensor was 1.2 V and response time 5 s. Biosensor was evaluated in terms of linear concentration range of 1.0–100 and 100–500 mg dL^{-1}, respectively. The detection limit was 1.0 μg mL^{-1}, and the analytical recovery of added concentration was 95% and 96% at 50 and 100 mg dL^{-1}, respectively. The coefficients of variation were 2% and 2%, respectively. The storage stability of working electrode was 90 days.

5.3 Biosensor Based on Immobilization of Nanoparticles of Hb

An amperometric biosensor was constructed by immobilizing Hb ENPs onto the Au electrode (Narwal et al., 2017) (Fig. 12). The working electrode

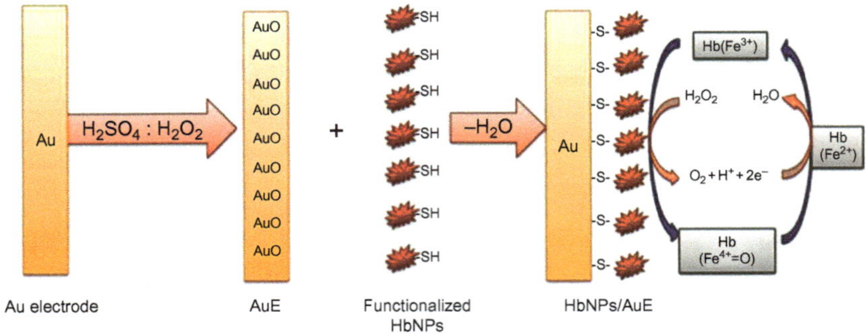

Fig. 12 Schematic representation and electrochemical reaction involved in the fabrication of HbNPs onto Au electrode for the detection of H_2O_2 in the serum sample. *From Narwal, V., Yadav, N., Thakur, M., & Pundir, C. S. (2017). An amperometric H_2O_2 biosensor based on hemoglobin nanoparticles immobilized onto a gold electrode.* Bioscience Reports, 37, *1–10.*

was characterized by cyclic voltammetry (CV), SEM, FTIR spectroscopy, and electrochemical impedance spectroscopy (EIS). The NPs of copper were characterized by TEM and UV-spectroscopy. HbNPs/AuE gave good response within 2.5 s, at pH 6.5 and temperature 30°C when operated at $20\,\mathrm{mV\,s^{-1}}$. The biosensor was evaluated in terms of its detection limit 1.0 μ*M*, linear range was 1–1200 μ*M*, and the analytical recovery of biosensor was 99% and 98%. The coefficients of variation were 3% and 4%, respectively. The storage stability of working electrode was 90 days.

5.4 Biosensor Based on Immobilized HbNPs for the Determination of Acrylamide in Processed Foods

HbNPs were prepared by the desolvation method and characterized by TEM, UV–visible spectroscopy, FTIR, XRD, and AFM (Yadav et al., 2018). These HbNPs were immobilized covalently onto Au electrode (Fig. 13). The Au electrode was characterized by SEM and EIS. HbNPs/AuE was optimized in terms of its pH 5.0, and working potential 0.26 V at 20°C. The analytical recovery of biosensor was 99% and 98% at 5 and 10 m*M* acrylamide, respectively. The coefficients of variation were 4% and 5%, respectively. The proposed biosensor was used for the detection of acrylamide in processed foods.

5.5 Biosensor Based on Immobilized HRP Nanoparticles

Liu et al. (2005) have constructed an improved amperometric biosensor by immobilizing the HRPNPs covalently onto Au electrode (Fig. 14). The

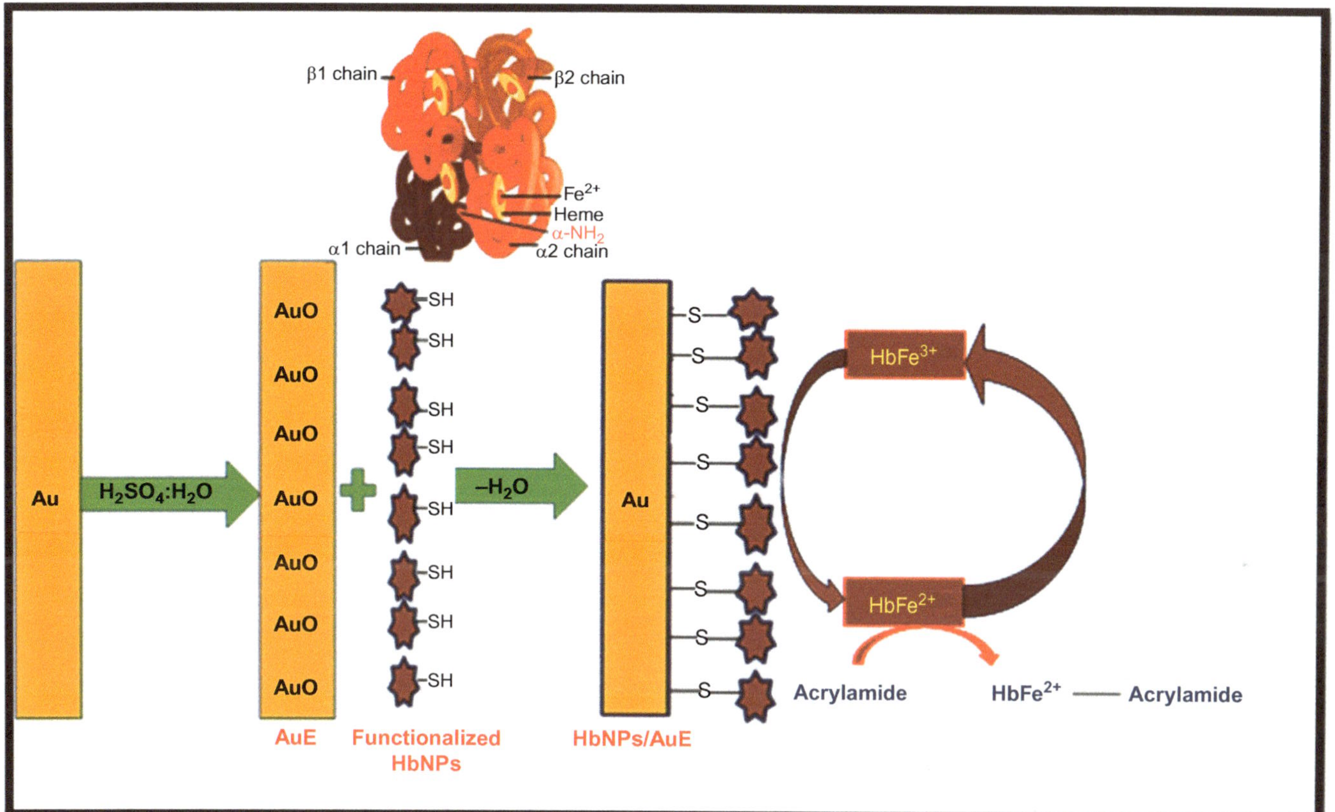

Fig. 13 Schematic representation and electrochemical reaction involved in the fabrication of HbNPs onto Au electrode. *From Yadav, N., Chhillar, A.K., & Pundir, C.S. (2018). Preparation, characterization and application of haemoglobin nanoparticles for detection of acrylamide in processed foods. International Journal of Biological Macromolecules, 107, 1000–1013.*

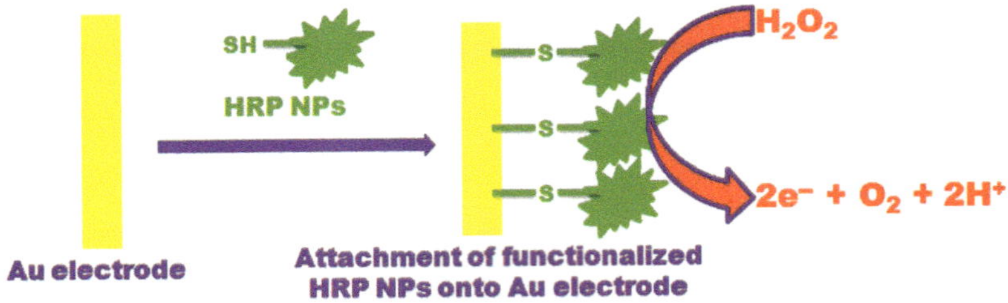

Fig. 14 Schematic representation of attachment of functionalized horseradish peroxidase (HRP) nanoparticles onto Au electrode (Liu et al., 2005).

performance of the biosensor was optimum at pH 7.4 when operated at 0.16 V. The linear range of HRPNPs/AuE was 1–9 μ*M*, detection limit was 0.1 μ*M*, and coefficient of variation was 3%.

5.6 Biosensor Based on Immobilized GOD Nanoparticles

The biosensor was fabricated by using GOD NPs (Kundu et al., 2012; Sharma et al., 2011) (Fig. 15). The biosensor was used for the electrochemical quantification of H_2O_2 produced from glucose. GOD NPs/AuE worked optimally at pH 7.4, and 27°C, with a linear range of 1–35 mg dL^{-1}. The working range of the biosensor was 35 mg dL^{-1}. The LOD of GOD NPs/AuE was 18 μ*M*.

5.7 Biosensor Based on Immobilized ChOx Nanoparticles

An improved amperometric biosensor was fabricated by immobilizing ChOx NPs onto Au electrode (Fig. 16) (Chawla et al., 2013). The ChOxNPs/AuE worked best at pH 6.0, and 35°C, when operated at 0.27 V vs Ag/AgCl. The working range of ChOxNPs/AuE was 12.5–700 mg dL^{-1}, with a detection limit of 1.56 mg dL^{-1}. The storage stability of biosensor was 90 days at 4°C.

5.8 Biosensor Based on Immobilized Uricase Nanoparticles

Chauhan et al. (2014) constructed a highly sensitive biosensor for the determination of uric acid in blood or urine (Fig. 17). The optimum performance of the biosensor was observed within 7 s, at pH 8.5, and 40°C at a potential of 0.2 V. The uricase NPs/AuE had a linear range of 0.005–0.8 m*M*, analytical recovery 100% and 99% at 10 and 29 mg L^{-1} uric acid, high sensitivity 0.003 mA μM^{-1} cm^{-2}, and stored for 210 days at 4°C. The precision of biosensor was 5.6% and 4.7%, respectively.

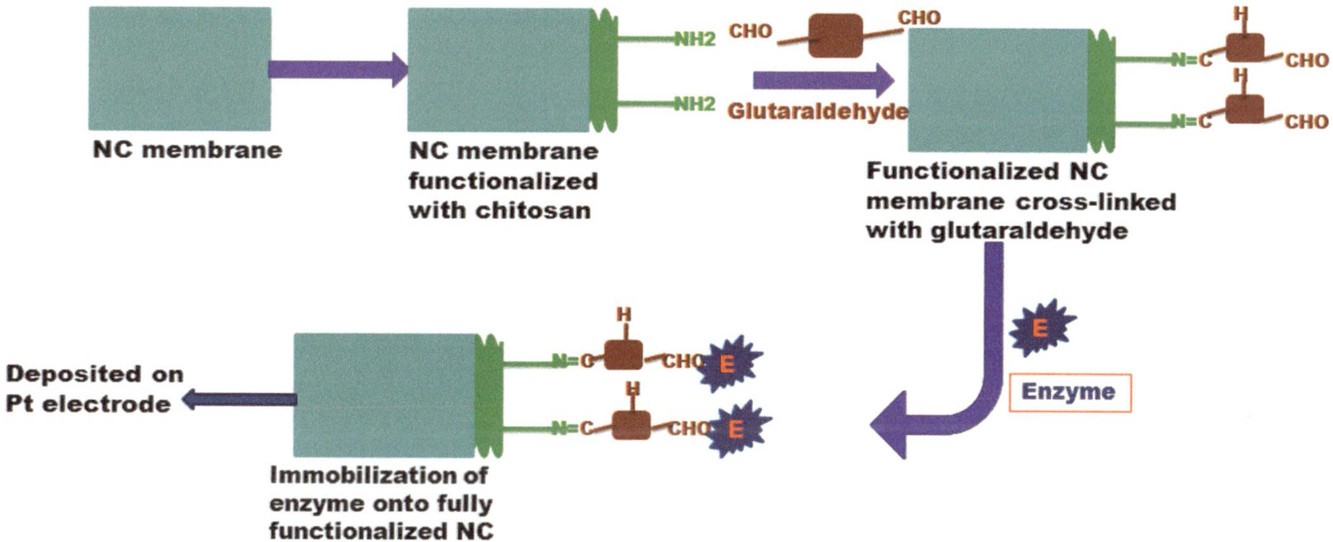

Fig. 15 Schematic representation of attachment of functionalized nitrocellulose (NC) membrane functionalized with chitosan and glutaraldehyde followed by immobilization of glucose oxidase (GOD) NPs and deposited onto Pt electrode (Sharma et al., 2011).

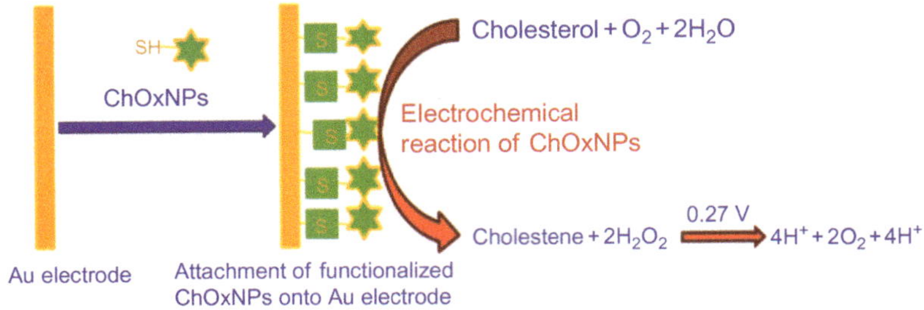

Fig. 16 Schematic representation of attachment of functionalized ChOxNPs and their electrochemical reaction onto Au electrode (Chawla et al., 2013).

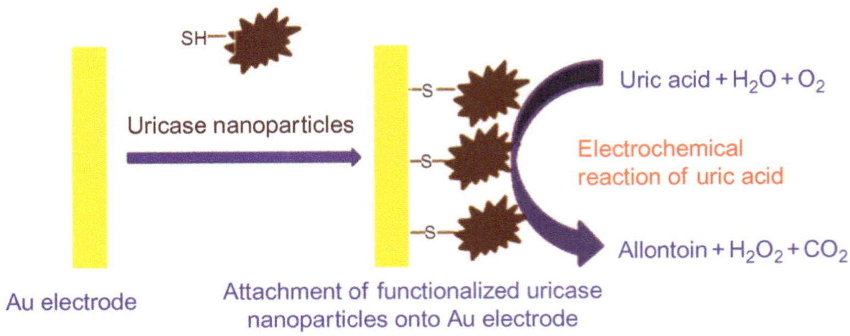

Fig. 17 Schematic representation of attachment of functionalized uricase nanoparticles and electrochemical reaction onto Au electrode (Chauhan et al., 2014).

6. DETAILED PROCEDURE FOR THE DEVELOPMENT OF HbNPs-BASED BIOSENSORS

Acrylamide is carcinogenic, neurotoxic, and reproductive toxicant, and hence, its detection is important. An improved amperometric acrylamide biosensor based on HbNPs is described here (Yadav et al., 2018).

6.1 Reagents

Human Hb from Sigma-Aldrich; acrylamide, potassium hexacyanoferrate(II) trihydrate (Carrez I), acetic acid, zinc sulfate heptahydrate (Carrez II), and glutaraldehyde from Merck; cysteamine dihydrochloride, potassium ferrocyanide, potassium ferricyanide, and copper sulfate from Sisco Research Labs (Mumbai, India); and ethanol from Changshu Yangyuan Chemical (China) have been used. All other chemicals were of analytical reagent grade. Au wire purchased from local jeweler, different brands of biscuits, kurkure, crisps, fried cereals, and diverse bread items were purchased from a local market.

6.2 Instruments Used

A potentiostat/galvanostat (Autolab, AUT83785, Eco Chemie) with a three-electrode system consisting of a Au wire as a working electrode, Pt wire as a counter electrode, and Ag/AgCl as a reference electrode was used. Zeiss EV040 SEM, UV-spectrophotometer (Dynamica HALO DB-20, UK), TEM, and FTIR spectrophotometer (Thermo Scientific, USA) were used. Refrigerated centrifuge (Sigma 3-30K, Germany), Spectronic-20, electronic balance, and temperature-controlled water bath (NSW) were used.

6.3 Preparation and Testing of HbNP Biosensor

Prepare a solution of Hb by dissolving Hb powder (4 mg) in sodium phosphate buffer (2 mL; 0.02 *M*, pH 5.0). Add ethanol (4 mL) dropwise into Hb solution under constant stirring, followed by addition of glutaraldehyde (1.8 mL of 2.5%), and stir for a whole. Add glutaraldehyde (1.8 mL; 2.5%) into this preparation, followed by the addition of cysteamine dihydrochloride (0.12 g), and let it stir for 5–6 h. Then centrifuge the sample (4°C) at 1200 rpm for 10 min and separate the supernatant from the pellet. Dissolve the pellet 0.1 *M* sodium phosphate buffer (pH 7.0) and store at 4°C until it is used. Characterize the samples by UV and FTIR as described in Figs. 3 and 4.

Immobilize the HbNPs onto Au electrode and wash the Au wire with piranha solution (concentrated H_2SO_4 and H_2O_2, in 3:1 ratio), followed by polishing with silica gel. Sonicate the Au wire and wash with distilled water. Dip the Au wire into a suspension of HbNPs, and remove and rinse with 0.1 *M* sodium phosphate buffer (pH 7.0). Characterize the working electrode by SEM (Fig. 5).

Construct the acrylamide biosensor using HbNPs/Au as a working electrode, Ag/AgCl as a reference electrode, and Pt wire as a counter electrode connected to a potentiostat. Measure the current (in mA) as a function of time (in seconds) (Fig. 14). Record the CV of the working electrode in potentiostat galvanostat in the potential range of −0.750 to 0.500 V in 25 mL of 0.2 *M* sodium acetate buffer (pH 6.0) containing 0.1 mL, 3.5 μ*M* acrylamide at a scan rate of 20 mV s^{-1}. Record the response of the biosensor (in current) for acrylamide in the range of 0.1–100 n*M*.

Test the biosensor to measure acrylamide in real samples with various concentrations of processed foods. Crush the samples to form a fine powder and dissolve a known amount of food powder in 100 mL distilled water, placed at room temperature for 20 min to swell to form a suspension.

After, 1 h of stirring at 60°C, centrifuge the suspension at 4500 rpm, at 4°C for 20 min. Remove the supernatant and test for acrylamide concentration, under standard conditions of 25 mL sodium acetate buffer, pH 5.0 at 20°C and by applying a potential of 0.26 V. Add the sample (100 µL) to 25 mL sodium acetate buffer (pH 5.0) and measure the current at 0.26 V at a scan rate of 20 mV s^{-1} vs Ag/AgCl. Determine the acrylamide content in the sample from a standard curve. Evaluate the biosensor by investigating its various parameters such as linearity (0.05–100 n*M*), LOD (0.1 n*M*)%, analytical recovery (99% and 98%), and precision (3.85% and 4.67%), and correlate with HPLC analysis.

7. CONCLUSIONS AND FUTURE PERSPECTIVE

Enzyme sensors are reported here, and these sensors are simple, fast, sensitive, and used for the detection of various analytes. ENPs-based biosensors have exhibited good analytical performance in terms of detection limit and shelf life. In future, ENPs could be employed for designing biosensors, biomedical devices, biofuel cells, and enzyme-based reactors. These functionalized ENPs could be used as tags for bioaffinity assays of protein and DNA. Such biosensor technology may open new opportunities for biosensors, clinical diagnosis, and medical management.

CONFLICT OF INTEREST

All authors of this chapter have no conflict of interest.

REFERENCES

Aggarwal, V., Malik, J., Prashant, A., Jaiwal, P. K., & Pundir, C. S. (2016). Amperometric determination of serum total cholesterol with nanoparticles of cholesterol esterase and cholesterol oxidase. *Analytical Biochemistry, 500*, 6–11.

Batra, B., Lata, S., Sharma, M., & Pundir, C. S. (2012). An acrylamide biosensor based on immobilization of haemoglobin onto multiwalled carbon nanotubes/copper nanoparticles/polyaniline hybrid film. *Analytical Biochemistry, 433*, 210–217.

Blanchette, C., Lacayo, C. I., Fischer, N. O., Hawang, M., & Thelen, M. P. (2012). Enhanced cellulose degradation using cellulose nanospheres. *PLoS One*, 7, 42–116.

Cammann, K. (1977). Biosensors based on ion-selective electrodes. *Fresenius Journal of Analytical Chemistry, 287*, 1–9.

Cete, S., Yaşar, A., & Arslan, F. (2006). An amperometric biosensor for uric acid determination prepared from uricase immobilized in polypyrrole film. *Artificial Cells, Blood Substitutes, and Immobilization Biotechnology, 34*(3), 367–380.

Chauhan, N., Kumar, A., & Pundir, C. S. (2014). Construction of an uricase nanoparticles modified Au electrode for amperometric determination of uric acid. *Applied Biochemistry and Biotechnology, 174*, 1683–1694.

Chawla, S., Rawal, R., Sonia, R., & Pundir, C. S. (2013). Preparation of cholesterol oxidase nanoparticles and their application in amperometric determination of cholesterol. *Journal of Nanoparticle Research, 15*, 1934–1943.
Chen, Y., Xiao, L., Liu, Y., Li, X., Zhanh, J., & Shu, Y. (2014). A lipase-based electrochemical biosensor for target DNA. *Microchimica Acta, 181*, 615–621.
Coester, C. J., Langer, K., & Von, B. (2000). Gelatin nanoparticles by two step desolvation: A new preparation method, surface modification and cell uptake. *Journal of Microencapsulation*, (2), 187–193.
Dagar, K., & Pundir, C. S. (2017). An improved amperometric L-lactate biosensor based on covalent immobilization of microbial lactate oxidase onto carboxylated multiwalled carbon nanotubes/copper nanoparticles/polyaniline modified pencil graphite electrode. *Enzyme and Microbial Technology, 96*, 177–186.
Deshapriya, I., Stromer, B. S., Kim, C. S., Patel, V., Gutkind, J. S., & Kumar, C. V. (2015). Novel, protein-based nanoparticles: Improved half-lives, retention of protein structure and activities. *Bioconjugate Chemistry, 26*, 396–404.
Elif, B. B., & Mustafa, S. K. (2015). Applications of commercial biosensors in clinical, food, environmental and biothreat/biowarfare analyses. *Analytical Biochemistry, 478*, 107–120.
Ezpeleta, I., Irchae, J. M., Stainmesses, S., Chabenat, C., Popineau, Y., & Orecchionic, A. M. (1999). Preparation of Ulex europaeus lectin-gliadin nanoparticle conjugates and their interaction with gastrointestinal mucus. *International Journal of Pharmaceutics, 191*, 25–32.
Freitas, D. D. S., Spencer, P. J., Vasao, R. C., & Abrahao-Neto, J. (2009). Biochemical and biopharmaceutical properties of PEGylated uricase. *International Journal of Pharmaceutics, 387*, 215–222.
Guo, M., Chen, J., Li, J., Nie, L., & Yao, S. (2004). Carbon nanotubes-based amperometric cholesterol biosensor fabricated through layer-by-layer technique. *Electroanalysis, 16*, 1–4.
Kalia, V., Goyal, L., & Pundir, C. S. (1999). Determination of serum glucose with alkyl amine glass bound glucose oxidase. *Chinese Journal of Biotechnology, 14*, 201–204.
Khosravi, A., Vossoughi, M., Sharokhian, S., & Alemzadeh, I. (2012). In *Synthesis and stability evaluation of HRP single enzyme nanoparticles. The proceedings of international conference on nanostructures (ICNS4)* (pp. 857–869).
Kim, J., & Grate, J. W. (2003). Single enzyme nanoparticles armored by a nanometer scale organic/inorganic network. *Nano Letters, 3*, 1219–1222.
Kundu, N., Yadav, S., & Pundir, C. S. (2012). Preparation and characterization of glucose oxidase nanoparticles and their application in DO metric determination of serum glucose. *Journal of Nanoscience and Nanotechnology, 13*, 1710–1716.
Liu, G., Lin, Y., Ostatna, V., & Wang, J. (2005). Enzyme nanoparticles based electronic biosensor. *Chemical Communications*, (27), 3481–3483.
Narwal, V., Yadav, N., Thakur, M., & Pundir, C. S. (2017). An amperometric H_2O_2 biosensor based on hemoglobin nanoparticles immobilized onto a gold electrode. *Bioscience Reports, 37*, 1–10.
Nikalje, A. P. (2015). Nanotechnology and its applications in medicine. *Medicinal Chemistry, 5*, 81–89.
Pundir, C. S. (2015). Introduction to enzyme and nanotechnology. In A. Cooper (Ed.), *Enzyme nanoparticles: Preparation, characterization, properties, and applications (micro and nanotechnology series)*. Oxford, UK: Elsevier.
Pundir, C. S., & Aggarwal, V. (2017). Amperometric triglyceride bionanosensor based on nanoparticles of lipase, glycerol kinase, glycerol-3-phosphate oxidase. *Analytical Biochemistry, 517*, 56–63.
Roco, M. C. (2003). Converging science and technology at the nanoscale: Opportunities for education and training. *Nature Biotechnology, 21*, 1247–1249.

Sailaja, A. K., Amareshwar, P., & Chakravarty, P. (2011). Different techniques used for the preparation of nanoparticles using natural polymers and their application. *International Journal of Pharmacy and Pharmaceutical Sciences*, *3*, 45–50.

Scheffel, U., Rhodes, B. A., Natrajan, T. K., & Wagner, H. N. (1972). Albumin microspheres for study of reticuloendothelial system. *Journal of Nuclear Medicine*, *13*, 498–503.

Sharma, S., Shrivastav, A., Gupta, N., & Srivastava, S. (2011). In *Amperometric biosensor: Increased sensitivity using enzyme nanoparticles. 2010 international conference on nanotechnology and biosensors, IPCBEE (Vol. 2)* (pp. 24–26).

Shukla, S. K., Deshpande, S. R., Shukla, S. K., & Tiwari, A. (2012). Fabrication of a tunable glucose biosensor based on zinc oxide/chitosan-graft-poly(vinylalcohol) core-shell nanocomposite. *Talanta*, *99*, 283–287.

Shukla, S. K., Mishra, A. K., Mamba, B. B., & Omotayo, A. A. (2014). Zirconia-poly (propylene imine) dendrimer nanocomposite based electrochemical urea biosensor. *Enzyme and Microbial Technology*, *66*, 48–55.

Silva, G. A. (2004). Introduction to nanotechnology and its applications to medicine. *World Neurosurgery*, *61*, 216–220.

Stromer, B. S., & Kumar, C. V. (2016). White-emitting protein nanoparticles for cell-entry and pH sensing. *Advanced Functional Materials*, *27*, 1603874–1603879.

Turner, A., Karube, I., & Wilson, G. S. (Eds.), (1987). *Biosensors: Fundamentals and applications*. Oxford, UK: Oxford University Press p. V.

Valipour, M. (2015). *Handbook of environmental engineering problems*. Foster City, CA: OMICS Group eBook, pp. 1–77.

Varma, R., & Nene, S. (2003). Biosynthesis of cholesterol oxidase by Streptomyces lavendulae NCIM 2421. *Enzyme and Microbial Technology*, *33*, 286–291.

Wang, X., Piro, B., Reisberg, S., Anquetin, G., Rocquigny, H. D., Jiang, P., et al. (2014). Direct, reagentless electrochemical detection of the BIR3 domain of X-linked inhibitor of apoptosis protein using a peptide-based conducting polymer sensor. *Biosensors & Bioelectronics*, *61*, 57–62.

Yadav, N., Chhillar, A. K., & Pundir, C. S. (2018). Preparation, characterization and application of haemoglobin nanoparticles for detection of acrylamide in processed foods. *International Journal of Biological Macromolecules*, *107*, 1000–1013.

FURTHER READING

Bahadir, E. B., & Mustafa, S. K. (2015). Applications of commercial biosensors in clinical, food, environmental and biothreat/biowarfare analyses. *Analytical Biochemistry*, *478*, 107–120.

Choudhary, M., Shukla, S. K., Taher, A., Siwal, S., & Kaushik, M. (2014). Organic–inorganic hybrid supramolecular assembly: An efficient platform for nonenzymatic glucose sensor. *ACS Sustainable Chemistry & Engineering*, *2*, 2852–2858.

CHAPTER SEVEN

Encapsulation of Microorganisms, Enzymes, and Redox Mediators in Graphene Oxide and Reduced Graphene Oxide

Orr Schlesinger, Lital Alfonta[1]
Department of Life Sciences and Ilse Katz Institute for Nanoscale Science and Technology, Ben-Gurion University of the Negev, Beer-Sheva, Israel
[1]Corresponding author: e-mail address: alfontal@bgu.ac.il

Contents

Abstract

Graphene oxide (GO) and reduced graphene oxide (rGO) were demonstrated in the past decade as biocompatible carbon-based materials that could be efficiently used in bioelectrochemical systems (BESs). Specifically, for redox enzyme encapsulation in order to improve electron communication between enzymes and electrodes. The addition of GO to different solvents was shown to cause gelation while still allowing small molecule diffusion through its gel-like matrix. Taking the combination of these traits together, we decided to use GO hydrogels for the encapsulation of enzymes displayed on the surface of yeast in anodes of microbial fuel cells. During our studies we have followed the changes in the physical characteristics of GO upon encapsulation of yeast cells displaying glucose oxidase in the presence of glucose and noted that GO is being rapidly reduced to rGO as a function of glucose concentrations. GO reduction under these conditions served as a proof of electron communication between the surface-displayed

Methods in Enzymology, Volume 609
ISSN 0076-6879
https://doi.org/10.1016/bs.mie.2018.05.008

enzymes and GO. Hence, we set out to study this phenomenon by the encapsulation of a purified glucose dehydrogenase (in the absence of microbial cells) in rGO where improved electron transfer to the electrode could be observed in the presence of phenothiazone. In this chapter, we describe how these systems were technically constructed and characterized and how a very affordable matrix such as GO could be used to electrically wire enzymes as a good replacement for expensive mediator containing redox active polymers commonly used in BESs.

ABBREVIATIONS

BES bioelectrochemical system
ET electron transfer
GCE glassy carbon electrode
GDH glucose dehydrogenase
GOx glucose oxidase
PVDF polyvinylidene difluoride
SDCAA synthetic dextrose casamino acids
SGCAA synthetic galactose casamino acids
YSD yeast surface display

1. INTRODUCTION

Since the renewed interest in graphene in 2004 (Novoselov, 2004) and its derivative materials: graphene oxide (GO) (Dikin et al., 2007) and reduced graphene oxide (rGO) (Gómez-Navarro et al., 2007), interactions between these carbon-based sp^2- and sp^3-hybridized materials and biological molecules were of a great interest (Lu, Yang, Zhu, Chen, & Chen, 2009; Salas, Sun, Lüttge, & Tour, 2010; Zhou, Zhai, & Dong, 2009). The interest lies in the fact that GO and rGO can serve as a convenient and a cheap matrix for the encapsulation of proteins (Wang, Zhou, Zhang, Boey, & Zhang, 2009), microorganisms (Mohanty & Berry, 2008), cells (Wang et al., 2010), and viruses (Chen et al., 2012) for multiple applications such as biosensing (Kuila et al., 2011), bioelectrochemical systems (BESs) (Teymourian, Salimi, & Khezrian, 2013; Unnikrishnan, Palanisamy, & Chen, 2013), drug delivery, (Liu, Robinson, Sun, & Dai, 2008; Sun et al., 2008), stem cells differentiation (Lee et al., 2011), and more. Although inorganic, GO is graphene that contains oxygen on its surface in the form of epoxides, hydroxyls, and carboxylic groups. GO has shown biocompatibility and allowed adhesion of biological molecules, specifically proteins to its irregular/unordered sheets (Liu, Fu, Yuan, Li, & Deng, 2010), a characteristic that was not observed with the insoluble graphene (Russo & Passmore, 2014). In addition, due to its increased hydrophilicity compared to graphene, it is water soluble, hence allows facile work with biologically compatible solutions (Liu, Yu, Zeng, Miao, & Dai, 2010). Our experience with the encapsulation of enzymes (Ravenna et al., 2015) or microbially surface-displayed enzymes (Bahartan et al., 2012) in GO has shown that not only is it a biocompatible material but also redox enzymes can reduce (depending on the redox potential of the enzyme) GO to rGO in the presence of an appropriate substrate. Once GO has been transformed to rGO it enables more facile electron transfer (ET) (Iwase, Ng, Ishiguro, Kudo, & Amal, 2011), hence aids in the electron communication between a redox enzyme and the electrode on which it is immobilized. We hypothesized that the unique combination of graphene ET properties and the oxygen-based functional groups that exist in GO and rGO allows for the combination of properties needed in enzyme-based BESs. Herein, we give several examples and describe the methodology of enzyme encapsulation in GO gel or GO film, either for a free purified enzyme or a surface-displayed enzyme as well as the enzymatically mediated reduction of GO. In the example that we show in this report, yeast displaying glucose oxidase (GOx) on their surface using the yeast surface display (YSD) system were encapsulated in a GO hydrogel. The ability of modified yeast to reduce GO to rGO by glucose assimilation while maintaining viability was tested in time and deemed suitable for biofuel-cell applications (Bahartan et al., 2012). Reduction of GO to rGO by the surface-displaying yeast cells was compared to the ability of *Shewanella oneidensis* MR1 (*S. oneidensis*) an exoelectrogenic bacteria capable of electron respiration to external inorganic electron acceptors, to reduce GO to rGO and have shown superior abilities of ET to GO in the presence of glucose. In the absence of glucose no apparent ET between yeast cells and GO could be detected. From these results it could be concluded that the surface-displayed enzyme GOx can efficiently deliver electrons to GO upon oxidation of glucose (Fig. 1A). Another system that we have built and reported was a biocomposite of rGO film modified with phenothiazone that was generated in situ from phenothiazine by a redox reaction between phenothiazine and GO during which phenothiazine was oxidized and GO was reduced to rGO (Ravenna, 2015). In this composite, the glucose-oxidizing enzyme, glucose dehydrogenase (GDH) was entrapped in GO. Hence, a biocatalytically glucose-oxidizing active rGO film has resulted (Fig. 1B). The biosensor that was made of this unique biocomposite was highly specific

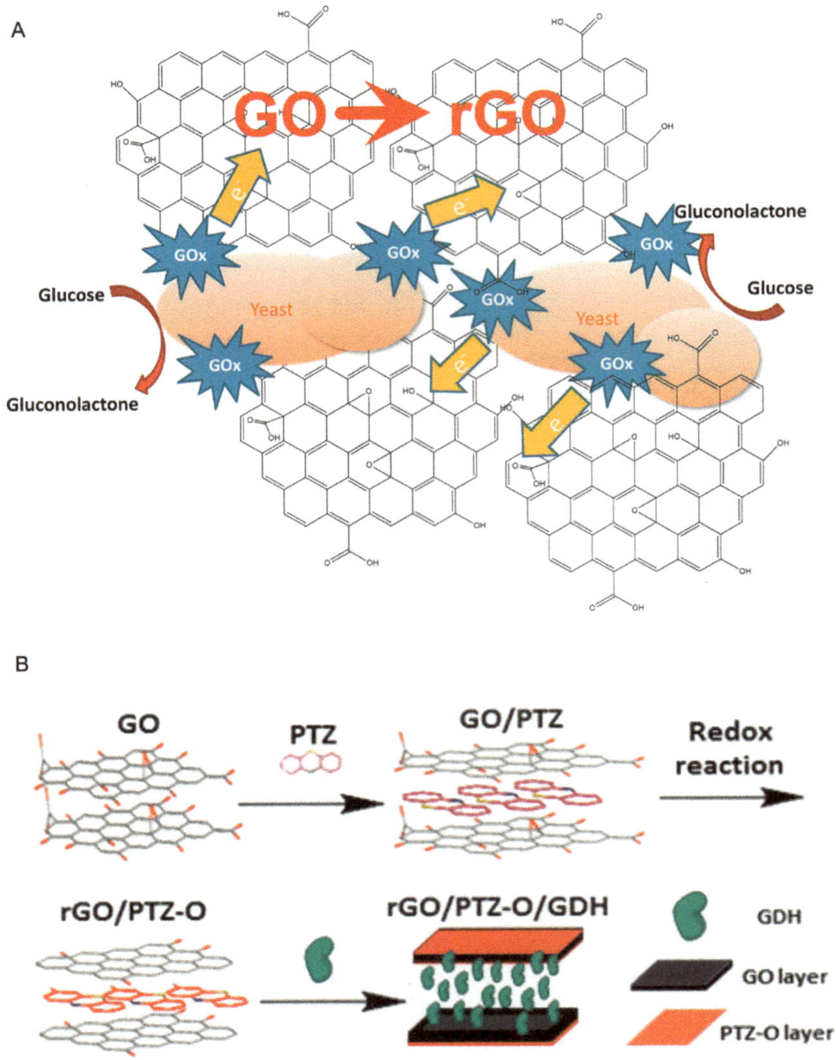

Fig. 1 (A) GOx displayed on the surface of yeast enables the enzyme to directly reduce GO into rGO with electrons gained by oxidation of glucose into gluconolactone. (B) GO is reduced to rGO by redox reaction with phenothiazine. The resultant phenothiazone enables direct electronic communication between rGO and encapsulated GDH. *Panel B was adapted with permission from Ravenna, Y., Xia, L., Gun, J., Mikhaylov, A. A., Medvedev, A. G., Lev, O., et al. (2015). Biocomposite based on reduced graphene oxide film modified with phenothiazone and flavin adenine dinucleotide-dependent glucose dehydrogenase for glucose sensing and biofuel cell applications.* Analytical Chemistry, *87(19), 9567–9571. Copyright (2015) American Chemical Society.*

to glucose due to the low redox potential of the biocomposite that is a result of the combined presence of phenothiazone and rGO. An enzyme-based anode of a biofuel cell that was constructed from an electrode coated with this biocomposite was very efficient with a very high power output due to the large surface area of the active rGO and high enzyme loading.

2. METHODS

2.1 Synthesis of GO

GO was prepared from inexpensive, nontoxic, and abundant raw materials (Sladkevich et al., 2012). The oxidation of graphene thin sheets in order to introduce oxygen-containing functional groups enabled the material to form a hydrogel in both aqueous solutions and organic solvents at very low concentrations (often less than 1% w/v) (Bahartan et al., 2012; Xia, Ravenna, & Alfonta, 2015). Many GO synthesis methods were created since GO was first synthesized in 1859 by Brodie who has oxidized Ceylon graphite using potassium chlorate and fuming nitric acid (Brodie, 1859; Staudenmaier, 1898). To date, the most commonly used GO synthesis protocols are based on a modified Hummers method (Hummers & Offeman, 1958). The Hummers method uses oxidation mixture of concentrated sulfuric acid and potassium permanganate in order to produce GO from pristine graphene sheets. This method was utilized in multiple recent publications to produce various composite materials containing GO: GO containing metal nanoparticles (Atar, Eren, Yola, Karimi-Maleh, & Demirdögen, 2015; Teymourian et al., 2013; Zhou, Huang, et al., 2009), GO functionalization with organic or metallic compounds (Bai, Li, Wang, & Shi, 2010; Dreyer, Park, Bielawski, & Ruoff, 2010; Huang, Bai, Li, & Shi, 2011; Iwase et al., 2011; Ravenna, 2015), and encapsulation of enzymes (Liu, Guo, Duan, Dong, & Gao, 2017; Liu, Yu, et al., 2010; Zhang, Zhang, et al., 2010) and microorganisms (Bahartan et al., 2012; Salas et al., 2010; Xia et al., 2015).

Equipment

- Hot plate with a magnetic stirrer
- Ice bath
- Ultrasonic bath
- Filter paper

Buffers and reagents

- Exfoliated graphite (xGnP, e.g., Sigma-Aldrich product # 900394)
- Potassium persulfate ($K_2S_2O_8$)

- Phosphorus pentoxide (P_2O_5)
- Concentrated sulfuric acid (H_2SO_4)
- Potassium permanganate ($KMnO_4$)
- 30% Hydrogen peroxide (H_2O_2)
- 1:10 Diluted HCl

Procedure

1. Add 1 g of exfoliated graphite to a solution of 1.67 g $K_2S_2O_8$ and 1.67 g P_2O_5 in 8 mL concentrated H_2SO_4.
2. Heat the mixture to 80°C for 4.5 h on a hot plate.
3. After cooling to RT, dilute the mixture with 350 mL deionized water followed by filtering.
4. Wash the preoxidized material with deionized water and dry overnight at 60–70°C.
5. Redisperse the dried material in 40 mL concentrated H_2SO_4 in an ice bath.
6. Slowly add 5 g of potassium permanganate ($KMnO_4$) while stirring the solution to avoid overheating.
7. Stir for 2 h at 35°C.
8. Slowly dilute with 80 mL of deionized water in an ice bath and stir for additional 2 h.
9. Add 200 mL of deionized water followed by 6 mL of 30% hydrogen peroxide to react with the excess of permanganate. The color of the solution will turn yellow.
10. Filter the product and wash with 100 mL diluted HCl to remove metal impurities.
11. Wash with 300 mL deionized water and dialyze against deionized water to remove any residual acid.
12. Disperse the GO in water by sonication in an ultrasonic bath for 2 h.
13. The final GO concentration in suspension should be 0.99 g per 100 mL solution (~1% w/v).

Notes

Potassium persulfate, phosphorus pentoxide, sulfuric acid, potassium permanganate, hydrogen peroxide, and hydrochloric acid are highly corrosive and therefore appropriate safety measures should be taken! Avoid breathing graphite dust and fumes that may be generated during this procedure!

2.2 Encapsulation of Yeast and Bacteria in GO Hydrogel

GO is very biocompatible compared to other graphene-based materials such as single-walled carbon nanotubes which were shown to damage cells

(Smart, Cassady, Lu, & Martin, 2006). GO biocompatability allows for the encapsulation of microorganisms inside the hydrogel matrix. Encapsulation of microorganisms in the microenvironment of the hydrogel was shown to preserve the viability of the microorganisms and allowed the diffusion of small molecules such as nutrients through the gel matrix and reach the cells (Fig. 2). Moreover, encapsulated microorganisms were shown to interact with the graphene matrix either directly or by using small molecules as mediators (Bahartan et al., 2012; Wang, Qian, Saltikov, Jiao, & Li, 2011).

Equipment

- Tabletop centrifuge

Buffers and reagent

- 100 m*M* phosphate buffer pH 7.0 (or growth medium or other appropriate buffer)
- 1% w/v GO suspension

Procedure

1. Inoculate a single colony of yeast or bacteria in the appropriate media overnight until it reaches an OD_{600} of 1.0. (Equals to ca. 10^7 yeast cells/mL or 10^{10} bacterial cells/mL.) See culture protocol for surface-displayed enzymes in *Saccharomyces cerevisiae* (*S. cerevisiae*) at the notes section of this protocol.
2. Centrifuge at 13,000 × *g* for 1 min for a 1.5 mL sample or 2500 × *g* for 10 min for larger volumes.

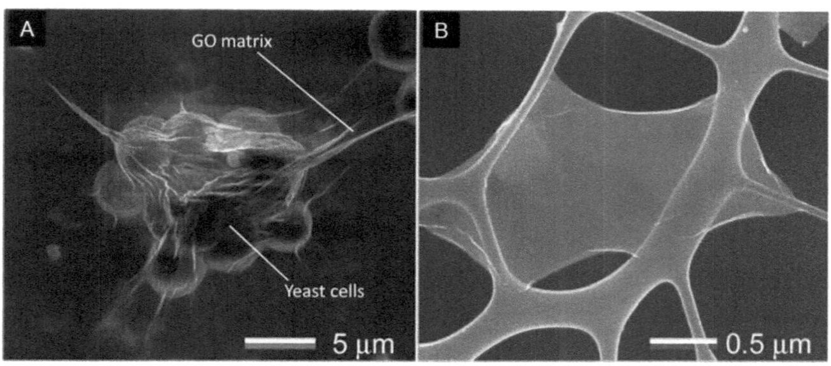

Fig. 2 (A) SEM micrograph of a cluster of *S. cerevisiae* cells in GO wrapping. Graphene connections between proximal, as well as between more distal cells (e.g., on the right) are apparent. (B) SEM micrograph of GO sheets on a lacey carbon support. The multi-layered sheets and the folded GO can be distinguished by different shadings. *Adapted from Bahartan, K., Gun, J., Sladkevich, S., Prikhodchenko, P. V., Lev, O., & Alfonta, L. (2012). Encapsulation of yeast displaying glucose oxidase on their surface in graphene oxide hydrogel scaffolding and its bioactivation.* Chemical Communications, *48(98), 11957–11959, with permission of The Royal Society of Chemistry.*

3. Discard supernatant and resuspend the pellet with half the initial culture volume in 100 m*M* phosphate buffer pH 7 or other buffer or fresh growth medium that is suitable for the microorganism.
4. Suspend 1% GO in the same buffer or medium and combine the cells and GO suspension in equal volumes (1:1), so the final concentration of the cells is the same as in the original culture (ca. 10^7 yeast cells/mL or 10^{10} bacterial cells/mL).

Notes

For enzymes display on yeast surface, the *S. cerevisiae* EBY100 strain should be used. Yeast cultures are grown using synthetic dextrose casamino acids (SDCAA) medium (20 g/L glucose, 6.7 g/L yeast nitrogen base without amino acids, 5 g/L bacto-casamino acids, 6.8 g/L $Na_2HPO_4 \cdot 12H_2O$, 9.6 g/L $NaH_2PO_4 \cdot H_2O$). Induction can be performed using 90% synthetic galactose casamino acids (SGCAA) medium (SDCAA with galactose instead of glucose), 10% SDCAA, supplemented with 3.6 m*M* δ-aminolevulinic acid and 0.2 m*M* ferric citrate (Szczupak, Kol-Kalman, & Alfonta, 2012).

2.3 Encapsulation of Enzymes in GO

Enzymes are very specific, versatile, and efficient biocatalysts (Majewski et al., 2017). The entrapment or encapsulation of enzymes is generally used in order to prevent them from leaking out to the medium, while still allowing an access of the substrate that diffuses freely through the matrix to the enzyme. Enzyme immobilization and encapsulation have also showed in some cases higher stability compared to the free enzyme (Ackermann, Guschin, Eckhard, Shleev, & Schuhmann, 2010; Pita et al., 2011). The encapsulation of enzymes in a hydrogel matrix enables control of the micro-environment in the hydrogel which can be different from conditions in the bulk of the solution, optimizing the environment for enzymatic activity (Betancor & Luckarift, 2008; Zhang, Zhang, & McClements, 2017). Redox enzymes catalyze the reduction and oxidation of many different molecules, extracting or donating electrons in the process. There is especially a great promise in encapsulation of redox enzymes in GO because it is inherently semiconductive and biocompatible. The encapsulation of redox enzymes in a conductive/semiconductive matrix, with which they can communicate electrically, enables efficient direct ET, without loss in enzymatic activity. In order to further improve the communication between the GO matrix and the enzyme, a redox mediator may be coencapsulated along with the enzyme.

Buffers and reagents

- 1% GO suspension
- Buffer compatible with the desired enzyme to be encapsulated
- 1–10 mg/mL enzyme solution dissolved in the appropriate buffer (concentrations should be optimized according to enzyme-specific activity).

Procedure

1. Centrifuge the GO suspension and decant the clear water layer
2. Add the same amount of buffer to produce an equilibrated 1% GO suspension
3. Dissolve the enzyme in the desired buffer at 1–10 mg/mL
4. Mix GO suspension with enzyme solution in equal volumes

Notes

- GO suspension may also be modified with organic molecules or redox mediators in order to improve bioelectrocatalysis or stability of the encapsulated enzymes.
- The final concentrations should be optimized for each enzyme/mediator individually.

2.4 Preparation of Mediator/Nanoparticle-Modified GO Hydrogel

Encapsulation of small redox active molecules inside the hydrogel matrix introduces reversible electrochemical behavior to GO (Ravenna, 2015). Many enzymes or microorganisms are not able to directly communicate with the GO matrix due to the insulating nature of the protein matrix or the cell membrane (Ray et al., 2013). In order to overcome this limitation, small molecules can be used as electron shuttles between the active sites of enzymes and the GO matrix. The redox mediator can help facilitate and improve the electronic communication between graphene sheets of GO and the cells or enzyme catalysts. Many redox mediators are comprised of a system of conjugated aromatic rings (e.g., phenothiazine, ABTS, quinones), which can improve adsorption of the mediator to the graphene sheets of GO by π–π interactions (Zhang et al., 2007). Other molecules can be conjugated to polyaromatic moieties such as pyrene and anthracene in order to improve their adsorption to graphene (Georgakilas et al., 2016). Small molecules or nanoparticles can also be used in GO in order to introduce new optical (Zhou, Huang, et al., 2009), chemical (Brownson & Banks, 2010), and physical (Teymourian et al., 2013) properties and in order to catalyze chemical/biochemical reactions.

Equipment
- Magnetic stirrer
- Tabletop centrifuge
- Dialysis membrane

Buffers and reagents
- 1% w/v GO suspension
- Mediator solution or nanoparticle suspension in water or organic solvent

Procedure
1. This procedure may be used with any redox mediator such as phenothiazine, 2,2′-azino-bis(3-ethylbenzothiazoline-6-sulfonic acid) (ABTS), and various quinones.
2. Centrifuge 35 mL of 1% w/v GO suspension until a separation of GO colloid and clear water occurs.
3. Discard the clear water layer (approximately 20 mL) and add the same volume of mediator solution in water or organic solvent (for phenothiazine 100 mg were dissolved in acetonitrile).
4. Incubate the mixture overnight at RT with constant stirring.
5. Centrifuge and dialyze the solid fraction against deionized water to remove any residual mediator. Replace the diluent in the dialysis vessel twice a day for 2 weeks.
6. Instead of dialysis, the GO/mediator composite can be washed several times with water. Centrifuge the GO/mediator suspension for 20 min or until separation of GO colloid and the water layer. Remove the clear aqueous layer and add the same volume of deionized water. Stir for 30 min at RT. Repeat the wash steps at least five times to remove any excess of mediator.

2.5 Electrode Modification With GO Hydrogel

GO is used to modify various types of electrodes in order to facilitate the harvest of electrons produced by the biocatalysts encapsulated in the hydrogel matrix. The electrode reinforces and protects the fragile GO matrix from breaking by providing a stable solid support. For electrode modification with GO, carbon-based materials should be used such as glassy carbon and graphite plates as rigid supports (Bahartan et al., 2012; Wang et al., 2009). In order to create more foldable, flexible electrodes, carbon cloth, or carbon paper should be used (Dikin et al., 2007; Liu et al., 2012). In addition, carbon/graphene-based materials are superior supports for GO-based materials due to π–π stacking interactions between the electrode support and GO sheets.

2.5.1 Modification of Glassy Carbon Electrodes With GO and GO Composites

Glassy carbon electrodes (GCEs) are very inert both chemically and electrochemically and are relatively reproducible. However, they are quite expensive, therefore are generally used for analytical purposes only (Ravenna, 2015; Unnikrishnan et al., 2013). A GO film can be deposited on the electrode surface in order to reduce or analyze the GO film electrochemically (Fig. 3).

Equipment

- GCE (3 mm diameter, ALS, Tokyo, Japan)
- 0.05 μm alumina slurry and a polishing pad
- Ultrasonic bath

Buffers and reagents

- 1% w/v GO or GO composite suspension

Procedure

1. Polish the GCE with 0.05 μm alumina slurry, rinse with deionized water, and sonicate in ultrasonic bath for 5 min.
2. Dry the electrode under a stream of pure nitrogen.
3. Pipette 10 μL of the GO mix onto the clean electrode and incubate at 4°C for at least 12 h to form a dry GO film.
4. Equilibrate the electrode in a desired electrolyte solution at RT for at least 30 min prior to analysis.

2.5.2 Filtration–Evaporation Method for Carbon Cloth–GO Electrode Fabrication

As GO hydrogel is a soft and flexible material it enables the generation of paper or foil-like electrodes that can be used as chemical filters, battery or electrical components, flexible biosensor electrodes, etc. (Dikin et al., 2007; Li, Müller, Gilje, Kaner, & Wallace, 2008). Carbon cloth is a good scaffold for a GO-based electrode because it reinforces the hydrogel while maintaining its flexibility, thermal, and electrical conductivity. Microorganisms, enzymes, and small molecules can be encapsulated in GO to create a biocatalytic electrode with high selectivity, specificity, and efficiency (Ravenna, 2015) (Fig. 3). Below is a modified version of the GO paper preparation protocol from Dikin et al. (2007).

Equipment

- Vacuum filtration apparatus with sintered glass as support (such as Starlab Scientific product # SLGSF05002)
- Vacuum pump

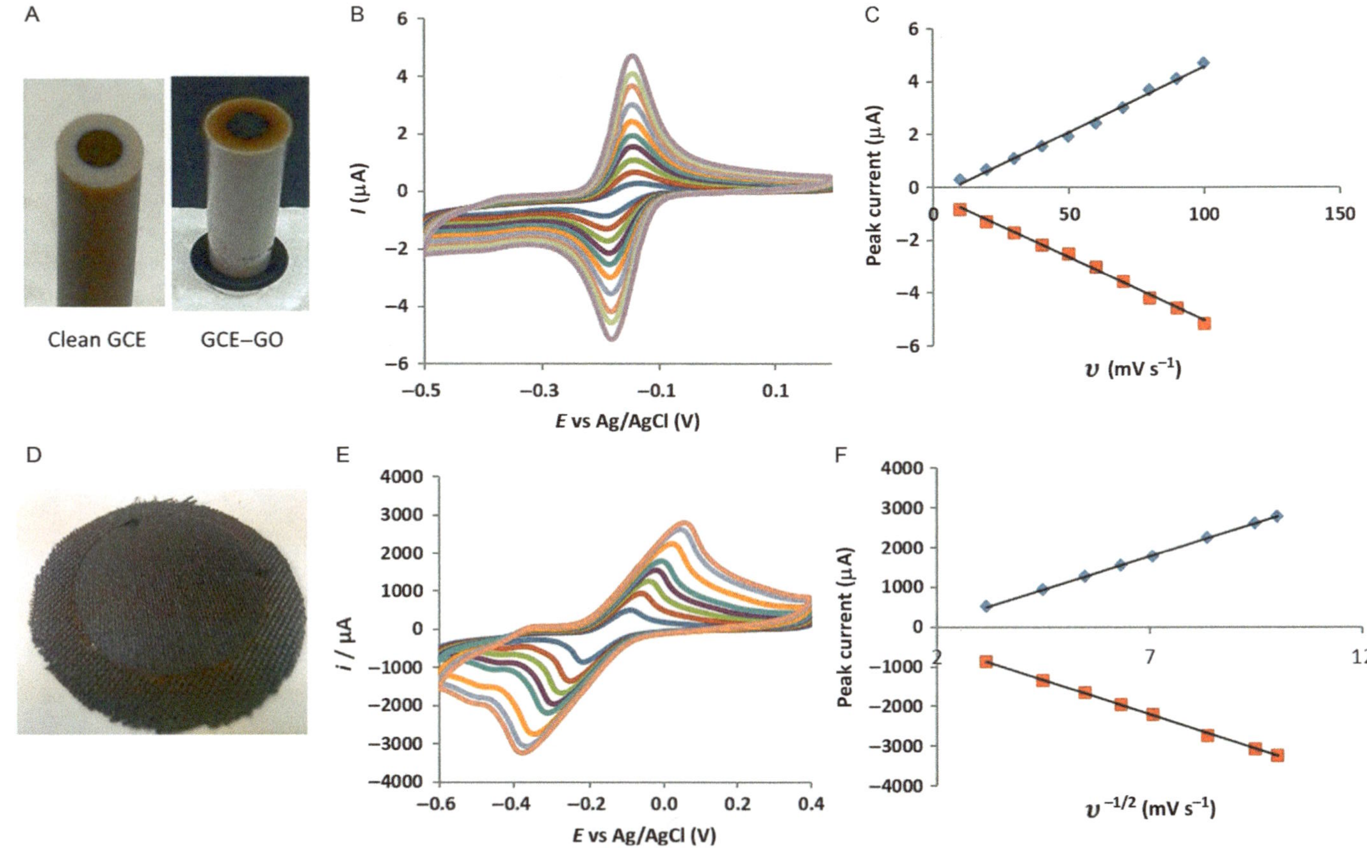

Fig. 3 (A) Image of clean GCE and GO/2,3-dichloro-1,4-naphthoquinone film on GCE; Peak current dependence on scan rate for GO/dCNQ film fabricated on a GCE (B), (C) or on a carbon cloth using filtration–evaporation method (E), (F); (D) image of GO/dCNQ film on carbon cloth. Scan rates: 10–100 mV/s. Graphite rod and Ag/AgCl (3 *M* KCl) were used as auxiliary and reference electrodes, respectively. *Figure was adapted with permission from Ravenna, Y. (2015). Graphene oxide-based biocomposite for highly efficient biosensors and biofuel cells. Ben-Gurion University of the Negev, M.Sc. Thesis.*

Buffers and reagents

- Hydrophilic carbon cloth (ELAT®) (e.g., Fuel cell store product # 1591002)
- *Polyvinylidene difluoride* (PVDF) membrane (0.2 μm pore size)
- 1% w/v GO suspension (can also be achieved with GO suspension containing enzymes, mediators, bacteria, or yeast cells)

Procedure

1. Assemble the filtration apparatus.
2. Cut the PVDF membrane to size and place onto the sintered glass (47 mm diameter).
3. Place the hydrophilic carbon cloth on the PVDF membrane and carefully pour 10 mL of the GO suspension (composite GO suspension with microorganisms, enzymes, or mediators may also be used).
4. Connect the vacuum pump.
5. Leave the vacuum on for approximately 12 h, until the film is dry.
6. Peel off and remove the carbon cloth—GO composite from the filtration apparatus.

Notes

Standard, untreated carbon cloth can be treated for increased hydrophilicity. Incubate the carbon cloth in concentrated nitric acid (HNO_3) at 90°C for 1 h (Warning: warm concentrated nitric acid is extremely corrosive! work with coaution). Then wash the carbon cloth thoroughly with water and dry at 60°C for 24 h. Keep in a desiccator until use. Other methods for carbon cloth treatment can be found in Polovina, Babić, Kaluderović, and Dekanski (1997).

2.6 Reduction of GO to rGO

GO is comprised of planar graphene sheets at random sizes, decorated with hydroxyl, epoxide, diol, and ketone groups which make it more hydrophilic and enables it to form the hydrogel matrix (Becerril et al., 2008). On the other hand, these functional groups disrupt the sp^2 bonds hybridization of graphene and thus π-bonding network that characterize graphene materials (Dreyer et al., 2010). The disruption of the π network of graphene during the oxidation and production of GO leads to a significant reduction in its electrical conductivity. By reducing GO, the π-bonds' delocalized electrons can be partially recovered, regaining some of the thermal, electrical, and mechanical properties of graphene. The reduction of GO to reduced GO (rGO) can be achieved by thermal (McAllister et al., 2007;

Schniepp et al., 2006), chemical (Fan et al., 2008; Stankovich et al., 2007; Zhang, Yang, et al., 2010), electrochemical (Wang et al., 2009; Zhou, Wang, et al., 2009), and bioelectrochemical (Bahartan et al., 2012; Salas et al., 2010; Wang et al., 2011) means. The caveat of the thermal and chemical GO reduction methods is that they utilize high temperatures, extreme pH, and harsh-reducing agents (e.g., hydrazine). These are not compatible with microorganisms and enzymes, thus when working with biological materials in GO, mild electrochemical or bioelectrochemical reduction methods are preferred.

2.6.1 Electrochemical Reduction

As mentioned earlier, GO can be reduced electrochemically using an electrode. This process can be performed in a biocompatible manner by using mild conditions (temperature, pH, aqueous solution, etc.) and with relatively low redox potentials and currents. The advantage of this method is that it is fully controlled by a potentiostat and it is quite rapid (depending on sample size). However, for industrial or preparative scale needed for production of large electrodes used in biosensors or fuel cells, electrochemical reduction may become ineffective. The deposition of rGO on electrodes is likely to make bulk electrochemical reduction difficult on a preparative scale (Dreyer et al., 2010).

Equipment

- GO-modified electrode
- Potentiostat

Buffers and reagents

- Electrolyte buffer solution (e.g., 100 m*M* phosphate buffer pH 7.0)

Procedure

- Bias the potential of the GO-modified electrode to −0.85 V vs Ag/AgCl reference electrode for 200 s (ca. −1.05 V vs NHE)

Notes

The potential and duration of reduction were optimized to minimize damage or activity loss to enzymes or microorganisms encapsulated in the GO hydrogel. This also depends on the redox potential of the enzyme and the nature of the enzyme and may vary between enzymes.

2.6.2 Reduction of GO to rGO by Biocatalysis

GO can be bioelectrochemically reduced under mild conditions. The reduction of GO by an encapsulated biocatalyst, without introduction of electric current, heat, or chemical reductants, improves the chances of microbial cell survival and maintains enzymatic activity. Electrogenic microorganisms such as *S. oneidensis* are able to transfer electrons produced in the cytoplasm through the cellular membrane to an electron acceptor in the outside medium. This makes them good candidates as catalysts for GO reduction. Other non-electrogenic microorganisms can be engineered to express redox enzymes on the cell surface using surface display systems developed for various microorganisms (Gai & Wittrup, 2007; Maurer, Jose, & Meyer, 1997). This enables the expression of various redox enzymes, utilizing various fuels as electron sources, while avoiding the electron and material transfer barrier posed by the cellular membrane (Amir et al., 2013; Fishilevich, Amir, Fridman, Aharoni, & Alfonta, 2009; Gal, Schlesinger, Amir, & Alfonta, 2016; Szczupak et al., 2012). Surface displayed, as well as free GOx, encapsulated in GO hydrogel, may be used as catalysts for GO reduction (Fig. 1A) (Bahartan et al., 2012). In some cases, chemical mediators are coencapsulated with the enzyme in order to facilitate and improve ET efficiency to GO (Ravenna, 2015). The reduction of GO directly by encapsulated biocatalyst may present a practical, cost effective alternative for preparative scale reduction of GO to rGO as no outside current, expensive chemicals, or heat are needed.

2.6.3 GO Reduction by Encapsulated GOx or GOx-Displaying Yeast

Buffers and reagents

- 100 m*M* phosphate buffer supplemented with 100 m*M* glucose for suspension of yeast surface displayed or free GOx
- GO suspension

Procedure

1. Culture the yeast and induce the surface-displayed expression of GOx using SDCAA and SGCAA growth media according to a protocol from Fishilevich et al. (2009).
2. Suspend the cells or the free enzyme in a solution of 100 m*M* glucose.
3. Aseptically encapsulate the yeast displayed or free GOx in GO according to previous protocols. The final number of yeast cells in the hydrogel should be 10^7 cell/mL. The recommended range for the final concentration of free enzyme in the GO hydrogel is between 5 and 50 mU/mL. The final concentration of glucose in the hydrogel, assuming a mixing ratio of 1:1 between GO:GOx/yeast suspension should be ca. 50 m*M*.
4. Incubate the hydrogel at room temperature for 1–5 days, allowing GOx to reduce the GO by oxidizing the glucose present in the medium.

Notes

- Reduction progress can be monitored in samples of GO by X-ray photoelectron spectroscopy (XPS) or Raman spectroscopy (Bahartan et al., 2012) (see Fig. 4).
- For surface-displayed enzymes expression in yeast, the *S. cerevisiae* EBY100 strain should be used. Yeast cultures are grown using SDCAA medium (20 g/L glucose, 6.7 g/L yeast nitrogen base without amino acids, 5 g/L bacto-casamino acids, 6.8 g/L $Na_2HPO_4 \cdot 12H_2O$, 9.6 g/L $NaH_2PO_4 \cdot H_2O$). Induction can be achieved using 90% SGCAA medium (SDCAA with galactose instead of glucose), 10% SDCAA, supplemented with 3.6 m*M* δ-aminolevulinic acid and 0.2 m*M* ferric citrate (Szczupak et al., 2012).

2.6.4 GO Reduction by Encapsulated S. oneidensis

Equipment

- Tabletop centrifuge

Buffers and reagents

- *S. oneidensis* culture at $OD_{600} = 1$
- GO suspension

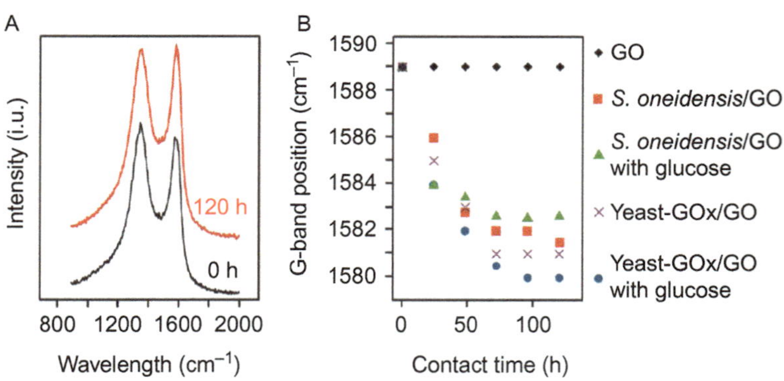

Fig. 4 Characteristic features of the Raman D and G bands as a function of contact time of the GO and the different microorganisms. (A) Raman spectra collected from the sampled Yeast-GOx/GO supplemented with glucose immediately after preparation and after 120 h of GO exposure to Yeast-GOx and glucose. (B) Peak intensity ratios ID/IG (obtained from Raman spectra) vs the contact time between GO and the solution. *Adapted from Bahartan, K., Gun, J., Sladkevich, S., Prikhodchenko, P. V., Lev, O., & Alfonta, L. (2012). Encapsulation of yeast displaying glucose oxidase on their surface in graphene oxide hydrogel scaffolding and its bioactivation.* Chemical Communications, *48(98), 11957–11959, with permission from The Royal Society of Chemistry.*

Procedure

1. Prepare *S. oneidensis* overnight cultures in tryptic soy broth at 30°C with continuous shaking (250 rpm).
2. After reaching $OD_{600nm} = 1.0$ (equivalent to 10^{10} cells per mL), centrifuge the cells and discard the used medium.
3. Resuspend the cell pellet with half the original culture volume of fresh soy broth.
4. Mix the cell suspension with the GO at a ratio of 1:1 according to a previous protocol.
5. Incubate the hydrogel at room temperature for 1–5 days, allowing *S. oneidensis* to reduce GO.

Notes

- Reduction progress can be monitored in samples of GO by XPS or Raman spectroscopy (Bahartan et al., 2012) (see Fig. 4).

2.7 Electrochemical Characterization of Modified Electrodes

GO composite-modified electrodes can be characterized using various electrochemical techniques. The modified electrode is used as a working electrode in a standard three-electrode configuration along with a reference (e.g., Ag/AgCl, calomel) and a counter electrode. Biocatalytic behavior, electrochemical response, electrical impedance, and other characteristics of the modified electrode can be analyzed using various electrochemical methods: Cyclic voltammetry can be performed by scanning the potential window where enzymatic catalysis/redox mediator activity is expected (Fig. 3B) (Bahartan et al., 2012; Ravenna, 2015). The evolution of a bioelectrocatalytic current can be observed and monitored in the presence of the relevant substrate, provided that the enzyme redox active site is properly wired. Correlations between substrate concentrations and measured currents can be made, thus quantification of substrate concentrations may be achieved. Also, the nature of the ET process can be determined by repeating the analysis at various scan rates. By plotting the peak current against the scan rate it can be determined whether the reaction is diffusion limited or not (Fig. 3C) (Ravenna, 2015). Amperometric methods can be utilized to analyze the response of the electrode to the addition of various substrates (for GO-enzyme composite) or compounds (Ravenna, 2015). Electrochemical impedance spectroscopy can be used to determine the electrical impedance and other electronic characteristics of the modified electrodes (Casero et al., 2012). ET rates between the electrode and the GO composite can be analyzed using cyclic voltammetry according to Laviron's theory (Laviron, 1979), or using a more direct, chronoamperometric method (Finklea & Hanshew, 1992; Schlesinger, Pasi, Dandela, Meijler, & Alfonta, 2018).

3. CONCLUDING REMARKS

In this chapter we have shown all useful techniques and methods whether previously published or yet unpublished, used by us to modify electrodes with GO or rGO that encapsulated engineered enzyme surface-displaying yeast, native electrogenic bacteria, or purified redox enzymes, nonspecifically wired through GO, rGO, or redox mediator modified rGO. GO was shown to be an effective semiconducting scaffolding for yeast-displaying GOx on their surface, while maintaining stable power output and yeast viability for at least 15 days. When compared to exoelectrogens, the engineered microorganisms have displayed superior stability over time. Demonstrating the potential of GO/rGO as a nonspecific-wiring matrix for nonelectrogenic-engineered yeast cells in BESs. (Bahartan et al., 2012).

When GO/rGO exists in the hydrogel form, we have noticed diffusional limitations of substrates, we have thus improved these diffusion limitation by developing two methods of fabrication of thin films with GO for enzyme encapsulation along with film formation. We have used a filtration–evaporation method (shown in this chapter), in order to generate a high surface area film on different electrode supports. Using bioelectrocatalysis, we could successfully demonstrate the presence of GDH in the generated film of rGO deposited on a GCE. When phenothiazine was adsorbed in GO, we have noticed a very interesting catalytic process. Phenothiazine was gradually oxidized to the low redox potential mediator phenothiazone by GO while at the same time GO was reduced to rGO. This has generated a biocomposite which was conductive for a suitable and efficient mediation of electrons from glucose-oxidizing enzymes such as GDH. Performance of an anode in a bioelectrochemical cell and as a glucose-sensing device was very promising with high sensitivity, selectivity, and stability. Direct ET has been achieved for GDH encapsulated in a phenothiazone/rGO film (Ravenna et al., 2015).

The ability of rGO to encapsulate enzymes and to form a film contributes to its appeal in future biosensing applications. In addition, the low-operating potential of the systems as well as improved communication of the enzyme active site with rGO or with an adsorbed redox mediator enables the composite to serve as an efficient bioanode and a stable and robust BES.

ACKNOWLEDGMENTS

We would like to thank Prof. Ovadia Lev and Dr. Jenny Gun for a fruitful collaboration in the various projects related to the use of GO and rGO.

FUNDING

This research was supported by the Israel Science Foundation (Grant number 232/13).

REFERENCES

Ackermann, Y., Guschin, D. A., Eckhard, K., Shleev, S., & Schuhmann, W. (2010). Design of a bioelectrocatalytic electrode interface for oxygen reduction in biofuel cells based on a specifically adapted Os-complex containing redox polymer with entrapped Trametes hirsuta laccase. *Electrochemistry Communications*, *12*(5), 640–643. https://doi.org/10.1016/j.elecom.2010.02.019.

Amir, L., Carnally, S. A., Rayo, J., Rosenne, S., Melamed Yerushalmi, S., Schlesinger, O., et al. (2013). Surface display of a redox enzyme and its site-specific wiring to gold electrodes. *Journal of the American Chemical Society*, *135*(1), 70–73. https://doi.org/10.1021/ja310556n.

Atar, N., Eren, T., Yola, M. L., Karimi-Maleh, H., & Demirdögen, B. (2015). Magnetic iron oxide and iron oxide@gold nanoparticle anchored nitrogen and sulfur-functionalized reduced graphene oxide electrocatalyst for methanol oxidation. *RSC Advances*, *5*(33), 26402–26409. https://doi.org/10.1039/C5RA03735B.

Bahartan, K., Gun, J., Sladkevich, S., Prikhodchenko, P. V., Lev, O., & Alfonta, L. (2012). Encapsulation of yeast displaying glucose oxidase on their surface in graphene oxide hydrogel scaffolding and its bioactivation. *Chemical Communications*, *48*(98), 11957–11959. https://doi.org/10.1039/c2cc36959a.

Bai, H., Li, C., Wang, X., & Shi, G. (2010). A pH-sensitive graphene oxide composite hydrogel. *Chemical Communications*, *46*(14), 2376–2378. https://doi.org/10.1039/c000051e.

Becerril, H. A., Mao, J., Liu, Z., Stoltenberg, R. M., Bao, Z., & Chen, Y. (2008). Evaluation of solution-processed reduced graphene oxide films as transparent conductors. *ACS Nano*, *2*(3), 463–470. https://doi.org/10.1021/nn700375n.

Betancor, L., & Luckarift, H. R. (2008). Bioinspired enzyme encapsulation for biocatalysis. *Trends in Biotechnology*, *26*(10), 566–572. https://doi.org/10.1016/j.tibtech.2008.06.009.

Brodie, B. C. (1859). On the atomic weight of graphite. *Philosophical Transactions of the Royal Society of London*, *149*(0), 249–259. https://doi.org/10.1098/rstl.1859.0013.

Brownson, D. A. C., & Banks, C. E. (2010). Graphene electrochemistry: An overview of potential applications. *The Analyst*, *135*(11), 2768–2778. https://doi.org/10.1039/c0an00590h.

Casero, E., Parra-Alfambra, A. M., Petit-Domínguez, M. D., Pariente, F., Lorenzo, E., & Alonso, C. (2012). Differentiation between graphene oxide and reduced graphene by electrochemical impedance spectroscopy (EIS). *Electrochemistry Communications*, *20*, 63–66. https://doi.org/10.1016/J.ELECOM.2012.04.002.

Chen, L., Zhang, X., Zhou, G., Xiang, X., Ji, X., Zheng, Z., et al. (2012). Simultaneous determination of human enterovirus 71 and Coxsackievirus B3 by dual-color quantum dots and homogeneous immunoassay. *Analytical Chemistry*, *84*(7), 3200–3207. https://doi.org/10.1021/ac203172x.

Dikin, D. A., Stankovich, S., Zimney, E. J., Piner, R. D., Dommett, G. H. B., Evmenenko, G., et al. (2007). Preparation and characterization of graphene oxide paper. *Nature, 448*(7152), 457–460. https://doi.org/10.1038/nature06016.

Dreyer, D. R., Park, S., Bielawski, C. W., & Ruoff, R. S. (2010). The chemistry of graphene oxide. *Chemical Society Reviews, 39*(1), 228–240. https://doi.org/10.1039/B917103G.

Fan, X., Peng, W., Li, Y., Li, X., Wang, S., Zhang, G., et al. (2008). Deoxygenation of exfoliated graphite oxide under alkaline conditions: A green route to graphene preparation. *Advanced Materials, 20*(23), 4490–4493. https://doi.org/10.1002/adma.200801306.

Finklea, H. O., & Hanshew, D. D. (1992). Electron-transfer kinetics in organized thiol monolayers with attached pentaammine(pyridine)ruthenium redox centers. *Journal of the American Chemical Society, 114*(9), 3173–3181. https://doi.org/10.1021/ja00035a001.

Fishilevich, S., Amir, L., Fridman, Y., Aharoni, A., & Alfonta, L. (2009). Surface display of redox enzymes in microbial fuel cells. *Journal of the American Chemical Society, 131*(34), 12052–12053. https://doi.org/10.1021/ja9042017.

Gai, S. A., & Wittrup, K. D. (2007). Yeast surface display for protein engineering and characterization. *Current Opinion in Structural Biology, 17*(4), 467–473. https://doi.org/10.1016/j.sbi.2007.08.012.

Gal, I., Schlesinger, O., Amir, L., & Alfonta, L. (2016). Yeast surface display of dehydrogenases in microbial fuel-cells. *Bioelectrochemistry, 112*, 53–60. https://doi.org/10.1016/j.bioelechem.2016.07.006.

Georgakilas, V., Tiwari, J. N., Kemp, K. C., Perman, J. A., Bourlinos, A. B., Kim, K. S., et al. (2016). Noncovalent functionalization of graphene and graphene oxide for energy materials, biosensing, catalytic, and biomedical applications. *Chemical Reviews, 116*(9), 5464–5519. https://doi.org/10.1021/acs.chemrev.5b00620.

Gómez-Navarro, C., Weitz, R. T., Bittner, A. M., Scolari, M., Mews, A., Burghard, M., et al. (2007). Electronic transport properties of individual chemically reduced graphene oxide sheets. *Nano Letters*, 7(11), 3499–3503. https://doi.org/10.1021/nl072090c.

Huang, C., Bai, H., Li, C., & Shi, G. (2011). A graphene oxide/hemoglobin composite hydrogel for enzymatic catalysis in organic solvents. *Chemical Communications, 47*(17), 4962–4964. https://doi.org/10.1039/c1cc10412h.

Hummers, W. S., & Offeman, R. E. (1958). Preparation of graphitic oxide. *Journal of the American Chemical Society, 80*(6), 1339. https://doi.org/10.1021/ja01539a017.

Iwase, A., Ng, Y. H., Ishiguro, Y., Kudo, A., & Amal, R. (2011). Reduced graphene oxide as a solid-state electron mediator in Z-scheme photocatalytic water splitting under visible light. *Journal of the American Chemical Society, 133*(29), 11054–11057. https://doi.org/10.1021/ja203296z.

Kuila, T., Bose, S., Khanra, P., Mishra, A. K., Kim, N. H., & Lee, J. H. (2011). Recent advances in graphene-based biosensors. *Biosensors and Bioelectronics, 26*(12), 4637–4648. https://doi.org/10.1016/j.bios.2011.05.039.

Laviron, E. (1979). General expression of the linear potential sweep voltammogram in the case of diffusionless electrochemical systems. *Journal of Electroanalytical Chemistry and Interfacial Electrochemistry, 101*(1), 19–28.

Lee, W. C., Lim, C. H. Y. X., Shi, H., Tang, L. A. L., Wang, Y., Lim, C. T., et al. (2011). Origin of enhanced stem cell growth and differentiation on graphene and graphene oxide. *ACS Nano, 5*(9), 7334–7341. https://doi.org/10.1021/nn202190c.

Li, D., Müller, M. B., Gilje, S., Kaner, R. B., & Wallace, G. G. (2008). Processable aqueous dispersions of graphene nanosheets. *Nature Nanotechnology, 3*(2), 101–105. https://doi.org/10.1038/nnano.2007.451.

Liu, J., Fu, S., Yuan, B., Li, Y., & Deng, Z. (2010). Toward a universal "adhesive nanosheet" for the assembly of multiple nanoparticles based on a protein-induced reduction/decoration of graphene oxide. *Journal of the American Chemical Society, 132*(21), 7279–7281. https://doi.org/10.1021/ja100938r.

Liu, H., Guo, K., Duan, C., Dong, X., & Gao, J. (2017). Hollow TiO_2 modified reduced graphene oxide microspheres encapsulating hemoglobin for a mediator-free biosensor. *Biosensors and Bioelectronics*, *87*, 473–479. https://doi.org/10.1016/j.bios.2016.08.089.

Liu, J., Qiao, Y., Guo, C. X., Lim, S., Song, H., & Li, C. M. (2012). Graphene/carbon cloth anode for high-performance mediatorless microbial fuel cells. *Bioresource Technology*, *114*, 275–280. https://doi.org/10.1016/j.biortech.2012.02.116.

Liu, Z., Robinson, J. T., Sun, X., & Dai, H. (2008). PEGylated nanographene oxide for delivery of water-insoluble cancer drugs. *Journal of the American Chemical Society*, *130*(33), 10876–10877. https://doi.org/10.1021/ja803688x.

Liu, Y., Yu, D., Zeng, C., Miao, Z., & Dai, L. (2010). Biocompatible graphene oxide-based glucose biosensors. *Langmuir*, *26*(9), 6158–6160. https://doi.org/10.1021/la100886x.

Lu, C., Yang, H., Zhu, C., Chen, X., & Chen, G. (2009). A graphene platform for sensing biomolecules. *Angewandte Chemie*, *121*(26), 4879–4881. https://doi.org/10.1002/ange.200901479.

Majewski, M. B., Howarth, A. J., Li, P., Wasielewski, M. R., Hupp, J. T., & Farha, O. K. (2017). Enzyme encapsulation in metal–organic frameworks for applications in catalysis. *CrystEngComm*, *19*(29), 4082–4091. https://doi.org/10.1039/C7CE00022G.

Maurer, J., Jose, J., & Meyer, T. F. (1997). Autodisplay: One-component system for efficient surface display and release of soluble recombinant proteins from Escherichia coli. *Journal of Bacteriology*, *179*(3), 794–804.

McAllister, M. J., Li, J.-L., Adamson, D. H., Schniepp, H. C., Abdala, A. A., Liu, J., et al. (2007). Single sheet functionalized graphene by oxidation and thermal expansion of graphite. *Chemistry of Materials*, *19*(18), 4396–4404. https://doi.org/10.1021/cm0630800.

Mohanty, N., & Berry, V. (2008). Graphene-based single-bacterium resolution biodevice and DNA transistor: Interfacing graphene derivatives with nanoscale and microscale biocomponents. *Nano Letters*, *8*(12), 4469–4476. https://doi.org/10.1021/nl802412n.

Novoselov, K. S. (2004). Electric field effect in atomically thin carbon films. *Science*, *306*(5696), 666–669. https://doi.org/10.1126/science.1102896.

Pita, M., Gutierrez-Sanchez, C., Olea, D., Velez, M., Garcia-Diego, C., Shleev, S., et al. (2011). High redox potential cathode based on laccase covalently attached to gold electrode. *The Journal of Physical Chemistry C*, *115*(27), 13420–13428. https://doi.org/10.1021/jp203643h.

Polovina, M., Babić, B., Kaluderović, B., & Dekanski, A. (1997). Surface characterization of oxidized activated carbon cloth. *Carbon*, *35*(8), 1047–1052. https://doi.org/10.1016/S0008-6223(97)00057-2.

Ravenna, Y. (2015). *Graphene oxide-based biocomposite for highly efficient biosensors and biofuel cells*. Ben-Gurion University of the Negev.

Ravenna, Y., Xia, L., Gun, J., Mikhaylov, A. A., Medvedev, A. G., Lev, O., et al. (2015). Biocomposite based on reduced graphene oxide film modified with phenothiazone and flavin adenine dinucleotide-dependent glucose dehydrogenase for glucose sensing and biofuel cell applications. *Analytical Chemistry*, *87*(19), 9567–9571. https://doi.org/10.1021/acs.analchem.5b02949.

Ray, S., Chand, S., Zhang, Y., Nussbaum, S., Rajeshwar, K., & Perera, R. (2013). Implications of active site orientation in myoglobin for direct electron transfer and electrocatalysis based on monolayer and multilayer covalent immobilization on gold electrodes. *Electrochimica Acta*, *99*, 85–93. https://doi.org/10.1016/j.electacta.2013.03.080.

Russo, C. J., & Passmore, L. A. (2014). Controlling protein adsorption on graphene for cryo-EM using low-energy hydrogen plasmas. *Nature Methods*, *11*(6), 649–652. https://doi.org/10.1038/nmeth.2931.

Salas, E. C., Sun, Z., Lüttge, A., & Tour, J. M. (2010). Reduction of graphene oxide via bacterial respiration. *ACS Nano*, *4*(8), 4852–4856. https://doi.org/10.1021/nn101081t.

Schlesinger, O., Pasi, M., Dandela, R., Meijler, M. M., & Alfonta, L. (2018). Electron transfer rate analysis of a site-specifically wired copper oxidase. *Physical Chemistry Chemical Physics*, *20*(9), 6159–6166. https://doi.org/10.1039/c8cp00041g.

Schniepp, H. C., Li, J.-L., McAllister, M. J., Sai, H., Herrera-Alonso, M., Adamson, D. H., et al. (2006). Functionalized single graphene sheets derived from splitting graphite oxide. *The Journal of Physical Chemistry B*, *110*(17), 8535–8539. https://doi.org/10.1021/jp060936f.

Sladkevich, S., Gun, J., Prikhodchenko, P. V., Gutkin, V., Mikhaylov, A. A., Medvedev, A. G., et al. (2012). The formation of a peroxoantimonate thin film coating on graphene oxide (GO) and the influence of the GO on its transformation to antimony oxides and elemental antimony. *Carbon*, *50*(15), 5463–5471.

Smart, S. K., Cassady, A. I., Lu, G. Q., & Martin, D. J. (2006). The biocompatibility of carbon nanotubes. *Carbon*, *44*(6), 1034–1047. https://doi.org/10.1016/j.carbon.2005.10.011.

Stankovich, S., Dikin, D. A., Piner, R. D., Kohlhaas, K. A., Kleinhammes, A., Jia, Y., et al. (2007). Synthesis of graphene-based nanosheets via chemical reduction of exfoliated graphite oxide. *Carbon*, *45*(7), 1558–1565. https://doi.org/10.1016/j.carbon.2007.02.034.

Staudenmaier, L. (1898). Verfahren zur Darstellung der Graphitsäure. *Berichte der Deutschen Chemischen Gesellschaft*, *31*(2), 1481–1487. https://doi.org/10.1002/cber.18980310237.

Sun, X., Liu, Z., Welsher, K., Robinson, J. T., Goodwin, A., Zaric, S., et al. (2008). Nano-graphene oxide for cellular imaging and drug delivery. *Nano Research*, *1*(3), 203–212. https://doi.org/10.1007/s12274-008-8021-8.

Szczupak, A., Kol-Kalman, D., & Alfonta, L. (2012). A hybrid biocathode: Surface display of O_2-reducing enzymes for microbial fuel cell applications. *Chemical Communications*, *48*(1), 49–51. https://doi.org/10.1039/C1CC16207A.

Teymourian, H., Salimi, A., & Khezrian, S. (2013). Fe_3O_4 magnetic nanoparticles/reduced graphene oxide nanosheets as a novel electrochemical and bioelectrochemical sensing platform. *Biosensors and Bioelectronics*, *49*, 1–8. https://doi.org/10.1016/j.bios.2013.04.034.

Unnikrishnan, B., Palanisamy, S., & Chen, S.-M. (2013). A simple electrochemical approach to fabricate a glucose biosensor based on graphene–glucose oxidase biocomposite. *Biosensors and Bioelectronics*, *39*(1), 70–75. https://doi.org/10.1016/j.bios.2012.06.045.

Wang, Y., Li, Z., Hu, D., Lin, C.-T., Li, J., & Lin, Y. (2010). Aptamer/graphene oxide nanocomplex for in situ molecular probing in living cells. *Journal of the American Chemical Society*, *132*(27), 9274–9276. https://doi.org/10.1021/ja103169v.

Wang, G., Qian, F., Saltikov, C. W., Jiao, Y., & Li, Y. (2011). Microbial reduction of graphene oxide by Shewanella. *Nano Research*, *4*(6), 563–570. https://doi.org/10.1007/s12274-011-0112-2.

Wang, Z., Zhou, X., Zhang, J., Boey, F., & Zhang, H. (2009). Direct electrochemical reduction of single-layer graphene oxide and subsequent functionalization with glucose oxidase. *The Journal of Physical Chemistry C*, *113*(32), 14071–14075. https://doi.org/10.1021/jp906348x.

Xia, L., Ravenna, Y., & Alfonta, L. (2015). Layer-by-layer assembly of a redox enzyme displayed on the surface of elongated bacteria into a hierarchical artificial biofilm based anode. *Chemical Communications*, *51*(13), 2633–2636. https://doi.org/10.1039/C4CC09781E.

Zhang, Y., Liu, C., Shi, W., Wang, Z., Dai, L., & Zhang, X. (2007). Direct measurements of the interaction between pyrene and graphite in aqueous media by single molecule force spectroscopy: Understanding the pi-pi interactions. *Langmuir: The ACS Journal of Surfaces and Colloids*, *23*(15), 7911–7915. https://doi.org/10.1021/la700876d.

Zhang, J., Yang, H., Shen, G., Cheng, P., Zhang, J., & Guo, S. (2010). Reduction of graphene oxide via L-ascorbic acid. *Chemical Communications*, *46*(7), 1112–1114. https://doi.org/10.1039/B917705A.

Zhang, Z., Zhang, R., & McClements, D. J. (2017). Lactase (β-galactosidase) encapsulation in hydrogel beads with controlled internal pH microenvironments: Impact of bead characteristics on enzyme activity. *Food Hydrocolloids*, *67*, 85–93. https://doi.org/10.1016/j.foodhyd.2017.01.005.

Zhang, J., Zhang, F., Yang, H., Huang, X., Liu, H., Zhang, J., et al. (2010). Graphene oxide as a matrix for enzyme immobilization. *Langmuir*, *26*(9), 6083–6085. https://doi.org/10.1021/la904014z.

Zhou, X., Huang, X., Qi, X., Wu, S., Xue, C., Boey, F. Y. C., et al. (2009). In situ synthesis of metal nanoparticles on single-layer graphene oxide and reduced graphene oxide surfaces. *The Journal of Physical Chemistry C*, *113*(25), 10842–10846. https://doi.org/10.1021/jp903821n.

Zhou, M., Wang, Y., Zhai, Y., Zhai, J., Ren, W., Wang, F., et al. (2009). Controlled synthesis of large-area and patterned electrochemically reduced graphene oxide films. *Chemistry—A European Journal*, *15*(25), 6116–6120. https://doi.org/10.1002/chem.200900596.

Zhou, M., Zhai, Y., & Dong, S. (2009). Electrochemical sensing and biosensing platform based on chemically reduced graphene oxide. *Analytical Chemistry*, *81*(14), 5603–5613. https://doi.org/10.1021/ac900136z.

CHAPTER EIGHT

Chemical and Biochemical Approach to Make a Perfect Biocatalytic System on Carbonaceous Matrices

Paulina Bolibok*, Katarzyna Roszek†, Marek Wiśniewski*,[1]
*Faculty of Chemistry, Physicochemistry of Carbon Materials Research Group, Nicolaus Copernicus University in Toruń, Toruń, Poland
†Department of Biochemistry, Faculty of Biology and Environmental Protection, Nicolaus Copernicus University in Toruń, Toruń, Poland
[1]Corresponding author: e-mail address: marek.wisniewski@umk.pl

Contents

Abstract

Enzymatic processes are widely used in food industry, pharmacy, cosmetic and household chemistry, and medicine. However, the common and efficient application of the biological catalysts is limited by a number of factors that influence enzymes activity. One of the most frequent methods to improve the biocatalysts' properties is immobilization. This chapter presents a recent overview of our attempts to obtain the perfect biocatalytic system. The experimental approach, proposed in this chapter, includes the critical points like: the choice of adequate immobilization method, most suitable carrier, determination of enzyme kinetic parameters, stability, and toxicity of obtained systems. As carbon materials including graphene-derived materials offer unique properties and a plenty of different modifications, these parameters seem to be of decisive importance

Methods in Enzymology, Volume 609
ISSN 0076-6879
https://doi.org/10.1016/bs.mie.2018.05.015

to understand chemistry of complex systems. Consideration of all the mentioned requirements lead us to the conclusion that graphene oxide could be the best candidate for support in perfect biocatalytic systems.

Highlights

- Graphene oxide is proposed to be the perfect matrix for immobilization of enzymes
- The specific, strong interactions of proteins with surface oxygen functionalities appeared
- The slight conformational changes increase the catalytic activity as well as pH and temperature resistance

1. INTRODUCTION

Enzymes are known from their unique catalytic and physicochemical properties, and therefore they are used as biocatalysts in complex chemical processes. The growing interest in biocatalysis and its abundant applications in many branches of life, industry, and science are also indisputably linked with specific properties of enzymes. At present, enzymatic processes are widely used in food industry, pharmacy, cosmetic and household chemistry, and medicine (Cowan & Fernandez-Lafuente, 2011; Wang, Wan, Liu, Huang, & Xu, 2009). The potency of enzymes is used in production of biosensors and biofuel cells, as well as in proteomic analysis. Biocatalysis is also extensively studied in numerous research projects, resulting in the growing number of articles providing the current information (Fig. 1).

The wide application of the biological catalysts is limited by a number of factors, including:

- *mild conditions* of enzyme-mediated reactions and processes (Eijsink et al., 2004), while many chemical processes require nonphysiological conditions,
- *high costs* of enzymatic processes due to inability of catalyst effective separation from obtained products and its reuse (Asgher, Shahid, Kamal, & Iqbal, 2014; Betancor & Luckarift, 2008; Cao, Van Langen, & Sheldon, 2003; Sheldon, 2007),
- *lack of stability* during action as well as during storage, enzymes requires special conditions (Huisman & Collier, 2013; Rekuć, Bryjak, Szymańska, & Jarzębski, 2009; Sanchez-Otero, Valerio-Alfaro, Garcia-Galindo, & Oliart-Ros, 2008; Wang, Wan, et al., 2009; Wang, Yang, et al., 2009). Biocatalysts with low stability have to be changed frequently and therefore, generate higher costs (Eijsink et al., 2004).

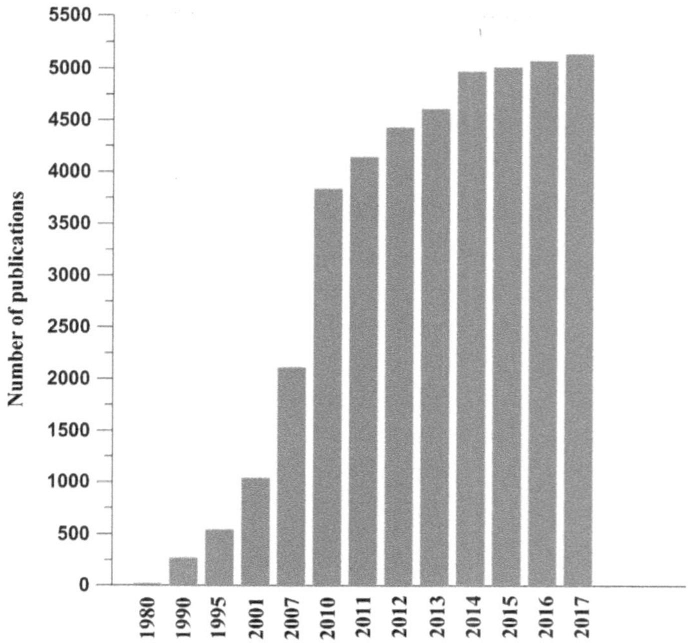

Fig. 1 The number of publications on "biocatalysis" according to Scopus database.

The reasons enumerated above have promoted plentiful studies to improve the enzymes stability and resistance, together with maintaining their considerable catalytic activities. Enzymes isolated from extremophiles or genetically modified organisms are the first choice solution to overcome their instability, e.g., at higher temperatures. The engineered enzymatic proteins are modified in terms of specificity, activity, selectivity, and stability at extreme conditions. However, extremophile-derived enzymes are rather expensive while genetic and protein modifications do not solve all problems limiting the use of enzymes at high-scale industrial applications, e.g., the issue of biocatalysts recovery and reuse (Eijsink et al., 2004; Sheldon, 2007). The good solubility of enzymes in water environment can be a serious obstacle, limiting their repeated use (Betancor & Luckarift, 2008).

At the heart of intensely explored technologies for enzyme improvement is immobilization. This strategy is widely described as beneficial in terms of enzyme properties modification (Rekuć et al., 2009; Sheldon, 2007; Tang, Saquing, Sarin, Kelly, & Khan, 2014; Wu, Yuan, & Sheng, 2005). The commonly listed advantages of immobilization are easy enzyme recovery and reuse, stable activity in numerous work cycles, enzyme stability at extreme temperatures, pH, as well as in the presence of organic solvents. A large selection of carriers (natural, synthetic, mixed) and a variety of techniques (adsorption, inclusion, microencapsulation, covalent bonding, cross-linking) provide the possibility of targeted modification of the catalytic and physicochemical properties of enzymes.

1.1 Methods of Immobilization

An unambiguous categorization of immobilization methods is a very difficult challenge. In literature, one can find more or less exhaustive classifications. The simplest of them is the division into methods with or without a solid support (Cao et al., 2003) (Fig. 2). Regardless of the presence of the enzyme carrier, all the methods aim to modify enzyme properties such as: activity and substrate affinity, resistance to thermal denaturation, and stability in the broad spectrum of pH, targeted enantioselectivity, and extended half-life. In the recent years, immobilization has gained a new meaning and

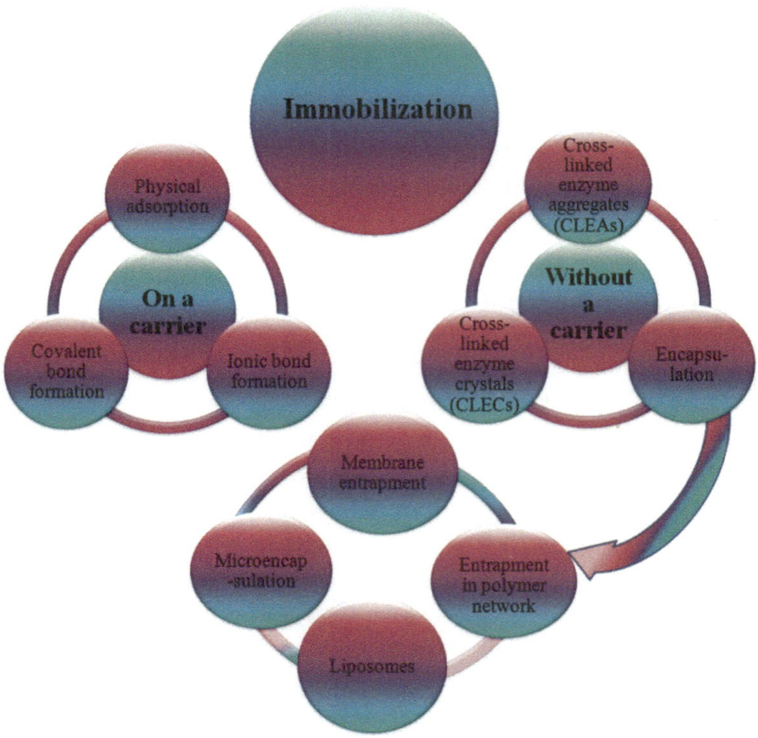

Fig. 2 Schematic representation of immobilization methods.

usage in enzymatic technology, since growing attention has been paid to the modifications of biocatalysts' physicochemical and kinetic properties. These goals can be achieved by enzyme immobilization on appropriately synthesized and/or functionalized supports (Brena & Batista-Viera, 2006; Es, Vieira, & Amaral, 2015; Homaei, Sariri, Vianello, & Stevanato, 2013).

Protein adsorption on the carrier surface is a very attractive application of a variety of different materials. It is the mildest method of immobilization, and therefore gives the greatest opportunity to preserve the native and active form of the biological particle. The immobilization process does not involve chemical reactions, but only the spontaneous formation of van der Waals, hydrophobic, or electrostatic interactions between the protein and the carrier. These interactions allow to protect the enzyme active center (Bhakta, Evans, Benavidez, & Garcia, 2015; Liu, 2016). In addition, the attractiveness of the technique is increased by simplicity, low cost, and the reversibility of the adsorption process (Sanchez-Otero et al., 2008). The last advantage means that after deactivation of the enzyme and separation of the product, both the carrier and the enzyme can be recovered and reused (Betancor & Luckarift, 2008).

Unfortunately, this method has one major disadvantage. The immobilized protein is in equilibrium with the solution and the enzyme to surface binding is weak, therefore the desorption of the enzyme cannot be negligible. Moreover, the enzymatic protein desorption can increase when the pH of the solution changes or when introducing competing molecules (e.g., detergents and others) into solution (Betancor & Luckarift, 2008; Liu, 2016; Sheldon, 2007; Sheldon & van Pelt, 2013). The interactions between enzyme and carrier can be strengthened (Bhakta et al., 2015), inter alia by cross-linking with glutaraldehyde treatment (Liu, 2016), to minimize desorption.

In general, the process of adsorption includes several stages of varying duration. At first, enzyme is adsorbed from basic solution to the interphase, then attached to the carrier surface, relaxed, and finally detached from the surface and desorbed back into the solution (Bhakta et al., 2015). As the adsorption progresses, more enzyme molecules adhere to the first layer of the enzymatic protein. From the perspective of further utilization of the emerging structures, it is most efficient to obtain an adsorbate monolayer. Despite the aforementioned simplicity of the process, many factors influence its course. These include, among others, temperature, pH of the solution, the size of the adsorbed protein (the smaller the particles, the better adsorption), the shape of both carrier and catalyst, and the ability to form hydrogen bonds (Bhakta et al., 2015).

The combination of enzyme and carrier through covalent binding, in the literature is referred to as the strongest bond that can be produced between a protein and a matrix (Wang & Hsieh, 2008). The covalent connection between enzyme and carrier can be achieved with two different methods involving appropriate chemical reactions. The first of these methods is direct binding of the enzyme to the carrier by means of a covalent bond. A much more effective method is the binding of the catalyst using a cross-linking reagent, which on the one hand binds to the enzyme and to the carrier surface, on the other (Asgher et al., 2014; Rekuć et al., 2009).

There is no agreement as to the safety of enzyme immobilization through covalent bonding. Conflicting sets of results can be found in the literature; some of them confirm that it is a very safe method, and thanks to the use of this technique, enzyme molecules become stiffer, which prevents molecular motions that lead to conformational changes and inactivation of catalysts (Betancor & Luckarift, 2008). On the other hand, this method often requires the use of toxic chemicals, such as carbodiimides. There are literature reports proving that in the presence of carbodiimides the enzymatic activity decreases, which in turn reduces the efficiency of industrial processes (Harris, Epting, & Kelly, 2010). Moreover, due to the strength of the binding formed, it is not possible to recover the carrier in the incident of enzyme deactivation (Sheldon & Van Pelt, 2013).

The most characteristic properties of different immobilization methods are summarized and compared in Table 1. The advantages and disadvantages of enzymes are described in terms of their applicative potential in technological processes. Even a superficial analysis clearly shows that the most valuable complexes are formed with biospecific as well as ionic interactions, due to their strength, specificity, versatility, and a weak influence on protein conformation. It is worth to notice that despite the commonly described "pure" physical and chemical immobilization methods, also strong physical adsorption and/or weak chemical interactions are possible.

1.2 Examples of the Most Promising Immobilization Matrices

Regardless of the immobilization method, growing attention has been paid to the possibility of modifying physicochemical and kinetic properties of biocatalysts through immobilization on tailored substrates. One can envision, that the choice of the carrier on which the enzyme is to be immobilized, is the key stage of the process. The interaction between the protein and substrate affects not only the immobilization efficiency

Table 1 The Properties of Physical and Chemical Methods of Enzyme Immobilization

Method	Immobilization Mechanism	Bonding Character	Influence on the Enzyme Structure	Reversibility	Substrates
Adsorption	Physical adsorption	Weak physical interactions (van der Waals, hydrophobic)	No	Yes	High- and low-molecular
Inclusion	Inclusion in polymer matrix				Low-molecular
Microencapsulation	Core–shell				
Strong physical adsorption	Strong physical/weak chemical adsorption	Ion-, biospecific-, electrostatic-, affinity interactions	Yes/No	Yes/No	High- and low-molecular
Cross-linking	Cross-linking	Strong chemical (covalent)	Yes	No	Low-molecular
Covalent bonding	Strong covalent interactions				High- and low-molecular

but also the thermal and chemical stability of the biocatalyst (Bernal, Escobar, Wilson, Illanes, & Mesa, 2014; Wang, Wan, et al., 2009; Wang, Yang, et al., 2009) or even its enzymatic activity and kinetics (Bolibok, Wiśniewski, Roszek, & Terzyk, 2017). Although the first reports on the enzyme immobilization on solid supports date from 1950 and many materials have been analyzed so far in terms of the possibility of their use in biocatalysis, one single universal substrate has still not been found (Prlainović et al., 2016).

The intended use of carrier–enzyme biocatalytic systems imposes a variety of requirements on the substrate. From the biotechnological point of view, the most commonly desired features of the carrier are (Ansari & Husain, 2012; Bernal et al., 2014; Cipolatti et al., 2014):

- large surface area
- mechanical strength
- biocompatibility
- possibility to regenerate
- bacterial resistance
- easy to obtain/synthesize
- low cost

The support serves to stabilize the enzyme, so it should have the largest possible surface area available for the biocatalyst. Besides its surface area, also volume, shape, pore size, and stability at different conditions influence the carrier final application. The cost of the support is particularly important in relation to the overall costs of the process. Significant savings are possible, e.g., when the support may regenerate and can be used repeatedly. The only exception may be the small-scale production of high-quality products (e.g., for medical applications), where expensive carriers are acceptable (Es et al., 2015; Liu, 2016). The number of supports that satisfy all the above-mentioned requirements is still rather modest, although rapidly growing.

The most essential point of immobilizing the enzyme on any support is the interaction between functional groups of the biocatalyst and matrix. Enzymatic proteins consist of amino acids with different functional groups in their side chains, e.g., amino group of lysine, thiol group of cysteine, or hydroxyl group of serine. These groups are involved in interactions with the functional groups of the matrix, forming covalent bond (Es et al., 2015). Group affinity and type of bonds determine the durability and reversibility of the biocatalytic system. A reversible adsorption system with predominantly weak physical bonds guarantees less stability than that created by strong covalent bonds. In some cases, there is no possibility to link the enzyme directly to the matrix and an additional compound is required, e.g., carbodiimide. Two functional groups of the linker enable simultaneous interaction with the enzyme and the matrix. Some substrates are enriched with additional, e.g., magnetic, compounds such as iron oxide. Magnetic supports enable the recovery of the biocatalytic system from the reactor by applying only magnetic field without decanting, centrifugation, or filtration. Other additives may be manganese dioxide or titanium dioxide, which removes the hydrogen peroxide produced by oxidases, thus protecting the enzyme from oxidative inactivation (Zucca & Sanjust, 2014).

In order to efficiently bind enzymatic protein and improve its physicochemical properties, the matrix can be activated and modified. Of all possible matrices, certain carbon structures have the ability to smoothly modify the surface by adding or removing appropriate functional groups. For example, covering the carbon matrix with functionalities increasing its surface charge (predominantly at high ionic strength conditions) reduces effectively the desorption of the bound enzyme.

Carbon nanomaterials are increasingly being used in biocatalysis according to the ability of shape and size control, low prices compared to most other raw materials, as well as relatively high biocompatibility (Bernal et al., 2014; de Poulpiquet, Ciaccafava, & Lojou, 2014). Among numerous carbonaceous structures the single- and multiwalled carbon nanotubes (Filip, Sefćovićova, Gemeiner, & Tkac, 2013; Karachevtsev et al., 2011; Krishnan & Armstrong, 2012; Lee, Shin, Kang, Park, & Kim, 2011a, 2011b; Miyake, Yoshino, Yamada, Hata, & Nishizawa, 2011; Prlainović et al., 2016; Zebda et al., 2011), soot (de Poulpiquet et al., 2014), carbon nanoparticles (Szot, Watkins, Bull, Marken, & Opallo, 2010), graphene (Liang et al., 2013; Zheng et al., 2010), mesoporous carbon foams (Guo, Hu, Lou, & Li, 2010; Wang, Yang, et al., 2009), and nanodiamonds (Wei, Zhang, Lu, & Yang, 2010) deserve special attention.

Nanodiamonds are particularly interesting due to their possible biomedical applications. Their representative features include high stability, chemical resistance, nontoxic properties to biological organisms, and biocompatibility. The rapidly growing number of literature data proves the nanodiamonds' ability to conjugate with biologically active molecules like enzymes, DNA (Ushizawa et al., 2002), or hormones (Cheng et al., 2007). Wei et al. (2010) have immobilized trypsin on the nanodiamonds surface. The immobilized enzyme exhibited higher catalytic activity, stability over time and resistance to acids, bases, and ethanol. Furthermore, immobilization of β-galactosidase on the glutaraldehyde-modified support resulted in 95% catalytic activity of native enzyme (Ansari et al., 2015). According to the authors, this is the best efficiency achieved and the slight decrease in the catalytic properties was explained with low active center availability for the substrate.

An important factor affecting the possibility of using the enzyme on an industrial scale is its stability over time and under various conditions such as as temperature and pH. Ansari et al. (2015) showed that β-galactosidase immobilized on the surface of nanodiamonds is characterized by a broad spectrum of both pH and temperature resistance. Owing to the above-mentioned properties and to the possibility of easy enzyme separation from the reaction mixture, the obtained catalytic systems proved to be very attractive for large-scale applications. However, nanodiamonds have one disadvantage that cannot be ignored—their high cost.

Among numerous obtained carbonaceous nanomaterials, carbon nanotubes (CNT) are a promising immobilization support since their structure allows the easy recovery of the carrier from other reagents by simple filtration (Prlainović et al., 2016). In addition, CNT compared with other nanoparticles exhibit environmental stability, effortless synthesis, and ability to acquire high amounts of protein on their surface.

A confirmation of these advantages can be found in literature, from which it results that nanotubes are the most popular material for enzyme immobilization. However, we assume that this is not a perfect immobilization matrix. CNT were described as a beneficial support for such enzymes as catalase, lipase, β-galactosidase, and glucose oxidase (Mohamad et al., 2015; Vilian, Chen, & Lou, 2014; Zhang, Luo, & Chen, 2013). Zhang and colleagues studied kinetic parameters of catalase immobilized on single- and multiwalled carbon nanotubes (Vilian et al., 2014). The major differences can be noticed in the efficiency of enzyme adsorption. The enzymatic protein was immobilized most efficiently on oxidized single-walled carbon nanotubes. This is explained by the presence of large number of oxygen functional groups (OH, COOH, etc.) on the surface of the carrier. The smallest amount of catalyst was adsorbed on the surface of multiwalled carbon nanotubes, which may be due to the presence of a considerable number of micropores. However, regardless of the amount of immobilized catalase, each sample was characterized by lower activity than the native enzyme. On the other hand, the resulting biocatalytic systems showed higher affinity (lower K_m values) to the substrate than the nonimmobilized catalase. The above results are elucidated by changes in the protein structure, which has been proved by the FTIR analysis (Zhang et al., 2013). Another set of results was presented by Ansari et al. The authors focused on β-galactosidase immobilization on multiwalled CNT (Ansari, Satar, Chibber, & Khan, 2013). The biocatalytic system was considerably less active than native protein, and the enzyme specificity decreased. On the other hand, the obtained complexes were characterized by more stable β-galactosidase activity in a wide range of both pH and temperature, than the native enzyme. According to the authors, the obtained biocatalytic systems can be successfully used in industrial scale processes, however, further research and analysis is necessary. According to our research experience, Ansari's study (Ansari et al., 2013) suffers from a lack of biocatalytic systems biocompatibility evaluation which limits their biomedical applications. This issue will be discussed later.

Graphene and graphene oxide (GO), together with reduced graphene oxide and graphene nanosheets, belong to another group of carbonaceous nanomaterials. Graphene is a single layer of carbon atoms packed into a dense hexagonal network similar to honeycomb. Compared with carbon nanotubes, graphene-based materials can provide a larger surface area and better dispersibility in most solvents. In addition to the wide applications in electronics and semiconductor industry, graphene, due to the large surface, can be used to immobilize biologically active molecules, including enzymes or anticancer drugs (Cipolatti et al., 2014). Graphene is also applicable in biosensors. Regrettably, these applications represent a still poorly developed part of nanomedicine, which challenges scientists and physicians (Feng & Liu, 2011).

Graphene oxide, which may be a graphene precursor, attracts a huge research interest of the scientific community. We postulate that this material is a good support for proteins and/or enzymes. It is described as highly stable support with large surface and oxygen functional groups on it. Therefore, immobilization method does not require surface modification (Li, Wen, Shi, & Zheng, 2016; Zhou, Jiang, Gao, Zhao, Ma, et al., 2012). Moreover, until now, GO was used as a matrix for enzyme immobilization in biosensors, biofuel cells, and in targeted drug delivery systems (Hernandez-Cancel et al., 2015).

Despite the enormous potential of graphene oxide, literature reports on its impact on the catalytic properties of immobilized molecules are still insufficient. The published results are difficult to compare with each other because they either refer to different molecules or the material was obtained by different methods. Research on bilirubin oxidase immobilized on GO sheets allowed to determine the effect of chemical glycosylation and immobilization on the catalytic properties of the enzyme (Hernandez-Cancel et al., 2015). It was shown that the glycosylation process decreases catalytic properties of the enzyme, while increasing its thermal stability due to conformational changes of the catalyst. After immobilization of the glycosylated enzyme, the above changes were even more evident. According to the authors, the increased stability of the immobilized enzyme favors such complexes for application in nanobiotechnology (Hernandez-Cancel et al., 2015). Zhou et al. reported that also glucose oxidase immobilization declines enzyme catalytic properties. V_{max} value decreases as compared to the native catalyst, and K_m value increases with lower affinity for glucose (Zhou, Jiang, Gao, Zhao, Ma, et al., 2012). These differences can be explained with the conformational changes in the protein structure. Moreover, the immobilized glucose oxidase is more stable than its native form. The positive effect of enzyme immobilization on the graphene oxide surface on the stability of its catalytic properties has also been demonstrated by Zhang and collaborators. Immobilized horseradish peroxidase (HRP) retained its activity over a wide pH range. In addition, it exhibited improved stability over time and better resistance to high temperatures than the native enzyme. However, the publication lacks results on the influence of immobilization on the kinetic parameters of HRP (Zhang et al., 2010). Wei and Ge (2013) and their study focused on the GO influence on conformation and activity of immobilized enzyme. The enzyme on GO surface exhibited conformational changes: the percentage of α-helix structure decreased, while β-sheets increased. The protein structure is therefore relaxed and its catalytic activity declines. The changes depend on the support used and the immobilization (adsorption) time. Another explanation of the reduced enzyme activity is the possibility that GO interacts with functional groups directly in the catalytic center, however, the mechanism of this phenomenon was unknown.

Besharati Vineh, Saboury, Poostchi, Rashidi, and Parivara (2018) reported that HRP immobilization on the reduced GO surface, obtained by modified Hummer's method (Lee et al., 2014), positively affects the properties of the enzymatic protein, i.e., increases the catalytic efficiency. The formed covalent bonds result in higher substrate affinity and higher activity of immobilized enzyme compared to its native form. It was suggested that the observed phenomena was due to changes in the secondary structure of the protein. The immobilization on RGO caused a reduction in the amount of α-helices and parallel increase in the percentage of β-structures. The conformational changes influence enzyme catalytic properties and stability at varying environmental conditions. Adsorbed HRP exhibits high catalytic activity over a wider pH and temperature range as compared to the native form. The authors emphasize the fact that immobilized HRP maintains its activity at a satisfactory level even through 10 catalytic cycles. Authors also point to the strength of the covalent bonding, which prevents the loss of the enzyme during rinsing the enzyme–matrix conjugates between cycles. The activity reduction after 10 cycles is explained by the accumulation of products rather than inactivation.

The immobilization of HRP on the graphene-derived surfaces was studied by Zhang, Chen, Alvarez, and Chen (2015). Based on circular dichroism spectra, the authors proved that adsorption in this case is very strong and initiates changes in the secondary structure of the protein. The authors conclude that the stronger is the interaction, the more considerable conformational changes it triggers.

Hermanová, Zarevúcká, Bouˇsa, Mikulics, and Sofer (2016) proved high potential of physical conjugation of lipase with GO for use in the pharmaceutical or food industry. The authors have demonstrated increased resistance of obtained systems to organic solvents as well as high temperatures. Mathesh et al. (2016) focused on determining the effect of hydrophobicity of the carrier surface on conformational changes in secondary protein structure. Authors supported their experiment with computer simulations, which are not very common in this area. They concluded that GO reduction leads to diminishing the amount of oxygen groups, and therefore, increasing its hydrophobicity. Immobilization of the protein on hydrophobic surface leads to a decrease in the number of α-helices, followed by relaxing the conformation, and subsequent increase in enzyme activity.

Physical adsorption of glucose oxidase on GO surface decreased the enzyme sensitivity to temperature and pH changes (Zhou, Jiang, Gao, Zhao, & Ma, 2012). Parallel, with the improvement of protein stability, the specificity of the enzyme to the substrate decreases. The authors indicate that surface oxygen groups of GO make it a perfect material serving as a matrix for mild but effective immobilization through physical adsorption, which will not cause significant conformational changes in the structure of immobilized proteins.

Zhou, Jiang, Gao, Zhao, Ma, et al. (2012) compared changes in catalytic properties of glucose oxidase as well as enzyme stability caused by physical and chemical immobilization. It was not a surprise that more stable and resistant to changing environmental conditions was the enzyme immobilized via covalent bond. In both cases, the immobilized enzyme showed a higher K_m value than the native one, resulting in a decreased affinity to glucose (Zhou, Jiang, Ma, He, & Gao, 2015). The authors explained the observed phenomena by the occurrence of conformational changes in the secondary structure of the protein, but they did not try to explore this issue. It was explained in the work of Bolibok et al. (2017) in which the authors correlate changes in the secondary protein structure with differences in its catalytic properties. These changes strongly depend on the amount of enzyme adsorbed on the GO surface. At low surface coverage, there is a substantial decrease in the number of α-helices and a simultaneous increase in β-turn content. It results in relaxing the protein structure and in increasing its activity. Oppositely, higher concentration of enzyme on the carrier surface allows to obtain biocatalytic systems with activity similar to the native enzyme, however, with much better stability.

The group of Srivastava (Das, Talat, Srivastava, & Kayastha, 2018; Srivastava, Singh, Talat, Srivastava, & Kayastha, 2014) examined the adsorption of β-amylase on functionalized GO surface. They proved that enzyme has upgraded thermal and pH tolerance as well as the possibility of multiple reuse with maintaining a reasonable catalytic activity. Also Pang, Li, and Zhang (2015) determined the effect of laccase immobilization (physical adsorption) on the surface of various carbon nanomaterials. Authors showed that only in the case of the enzyme immobilized on GO, the value of K_m remained at a level comparable to the one obtained for the native protein. For other tested materials, these values have increased many times, which indicates a lower catalyst affinity to the substrate. Similar to other reports, the observed phenomena are explained by conformational changes in the protein structure.

Kishore, Talat, Srivastava, and Kayastha (2012) demonstrated that β-galactosidase immobilized on GO surface significantly improves its catalytic properties. The conjugated enzyme was characterized by a lower K_m value (higher affinity to lactose) and higher catalytic activity. Similar results were obtained by Singh et al. (2014) and Satar, Ismail, Rehan, and Ansari (2016) who concluded that immobilization triggered also widening of the pH range tolerance. Conjugated enzyme activity was very high in slightly acidic environment, while in the basic one, its activity was even higher as compared to the native protein. The authors explained it with specific interactions between the protein and the matrix.

Table 2 Fitted Parameters of Bimodal Langmuir–Freundlich Equation (R^2 Is the Value of Determination Coefficient Showing the Quality of the Fit of Theoretical Model to Experimental Data)

Sample	K_1 (mL/mg)	n_1	K_2 (mL/mg)	n_2	R^2
GO	1.170	0.417	0.158	2.518	0.9998

According to the recent results from our laboratory, binding efficiency on the graphene oxide surface is independent of enzyme class but depends rather on protein to matrix mass ratio (Bolibok et al., 2017). At lower coverage of GO surface, the enzyme kinetic parameters increase drastically, whereas at higher surface protein concentrations the values decrease to the level of native protein. The same correlations were observed for catalase, β-galactosidase, and adenylate kinase (Hetmann et al., 2018). Moreover, the storage stability of obtained biocatalytic systems rises for all studied enzymes as compared with the native enzymatic protein. We assume that strong physical adsorption and/or weak chemical interactions (see Table 2) are crucial for specific and beneficial properties of graphene oxide-based biocatalytic systems. However, the precise mechanism has to be further elucidated.

1.3 Safety Assessment of Biocatalytic System Application

Graphene-related nanomaterials have emerged as promising supports for a wide range of technological and biomedical applications. One of the main requirements to be met by perfect biocatalytic system is its biocompatibility. For many immobilization matrices, there are gaps in our knowledge concerning this issue. To ensure the safe use of the studied systems in biomedicine, the extensive safety assessment regarding their toxicity to human organism as well as to the environment is mandatory.

The most commonly used cytotoxicity assays to evaluate the toxicity of carbonaceous nanomaterials are MTT test, neutral red uptake, LDH membrane integrity assay, generation of reactive oxygen species (ROS), cell adhesion and morphology, apoptosis assay, and genotoxicity (Seabra, Paula, de Lima, Alves, & Durán, 2014). Genotoxicity means the injury of cell genetic material at a chromosome level, a molecular or at the DNA base level. If the damage is irreversible, it can result in cell death or carcinogenesis.

From the carbonaceous materials mentioned above, the worst reputation in the literature is attributed to CNT, what mostly excludes them from applications in the living organisms. Many authors reported that CNT show a significant toxicity in vitro, as tested on variety of different human and animal cells (Werengowska-Ciećwierz et al., 2015; Zhu et al., 2012).

Oppositely, nanodiamonds and graphene oxide exhibit no apparent cytotoxicity, do not interfere with the gene or protein expression and their in vivo toxicity is much smaller than that of CNT (Werengowska-Ciećwierz et al., 2015; Zhu et al., 2012). On the other hand, numerous studies devoted to the evaluation of in vitro and in vivo toxicity, frequently show contradictory results, therefore generalized conclusions must be avoided (Seabra et al., 2014). Safety risk associated with graphene-derived materials depends on several parameters such as the specific material type and dimensions. For example, the higher is graphitization degree, the higher toxicity should be expected. Systematic investigations should be carefully performed to correlate each of the GO properties to the biological event induced and to asses toxicity profile.

2. METHODS

We believe that GO fulfills all the requirements of a good support, thus the below-described methods concern GO, however, could be successfully applied to other supports. We have also focused the methodology on catalase as the model enzyme.

2.1 Synthesis of Graphene Oxide

One of the most frequently used methods of GO synthesis is the modified Hummers method (Krishnamoorthy, Veerapandian, Yun, & Kim, 2013; Remyamol, John, & Gopinath, 2013).

1. Add 50 mL of concentrated H_2SO_4 to graphite flakes (0.175 g).
2. Slowly add $KMnO_4$ powder (2.25 g) to the suspension.
3. Keep the reaction mixture at 25°C and stir for 24 h.
4. Add 5 mL of 30% H_2O_2 in order to stop the oxidation process.
5. Centrifuge the mixture at 8000 × *g* at least, for 5 min and the remaining solid material wash several times with 200 mL of 30% HCl, and subsequently with 500 mL of water. After each wash, the mixture should be centrifuged (15,000 × *g* for 20 min).
6. Freeze-dry the final resulting material for at least 24 h.
7. Before using, dilute GO in deionized water and ultrasonicate for 60 min to obtain the GO solution.

2.2 Immobilization of Catalase on Graphene Oxide and Adsorption Reversibility

For the experimental procedures catalase (EC 1.11.1.6, from bovine liver, tetramer consisting of four equal subunits each of 60kDa) should be dissolved in appropriate concentration in 50mmol/LPBS buffer, pH 7.4.

1. Perform the adsorption of catalase (Cat) at 4°C in the initial concentration range 0.1–5 mg_{cat}/mL and the constant concentration of GO 0.1 mg/mL.
2. After 48h of equilibration, centrifuge each solution at 5000 × *g* for 10min.
3. Measure the concentration of protein in supernatant spectrophotometrically at 280nm, while precipitates should be subjected to enzymatic activity assay.
4. Perform the enzyme desorption: after centrifugation of adsorbed and equilibrated catalytic systems, add fresh portion of PBS to the precipitates.
5. Determine the amount of protein in the supernatant every 3 days for as long as you need.

GO due to the large amount of surface oxygen species possesses drastically higher adsorptive capacity than other carbon materials (Bolibok et al., 2017). Bearing in mind that dry GO measured surface area is lower than 1 m^2/g (based on N_2 low temperature adsorption), it is obvious that after dispersion in water the graphene layers are separated facilitating such high-adsorption ratio of proteins.

Usually the experimental data fit well by the bimodal Langmuir–Freundlich equation. This simple model was presented as applicable to the description of different adsorption data, see e.g., Huang et al. (2014). This model is represented by expansion of the equation proposed by Jeppu and Clement (2012) and Umpleby, Baxter, Chen, Shah, and Shimizu (2001) to bimodal form:

$$Q_{eq} = \frac{A_1 (K_1 C)^{\frac{1}{n_1}}}{1 + (K_1 C)^{\frac{1}{n_1}}} + \frac{A_2 (K_2 C)^{\frac{1}{n_2}}}{1 + (K_2 C)^{\frac{1}{n_2}}} \quad (1)$$

$$A_x = w_x Q_{m,x}; \quad x = 1, 2 \quad (2)$$

where Q_{eq} is the amount adsorbed at equilibrium (mg_{Cat}/mg_C); Q_m is the maximum adsorbed capacity of the system (mg_{Cat}/mg_C); w_x is the weight; C is the protein concentration in solution at equilibrium (mg/mL); K_x ($x = 1,2$) is the affinity constant for adsorption (mL/mg); and n is the index of heterogeneity.

The fitted values of K and n are summarized in Table 2. The constant K is related to the affinity between the adsorbate and the adsorbent. The n parameter is the site heterogeneity index. Value of $n=1$ suggests non-interacting sites, while $0<n<1$ the positive cooperativity, whereas $n>1$ negative cooperativity is expected during adsorption process.

The left part of the table (n_1, K_1) indicates the sites with higher carbon surface to protein affinity. For GO the positive cooperativity is observed. Oppositely, lower carbon-protein affinity for other carbons is described in the literature Bolibok et al. (2017). For GO, $n_2>1$ indicates that negative cooperative adsorption takes place, probably due to steric hindrance caused by multilayer adsorption rather than to nonfavorable electrostatic interactions.

2.3 Enzymatic Activity and Stability

The easiest way to determine the enzymatic activity of Catalase (Cat) is to use H_2O_2 as substrate and incubate at 25°C.

1. Dose of 0.1 mL of Cat solution in PBS (75 μg/mL) add to 2.9 mL H_2O_2 (in concentration range 1–100 mmol/L).
2. Monitor spectrophotometrically the changes in the absorbance at 240 nm continuously during 10 min reaction.
3. In order to determine the storage stability of native and immobilized Cat, store the tested samples at 4°C in 50 mmol/L PBS buffer, pH 7.4 for 21 days.
4. The aliquots from each preparation (0.1 mL) should be taken in triplicates at established time points and analyzed for the remaining activity.
5. The activity of native Cat, determined on the first day should be taken as control (100%) for the calculation of remaining activity.

2.4 The Secondary Structure

The secondary structure is usually determined using circular dichroism, however, we believe that the infrared techniques could be equally useful, if properly performed. The analysis of the amide I band at approximately 1700–1600 cm^{-1} gives information on possible conformational changes of enzyme resulting from the immobilization on GO surface (Tzialla et al., 2010). This band consists of several overlapping components that should be recognized throughout analysis of the second derivative or self-deconvolution. The bands appearing could be assigned to different secondary structure elements (α-helix, β-sheets, β-turns, random coil)

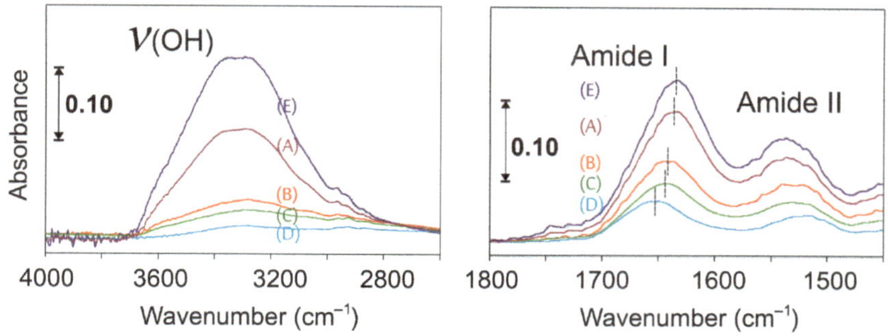

Fig. 3 The OH stretching and amide I and II regions of the FTIR spectra of catalase immobilized on GO; (A)—as received after lyophilization and after vacuum treatment for (B) 15 min, (C) 30 min, (D) 60 min, (E) after H_2O adsorption.

(Natalello, Ami, Brocca, Lotti, & Doglia, 2005; Zhao, Du, Li, & Yang, 2012). Next, the spectral decomposition of native and immobilized enzyme can be performed showing some major changes in the secondary structures.

One of the key problems in the infrared techniques is the presence of water. Its presence disrupts drastically the spectral analysis. As an example, Fig. 3 presents the spectra of immobilized catalase on GO (1/1 m/m). The spectral changes caused by water desorption are the blue-shifting of the amide I band which could be interpreted as α-helix and random coil formation. However, the real reason responsible for the observed phenomena is the diminishing in the intensity of HOH-bending-band at ca. 1640 cm^{-1}. It is proved through red-shifting of amide I band after water readsorption.

2.5 Cytotoxicity Evaluation of Carbon Nanostructures

1. Grow the chinese hamster ovary (CHO) cells to obtain subconfluent monolayer in F-12 medium containing 10% fetal bovine serum, in a CO_2 incubator at 37°C in humidified atmosphere containing 5% CO_2.
2. Twenty four hours before the experiment, seed 5×10^4 cells (in a volume of 25 μL) to each well of a 12-well culture plate.
3. After cell attachment (about 24 h) add GO to the growing cells in the final concentrations of 1, 10, 50, 250, 500, and 1000 μg mL^{-1}, respectively, and incubate for the next 24 h.
4. Perform MTT test based on the ability to reduce 3-(4,5-dimethylthiazol-2-yl)-2,5-diphenyltetrazoliumbromide (MTT) by mitochondrial dehydrogenases in triplicate for assessing cell metabolic activity and viability.

5. Discard the culture medium from tested cells. Prepare the MTT solution ($1\,mg\,mL^{-1}$) in culture medium, add aliquots of 1 mL to every well. Incubate from 20 to 60 min (dependent on the cell density).
6. Discard the MTT solution and add 1 mL DMSO to dissolve the obtained formazan crystals.
7. Measure spectrophotometrically at 570 nm. The viability of cells growing without any supplementation should be taken as a 100% control.

2.6 In Vitro Antioxidant Activity of Biocatalytic Systems

CHO cells could be used as a model to test the functionality and biological activity of obtained Cat/GO complexes, expressed as their antioxidant capacity in the in vitro model. The antioxidant capacity could be defined as the ability to improve viability of CHO cells exposed to ROS.

1. Twenty four hours before the experiment, seed 5×10^4 cells (in a volume of 25 μL) to each well of a 12-well culture plate.
2. After cell attachment (about 24 h), add the tested biocatalytic systems (Cat/GO complexes) to the growing CHO cells in concentrations of $250\,ng\,mL^{-1}$ and $2.5\,\mu g\,mL^{-1}$ together with $500\,\mu M$ H_2O_2 or $100\,\mu M$ tBuOOH and incubate for the next 24 h.
3. Use the MTT test to assess the viability of cells.

3. SUMMARY AND PERSPECTIVES

In general, numerous studies have been performed on graphene oxide applicability for immobilization, and undoubtedly, GO deserves its position as a candidate for immobilization of different enzymes. Specific, strong interactions of proteins with surface oxygen functionalities cause slight conformational changes and thus increase the catalytic activity as well as pH and temperature resistance. Moreover, prepared biocatalytic systems are more stable during working cycles and storage. Their biocompatibility is not to be underestimated, when biomedical application is foreseen.

However, such easily prepared material is in parallel very complex in nature, therefore some caution is needed in interpreting the results. Some researchers do not entirely realize what specific kind of GO-derived material and interactions are they working with. For example, inserting the quartz tube with sample into a tube furnace preheated to 1050°C and keeping it for 30 s can exfoliate the GO samples, due to fast water evaporation, however, it does not cause the formation of graphene. Analogously, bath in ethanol does not form—OH functionalities. Therefore, even when the chemical bonding was intended to form, in fact, simple physical adsorption of enzyme on the modified surface could occur. Undoubtedly, this issue will gain more attention and more detailed approach in future studies.

In summary, taking into account the existing literature data as well as our results, we postulate that GO is tested here as an almost perfect immobilization matrix for all purposes. The only disadvantage, which nowadays should be intensely studied, is the removal of GO-based catalytic systems from the reaction chamber. A reasonable solution seems to synthesize the corresponding magnetic composites.

REFERENCES

Ansari, S. A., & Husain, Q. (2012). Potential applications of enzymes immobilized on/in nano materials: A review. *Biotechnology Advances*, *30*(3), 512–523.

Ansari, S. A., Satar, R., Chibber, S., & Khan, M. J. (2013). Enhanced stability of *Kluyveromyces lactis* β galactosidase immobilized on glutaraldehyde modified multiwalled carbon nanotubes. *Journal of Molecular Catalysis B: Enzymatic*, *97*, 258–263.

Ansari, S. A., Satar, R., Zaidi, S. K., Naseer, M. I., Karima, S., Alqahtani, M. H., et al. (2015). Nanodiamonds as an effective and novel matrix for immobilizing β-galactosidase. *Food and Bioproducts Processing*, *95*, 298–303.

Asgher, M., Shahid, M., Kamal, S., & Iqbal, H. M. N. (2014). Recent trends and valorization of immobilization strategies and ligninolytic enzymes by industrial biotechnology. *Journal of Molecular Catalysis B: Enzymatic*, *101*, 56–66.

Bernal, C., Escobar, S., Wilson, L., Illanes, A., & Mesa, M. (2014). Carbonaceous siliceous composite materials as immobilization support for lipase from *Alcaligenes* sp.: Application to the synthesis of antioxidants. *Carbon*, *74*, 96–103.

Besharati Vineh, M., Saboury, A. A., Poostchi, A. A., Rashidi, A. M., & Parivara, K. (2018). Stability and activity improvement of horseradish peroxidase by covalent immobilization on functionalized reduced graphene oxideand biodegradation of high phenol concentration. *International Journal of Biological Macromolecules*, *106*(2018), 1314–1322.

Betancor, L., & Luckarift, H. R. (2008). Bioinspired enzyme encapsulation for biocatalysis. *Trends in Biotechnology*, *26*(10), 566–572.

Bhakta, S. A., Evans, E., Benavidez, T. E., & Garcia, C. D. (2015). Protein adsorption onto nanomaterials for the development of biosensors and analytical devices: A review. *Analytica Chimica Acta*, *872*, 7–25.

Bolibok, P., Wiśniewski, M., Roszek, K., & Terzyk, A. P. (2017). Controlling enzymatic activity by immobilization on graphene oxide. *Naturwissenschaften*, *104*, 36.

Brena, B. M., & Batista-Viera, F. (2006). Immobilization of enzymes. In J. M. Guisán (Ed.), *Immobilization of enzymes and cells, methods in biotechnology series: Vol. 22.* (pp. 15–30). Humana Press.

Cao, L., Van Langen, L., & Sheldon, R. A. (2003). Immobilised enzymes: Carrier-bound or carrier-free? *Current Opinion in Biotechnology*, *14*(4), 387–394.

Cheng, C.-Y., Perevedentseva, E., Tu, J.-S., Chung, P.-H., Cheng, C.-L., Liu, K.-K., et al. (2007). Direct and in vitro observation of growth hormone receptor molecules in A549 human lung epithelial cells by nanodiamond labelling. *Applied Physics Letters*, *90*, 163903.

Cipolatti, E. P., Silva, M. J. A., Klein, M., Feddern, V., Feltes, M. M. C., Oliveira, J. V., et al. (2014). Current status and trends in enzymatic nanoimmobilization. *Journal of Molecular Catalysis B: Enzymatic*, *99*, 56–67.

Cowan, D. A., & Fernandez-Lafuente, R. (2011). Enhancing the functional properties of thermophilic enzymes by chemical modification and immobilization. *Enzyme and Microbial Technology*, *49*(4), 326–346.

Das, R., Talat, M., Srivastava, O. N., & Kayastha, A. M. (2018). Covalent immobilization of peanut β-amylase for producing industrial nanobiocatalysts: A comparative study of kinetics, stability and reusability of the immobilized enzyme. *Food Chemistry*, *245*, 488–499.

de Poulpiquet, A., Ciaccafava, A., & Lojou, E. (2014). New trends in enzyme immobilization at nanostructured interfacesfor efficient electrocatalysis in biofuel cells. *Electrochimica Acta*, *126*, 104–114.

Eijsink, V. G. H., Bjørk, A., Gaseidnes, S., Sirevag, R., Synstad, B., van den Burg, B., et al. (2004). Rational engineering of enzyme stability. *Journal of Biotechnology*, *113*(1–3), 105–120.

Es, I., Vieira, J. D. L., & Amaral, A. C. (2015). Principles, techniques, and applications of biocatalyst immobilization for industrial application. *Applied Microbiology and Biotechnology*, *99*(5), 2065–2082.

Feng, L., & Liu, Z. (2011). Graphene in biomedicine: Opportunities and challenges. *Nanomedicine*, *6*(2), 317–324.

Filip, J., Sefćovićova, J., Gemeiner, P., & Tkac, J. (2013). Electrochemistry of bilirubin oxidase and its use in preparation of a low cost enzymatic biofuel cell based on a renewable composite binder chitosan. *Electrochimica Acta*, *87*, 366–374.

Guo, C. X., Hu, F. P., Lou, X. W., & Li, C. M. (2010). High-performance biofuel cell made with hydrophilic ordered mesoporous carbon as electrode material. *Journal of Power Sources*, *195*(13), 4090–4097.

Harris, J. M., Epting, K. L., & Kelly, R. M. (2010). N-terminal fusion of a hyperthermophilic chitin-binding domain to xylose isomerase from *Thermotoga neapolitana* enhances kinetics and thermostability of both free and immobilized enzymes. *Biotechnology Progress*, *26*(4), 993–1000.

Hermanová, S., Zarevúcká, M., Bouˇsa, D., Mikulics, M., & Sofer, Z. (2016). Lipase enzymes on graphene oxide support for high-efficiency biocatalysis. *Applied Materials Today*, *5*, 200–208.

Hernandez-Cancel, G., Suazo-Davila, D., Ojeda-Cruzado, A. J., Garcia-Torres, D., Cabrera, C. R., & Griebenow, K. (2015). Graphene oxide as a protein matrix: Influence on protein biophysical properties. *Journal of Nanobiotechnology*, *13*, 70.

Hetmann, A., Wujak, M., Bolibok, P., Zięba, W., Wiśniewski, M., & Roszek, K. (2018). Novel biocatalytic systems for maintaining the nucleotide balance based on adenylate kinase immobilized on carbon nanostructures. *Materials Science and Engineering: C*, *88*, 130–139.

Homaei, A. A., Sariri, R., Vianello, F., & Stevanato, R. (2013). Enzyme immobilization: An update. *Journal of Chemical Biology*, *6*(4), 185–205.

Huang, W., Zhou, X., Xia, Q., Peng, J., Wang, H., & Li, Z. (2014). Preparation and adsorption performance of GrO@Cu-BTC for separation of CO2/CH4. *Industrial & Engineering Chemistry Research*, *53*, 11176–11184.

Huisman, G. W., & Collier, S. J. (2013). On the development of new biocatalytic processes for practical pharmaceutical synthesis. *Current Opinion in Chemical Biology*, *17*(2), 284–292.

Jeppu, G. P., & Clement, T. P. (2012). A modified Langmuir-Freundlich isotherm model for simulating pH-dependent adsorption effects. *Journal of Contaminant Hydrology*, *129–130*, 46–53.

Karachevtsev, V. A., Stepanian, S. G., Glamazda, A. Y., Karachevtsev, M. V., Eremenko, V. V., Lytvyn, O. S., et al. (2011). Noncovalent interaction of single-walled carbon nanotubes with 1-pyrenebutanoic acid succinimide ester and glucoseoxidase. *The Journal of Physical Chemistry C*, *115*(43), 21072–21082.

Kishore, D., Talat, M., Srivastava, O. N., & Kayastha, A. M. (2012). Immobilization of β-galactosidase onto functionalized graphene nano-sheets using response surface methodology and its analytical applications. *PLoS One*, 7(7), e40708.

Krishnamoorthy, K., Veerapandian, M., Yun, K., & Kim, S.-J. (2013). The chemical and structural analysis of graphene oxide withdifferent degrees of oxidation. *Carbon*, *53*, 38–49.

Krishnan, S., & Armstrong, F. A. (2012). Order-of-magnitude enhancement of an enzymatic hydrogen-air fuel cell based on pyrenyl carbon nanostructures. *Chemical Science*, *3*(4), 1015–1023.

Lee, K. H., Lee, B., Hwang, S.-J., Lee, J.-U., Cheong, H., Shin, K., et al. (2014). Large scale production of highly conductive reduced graphene oxide sheets by a solvent-free low temperature reduction. *Carbon*, *69*, 327–335.

Lee, J. Y., Shin, H. Y., Kang, S. W., Park, C., & Kim, S. W. (2011a). Improvement of electrical properties via glucose oxidase-immobilization by actively turning over glucose for an enzyme-based biofuel cell modified with DNA-wrapped single walled nanotubes. *Biosensors and Bioelectronics*, *26*(5), 2685–2688.

Lee, J. Y., Shin, H. Y., Kang, S. W., Park, C., & Kim, S. W. (2011b). Application of an enzyme-based biofuel cell containing a bioelectrode modified with deoxyribonucleic acid-wrapped single-walled carbon nanotubes to serum. *Enzyme and Microbial Technology*, *48*(1), 80–84.

Li, W., Wen, H., Shi, Q., & Zheng, G. (2016). Study on immobilization of (+) γ-lactamase using a new type of epoxy graphene oxide carrier. *Process Biochemistry*, *51*(2), 270–276.

Liang, B., Fang, L., Yang, G., Hu, Y., Guo, X., & Ye, X. (2013). Direct electron transfer glucose biosensor based on glucose oxidase self-assembled on electrochemically reduced carboxyl graphene. *Biosensors and Bioelectronics*, *43*, 131–136.

Liu, S. (2016). Enzymes. In S. Liu (Ed.), *Bioprocess engineering, kinetics, biosystems, sustainability, and reactor design* (pp. 324–371). Elsevier.

Mathesh, M., Luan, B., Akanbi, T., Weber, J. K., Liu, J., Barrow, C. J., et al. (2016). Opening lids: Modulation of lipase immobilization by graphene oxides. *ACS Catalysis*, *6*(7), 4760–4768.

Miyake, T., Yoshino, S., Yamada, T., Hata, K., & Nishizawa, M. (2011). Self-regulating enzyme-nanotube ensemble films and their application as flexible electrodes for biofuel cells. *Journal of the American Chemical Society*, *133*(13), 5129–5134.

Mohamad, N. R., Buang, N. A., Mahat, N. A., Jamalis, J., Huyop, F., Aboul-Enein, H. Y., et al. (2015). Simple adsorption of *Candida rugosa* lipase onto multi-walled carbon nanotubes for sustainable production of the flavor ester geranyl propionate. *Journal of Industrial and Engineering Chemistry*, *32*, 99–108.

Natalello, A., Ami, D., Brocca, S., Lotti, M., & Doglia, S. M. (2005). Secondary structure, conformational stability and glycosylation of a recombinant *Candida rugosa* lipase studied by Fourier-transform infrared spectroscopy. *The Biochemical Journal*, *385*, 511–517.

Pang, R., Li, M., & Zhang, C. (2015). Degradation of phenolic compounds by laccase immobilized on carbon nanomaterials: Diffusional limitation investigation. *Talanta*, *131*, 38–45.

Prlainović, N. Z., Bezbradica, D. I., Rogan, J. R., Uskoković, P. S., Mijin, D. Z., & Marinković, A. D. (2016). Surface functionalization of oxidized multi-walled carbon nanotubes: *Candida rugosa* lipase immobilization. *Comptes Rendus Chimie*, *19*(3), 363–370.

Rekuć, A., Bryjak, J., Szymańska, K., & Jarzębski, A. B. (2009). Laccase immobilization on mesostructured cellular foams affords preparations with ultra high activity. *Process Biochemistry*, *44*(2), 191–198.

Remyamol, T., John, H., & Gopinath, P. (2013). Synthesis and nonlinear optical properties of reduced graphene oxide covalently functionalized with polyaniline. *Carbon*, *59*, 308–314.

Sanchez-Otero, M. G., Valerio-Alfaro, G., Garcia-Galindo, H. S., & Oliart-Ros, R. M. (2008). Immobilization in the presence of Triton X-100: Modifications in activity and thermostability of *Geobacillus thermoleovorans* CCR11 lipase. *Journal of Industrial Microbiology & Biotechnology*, *35*(12), 1687–1693.

Satar, R., Ismail, S. A., Rehan, M., & Ansari, S. A. (2016). Elucidating the binding efficacy of β-galactosidase on graphene by docking approach and its potential application in galacto-oligosaccharide production. *Bioprocess and Biosystems Engineering*, *39*(5), 807–814.

Seabra, A. B., Paula, A. J., de Lima, R., Alves, O. L., & Durán, N. (2014). Nanotoxicity of graphene and graphene oxide. *Chemical Research in Toxicology*, *27*(2), 159–168.

Sheldon, R. A. (2007). Enzyme immobilization: The quest for optimum performance. *Advanced Synthesis & Catalysis*, *349*(8–9), 1289–1307.

Sheldon, R. A., & Van Pelt, S. (2013). Enzyme immobilisation in biocatalysis: Why, what and how. *Chemical Society Reviews*, *42*, 6223–6235.

Singh, N., Srivastava, G., Talat, M., Raghubanshi, H., Srivastava, O. N., & Kayastha, A. M. (2014). *Cicer*α-galactosidase immobilization onto functionalized graphene nanosheets using response surface method and its applications. *Food Chemistry*, *142*, 430–438.

Srivastava, G., Singh, K., Talat, M., Srivastava, O. N., & Kayastha, A. M. (2014). Functionalized graphene sheets as immobilization matrix for fenugreek β-amylase: Enzyme kinetics and stability studies. *PLoS One*, *9*(11), e113408.

Szot, K., Watkins, J. D., Bull, S. D., Marken, F., & Opallo, M. (2010). Three dimensional film electrode prepared from oppositely charged carbon nanoparticles as efficient enzyme host. *Electrochemistry Communications*, *12*(6), 737–739.

Tang, C., Saquing, C. D., Sarin, P. K., Kelly, R. M., & Khan, S. A. (2014). Nanofibrous membranes for single-step immobilization of hyperthermophilic enzymes. *Journal of Membrane Science*, *472*, 251–260.

Tzialla, A. A., Pavlidis, I. V., Felicissimo, M. P., Rudolf, P., Gournis, D., & Stamatis, H. (2010). Lipase immobilization on smectite nanoclays: Characterization and application to the epoxidation of α-pinene. *Bioresource Technology*, *101*, 1587–1594.

Umpleby, R. J., Baxter, S. C., Chen, Y., Shah, R. N., & Shimizu, K. D. (2001). Characterization of molecularly imprinted polymers with the Langmuir – Freundlich isotherm. *Analytical Chemistry*, *73*, 4584–4591.

Ushizawa, K., Sato, Y., Mitsumori, T., Machinami, T., Ueda, T., & Ando, T. (2002). Covalent immobilization of DNA on diamond and its verification by diffuse reflectance infrared spectroscopy. *Chemical Physics Letters*, *351*(1–2), 105–108.

Vilian, A. T. E., Chen, S.-M., & Lou, B.-S. (2014). A simple strategy for the immobilization of catalase on multi-walled carbon nanotube/poly (L-lysine) biocomposite for the detection of H_2O_2 and iodate. *Biosensors and Bioelectronics*, *61*, 639–647.

Wang, Y., & Hsieh, Y.-L. (2008). Immobilization of lipase enzyme in polyvinyl alcohol (PVA) nanofibrous membranes. *Journal of Membrane Science*, *309*(1–2), 73–81.

Wang, Z.-G., Wan, L.-S., Liu, Z.-M., Huang, X.-J., & Xu, Z.-K. (2009). Enzyme immobilization on electrospun polymer nanofibers: An overview. *Journal of Molecular Catalysis B: Enzymatic*, *56*(4), 189–195.

Wang, K., Yang, H., Zhu, L., Ma, Z., Xing, S., Lv, Q., et al. (2009). Direct electron transfer and electrocatalysis of glucose oxidase immobilized on glassy carbon electrode modified with Nafion and mesoporous carbon FDU-15. *Electrochimica Acta*, *54*(20), 4626–4630.

Wei, X.-L., & Ge, Z.-Q. (2013). Effect of graphene oxide on conformation and activity of catalase. *Carbon*, *60*, 401–409.

Wei, L., Zhang, W., Lu, H., & Yang, P. (2010). Immobilization of enzyme on detonation nanodiamond for highly efficient proteolysis. *Talanta*, *80*(3), 1298–1304.

Werengowska-Ciećwierz, K., Wiśniewski, M., Terzyk, A. P., Roszek, K., Czarnecka, J., Bolibok, P., et al. (2015). Conscious changes of carbon nanotubes cytotoxicity by

manipulation with selected nanofactors. *Applied Biochemistry and Biotechnology*, *176*(3), 730–741.
Wu, L., Yuan, X., & Sheng, J. (2005). Immobilization of cellulase in nanofibrous PVA membranes by electrospinning. *Journal of Membrane Science*, *250*(1–2), 167–173.
Zebda, A., Gondran, C., Goff, A. L., Holzinger, M., Cinquin, P., & Cosnier, S. (2011). Mediatorless high-power glucose biofuel cells based on compressed carbon nanotube-enzyme electrodes. *Nature Communications*, *2*, 370.
Zhang, C., Chen, S., Alvarez, P. J. J., & Chen, W. (2015). Reduced graphene oxide enhances horseradish peroxidase stability by serving as radical scavenger and redox mediator. *Carbon*, *94*, 531–538.
Zhang, C., Luo, S., & Chen, W. (2013). Activity of catalase adsorbed to carbon nanotubes: Effects of carbon nanotube surface properties. *Talanta*, *113*, 142–147.
Zhang, F., Zheng, B., Zhang, J., Huang, X., Liu, H., Guo, S., et al. (2010). Horseradish peroxidase immobilized on graphene oxide: Physical properties and applications in phenolic compound removal. *The Journal of Physical Chemistry C*, *114*(18), 8469–8473.
Zhao, H.-Z., Du, Q., Li, Z.-S., & Yang, Q.-Z. (2012). Mechanisms for the direct electron transfer of cytochrome c induced by multi-walled carbon nanotubes. *Sensors*, *12*, 10450–10462.
Zheng, W., Zhao, H. Y., Zhang, J. X., Zhou, H. M., Xu, X. X., Zheng, Y. F., et al. (2010). A glucose/O2 biofuel cell base on nanographene platelet-modified electrodes. *Electrochemistry Communications*, *12*, 869–871.
Zhou, L., Jiang, Y., Gao, J., Zhao, X., & Ma, L. (2012). Graphene oxide as a matrix for the immobilization of glucose oxidase. *Applied Biochemistry and Biotechnology*, *168*(6), 1635–1642.
Zhou, L., Jiang, Y., Gao, J., Zhao, X., Ma, L., & Zhou, Q. (2012). Oriented immobilization of glucose oxidase on graphene oxide. *Biochemical Engineering Journal*, *69*, 28–31.
Zhou, L., Jiang, Y., Ma, L., He, Y., & Gao, J. (2015). Immobilization of glucose oxidase on polydopamine functionalized graphene oxide. *Applied Biochemistry and Biotechnology*, *175*(2), 1007–1017.
Zhu, Y., Li, J., Li, W., Zhang, Y., Yang, X., Chen, N., et al. (2012). The biocompatibility of nanodiamonds and their application in drug delivery systems. *Theranostics*, *2*(3), 302–312.
Zucca, P., & Sanjust, E. (2014). Inorganic materials as supports for covalent enzyme immobilization: Methods and mechanisms. *Molecules*, *19*(9), 14139–14194.

CHAPTER NINE

Immobilization of a Mesophilic Lipase on Graphene Oxide: Stability, Activity, and Reusability Insights

Nalok Dutta, Malay K. Saha[1]
ICMR-National Institute of Cholera and Enteric Diseases, Kolkata, West Bengal, India
[1]Corresponding author: e-mail address: sahamk@yahoo.com

Contents

Methods in Enzymology, Volume 609
ISSN 0076-6879
https://doi.org/10.1016/bs.mie.2018.05.009

Abstract

Catalyst-mediated bioprocessing at an industrial scale is dependent on the sustained stability and activity of the biocatalyst. Here, we demonstrate that strategy employed to increase the stability and activity of a mesophilic lipase immobilized on graphene oxide (GO). The protein adsorption capacity of GO is much higher than of other large surface area carbonaceous materials. Its structure and physicochemical properties are reported beneficial for enzymatic activity modifications. A purified lipase from *Brevibacillus borstelensis* NLIP05 immobilized on GO showed remarkable increase in thermostability (at 95°C) over a broad alkaline pH range (pH 7–12) compared to the free enzyme. Thermodynamic analysis of the GO-lip showed decreases in K_m and activation energy (E_a) with increased V_{max} and deactivation energy (E_d) at both 45 and 95°C. The decrease in decay constant (k) coupled with the increase in $t_{1/2}$ value with temperature increase were salient features of the GO-lip system. This implies that GO-based immobilization conferred structural stability to the enzyme at higher temperature indicating chaperone like activity. Our findings support suitability of immobilized lipase on GO nanosupport for possible production of value-added materials with varied biological applications in the form of active pharmaceutical substances, synthetic building blocks, and effective synthesis of surfactants. The enhanced stability and activity of lipase indicate major application of GO on an industrial scale.

ABBREVIATIONS

free-lip free lipase
GO graphene oxide nanosupport
GO-lip lipase immobilized on graphene oxide nanosupport

1. INTRODUCTION

Lipase (triacylglycerol acylhydrolase EC 3.1.1.3) is an inexpensive biocatalyst which is capable of breaking down lipids with wide substrate specificity and high selectivity (Soler, Illanes, & Wilson, 2016). Lipases catalyze the hydrolysis of the ester bonds in triglycerides. Being insoluble in water, this reaction is catalyzed at the water–lipid interface as most lipases express higher catalytic activity at these interfaces than in aqueous solution. This increased lipase activity at the interface is termed as interfacial activation (Bastida et al., 1998; Fernandez-Lafuente, 2010). Studies involving structure of a typical lipase have shown that the active site of the lipase is covered by a flexible region called lid, which is composed of either one/two amphiphilic α-helices which blocks substrate access in the closed conformation. During the process of interfacial activation, the interaction with specific lipase substrate triggers opening of the lid to make the active site accessible (Bernal, Illanes, & Wilson, 2014; Secundo et al., 2006). Since interfacial activation has direct influence on the lipase activity, this property should be taken into account during the screening of lipase immobilization (Adlercreutz, 2013).

Different activation methods have been used to immobilize lipases to improve their catalytic activity. These include increasing the hydrophobicity of the supporter for immobilization (Chen, Liu, & Yang, 2017), ion-paired lipases (Adkins et al., 2008), surfactant-coated lipases (Kim et al., 2011), microemulsions (Maiti, Ghosh, & Das, 2011), and microemulsion-based organogels (Zhang, Wang, Zhou, He, & Yu, 2012; Zhang et al., 2011; Zhang, Zhang, et al., 2012).

Lipase offers a large number of potential applications in pharmaceuticals, food items, cosmetics, and the conversion of oil into fuel for biodiesel production (Duarte et al., 2015; Zhao, Qi, Yuan, Du, & Liu, 2015). Compared to enzymes from animals or plant sources, enzymes from microbial sources are more beneficial because of their possible high yields, diversity of catalytic activities, convenient production, and prompt growth on inexpensive media (Gurung, Ray, Bose, & Rai, 2013; Momsia & Momsia, 2013). Hence, we selected *B. borstelensis* lipase as our model enzyme for immobilization.

Nanomaterials and nanostructures generally provide large surface area, low mass transfer resistance, and high enzyme loading capability, which enables better interaction with the enzyme, resulting in increase in immobilization efficiency and enhancement in long-term storage and recycling ability of the enzyme. Nanomaterials can also be engineered to present multiple surface functional groups for interacting with biomolecules. A wide range of nanomaterials have been studied and their positive effect on enzyme was shown through immobilization process (Dwevedi et al., 2011; Dwevedi, Singh, Singh, Srivastava, & Kayastha, 2009; Konwarh, Karak, Rai, & Mukherjee, 2009; Momsia & Momsia, 2013).

Glutaraldehyde, a linear, 5-carbon dialdehyde, is a clear, colorless to pale straw-colored, pungent oily liquid that is soluble in all proportions in water and alcohol, as well as in organic solvents. It is mainly available as acidic aqueous solutions (pH 3.0–4.0), ranging in concentration from less than 2% to 70% (w/v). Glutaraldehyde has had great success because of its commercial availability and low cost in addition to its high reactivity. It reacts rapidly with amine groups at around neutral pH (Okuda, Urabe, Yamada, & Okada, 1991) and is more efficient than other aldehydes in generating thermally and chemically stable crosslinks (Nimni, Cheung,

Strates, Kodama, & Sheikh, 1987). Glutaraldehyde is one of the most widely used reagents in the design of biocatalysts. It is because of the fact that the treatment of enzymes immobilized on supports is nonreactive with glutaraldehyde (only supports inter- and intramolecular crosslinkings) to prevent enzyme leakage and obtain some enzyme stabilization via crosslinking. Moreover, the crosslinking of enzymes adsorbed on aminated supports facilitates other enzymatic reactions/crosslinking while the enzyme remains incorporated into the support. Glutaraldehyde facilitates intermolecular and intramolecular crosslinking. Glutaraldehyde was primarily used to preserve and fix tissues in combination with formaldehyde by the formation of intermolecular crosslinking. This was achieved utilizing the unique reactive capacity of glutaraldehyde to react with itself or with protein groups modified with a glutaraldehyde molecule (Kiernan, 2000). Of late, glutaraldehyde serves as one of the most potent crosslinker reagents in clinical, biomedical, and biotechnological applications. Several researchers have excellently and critically reviewed the importance and properties of glutaraldehyde in these applications (McGinley, 2012; Yan et al., 2015). Recent studies have demonstrated the utilization of glutaraldehyde preactivated supports for imparting stability to various enzymes in food industry due to their generally recognized as safe status (Contesini et al., 2013).

Nanomaterials serve as excellent supporting materials for enzyme immobilization, because they offer the ideal characteristics for balancing the key factors that determine biocatalysts efficiency. But, there are some salient disadvantages of the process which include (i) cost of the fabrication process: as some nanoparticles possess high costs (gold and nanodiamond) it is not profitable for industrial application. Large-scale simulation of the lab methodology and reaction medium separation after bioprocessing remains a concern for a large number of methods.

Various graphene-based nanomaterials have been used to fabricate functionalized biosystems integrated with nucleic acids, peptides, proteins, and even cells. Graphene is a free standing 2D crystal with one atom thickness. This allotrope of carbon comprises layers of six atom rings in a honey comb network and can be conceptually viewed as a true planar atomic macromolecule. Graphene also gets benefitted by biological modification which improves its biocompatibility, solubility, and selectivity. Proteins can exfoliate and modify graphene oxide due to presence of various functional groups, through physical adsorption or chemical bonding (Wang, Li, Wang, Li, & Lin, 2011; Wang, Ruan, et al., 2011). Being, composed of carbon atoms, it does not alter the native biochemical properties of attached biomolecules significantly. Recently, material based on graphene have found use in technical applications such as biofuel cells, biosensing material, drug delivery, and catalysis (Gu et al., 2011; Stankovich, Piner, Nguyen, & Ruoff, 2006; Stoller, Park, Zhu, An, & Rouff, 2008; Zhang, Tang, Fu, & Xu, 2010; Zhang, Zhang, et al., 2010). Issue of biocompatibility of graphene oxide has been studied on animal model and its nontoxic effect under low dose administration was demonstrated by Wang, Li, et al. (2011) and Wang, Ruan, et al. (2011). Their studies showed that GO with dose less than 20 μg/mL did not exhibit toxicity to human fibroblast cells whereas the dose exceeding 50 μg/mL triggered decreasing cell adhesion promoting apoptosis. Liu et al. (2014) showed that the size of the GO sheets affected their toxic property. The multiple-dose exposure to larger forms of GO induced a serious lymphocyte infiltration around the bronchioles in the lungs, as well as obvious renal damage. The multiple-dose exposure to smaller forms of GO did not significantly induce lung or kidney damages, but induced a liver index increase and an inflammatory cell infiltration.

GO, due to its varied flexibility in biological applications as nanosupport has found its use in production of biocatalysts, biosensors, and drug delivery vehicles (Kuila et al., 2011). GO provides large surface area for enzyme immobilization and maintains its integrity in different aqueous suspensions (Zhang, Tang, et al., 2010; Zhang, Wang, et al., 2012; Zhang, Zhang, et al., 2012, 2010; Zhao, Li, Jiang, Wang, & Mu, 2014). GO possesses a single-layered sp2 hybridized structure which is two dimensional in nature. Owing to its unique chemical and physical properties, it can be used for immobilization of organic biomacromolecules. The use of glutaraldehyde as a crosslinker for enzyme immobilization on graphene was found to confer increased structural stability to GO–enzyme complex which reflected in its increased activity and multiple usage (Sun, Feng, Hou, & Li, 2014). In this chapter, we have discussed a method devised for immobilization of an extracellular lipase obtained from *B. borstelensis* on GO.

2. MATERIALS AND METHODS

2.1 Materials

Single layer graphene oxide (SLGO) was purchased from mkNANO (80% carbon, 20% oxygen. Flake size: 0.5–5 μm. Thickness: Single atomic layer >80%. Color: Bright yellow. Purity: Carbon contents over 80%). PCR primers (forward and reverse) and silver nitrate for silver staining were obtained from Sigma–Aldrich (Germany). pNPP (para-Nitrophenylphosphate), agar powder, and glutaraldehyde were purchased from Himedia, India. All other reagents for bacterial culture media such as agar, and salts were purchased from SRL (India) and were of analytical grade. 20–200 kDa protein markers were obtained from Fermentas (Germany).

2.2 Methods

2.2.1 *Isolation of Enzyme Secreting Bacterial Strains*

The bacterial strain was isolated from Himalayan hot spring soil near Manikaran (Himachal Pradesh, India). Approximately 5 g of soil sample was added to sterile 0.9% saline water. The sample saline mix was incubated overnight at 37°C and the saline supernatant was used as the bacterial source. Lipolytic bacterial strains were isolated from the soil by using serial dilution technique on tributyrin agar containing (0.5% (w/v) peptone, 0.3% (w/v) yeast, 1% (v/v) tributyrin and 2% agar, pH 7.0). Pure cultures of the isolates were maintained on minimal media agar slants containing (1.5% yeast extract, 0.5% NaCl, 1% peptone and 2% agar, pH 7.0). Culture plates were incubated at 37°C. Colonies showing clear zones (due to hydrolysis of tributyrin) around them were picked out, purified on tributyrin agar plates, and transferred to agar slants. The isolate with higher lipolytic activity was used for further processes.

2.2.2 *Ribotyping Studies*

The lipolytic strain with the highest activity was coined as NLIP05.· We performed 16S rRNA gene sequencing using universal primers AGAGTTTGATCMTGGCTCAG and TACGGYACCTTGTTACGACTT. Genomic DNA was prepared from NLIP05 by sodium dodecyl sulfate proteinase K cetyltrimethylammonium bromide (CTAB) method. The 16S rRNA gene was partially amplified with the thermal cycler ABI 9700 (ABI, Foster City, USA). The PCR mixture consisted of 25 pmol of each primer, 10 ng of chromosomal DNA, 200 μM dNTPs, and 2.5 U of Taq polymerase in 50 µL of polymerase buffer. The PCR was carried out for 40 cycles of 95°C for 5 min, 52°C for 5 min, and 72°C for 2 min. The 16S rRNA gene fragment was sequenced with an ABI 3100 Genetic Analyzer and Blast program (www.ncbi.nlm.nih.gov/blast) was used to assess the DNA similarities. Sequencing reaction was performed by using the Big Dye terminator cycle sequencing kit V3.1 (Applied Biosystems, Foster City, USA).

2.2.3 Purification of Lipase

Lipase enzyme was purified from 100 mL culture of NLIP05.·Cell free supernatant was precipitated with 0%–80% saturation of ammonium sulfate followed by dialysis. All steps of the purification procedure were performed at 4°C. The dialyzed proteins (10.5 mL in volume; specific activity: 1204 units/mg) were loaded onto a phenyl Sepharose CL-4B (1.5 × 24 cm^2) column preequilibrated with 50 m*M* Tris–HCl buffer, pH 7.0, and allowed to equilibrate overnight. Each fraction was checked for enzyme activity as well as protein content. The enzyme was eluted with linear gradient of 1% (w/v) cholate in 50 m*M* Tris–HCl buffer, pH 7.0 with flow rate of 1 mL/min. All the fractions were checked for enzyme activity. The fractions containing lipase with the highest activity from affinity chromatography (8.5 mL in volume; specific activity: 3504 units/mg) was loaded on Sephadex G-100 column (40.0 cm × 1.2 cm) equilibrated with 50 m*M* Tris–HCl buffer, pH 7.0. The enzyme was then eluted with the same buffer with a flow rate of 1 mL/min.

Note: The dialysis membrane should be handled with utmost care for prevention from contamination. The clips to seal the dialysis bags should be cleaned properly and autoclaved to avoid any unwanted interferences.

Total protein estimation was performed following Lowry's method.

Procedure

1. Different dilutions of BSA solutions were prepared by mixing stock BSA solution (1 mg/mL) and water. The final volume was 5 mL. The BSA range was 0.05–1 mg/mL.
2. From different dilutions, 0.2 mL protein solution was pipetted out to different test tubes and 2 mL of alkaline copper sulfate reagent was added (analytical reagent).
3. This solution was incubated at room temperature for 10 min.
4. Then 0.2 mL of reagent Folin Ciocalteau solution was added to each tube and incubate for 30 min following measurement of optical density at 660 nm.
5. Absorbance against protein concentration was plotted to get a standard calibration curve.
6. Absorbance of unknown lipase was determined using the standard curve plotted earlier.

Specific activity of the enzyme samples were determined by dividing the total enzyme activity (units) by total protein (mg).

2.2.4 Enzyme Immobilization on GO

GO dispersion was prepared by dissolving (0.2–1.8) mg/mL of GO in 25 m*M* sodium phosphate buffer (pH 7.0) followed by treatment with glutaraldehyde (0.1–1.7) mL, (25% (v/v) in water), and kept in dark for 4 h at room temperature.

In the final test solution amounting to 10 mL in volume, GO dispersion constituted (0.2–1.8) mg/mL, glutaraldehyde constituted (0.1–1.7) mL, and the rest of the volume was made up with 25 m*M* sodium phosphate buffer (pH 7.0). After treatment and buffer wash, it was incubated with optimal lipase concentration (5 mL of purified protein with specific activity of 4500 units/mg) under dark condition for 22 h at 4°C·The final reaction volume of 10 mL was constituted by 1.1 mL glutaraldehyde; GO dispersion amounting to 1.2 mg/mL, 5 mL of enzyme, and the rest of the volume was made up with phosphate buffer. The immobilized enzyme was washed thoroughly with phosphate buffer. Each step of immobilization was followed by thorough washing with buffer and centrifugation at 8000 rpm for 10 min at 4°C.

Note: The preparation of GO dispersion should be carried out in amber colored containers as glutaraldehyde is light sensitive. Extreme care should be taken that the glovewears are clean throughout the process to avoid contamination.

2.2.5 Assay of Enzyme Activity of Free and Immobilized Lipase

The lipolytic activity of soluble and immobilized lipase was analyzed spectrophotometrically through a UV/vis JASCO spectrophotometer by evaluating the absorbance at 410 nm produced by the released p-nitrophenol. p-NPP was selected as the substrate for the determination of lipase activity. The substrate solution was prepared with p-NPP (150 m*M* in isopropanol), 150 mg Triton X-100, and 67.5 mL 25 m*M* Tris–HCl buffer, pH 7.0. The reaction mixture consisted of 2 mL substrate solution and 0.1 mL lipase solution. The reaction was carried out by incubating the reaction mixture at 37°C for 15 min in a water bath under continuous stirring. Thereafter, 0.25 mL of 0.1 (*M*) Na_2CO_3 was added to terminate the reaction. Lipolytic activity was determined, using substrate free blanks as control. The reaction rate was calculated from the slope of the absorbance curve, using software installed by the manufacturer (Beckman Coulter). One unit of lipase activity (U) is defined as the quantity of enzyme that causes the hydrolysis of 1 mmol of p-nitrophenol from p-NPP per min at analysis conditions.

2.2.6 Optimization of the Immobilization Process

To optimize the immobilization process of lipase on GO, the immobilization yield was calculated as:

$$\text{Immobilization yield\%} = \frac{\text{Enzyme content of immobilized enzyme(mg)} \times 100}{\text{Total enzyme used for immobilization(mg)}}$$

The loading capacity of GO was measured using the formula:

$$E_t = \frac{C_0 - C_t}{M} \times V$$

where C_0 and C_t denote amount of enzyme in the solution before and after adsorption, respectively; E_t is the enzyme adsorbed per time unit per amount of support; V is volume of solution; and M is the mass of the GO support (g).

2.2.7 Circular Dichroism (CD) Studies

The far CD spectra over the range of 190–250 nm obtained for the secondary structure of native lipase enzyme at 45°C, lipase at 95°C, GO-lip and GO at 45 degree and 95°C using a Chirascan spectropolarimeter (Applied Photosystem, UK). Enzyme concentration for the purified enzyme was maintained at 0.65 mg/mL for all observations. 1-mm path length cell and 1-nm bandwidth were maintained for every observation while carrying out the spectral analyses with a scan rate of 20 nm/min. A system of (50 m*M* sodium phosphate buffer, pH 8.0) was used as blank. Quartz cuvette with 1 cm optical path length was used. Molar ellipticity (θ, degree cm^2 dmol^{-1}) was measured every 2 s with an integration time of 50 nm min^{-1}. Three full wavelength scans were collected and averaged for each sample. The spectra were analyzed to characterize interaction between the GO and the enzyme. Secondary structures of lipase before and after GO immobilization were determined by comparing the molar ellipticity at 222 nm (θ 222) and 208 nm (θ 208). CD spectra for both the native enzyme and GO were recorded as a control. The spectrum of the CD buffer (50 m*M* sodium phosphate buffer, pH 8.0) was subtracted from the spectrum of the samples. Data obtained from CD were analyzed using the DICHROWEB website: http://www.dichroweb.cryst.bbk.ac.

2.2.8 Effect of Temperature and pH on Activity of GO-Lip and Untreated Lipase

The optimum temperature of GO-lip and untreated lipase were determined by carrying out the standard enzyme assay in sodium phosphate buffer at increasing temperatures, i.e., 37, 42, 45, 50, 60, 70, 80, 90, and 95°C. The substrate was preincubated with the enzyme at the said temperature for 25 min followed by the enzyme assay for each observation. To observe the effect of GO treatment on enzyme activities with decreasing temperature, optimal concentration of GO was added to the enzyme and preincubated for 50 min. The control system contained untreated enzyme. The effect of pH on GO-lip was studied in the pH range of 3.0–12.0 (citrate phosphate: pH 3.0–5.0, sodium phosphate: 5.5–7.0, Tris–HCl: 7.0–9.2, glycine buffer pH 9.2–12.0).

2.2.9 Stability of GO-Lip at Lower and Higher Temperatures

To measure the retention of enzyme activity at lower and higher temperatures, lipase immobilized on GO was incubated for 0–240 min at 45 and 95°C. The enzyme activities were determined as described in the earlier section after incubation. The results expressed were in comparison to the values of free enzyme systems.

3. ENZYME AGGREGATION ANALYSIS BY DYNAMIC LIGHT SCATTERING (DLS) STUDIES

DLS is a technique in physics that can be used to determine the size distribution profile of small particles in suspension or polymers in solution. A monochromatic light source, usually a laser, is shot through a polarizer and into a sample. The scattered light then goes through a second polarizer where it is collected by a photomultiplier and the resulting image is projected onto a screen. Upon analyzing the modulation of the scattered light intensity as a function of time, the hydrodynamic size of the particles, and particle agglomerates can be determined. Larger particles will diffuse slower than smaller particles, and the DLS instrument measures the time dependence of the scattered light, to generate a correlation function that can be mathematically linked to the particle size.

Study of enzyme aggregation by DLS was carried out to analyze whether GO can reduce protein aggregation at high temperature. Change of hydrodynamic radius was assessed through DLS Malvern Nano-ZS, ZEN 3600 DLS instrument. GO-lip and free-lip were heated up to 95°C and the hydrodynamic radius of all the samples were checked by aggregation signal.

3.1 Stability and Reusability of GO-Lip

To observe the reusability of GO-lip, the system was tested for enzyme activity of 12 repeated uses at a stretch with standard enzyme assay protocols. Storage stability of GO-lip was tested periodically for 160 days at 4°C.

3.2 Activation–Inactivation Parameters and Kinetics of Lipase Enzyme

The $K_m - V_{max}$, activation energy (E_a), and the activation/deactivation kinetics of both GO immobilized and free lipase enzyme systems were studied using the standard reaction mix and assay conditions as described in the earlier section. The substrate (pNPP) concentration used was from 0.015% to 1.25% to determine K_m and V_{max}. The assay temperature was case specific (45 or 95°C). The activation energies (E_a) of GO-lip and free lipase were calculated for the temperature range of 45–95°C (318–368 K) from the Arrhenius plots. Inactivation parameters comprising half-life ($t_{1/2}$), decay rate constant (k), energy of deactivation (E_d), enthalpy (ΔH), entropy (ΔS), and free energy change (ΔG) were obtained according to Ortega, de Diego, Rodríguez-Nogales, Perez-Mateos, and Busto (2004).

Note: Arrhenius energy of activation (E_a).

According to Arrhenius equation:

$$\ln k_1 = \frac{E_a}{RT} + Z_1 \tag{1}$$

where k_1 is the first-order reaction rate constant, T is the absolute temperature, R is the universal gas constant, Z_1 is the constant, and E_a is the Arrhenius energy of deactivation.

To measure the Arrhenius energy of activation (E_a), we calculated the specific activity values from 45 to 95°C at pH 7.0.

Note: Arrhenius energy of deactivation (E_d)

To calculate E_d value, first we treated the enzyme thermally at 45, 50, 60, 70, 80, and 95°C for up to 10 min.

The inactivation of enzyme is often described by a first-order kinetic model in which enzyme activity decreased log linearly as a function of time as described by the following equation:

$$\ln \frac{C_t}{C_0} = -k_t \tag{2}$$

where C_t is the activity at time t (min), C_0 is the activity at time zero, and k is the first-order inactivation rate constant.

Again, according to Arrhenius equation:

$$\ln k = -\frac{E_d}{RT} + Z_2 \quad (3)$$

where k is the first-order inactivation rate constant, T is the absolute temperature, R is the universal gas constant, Z_2 is the constant, and E_d is the Arrhenius energy of deactivation.

Half-life ($t_{1/2}$), inactivation rate constant (k), entropy (ΔS), Gibb's free energy of inactivation (ΔG), and enthalpy (ΔH) of lipase were determined at different temperatures.

(1) Half-life calculated from Eq. (4).

$$t_{1/2} = \ln \frac{2}{k/\min} \quad (4)$$

(2) Inactivation rate constant (k) calculated from Eq. (2).

(3) Enthalpy (ΔH) calculated from Eq. (5)

$$\Delta H = \Delta E_d - RT \quad (5)$$

(4) Gibb's free energy of inactivation (ΔG) calculated from Eq. (6)

$$\Delta G = -RT \ \ln(kh/KT) \quad (6)$$

where h (6.6262×10^{-34}J) is the Planck constant and K (1.2806×10^{-23}JK^{-1}) is the Boltzmann constant. k is the inactivation constant.

(5) Entropy (ΔS) calculated from Eq. (7)

$$\Delta S = (\Delta H - \Delta G)/T \quad (7)$$

4. RESULTS

4.1 Identification of the Bacterial Strain

The sequence of the partially amplified 16S rDNA fragments from NLIP05 was found to bear more than 98% similarity to *B. borstelensis* when compared to rDNA sequence database in GenBank. The GenBank accession number was allocated as KT783471.

4.2 Purification of Lipase From Bacterial Strains

The lipase containing fractions were pooled and assessed for protein content. The resulting enzyme was utilized for the characterization of the extracellular lipase. 61.28% recovery of lipase was achieved after affinity chromatography. The recovery of enzyme activity was about 45.17% after gel filtration (Table 1). A single band was obtained in SDS–PAGE analysis indicating complete purification of the enzyme. Compared to protein markers, the size of the purified enzyme was around 50 kDa (Fig. 1A). More than 8.75-fold purification was achieved with a final specific activity of 4500.4 units/mg (Table 1).

Table 1 Purification of Mesophilic Lipase From *Brevibacillus borstelensis* NLIP05

Purification Steps	Total Volume (mL)	Total Protein (mg)	Total Activity (Unit)	Specific Activity (Unit/mg)	Purification (Fold)	Yield (%)
Crude	100	29	13,476	514.3	1.0	100
(0–80)% $(NH_4)_2\ SO_4$	10.5	13.9	8258.7	1204	2.34	61.28
Sepharose CL-4B	8.5	7.4	6087.2	3504	6.813	45.17
Sephadex (G-100)	5.0	3.5	4680	4500.4	8.75	34.72

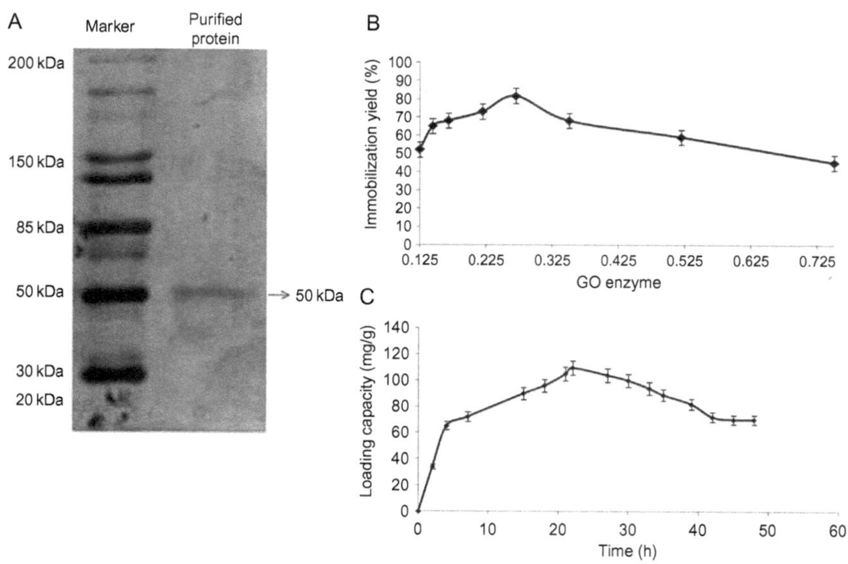

Fig. 1 (A) SDS–PAGE (12%) showing purification of lipase from *Brevibacillus borstelensis* NLIP05. (B) Optimization of enzyme immobilization yield (%) of lipase on GO. (C) Optimization of loading capacity of enzyme systems on GO as a function of time.

4.3 Immobilization of Lipase on GO

Purified lipase was coupled to the graphene oxide with the help of glutaraldehyde as a crosslinker. Enzyme activity was performed for each fraction of GO subjected to treatment with glutaraldehyde following incubation with optimum concentration of enzyme. This step was performed to ascertain the optimal concentration of glutaraldehyde needed for attachment of lipase to GO.

Thus, the optimum concentration of glutaraldehyde for crosslinking of lipase was found out to be 1.1 mL and graphene oxide concentration required for enzyme immobilization was found out to be 1.2 mg/mL (Table 2).

4.4 Optimal Immobilization of Lipase on GO

To achieve optimal immobilization of the enzyme systems on GO, the enzyme loading capacity on nanoparticles and the resultant immobilization yield was taken into account. As evident from the data illustrated in Fig. 1B, the immobilization yield increased concomitantly with the increase of GO to enzyme (w/w) ratio. When the amount of GO to enzyme was 27.5% (w/w), the maximum enzyme immobilization yield was found to be 81.6% (Fig. 1B). The minimum immobilization yield was found to be 45% for lipase when the ratio was 0.75 (Fig. 1B). From protein estimation, it was found that optimum amount of enzyme loading was 0.65 mg/mL for the enzyme systems. An optimal loading capacity of lipase per gram of the immobilization matrix was found out to be 109.6 mg after 22 h of the immobilization process (Fig. 1C).

4.5 CD Studies

The CD spectroscopy of purified lipase at 45°C showed a 4.67% of α-helix and 9.46% of β-strand per 100 residues (Fig. 2). Spectroscopic data for the free lipase at 95°C showed 2.47% α-helix content, 1.25-fold times the native enzyme at 50°C and 1.77% of β-strand, which was almost fourfold less than the native protease. GO-lip enzyme system at 95°C had a marked increase in β-strand content by ~9-fold compared to untreated enzyme system at 95°C. This result clearly indicates that β-strands are the most important contributor toward maintaining lipase enzyme activity.

4.6 Effect of Temperature and pH on Activity of GO-Lip and Untreated Lipase

In case of the purified lipase, the optimum temperature was 45°C which eventually became 95°C upon GO immobilization. The GO-lip activity

Table 2 Determination of Optimum Graphene Oxide and Glutaraldehyde Concentration for Enzyme Activity

GO Concentration (mg/mL)	Glutaraldehyde Concentration (mL)							
	0.1	0.3	0.7	0.9	1.1	1.3	1.5	1.7
	Enzyme Activity (Units/mL)							
0.2	30.9 ± 0.98	33.1 ± 1.99	29.3 ± 0.33	33.9 ± 1.41	52.1 ± 1.21	52.1 ± 2.33	50.9 ± 1.45	44.5 ± 1.26
0.4	28.6 ± 1.34	30.7 ± 1.08	34.6 ± 1.01	41.5 ± 1.51	58.7 ± 0.97	53.2 ± 3.21	53.9 ± 2.42	45.1 ± 1.08
0.6	31.3 ± 2.31	33.1 ± 2.01	40.8 ± 1.23	42.4 ± 0.42	69.8 ± 1.22	60.7 ± 1.98	55.6 ± 1.29	49.8 ± 1.57
0.8	34.6 ± 1.98	40.7 ± 2.03	42.3 ± 1.45	46.2 ± 1.33	72.4 ± 1.62	65.3 ± 1.45	60.9 ± 1.68	58.6 ± 0.62
1.0	47.8 ± 4.77	43.7 ± 0.99	44.5 ± 3.21	57.1 ± 2.47	79.4 ± 2.67	73.6 ± 2.67	63.2 ± 3.91	63.1 ± 2.08
1.2	45.7 ± 0.14	50.4 ± 1.98	52.8 ± 1.88	65.1 ± 1.45	80.9 ± 1.88	65.4 ± 2.12	63.9 ± 1.33	68.2 ± 1.56
1.4	46.8 ± 1.72	48.4 ± 0.76	55.6 + 1.91	64.8 ± 0.44	78.1 ± 2.09	70.8 ± 1.55	68.9 ± 0.21	59.6 ± 2.11
1.6	39.5 ± 1.08	40.4 ± 2.01	43.6 ± 2.34	55.1 ± 2.01	69.4 ± 1.41	63.2 ± 2.44	60.9 ± 2.11	52.5 ± 0.71
1.8	29.4 ± 3.1	31.5 ± 1.04	36.88 ± 1.56	45.2 ± 0.38	58.6 ± 0.34	60.7 ± 1.21	59.8 ± 1.87	49.7 ± 1.38

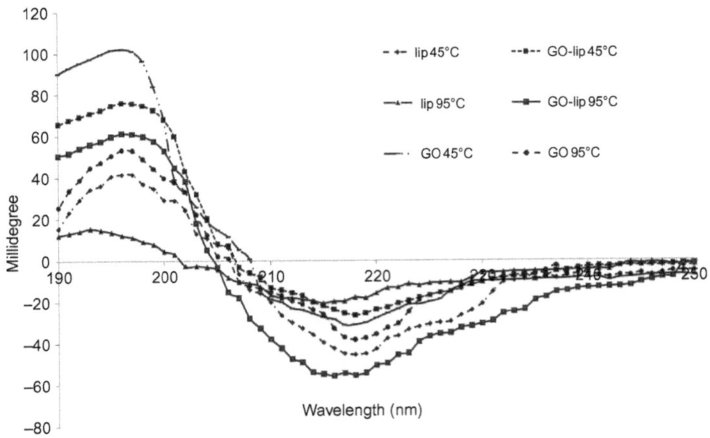

Fig. 2 Far-UV CD spectra (190–250 nm) of the purified lipase at 45 and 95°C, GO-lip at 45 degree and 95°C and GO at 45 degree and 95°C.

increased by 31.54%±1.18, 48.3%±1.8, 55.73%±2.21, 62.27%±2.23 at 50, 70, 80, and 95°C, respectively, compared to the free enzyme at 45°C (Fig. 3A). The free enzyme registered decreased activity beyond 50°C (Fig. 3C). The native lipase showed optimum activity at pH 7.0 (Fig. 3B). Upon GO immobilization, the immobilized enzyme showed increase in activity by 1.46-, 2.0-, 2.08-, and 3.38-fold compared to the free enzyme at pH 8, 9, 10, and 12, respectively (Fig. 3B).

4.7 GO-Lip Stability at Upper and Lower Temperature Extremes

At 95°C, GO-lip could retain 90% enzyme activity after 240 min whereas the free enzyme lost ~50% of its activity after 90 min (Fig. 3C). The standing enzyme activity upon incubation of the enzyme systems at 45°C showed that GO-lip could retain 81.5% enzyme activity after 240 min whereas the free lipase lost half of its initial activity after 90 min (Fig. 3D).

4.8 DLS Studies

DLS observations confirmed that GO inhibit protein aggregation at high temperature (Fig. 4A). Hydrodynamic radius analysis showed high intensity of lipase at 95°C between 100 and 1000 nm, which indicates extensive protein aggregation, whereas a good intensity for the control set (lipase at 45°C) was observed between 0.5 and 10 nm, which supports for monomer formation (Fig. 4A). A high intensity peak was observed for GO-lip at 95°C

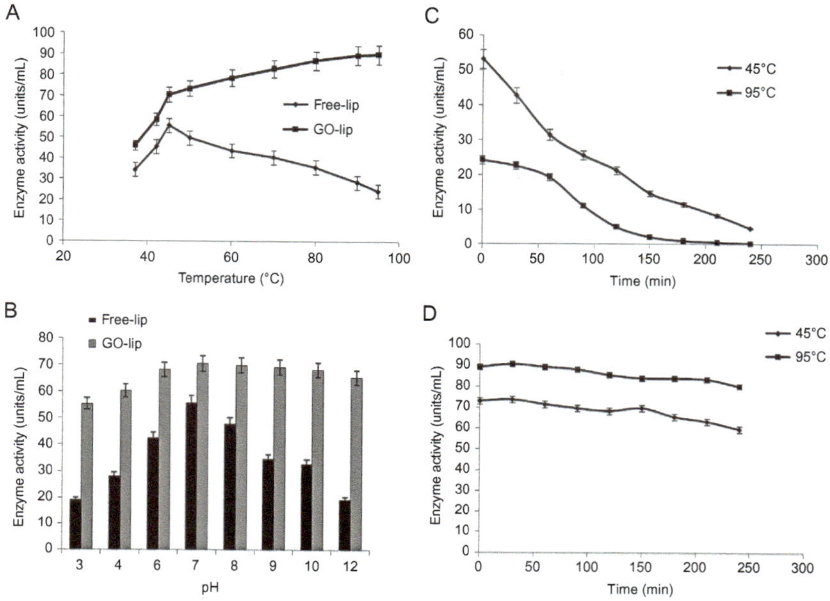

Fig. 3 (A) Temperature profile of free lipase and GO-lip. (B) Comparison of free lipase and GO-lip activity as a function of pH (3 – 12). (C) Activity profile of free lipase at 45 degree and 95°C as a function of time. (D) Activity profile of GO-lip at 45 degree and 95°C as a function of time.

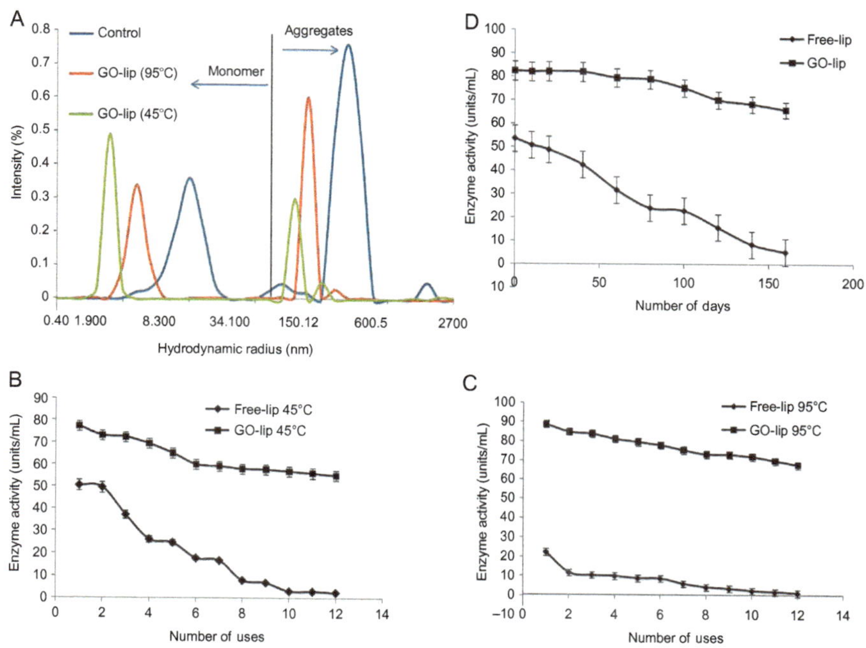

Fig. 4 (A) Study of hydrodynamic radius shift of GO-lip at 45 degree and 95°C. (B) Analysis of activity free lipase and GO-lip at 45 degree as a function of repeated usage. (C) Analysis of activity free lipase and GO-lip at 95 degree as a function of repeated usage. (D) Analysis of storage stability of free lipase and GO-lip at 4 degree as a function of number of days.

between 0 and 10 nm, almost similar to the control set at 45°C. A peak of considerable intensity of GO-lip was found in between 50 and 400 nm ranges at 95°C (Fig. 4A).

4.9 Stability and Reusability of GO-Lip

GO-lip could be reused for 12 repeated times retaining significant activity at both lower (71.17%; 45°C) and higher (76.4%; 95°C) temperatures whereas free lipase lost ~50% of activity after single use at 95°C (Fig. 4B and C). At 45°C, the free lipase lost its residual activity gradually after the third cycle retaining ~33% of its initial activity after the sixth use. GO immobilization of lipase via glutaraldehyde as crosslinker imparted flexibility to the temperature range for the enzyme activity. GO-lip could be stored at 4°C for more than 160 days with the enzyme retaining 79% activity whereas the free lipase could only retain 45% of its activity at the same temperature after 80 days (Fig. 4D).

4.10 Analysis of Activation–Inactivation Parameters and Kinetics of Lipase Enzyme

For the GO-lip, the K_m value was 0.00281 mmol/mL at 45°C with a V_{max} value of 73.4 units/mL (Table 2). For the free lipase at 45°C the K_m and the V_{max} values were 0.413 mmol/mL and 35.2 units/mL, respectively. At 95°C, for the GO-lip, the K_m value was 0.00132 mmol/mL with a V_{max} value of 113.4 units/mL (Table 2). On comparing these systems, we observed that there was 3.22-fold increase in V_{max} with 10.6 times decrease in K_m for the GO-lip at 95°C than the free lipase at 45°C. In contrast to this, the free lipase had 2.31-fold decreased V_{max} and 2.24-fold increased K_m at 95°C compared to the lipase at 45°C (Table 2). There was also an increase in the catalytic efficiency (k_{cat}/K_m) by 1.97-fold for the GO-lip at 95°C than the free lipase at 45°C (Table 3). The kinetics of inactivation of the GO-lip

Table 3 K_m, V_{max}, and E_a of Protease and GO-Lip Systems at 45 and 95°C

	45°C		95°C	
Activation Kinetics Parameters	**Lipase**	**GO-Lip**	**Lipase**	**GO-Lip**
K_m (mmol/mL)	0.413	0.00281	0.032	0.00132
V_{max} (Units/mL)	35.2	73.4	15.2	113.4
E_a (kJ mol^{-1})	29.6	12.7	44.28	9.8
k_{cat}/K_M (mM^{-1} s^{-1})	4912.2	11,036.2	351.45	9719.3

and free lipase enzyme systems were studied at a temperature range between 45 and 95°C (Table 4) for activity analysis at temperature extremes. Semi-logarithmic plots of residual activity vs time between 45 and 95°C for the systems were found to be linear. The plots suggested that free lipase was heat inactivated with first-order kinetics, whereas GO-lip was heat activated with first-order kinetics. For free lipase, a trend of increase in decay constant with increase in temperature was observed. At 45°C; decay constant (k), half-life ($t_{1/2}$), and deactivation energy (E_d) of free lipase were 0.013 min^{-1}, 53.3 min, and −42.29 kJ/mol, respectively (Table 4 and Fig. 5A). With increasing temperatures at 60, 70, 80, and 95°C, the corresponding half-lives of free lipase were observed to be 36.4, 22.35, 10.04, and 6.66 min, respectively. Whereas, GO-lip had 20.39-fold decreased k value of 0.0051 min^{-1} while $t_{1/2}$ (135.88 min) and E_d (11.009 kJ/mol) were increased by 20-fold and 53.305 units, respectively, w.r.t. free enzyme at 95°C (Table 4 and Fig. 5B).

The thermodynamic changes in the enzymatic parameters were summarized by the entropy–enthalpy map in presence and absence of GO. When GO was present in the system, a compensatory behavior was observed whereas the absence of GO triggered a critical change in the entropy–enthalpy profile (Fig. 5C). For GO-lip, the opposite signs of ΔH and ΔS suggested significant entropy–enthalpy compensation (Fig. 5D and Table 5). However this compensatory profile was not observed for the untreated enzyme set as it was operative in nature for the free enzyme (Table 5).

5. DISCUSSION

The high activity of the GO-lip system, as compared to free lipase is reflected in improved stability of the enzyme at higher temperatures which otherwise disrupts native structure and function of the enzyme (Dutta, Biswas, & Saha, 2016, 2017; Dutta, Mukhopadhyay, Dasgupta, & Chakrabarti, 2013, 2014). Enhancement in thermostability of enzyme immobilized on nanomaterials has been previously shown in earlier studies (Singh, Srivastava, Talat, Raghubanshi, & Srivastava, 2014). The GO-lip nanocomposite exhibited increased functionality at elevated temperatures due to the chaperonic nature of the GO where the enzyme activity is kept intact by guarding against unfavorable environment. Reusability of a biocatalyst is an important factor that considerably reduces the processing cost of any industrial merchandise. There are reports which show improved

Table 4 Effect of Variation in Kinetic Parameters for Immobilized GO-Lip and Free Lipase 318–368 K

		45°C		60°C		70°C		80°C		95°C		
Enzyme System: Lipase		**Decay Constant (*k*) (min^{-1})**	**Half-life ($t_{1/2}$) (min)**	**Decay Constant (*k*) (min^{-1})**	**Half-life ($t_{1/2}$) (min)**	**Decay Constant (*k*) (min^{-1})**	**Half-life ($t_{1/2}$) (min)**	**Decay Constant (*k*) (min^{-1})**	**Half-life ($t_{1/2}$) (min)**	**Decay Constant (*k*) (min^{-1})**	**Half-life ($t_{1/2}$) (min)**	**E_d ($kJ\,mol^{-1}$)**
GO	(−)	0.013 ± 1.22	53.3 ± 0.38	0.019 ± 2.1	36.47 ± 0.22	0.031 ± 1.7	22.35 ± 1.28	0.069 ± 0.14	10.04 ± 0.41	0.104 ± 0.11	6.66 ± 0.19	−42.296 ± 1.12
	(+)	0.009 ± 1.11	77.0 ± 0.14	0.0081 ± 0.92	85.55 ± 1.09	0.0072 ± 1.38	96.25 ± 1.15	0.0064 ± 1.0	108.2 ± 1.12	0.0051 ± 1.34	135.88 ± 1.12	11.009 ± 0.24

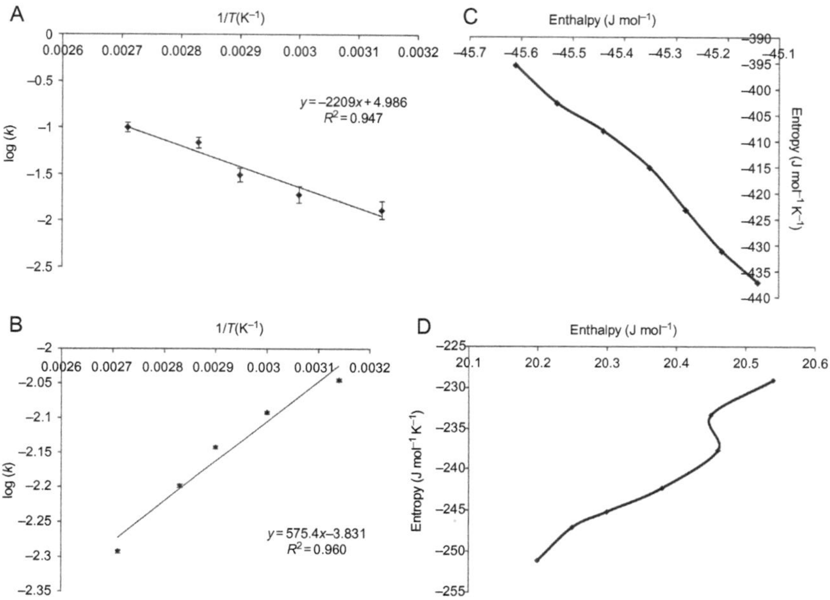

Fig. 5 (A) Arrhenius energy of deactivation (E_d) for free lipase (318–368 K). (B) Arrhenius energy of deactivation (E_d) for GO-lip (318–368 K). (C) Entropy–enthalpy map of free lipase (318–368 K). (D) Entropy–enthalpy map of GO-lip (318–368 K). All values are averages of results from triplicate trials; *error bars* indicate the SD values.

stability of the enzymes through supported attachment via covalent bonding (Dwevedi et al., 2009; Kumari & Kayastha, 2011). Immobilized enzyme with good reusability and prolonged stability provides an economical gain and can be used constantly in both batch and continuous reactors. Improvement in stability and reusability of enzymes following immobilization has been well documented (Dwevedi & Kayastha, 2009; Gaur, Pant, Jain, & Khare, 2006; Kishore & Kayastha, 2012). The K_m and V_{max} values of GO-lip at 45 and 95°C systems indicated that the enzyme system underwent a conformational change making it more efficient in catalytic turnover. The lipase from NLIP05 showed increased activity at 95°C when immobilized on GO which is indicative of the stability conferred to the enzyme by GO. The K_m, V_{max}, and E_a values for the enzyme indicate that substrate binding and catalytic efficiency were high at the temperature extremes.

We have devised a strategy of immobilizing lipase onto GO via glutaraldehyde as crosslinker. The enhanced enzyme activity at the upper temperature is a unique finding from this study.

The alteration in decay constant (k) with temperature, high deactivation energy (E_d), and the increase of $t_{1/2}$ value with approaching higher

Table 5 Effect of Variation in Kinetic Parameters for Immobilized GO-Lip and Free Lipase 318–368 K

		ΔH (kJ/mol)		ΔG (kJ/mol)		ΔS (J/mol/k)	
		GO		GO		GO	
Temperature (*k*)	**Enzymes**	(−)	(+)	(−)	(+)	(−)	(+)
318	Lip	−45.14 ± 1.6	20.54 ± 2.27	81.08 ± 2.12	92.14 ± 2.7	−437.2 ± 1.2	−229.1 ± 1.19
323		−45.21 ± 1.21	20.45 ± 2.43	85.66 ± 1.24	95.25 ± 3.5	−431.13 ± 2.4	−233.33 ± 1.19
333		−45.28 ± 1.44	20.46 ± 1.21	87.32 ± 3.22	98.93 ± 1.3	−423.25 ± 1.2	−237.7 ± 2.18
343		−45.35 ± 1.33	20.38 ± 2.25	88.74 ± 2.53	102.29 ± 0.9	−415.08 ± 2.59	−242.34 ± 1.22
353		−45.44 ± 0.91	20.30 ± 2.19	90.33 ± 0.98	106.19 ± 3.1	−408.0 ± 2.8	−245.18 ± 0.94
363		−45.53 ± 2.38	20.25 ± 1.72	93.49 ± 3.36	107.61 ± 2.4	−402.7 ± 2.1	−247.07 ± 1.73
368		−45.61 ± 0.99	20.20 ± 0.23	95.62 ± 1.23	110.72 ± 1.9	−395.4 ± 1.07	−251.09 ± 1.38

temperature of the GO-lip system implies that GO-based immobilization conferred structural stability to the enzyme at higher temperature indicating chaperone like activity.

6. CONCLUSION

In the current study, we have effectively optimized immobilization process of enzyme lipase on GO via glutaraldehyde as crosslinker. The high enzyme activity and storage stability of lipase indicate that industries can benefit following this process of enzyme immobilization. This methodology of immobilization could be easily implied to immobilization studies involving other enzymes for improved and stable end products.

ACKNOWLEDGMENT

The authors would like to thank all the lab members for their support.

FUNDING

This research did not receive any specific grant from funding agencies in the public, commercial, or not-for-profit sectors.

REFERENCES

Adkins, S. S., Hobbs, H. R., Benaissi, K., Johnston, K. P., Poliakoff, M., & Thomas, N. R. (2008). Stable colloidal dispersions of a lipase-perfluoropolyether complex in liquid and supercritical carbon dioxide. *The Journal of Physical Chemistry B*, *112*, 4760–4769.

Adlercreutz, P. (2013). Immobilisation and application of lipases in organic media. *Chemical Society Reviews*, *42*, 6406–6436.

Bastida, A., Sabuquillo, P., Armisen, P., Fernández-Lafuente, R., Huguet, J., & Guisán, J. M. (1998). A single step purification, immobilization, and hyperactivation of lipases via interfacial adsorption on strongly hydrophobic supports. *Biotechnology and Bioengineering*, *58*, 486–493.

Bernal, C., Illanes, A., & Wilson, L. (2014). Heterofunctional hydrophilic–hydrophobic porous silica as support for multipoint covalent immobilization of lipases: Application to lactulose palmitate synthesis. *Langmuir*, *30*, 3557–3566.

Chen, Z. M., Liu, L. L., & Yang, R. C. (2017). Improved performance of immobilized lipase by interfacial activation on Fe_3O_4 @PVBC nanoparticles. *RSC Advances*, *7*, 35169–35174.

Contesini, F. J., Figueira, J. A., Kawaguti, H. Y., Fernandes, P. C. B., Carvalho, P. O., Nascimento, M. G., et al. (2013). Potential applications of carbohydrases immobilization in the food industry. *International Journal of Molecular Sciences*, *14*, 1335–1369.

Duarte, S. H., del Peso Hernández, G. L., Canet, A., Benaiges, M. D., Maugeri, F., & Valero, F. (2015). Enzymatic biodiesel synthesis from yeast oil using immobilized recombinant Rhizopus oryzae lipase. *Bioresource Technology*, *183*, 175–180.

Dutta, N., Biswas, S., & Saha, M. K. (2016). Biophysical characterization and activity analysis of nano-magnesium supplemented cellulase obtained from a psychrobacterium following graphene oxide immobilization. *Enzyme and Microbial Technology*, *95*, 245–258.

Dutta, N., Biswas, S., & Saha, M. K. (2017). Nano-magnesium aided activity enhancement and biophysical characterization of a psychrophilic α-amylase immobilized on graphene oxide nanosupport. *Journal of Bioscience and Bioengineering, 124*, 15–22. https://doi.org/10.1016/j.jbiosc.2017.02.002.

Dutta, N., Mukhopadhyay, A., Dasgupta, A. K., & Chakrabarti, K. (2013). Nanotechnology enabled enhancement of enzyme activity and thermostability: Study on impaired pectate lyase from attenuated Macrophomina phaseolina in presence of hydroxyapatite nanoparticle. *PLoS ONE, 8*(5), e63567.

Dutta, N., Mukhopadhyay, A., Dasgupta, A. K., & Chakrabarti, K. (2014). Improved production of reducing sugars from rice husk and rice straw using bacterial cellulase and xylanase activated with hydroxyapatite nanoparticles. *Bioresource Technology, 153*, 269–277.

Dwevedi, A., & Kayastha, A. M. (2009). Optimal immobilization of β-galactosidase from pea (PsBGAL) onto Sephadex and chitosan beads using response surface methodology and its applications. *Bioresource Technology, 100*, 2667–2675.

Dwevedi, A., Routh, S. B., Yadav, A. S., Singh, A. K., Srivastava, O. N., & Kayastha, A. M. (2011). Response surface analysis of nano-ureases from *Canavalia ensiformis* and *Cajanus cajan*. *International Journal of Biological Macromolecules, 49*, 674–680.

Dwevedi, A., Singh, A. K., Singh, D. P., Srivastava, O. N., & Kayastha, A. M. (2009). Lactose nano-probe optimized using response surface methodology. *Biosensors & Bioelectronics, 25*, 784–790.

Fernandez-Lafuente, R. (2010). Lipase from Thermomyces lanuginosus: Uses and prospects as an industrial biocatalyst. *Journal of Molecular Catalysis B: Enzymatic, 62*, 197–212.

Gaur, R., Pant, H., Jain, R., & Khare, S. K. (2006). Galacto-oligosaccharide synthesis by immobilized Aspergillus oryzae β-galactosidase. *Food Chemistry, 97*, 426–430.

Gu, H., Yu, Y., Liu, X., Ni, B., Zhou, T., et al. (2011). Layer-by-layer self-assembly of functionalized graphene nanoplates for glucose sensing in vivo integrated with on-line microdialysis system. *Biosensors & Bioelectronics, 32*, 118–126.

Gurung, N., Ray, S., Bose, S., & Rai, V. (2013). A broader view: Microbial enzymes and their relevance in industries, medicine, and beyond. *BioMed Research International, 2013*, 329121.

Kiernan, J. A. (2000). Formaldehyde, formalin, paraformaldehyde and glutaraldehyde: What they are and what they do. *Microscopy Today, 1*, 8–12.

Kim, H., Choi, Y. K., Lee, J., Lee, E., Park, J., & Kim, M. J. (2011). Ionic-surfactant-coated *Burkholderia cepacia* lipase as a highly active and enantioselective catalyst for the dynamic kinetic resolution of secondary alcohols. *Angewandte Chemie, International Edition, 50*, 10944–10948.

Kishore, D., & Kayastha, A. M. (2012). Optimization of immobilization conditions for chick pea β-galactosidase (CpGAL) to alkylamine glass using response surface methodology and its applications in lactose hydrolysis. *Food Chemistry, 134*, 1650–1657.

Konwarh, R., Karak, N., Rai, S. K., & Mukherjee, A. K. (2009). Polymer-assisted iron oxide magnetic nanoparticle immobilized keratinase. *Nanotechnology, 20*, 225107.

Kuila, T., Bose, S., Khanra, P., Mishra, A. K., Kim, N. H., & Lee, J. H. (2011). Recent advances in graphene-based biosensors. *Biosensors & Bioelectronics, 26*, 4637–4648.

Kumari, A., & Kayastha, A. M. (2011). Immobilization of soybean (Glycine max) α-amylase onto chitosan and Amberlite MB-150 beads: Optimization and characterization. *Journal of Molecular Catalysis B: Enzymatic, 69*, 8–14.

Liu, J.-H., Wang, T.-C., Wang, H., Gu, Y., Xu, Y., Tang, H., et al. (2014). Biocompatibility of graphene oxide intravenously administrated in mice—Effects of dose, size and exposure protocols. *Toxicology Research, 4*, 83–91. https://doi.org/10.1039/C4TX00044G.

Maiti, S., Ghosh, M., & Das, P. K. (2011). Gold nanorod in reverse micelles: A fitting fusion to catapult lipase activity. *Chemical Communications, 47*, 9864–9866.

McGinley, H. R. (2012). Glutaraldehyde uses and counterfeits. *Water Technology, 5*, 30–32.
Momsia, T., & Momsia, P. (2013). A review on microbial lipase-versatile tool for industrial applications. *International Journal of Life Sciences Biotechnology and Pharma Research, 2*, 45–49.
Nimni, M. E., Cheung, D., Strates, B., Kodama, M., & Sheikh, K. (1987). Chemically modified collagen: A natural biomaterial for tissue replacement. *Journal of Biomedical Materials Research, 21*, 741–771.
Okuda, K., Urabe, I., Yamada, Y., & Okada, H. (1991). Reaction of glutaraldehyde with amino and thiol compounds. *Journal of Fermentation and Bioengineering, 71*, 100–105.
Ortega, N., de Diego, S., Rodríguez-Nogales, J. M., Perez-Mateos, M., & Busto, M. D. (2004). Kinetic behaviour and thermal inactivation of pectin lyase used in food processing. *International Journal of Food Science and Technology, 39*, 631–639.
Secundo, F., Carrea, G., Tarabiono, C., Gatti-Lafranconi, P., Brocca, S., Lotti, M., et al. (2006). The lid is a structural and functional determinant of lipase activity and selectivity. *Journal of Molecular Catalysis B: Enzymatic, 39*, 166–170.
Singh, N., Srivastava, G., Talat, M., Raghubanshi, H., & Srivastava, O. N. (2014). Cicer α-galactosidase immobilization onto functionalized graphene nanosheets using response surface method and its applications. *Food Chemistry, 142*, 430–438.
Soler, L., Illanes, A., & Wilson, L. (2016). Immobilization of Alcaligenes sp. lipase as catalyst for the transesterification of vegetable oils to produce biodiesel. *Catalysis Today, 259*, 177–182.
Stankovich, S., Piner, R., Nguyen, S., & Ruoff, R. (2006). Synthesis and exfoliation of isocyanate-treated graphene oxide nanoplatelets. *Carbon, 44*, 3342–3347.
Stoller, M., Park, S., Zhu, Y., An, J., & Rouff, R. S. (2008). Graphene-based ultracapacitors. *Nano Letters, 8*, 3498–3502.
Sun, X., Feng, Z., Hou, T., & Li, Y. (2014). Mechanism of graphene oxide as an enzyme inhibitor from molecular dynamics simulations. *ACS Applied Materials & Interfaces, 6*, 7153–7163.
Wang, Y., Li, Z., Wang, J., Li, J., & Lin, Y. (2011). Graphene and graphene oxide: Biofunctionalization and applications in biotechnology. *Trends in Biotechnology, 29*, 205–212.
Wang, K., Ruan, J., Song, H., Zhang, J., Wo, Y., et al. (2011). Biocompatibility of graphene oxide. *Nanoscale Research Letters, 6*, 8.
Yan, E. K., Caoa, H. L., Zhang, C. Y., Lu, Q. Q., Ye, Y. J., He, J., et al. (2015). Crosslinked protein crystals by glutaraldehyde and their applications. *RSC Advances, 5*, 26163–26174.
Zhang, Y., Tang, Z., Fu, X., & Xu, Y. (2010). TiO_2-graphene nanocomposites for gas-phase photocatalytic degradation of volatile aromatic pollutant: Is TiO_2-graphene truly different from other TiO_2-carbon composite materials? *ACS Nano, 4*, 7303–7314.
Zhang, W. W., Wang, N., Zhou, Y. J., He, T., & Yu, X. Q. (2012). Enhancement of activity and stability of lipase by microemulsion-based organogels (MBGs) immobilization and application for synthesis of arylethyl acetate. *Journal of Molecular Catalysis B: Enzymatic, 78*, 65–71.
Zhang, X., Yin, J., Peng, C., Hu, W., Zhu, Z., et al. (2011). Distribution and biocompatibility studies of graphene oxide in mice after intravenous administration. *Carbon, 49*, 986–995.
Zhang, Y., Zhang, J., Huang, X., Zhou, X., Wu, H., & Guo, S. (2012). Assembly of grapheme oxide–enzyme conjugates through hydrophobic interaction. *Small, 8*, 154–159.
Zhang, J., Zhang, F., Yang, H., Huang, X., Liu, H., Zhang, J., et al. (2010). Graphene oxide as a matrix for enzyme immobilization. *Langmuir, 26*, 6083–6085.
Zhao, F., Li, H., Jiang, Y., Wang, X., & Mu, X. (2014). Co-immobilization of multi-enzyme on control-reduced graphene oxide by non-covalent bonds: An artificial biocatalytic system for the one-pot production of gluconic acid from starch. *Green Chemistry, 16*, 2558–2565.

Zhao, X., Qi, F., Yuan, C., Du, W., & Liu, D. (2015). Lipase-catalyzed process for biodiesel production: Enzyme immobilization, process simulation and optimization. *Renewable and Sustainable Energy Reviews*, *44*, 182–197.

FURTHER READING

Zhang, W. W., Wang, N., Zhang, L., Wu, W. X., Hu, C. L., & Yu, X. Q. (2014). Effects of additives on lipase immobilization in microemulsion-based organogels. *Applied Biochemistry and Biotechnology*, *172*, 3128–3140.

CHAPTER TEN

A Simple Flow Reactor for Continuous Synthesis of Biographene for Enzymology Studies

Megan K. Puglia*, Murali Anuganti*, Yao Lin*,†, Challa V. Kumar*,‡,1

*Department of Chemistry, University of Connecticut, Storrs, CT, United States
†Polymer Program, Institute of Materials Science, University of Connecticut, Storrs, CT, United States
‡Department of Molecular and Cell Biology, University of Connecticut, Storrs, CT, United States
[1]Corresponding author: e-mail address: challa.kumar@uconn.edu

Contents

Abstract

The unique properties of graphene make it an intriguing platform for the attachment and enhancement of biological molecules, but it has yet to achieve its full potential in terms of biological applications. Single-layer graphene is expensive, making alternatives to this material highly desired for applications that require high-quality graphene in large quantities. In this context, we report a simple, environmentally friendly, nonlabor-intensive method for the synthesis of colloidal graphene suspensions of 3–5 layers, stabilized by bovine serum albumin, in water. The method involves a flow reactor designed to continually yield high-quality graphene colloids, synthesized,

Methods in Enzymology, Volume 609
ISSN 0076-6879
https://doi.org/10.1016/bs.mie.2018.08.015

purified, and optimized all in one setup. The flow reactor is able to produce colloidal graphene sheets on a multigram scale, and these colloids were characterized by Raman spectroscopy, electron microscopy, and zeta potential studies. The average size of the sheets is 0.16 μm^2, each consisting of 3–5 layers of graphene with little or no sp^3 defects. These graphene colloids stabilized by the protein were successfully used in protein kinetic studies as well as in surface plasmon resonance protein binding studies. The ease of synthesis of these high-quality graphene colloidal suspensions in water provides an exciting opportunity for biographene to be used on an industrial scale for electronic, thermal, and enzymology applications.

1. INTRODUCTION

Graphene, a carbon material discovered in 2004, is the name given to sheets of single carbon atoms connected in an infinite hexagonal pattern, such that each carbon atom is sp^2 bonded to three other carbon atoms (Novoselov et al., 2004). This carbon allotrope quickly rose to the forefront of research in materials science due to its unique thermal, electrical, and optical properties (Novoselov et al., 2005).

A sister to diamond and charcoal, this all-carbon, atomically thin material is biodegradable (Falcao Eduardo & Wudl, 2007; Sridhar, Lee, Chun, & Park, 2013), making it ideal for environmentally conscious and biological research. Graphene's unique combination of traits makes it highly intriguing for use in biomedical devices such as sensors, and has been proposed as a vehicle for drug delivery (McCallion, Burthem, Rees-Unwin, Golovanov, & Pluen, 2016; Pattammattel & Kumar, 2017; Yang, Asiri, Tang, Du, & Lin, 2013).

Graphene research is currently limited by the fact that it is tedious and expensive to make, often times involving high temperatures in the synthesis and necessitating very specific substrates (Bhuyan, Uddin, Islam, Bipasha, & Hossain, 2016). "Top-down" production methods have been integral in creating a platform for large-scale synthesis of graphene and other two-dimensional materials for use in laboratories (Paton et al., 2014). However, these methods regularly make graphene dispersions in organic solvents such as *N*-methyl-2-pyrrolidone and dimethyl sulfoxide, creating large obstacles for biological researchers looking to incorporate and study graphene (Johnson, Dobson, & Coleman, 2015; Pattammattel & Kumar, 2017).

Recently, methods for synthesizing high-quality graphene suspensions in water using shear force and bovine serum albumin (BSA) have been reported (Ahadian et al., 2015). One method allows for the use of a common kitchen blender and BSA in water. This simple method yielded production rates higher than 4 mg/mL/h for graphene suspensions in water, which were stable for up to 2 months (Pattammattel & Kumar, 2015). It was also subsequently shown that this method could also be used to make graphene suspensions using protein sera from a number of different species (Pattammattel et al., 2017). Graphene made in sera is ideal for biological applications due to its high stability as well as its inherent compatibility with biological systems.

BSA, a 66.4-kDa-sized carrier protein in the blood of cows, was chosen as the exfoliant for graphite powder for a number of reasons. BSA is an inexpensive by-product of the meat industry with a negative charge at neutral pH, which was shown to enhance exfoliation efficiency (Bah, Bekhit, Carne, & McConnell, 2013; Pattammattel & Kumar, 2015). BSA is thought to bind to the sp^2 carbons of graphene by the stacking of its aromatic residues onto the sheets as well as by adsorption of residues onto the surface (Gu et al., 2015). Additionally BSA has a number of carboxylic acid (99), as well as amine (82) groups that provide easy attachments sites for someone who wishes to do conjugation chemistry (Bujacz, Zielinski, & Sekula, 2014). Using BSA also helps to stabilize the graphene sheets in solution through noncovalent interactions between the hydrophobic segments of BSA and the graphene sheets, while the exposed hydrophilic segments of BSA allow for easy solubility in water (Ahadian et al., 2015). It is also suggested that the BSA helps to prevent the restacking of graphene sheets in solution and its attachment to the graphene sheets permits less harsh interactions between graphene and other biomolecules (Alava et al., 2013; Pattammattel & Kumar, 2015; Pattammattel et al., 2017, 2013).

Making water-soluble, biologically compatible graphene dispersions of high concentrations remains a laborious and challenging task. Here we report a novel flow reactor system for the synthesis of high concentrations of high-quality graphene dispersions. In a single reactor, the graphene is exfoliated into sheets from graphite flakes, purified of any unexfoliated graphite, and then separated based on size, all in one vessel. The graphene is coated with the minimal quantity of protein necessary to solubilize it in water and render it suitable for biological applications, as described later.

2. CONTINUAL SYNTHESIS OF HIGH-QUALITY BIOGRAPHENE

The proposed system (Scheme 1) has the ability to synthesize highly concentrated graphene suspensions in water, using BSA, with very minimal external effort necessary after the initial setup. The suspensions have been

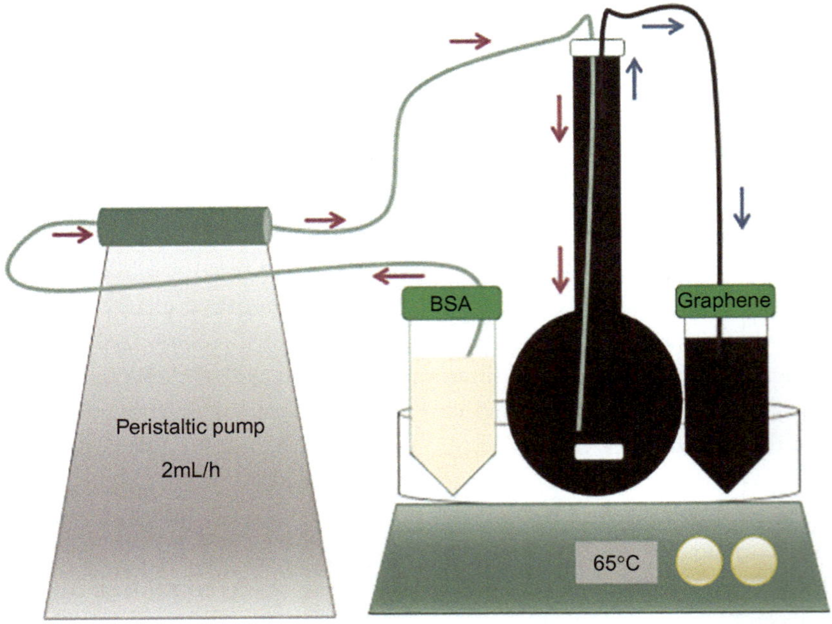

Scheme 1 Continual biographene flow setup. A round bottom flask connected to a long neck is filled with BSA solution in distilled water (20 mg/mL) and graphite flakes (100 mg/mL) is set up and stirred vigorously at 65°C. A BSA solution (*left*) (20 mg/mL) is flowed through a peristaltic pump at 2–3 mL/h into the bottom of the round bottom flask stirring at 65°C. Graphene solution at the top of the neck is then pushed out into the waiting collecting flask, also kept at 65°C.

characterized using Raman spectroscopy, transmission electron microscopy, UV–visible spectroscopy, and zeta potential measurements. The system requires only a stir bar, hot plate, and pump (peristaltic or syringe), all of which are items normally found in research laboratories. Similar to the method proposed by Pattammattel and Kumar (2015) a solution of BSA (20 mg/mL) in distilled water is added to the flask with 100 mg/mL graphite flakes. However, instead of using a kitchen blender for exfoliation, the solution is heated to 65°C and stirred vigorously, thus eliminating the need to turn machinery on and off multiple times (Fig. 1, Scheme 2).

This setup inherently eliminates the need for centrifugation after the exfoliating process due to the natural gravity separation caused by the elongated neck of the flask. All unexfoliated graphite will fall to the bottom of the flask, and few-layered graphene will float to the top of the flask and subsequently be the first to be pushed out of the exit tube. The collected graphene was analyzed using Raman spectroscopy (Fig. 2) and found to

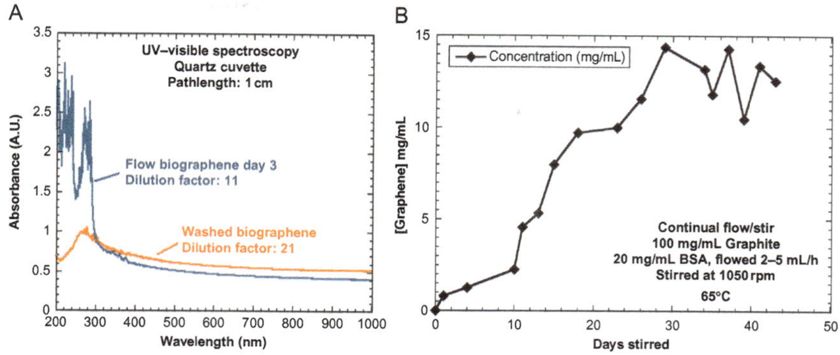

Fig. 1 (A) Typical UV–visible absorbance spectra of biographene (*blue*) and washed biographene (*orange*) diluted using distilled water. The absorbance at 660 nm was used to calculate the concentration using our lab's previously reported extinction coefficient (Pattammattel & Kumar, 2015). The large scattering around 280 nm seen in the biographene spectrum is due to the BSA in the sample. It can be seen that the washed biographene has a small peak at 280 nm due to the small amount of BSA bound to the surface. (B) Plot of graphene concentration (mg/mL), calculated using the extinction coefficient at 660 nm, vs days stirred in the flow reactor system at 65°C.

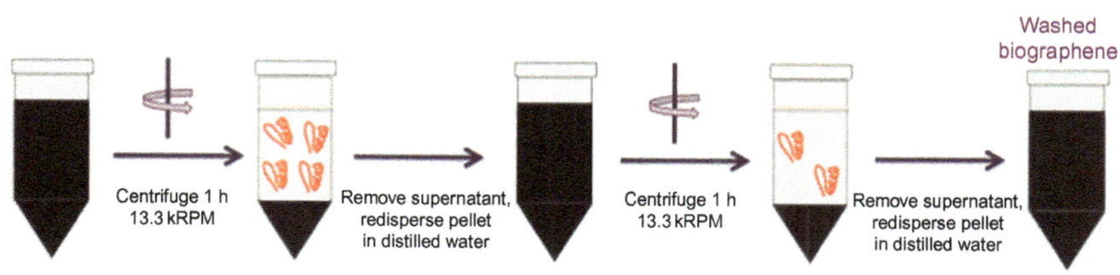

Scheme 2 Process for removal of free BSA in biographene solution. A biographene sample is centrifuged for 65 min at 13.3 kRPM and the resulting supernatant containing free BSA is removed. The pellet, containing graphene sheets bound to BSA, is then redispersed in distilled water. The centrifugation process is repeated and the second supernatant is removed. The resulting pellet is redispersed in distilled water again. Supernatants 1 and 2 can be tested for their [BSA] in order to confirm the total amount of BSA bound to the washed biographene. Washed biographene was found to contain less than 20% of the initial BSA in solution after the washing process.

have 3–4 layers per flake, with an average flake size of 0.16 μm^2. The size of these few-layered flakes was confirmed using transmission electron microscopy (Fig. 3).

The flow reactor also allows for the optimization of graphene as per the researcher's needs. If fewer layers per sheet are required, the neck attached to the round bottom flask may be extended. The concentration, as well as sample preparation time, can also be controlled by adjusting the flow rate of the

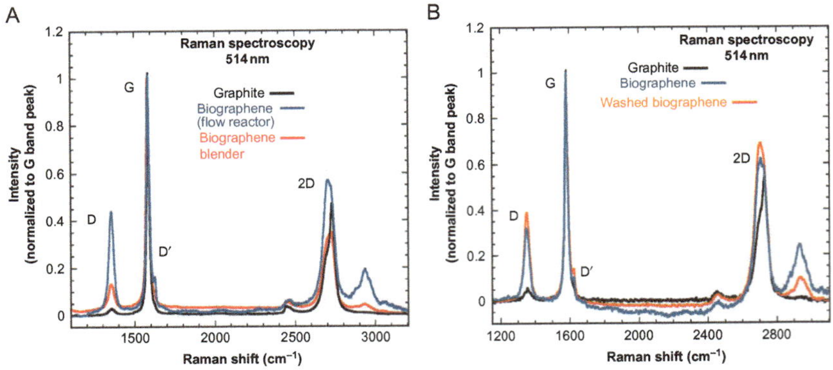

Fig. 2 (A) Raman spectra of biographene made in the flow reactor compared to the biographene made in a blender via the method reported by Pattammattel and Kumar (2015). (B) Raman spectra of biographene made in the flow reactor compared to biographene from the flow reactor that had all protein removed in the washing process discussed previously. A 514-nm excitation laser was used for all Raman measurements and each Raman band is normalized to the G band (1582 cm^{-1}).

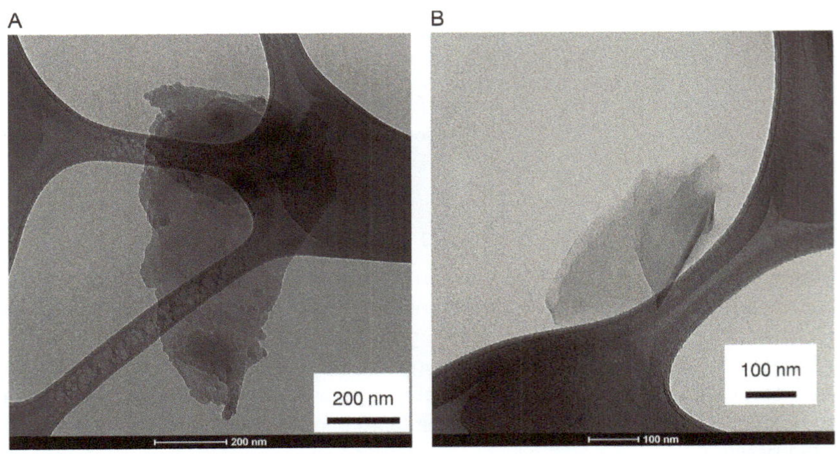

Fig. 3 Transmission electron microscopy images of a "washed" graphene sheet imaged after removing the free protein from the graphene using the aforementioned centrifugation method. (A) A jagged few-layered sheet, about 0.6 μm in length. (B) The sheet appears folded on itself and is transparent, indicating very few layers.

BSA solution into the bottom of the flask. In addition, higher or lower concentrations of protein solution can be flowed into the graphene depending on how much is desired. The BSA in the graphene solution can also be removed by centrifugation, in a case where free protein is not conducive to the experiments needed.

The method presented below is unique in that it is simplified to require very basic laboratory equipment (a stir bar and a hotplate) and minimal effort. Biological chemistry labs that may not be particularly familiar with top-down material exfoliation techniques can easily repeat this method. Additionally, the use of the protein in the exfoliation opens the door for research combining graphene with biological molecules without immediate incompatibility and could possibly even be amended to work with a protein of the researcher's choice. The method presented here also does not require an extensive purification process, which will be advantageous for labs wishing to extensively study graphene. The flow reactor method gives close to a 100% yield, as all of the graphite flakes added to the reactor are eventually synthesized into graphene with the continual flow of BSA solution.

3. PREPARATION AND ENZYMOLOGY APPLICATIONS

The biocatalytic opportunities afforded by graphene were not immediately realized upon its discovery in 2004. One major limitation was the fact that the hydrophobic sheets were not soluble in water and bio-incompatible chemicals were often used during synthesis, or to stabilize the sheets in solution. The synthesis of biographene (Ahadian et al., 2015; Jiang, Li, Yao, Shao, & Chen, 2016; Pattammattel & Kumar, 2015) has opened the door for a number of different applications in terms of proteins and enzyme studies (Pattammattel & Kumar, 2017; Pattammattel et al., 2017). Recently, biographene was used as a modification to a previously reported method allowing for the immobilization and stabilization of enzymes in cellulose fibers (Riccardi, Kasi, & Kumar, 2017; Riccardi et al., 2016). 1-Ethyl-3-(3-dimethylaminopropyl)carbodiimide (EDC) chemistry was used to entrap redox-active enzymes (horseradish peroxidase and glucose oxidase) in biographene-coated cellulose fibers. This allowed for enhanced enzyme stability as well as an increased relative activity. Specific samples made with the biographene showed a relative protein activity of up to 279%, compared to those without biographene. These improvements were attributed to graphene's ability to act as an electron carrier, as well as the additional attachments sites afforded by the BSA on the graphene surface. Additional enzyme attachments sites inhibit the denaturation of the enzyme by cradling and supporting the structure, and subsequently increasing the free energy of denaturation (Limbacher, Puglia, Riccardi, & Kumar, 2018).

Not all enzymology experiments would benefit from having an abundance of extra BSA in the graphene solution. For this reason we have developed a method to remove the free BSA from the biographene sheets, via centrifugation, to create what our group has deemed "washed biographene." The washed biographene sample has been analyzed with Raman spectroscopy, UV–visible spectroscopy, zeta potential measurements, and transmission electron microscopy.

This washed biographene sample has been used to coat a gold sensor chip for surface plasmon resonance studies in order to determine the direct binding of proteins to the washed biographene surface. One example is discussed later with the corresponding kinetic analysis of the binding. A better understanding of how proteins with different sizes, side chains, isoelectric points, and functions may react with the biographene surface is imperative information for further controlling the graphene–biomolecule interface for biological applications. A method for performing surface plasmon resonance experiments with washed biographene is described later.

4. METHODS

4.1 Flow Reaction Setup

Equipment

1. Stir plate
2. 500-mL Round bottom flask
3. Reflux condenser tube or long neck made with ground glass joint to fit round bottom flask perfectly
4. Septum to fit top of reflux condenser tube/neck piece
5. Tubing (1.6 mm diameter or larger)
6. Peristaltic pump (BIO-RAD, USA)

Materials

1. Graphite flakes (Millipore Sigma, USA, MKBW0432V)
2. BSA (Equitech-Bio, Inc., TX, BAH64-1366)
3. Distilled water

Preparation Method

1. BSA (20 mg/mL) was dissolved in distilled water and the pH was adjusted to 7.0 using HCl and added to the 500-mL round bottom flask enough to fill to the bottom of the flask neck.
2. *Tip*: You will want to measure the total volume of your round bottom flask and condenser setup beforehand and make a BSA solution for the entire volume, although it will not all be poured in at once.

3. Part of the BSA solution was stirred at 65–70°C in the open round bottom flask, while graphite flakes (100 mg/mL) were added in small increments, to avoid the stir bar getting buried in the flakes.
4. When the BSA graphene mixture is filled to just above the neck of the round bottom flask, attach the appropriate condenser tube for the flask, and finish pouring the rest of the BSA solution in from the top. Check to make sure there are no leaks in the system.
5. *Tip:* The setup is made much simpler by having a round bottom flask with an elongated neck made by a glass blowing expert, if a glass blowing service is available.
6. Puncture an appropriately sized septum with a needle (in two spots) and fit your tubing through. One of the tubes will be threaded through the septum just enough to reach the top of your liquid and the other should be threaded through so that it reaches the middle of the round bottom flask. Make sure you leave enough tubing on the other end to reach your pump system.
7. *Tip:* We used GO2 glue to seal any small leaks in the system. These usually occur where the tubing is piercing through the septum.
8. When the tubing has been attached to the top of the reflux condenser, attach your tubing to your peristaltic pump and load a 20-mg/mL BSA solution into the pump reservoir.
9. It is a good idea to pump the BSA solution into your system slowly at first in case there are any leaks in the system. Once everything is flowing smoothly, you can increase your flow rate.
10. It is not necessary to flow BSA through the solution right away, as it will take about 24 h for the reaction to produce higher concentrations of graphene.

4.2 Removing Free Protein From Biographene Solution

Equipment

1. Centrifuge (Fisher Scientific, USA)

Materials

1. Distilled water

Preparation Method

1. Centrifuge biographene at 13.3 kRPM for 65 min.
2. Remove clear supernatant (free BSA) from centrifuge tube, leaving only the graphene pellet at the bottom. Save the removed protein solution to later confirm its concentration.
3. Resuspend the graphene pellet in distilled water.
4. Centrifuge new suspensions at 13.3 kRPM for 65 min.

5. Remove clear supernatant from centrifuge tube, leaving only the graphene pellet at the bottom. Save the removed protein solution to later confirm its concentration.
6. Resuspend the graphene in distilled water.
7. At this point it is possible to further concentrate your graphene, if desired, by resuspending in a smaller volume of water.

Tip: If the graphene pellet is not dissolving well in water after centrifuging a second time, placing the graphene and water solution in a sonicator for 5–10 min can help breakup and redisperse the pellet into solution.

We have washed many different types of concentrations of graphene, made with different concentrations of BSA, and have consistently seen that at this speed 80%–90% of the BSA is always removed. The biographene also can be easily resuspended in the water, meaning that there is some residual BSA still bound to the sheets making them more hydrophilic. In a washed sample of biographene there is calculated to be 0.3 mg of protein for every 1 mg of graphene.

4.3 Determining Biographene Concentration UV–Visible Spectroscopy

Equipment

1. UV–vis spectrophotometer (Agilent Technologies, CA)
2. Quartz cuvette

Method

1. The flow reaction setup will synthesize, purify, and optimize your graphene solution all in one. Aliquots of the produced biographene solution can be taken out and tested using a UV–visible spectrophotometer to calculate the concentration of graphene being produced.
2. Measure the absorbance of the biographene solution at 660 nm after appropriately diluting the solution with distilled water so the absorbance is below 1.
3. The concentration can be calculated using the measured extinction coefficient of 3.98 mg^{-1} mL cm^{-1} (Pattammattel & Kumar, 2015).

4.4 Characterizing Graphene With Raman Spectroscopy

Equipment

1. Raman spectrometer, 514 nm Laser with 1 μm spacial resolution and 50 mW power (Renishaw System 2000)
2. Microscope slides (Globe Scientific Inc., NJ)

Method

1. Draw a ≈1.0 cm diameter circle with a wax pen on a clean microscope slide.
2. Drop-cast 50 μL of biographene (10 mg/mL) into your circle and let dry.
3. The Raman spectra can then be collected from 1100 to 3200 cm^{-1}, using a 514-nm laser Raman spectrometer.
4. A control spectra should also be collected of the unexfoliated graphene flakes used in the synthesis (from 1100 to 3200 cm^{-1}).
5. Each Raman band should be normalized by dividing the spectra so the G band (1582 cm^{-1}) has an intensity equal to 1 before analysis.
6. *Tip:* Tap the graphite flakes onto the microscope slide and press the mound down with a second glass slide to make imaging easier.
7. *Tip:* The protein in the biographene can become damaged from the 514-nm laser. To avoid burning of the protein in the sample, which can decrease the signal-to-noise ratio, you can reduce the laser power to 50% (25 mW) of the original source power. A range of reported laser powers used for collecting the Raman spectra of BSA is reported in the literature in order to minimize a decrease in the signal-to-noise ratio caused by perturbation of the sample from the laser (Biscar, Dhall, & Pennison, 1972; Bolton & Scherer, 1989; Lord & Yu, 1970).

4.5 Raman Spectroscopy Equations and Analysis

As can be seen from Fig. 2 both the flow biographene and the washed biographene (free protein removed) exhibited the characteristic Raman peaks at 1582 (G), 1622 (D′), 1353 (D), and 2726 (2D) cm^{-1}. The D′ band is a shoulder peak to the right of the G band. Using the normalized intensities (see Section 4.4, Method Step 5) of the 2D, D, and G peaks for both the graphene samples and bulk graphite, along with Eqs. (1) and (2), it was possible to estimate the number of layers per graphene sheet as well as lateral sheet size (Paton et al., 2014; Pattammattel & Kumar, 2015). Paton et al. developed these equations by correlating transmission electron microscopy, images showing sheet sizes, and atomic force microscopy images, giving number of layers per sheet, to Raman spectra for differently synthesized samples. Raman excitation wavelengths used for this pool of data included both 532.15 and 514 nm (Paton et al., 2014). The Raman spectra calculations for the flow reactor biographene suggest there are 3–4 layers per graphene sheet and the sheet size is about 0.16 μm^2.

In analyzing the Raman spectra it is also important to compare ID/ID′ ratios of the samples in order to determine levels of oxidation in the sample (Paton et al., 2014; Pattammattel & Kumar, 2015). For our graphene samples the ID/ID′ ratio was about 2.7 which suggests that our dispersed graphene sheets lack sp^3 defects and were not oxidized while in the flow reactor, or during the washing process. It has been reported previously that an ID/ID′ ratio below 3.5 implies the presence of edge defects on the sheets (Eckmann et al., 2012; Paton et al., 2014).

$$\text{Number of layers} = 10^{0.84M+0.45M^2} \cdot M = \frac{\left[I_{2\text{D peak}}/I_{2\text{D peak}-30\ \text{cm}^{-1}}\right]_{\text{graphene}}}{\left[I_{2\text{D peak}}/I_{2\text{D peak}-30\ \text{cm}^{-1}}\right]_{\text{graphite}}} \tag{1}$$

$$\text{Lateral sheet size} = \frac{k}{[I_{\text{D}}/I_{\text{G}}]_{\text{graphene}} - [I_{\text{D}}/I_{\text{G}}]_{\text{graphite}}} \tag{2}$$

*k was experimentally calculated to be 0.17 using transmission electron microscopy data as detailed in our previously published work (Pattammattel & Kumar, 2015).

4.6 Transmission Electron Microscopy

Equipment

1. FEI Tecnai T12 TEM

Materials

1. Ultrathin carbon film supported by a lacey carbon film on a 400 mesh copper grid, product no. 01824 (Ted Pella, Inc., CA)

Method

1. Dilute graphene sample with distilled water to 20 μg/mL.
2. Dropcast 12 μL of solution onto a copper mesh grid and wait 20 s.
3. Dab off liquid with corner of filter paper until there is just a sheen of liquid left on the grid.
4. Let dry, covered, overnight.
5. Image using an accelerating voltage of 120 kV.

4.7 Zeta Potential Studies

Equipment

1. Zeta Potential Analyzer (Brookhaven ZetaPlus, NY)

Materials

1. 100 mM KCl solution in water
2. Dilute HCl
3. Dilute NaOH

Method

1. Dilute graphene sample with distilled water so that it is translucent in color.
2. Adjust pH of solution to 7 using HCl or NaOH.
3. If the graphene solution is giving a low conductance reading, you can spike the solution with 100 μM KCl.

The zeta potential of the flow biographene solutions was found to be -36.7 ± 3.8 mV after sampling seven different samples at pH 7. The zeta potential of this sample reflects the negative charge on the protein at pH 7. The washed graphene samples had an average zeta potential value of -38.9 ± 4.1 mV.

4.8 Storage and Stability

Samples can be successfully stored at 4°C for more than 40 days. It may be necessary to stir older samples briefly before use. Due to the nature of the flow reaction it does not necessitate any further addition of graphite for more than 40 days after the initial setup.

4.9 SPR Studies of Protein Binding to Graphene

1. SPR 7000DC (Reichert Technologies)

Materials

1. 10 mM Sodium phosphate buffer pH 7.0
2. 500 μg/mL Biographene (free protein removed from solution)
3. Piranha solution (3:1 H_2SO_4/H_2O_2)
4. Plain gold SPR slides (Reichert Technologies)
5. Coupling oil (Reichert Technologies)
6. 500 μL Sample loop

Method

Making the necessary solutions

1. Degas the 10 mM sodium phosphate buffer overnight by gently stirring under vacuum.
2. Use degassed buffer to make the protein solutions of choice.
3. Filter all solutions that will go into the SPR tubing with a 0.2-μm filter.

4. Biographene solution should have all free protein removed as previously described in Section 4.2.

Making the SPR chips

1. Clean new SPR chips by submersing them in Piranha solution for 10 min and then thoroughly rinsing with water.
2. Dropcast 250 μL of 500 μg/mL washed biographene onto a new, clean SPR chip and allow to dry overnight.
   a. If reusing an old SPR chip, make sure it has been thoroughly cleaned.
3. After the washed biographene has dried, soak the chips in 10 m*M* sodium phosphate buffer (pH 7.0) overnight.
4. Remove the buffer and add fresh buffer the next morning, let soak all day. This will help to remove any protein bound to the graphene surface as well as any graphene not bound to the gold surface (Fig. 4).

Running SPR binding studies

1. Add clean chip with a drop of coupling oil into your SPR machine sensor holder. Make sure to appropriately clean the gasket before tightening the chip cover into place.
2. Check all tubing and connections.

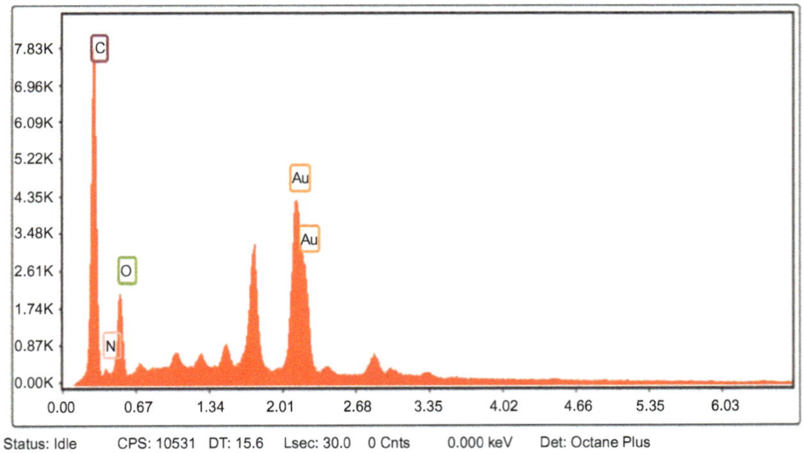

Fig. 4 Energy-dispersive X-ray spectroscopy spectrum of biographene on a gold SPR sensor chip. The Au-MLG sensor chip was soaked in sodium phosphate buffer (pH 7.0) overnight, twice, in order to remove any remaining BSA on the surface, as well as any unbound graphene. The nitrogen content was analyzed in order to determine the amount of protein remaining on the surface after the extensive washing procedure. The w/w percentages of the elements were 0.04%, 41.0%, and 8.80% for nitrogen, carbon, and oxygen, respectively. The nitrogen content is very low indicating that there is little to no protein remaining on the graphene surface, allowing for undeterred protein binding in the SPR studies.

3. Set infuse rate on the pump connected to SPR machine to 25 μL/min and allow degassed buffer to flow through your system. Make sure there are no air bubbles in your syringe pump or system.
4. Rinse 500 μL sample loop with ethanol, air, and water before injecting protein/sample solution.
5. Start by injecting 550 μL of a 1 μ*M* protein solution.
   a. Inject more protein solution than the length of the sample loop to make sure it is full.
6. Inject the protein solution for 600 s. The injection time can be adjusted according to the signal changes seen.
7. After the injection is finished, the sample loop can be cleaned by first evacuating the previous sample by pushing air through the system, then rinsing the system with distilled water, and then buffer.
8. Allow the signal to return to baseline.
9. A new sample can then be injected and tested after the signal has returned to baseline. If the protein has dissociated but the signal has not yet returned to the original baseline, sodium dodecyl sulfate (SDS) can be injected to regenerate the sensor surface.
10. Adjust the protein concentration and injection times according to the signal changes seen in order to get robust on and off rates.

4.10 Kinetic Analysis of SPR Binding Data

SPR binding data were collected using BSA on a plain gold sensor chip (Au sensor) as well as on a gold sensor chip coated with washed biographene (Au-MLG sensor) and the kinetic traces were plotted for analysis (Fig. 5). It can be seen in Fig. 5 that the biographene-coated sensor chip was much more sensitive to BSA binding than the regular gold sensor chip. The Au sensor showed a moderate signal for 1.66 g/mL of BSA, whereas a signal could be seen on the Au-MLG sensor for only 133 ng/mL of BSA.

From the traces in Fig. 5, k_a, k_d, and K_D were calculated for BSA biding to gold vs BSA binding to biographene. These parameters were estimated using the Langmuir adsorption kinetic model and the calculations were done with Eqs. (3)–(6) (Glaser, 1993). In these model equations "E" represents the binding protein (BSA), "S" is the material on the sensing chip (Au or Au-MLG), and "ES" is the conjugate formed by the interaction between the protein and the sensing surface.

$$\mathrm{E} + \mathrm{S} \underset{k_a}{\rightarrow} \mathrm{ES} \underset{k_d}{\rightarrow} \mathrm{E} + \mathrm{S} \tag{3}$$

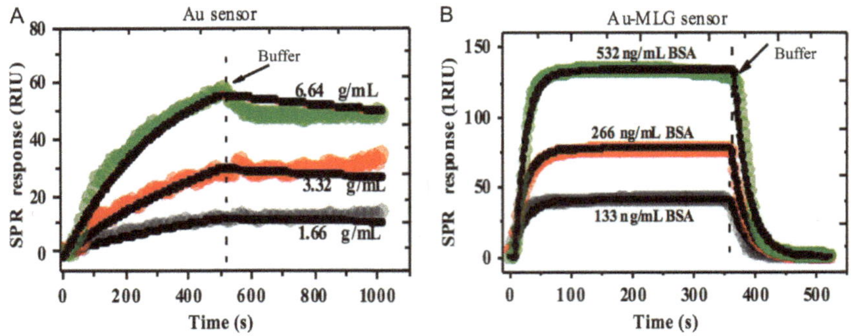

Fig. 5 (A) Kinetic traces of BSA binding to the plain gold surface. (B) Kinetic traces of BSA binding to biographene plated on gold surface. "MLG" stands for multilayer graphene. The data are shown in *black* and the global fitting is show in *green*, *red*, and *gray*. Kinetic parameters calculated from these binding study traces are shown in Table 1.

Table 1 Calculated Kinetic Parameters for BSA on a Biographene-Coated Gold SPR Chip Using a 1:1 Binding Model

Type of Sensor	k_a (M^{-1} s^{-1})	k_d (s^{-1})	K_D (M)
Au	$2.8 \pm 0.1 \times 10^4$	$3.2 \pm 0.5 \times 10^{-4}$	$1.1 \pm 0.2 \times 10^{-8}$
Au-MLG (biographene)	$6.1 \pm 0.3 \times 10^6$	$3.5 \pm 0.2 \times 10^{-2}$	$5.7 \pm 0.3 \times 10^{-9}$

$$\frac{d[ES]}{dt} = k_a[E][S] - k_d[ES] \tag{4}$$

$$\frac{d[S]}{dt} = -k_a[E][S] + k_d[ES] \tag{5}$$

The equilibrium dissociation constants (K_D) of the BSA to the Au sensor and to the Au-MLG sensor were determined using the adsorption rate (k_a) and the desorption rate (k_d) in Eq. (6).

$$K_D = k_d / k_a \tag{6}$$

The resulting K_D values calculated for the Au sensor and the Au-MLG sensor were $1.10 \times 10^{-8} M$ and $5.7 \times 10^{-9} M$, respectively. The lower K_D value estimated for the Au-MLG sensor implies a stronger affinity of the BSA to the biographene layer than to regular gold surface. This was expected due to the poor binding of water-soluble molecules to gold. The higher binding affinity of the BSA to the Au-MLG sensor also explains the signal enhancement seen for this sensor at lower BSA concentrations.

The SPR sensor surface was also able to be regenerated using SDS. After the dissociation of BSA on the Au-MLG sensor surface has finished,

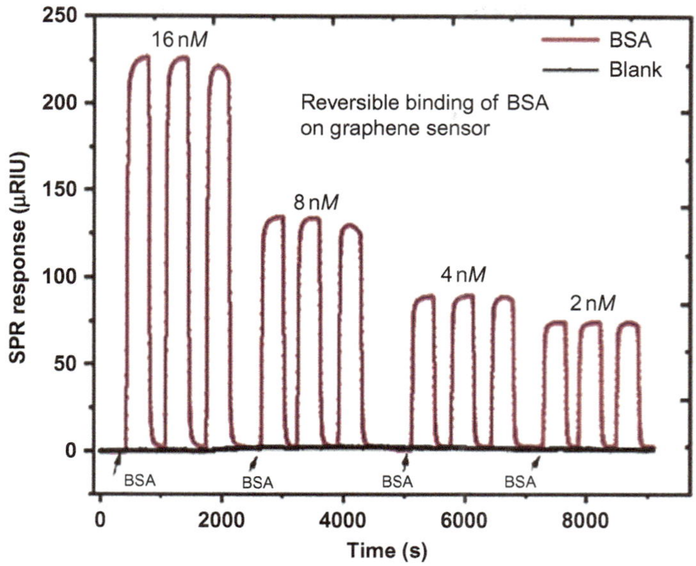

Fig. 6 An SPR response (μRIU) vs time (seconds) plot showing how the biographene-coated sensor chip can be regenerated using sodium dodecyl sulfate (SDS) and used multiple times over.

SDS can be injected into the system to remove any bound material remaining, and bring the signal back to baseline (Fig. 6). In this way the same sensor chip can be used multiple times to collect data and show the reversibility of protein binding. In Fig. 6, triplicates of BSA binding at each of four different concentrations were run, where the Au-MLG sensor chip was regenerated in between each protein injection. It can be seen that even after the graphene sensor is regenerated with SDS, the triplicate data have very little variance.

5. CONCLUSIONS

Here we report a facile method for synthesizing high-quality biographene dispersions that are stable in water for over 2 months at 4°C. This synthesis method has the potential to be scaled up to production levels reaching grams of biographene per day. The graphene sheets produced are 3–5 layers and have demonstrated the ability to enhance the enzymatic activity of a horseradish peroxidase–glucose oxidase enzyme cascade by acting as an electron carrier, as well as by cushioning the enzymes. Washed biographene has also been successfully used as a highly sensitive coating for protein binding studies using SPR spectroscopy. This high-yielding synthesis method could therefore overcome the current limitations for graphene to be used as a widespread tool in enzymology studies and biochemical research.

ACKNOWLEDGMENT

We thank the NASA and NSF for the funding for this project.

REFERENCES

Ahadian, S., Estili, M., Surya, V. J., Ramon-Azcon, J., Liang, X., Shiku, H., et al. (2015). Facile and green production of aqueous graphene dispersions for biomedical applications. *Nanoscale*, 7(15), 6436–6443. https://doi.org/10.1039/C4NR07569B.

Alava, T., Mann, J. A., Théodore, C., Benitez, J. J., Dichtel, W. R., Parpia, J. M., et al. (2013). Control of the graphene–protein interface is required to preserve adsorbed protein function. *Analytical Chemistry*, *85*(5), 2754–2759. https://doi.org/10.1021/ac303268z.

Bah, C. S. F., Bekhit, A. E.-D. A., Carne, A., & McConnell, M. A. (2013). Slaughterhouse blood: An emerging source of bioactive compounds. *Comprehensive Reviews in Food Science and Food Safety*, *12*(3), 314–331. https://doi.org/10.1111/1541-4337.12013.

Bhuyan, M. S. A., Uddin, M. N., Islam, M. M., Bipasha, F. A., & Hossain, S. S. (2016). Synthesis of graphene. *International Nano Letters*, *6*(2), 65–83. https://doi.org/10.1007/s40089-015-0176-1.

Biscar, J. P., Dhall, P., & Pennison, J. (1972). Raman behavior of bovine serum albumin. *Chemical Physics Letters*, *14*(5), 569–572. https://doi.org/10.1016/0009-2614(72)87209-9.

Bolton, B. A., & Scherer, J. R. (1989). Raman spectra and water absorption of bovine serum albumin. *The Journal of Physical Chemistry*, *93*(22), 7635–7640. https://doi.org/10.1021/j100359a021.

Bujacz, A., Zielinski, K., & Sekula, B. (2014). Structural studies of bovine, equine, and leporine serum albumin complexes with naproxen. *Proteins: Structure, Function, and Bioinformatics*, *82*(9), 2199–2208. https://doi.org/10.1002/prot.24583.

Eckmann, A., Felten, A., Mishchenko, A., Britnell, L., Krupke, R., Novoselov, K. S., et al. (2012). Probing the nature of defects in graphene by Raman spectroscopy. *Nano Letters*, *12*(8), 3925–3930. https://doi.org/10.1021/nl300901a.

Falcao Eduardo, H. L., & Wudl, F. (2007). Carbon allotropes: Beyond graphite and diamond. *Journal of Chemical Technology & Biotechnology*, *82*(6), 524–531. https://doi.org/10.1002/jctb.1693.

Glaser, R. W. (1993). Antigen-antibody binding and mass transport by convection and diffusion to a surface: A two-dimensional computer model of binding and dissociation kinetics. *Analytical Biochemistry*, *213*, 152–161.

Gu, Z., Yang, Z., Wang, L., Zhou, H., Jimenez-Cruz, C. A., & Zhou, R. (2015). The role of basic residues in the adsorption of blood proteins onto the graphene surface. *Scientific Reports*, *5*, 10873.

Jiang, X., Li, Z., Yao, J., Shao, Z., & Chen, X. (2016). One-step synthesis of soy protein/graphene nanocomposites and their application in photothermal therapy. *Materials Science and Engineering: C*, *68*, 798–804. https://doi.org/10.1016/j.msec.2016.07.034.

Johnson, D. W., Dobson, B. P., & Coleman, K. S. (2015). A manufacturing perspective on graphene dispersions. *Current Opinion in Colloid & Interface Science*, *20*(5), 367–382. https://doi.org/10.1016/j.cocis.2015.11.004.

Limbacher, M. R., Puglia, M. K., Riccardi, C. M., & Kumar, C. V. (2018). Interlocking enzymes in graphene-coated cellulose paper for increased enzymatic efficiency. In C. V. Kumar (Ed.), *Methods in enzymology*, Vol. 609 (pp. 1–22). Academic Press. https://doi.org/10.1016/bs.mie.2018.06.012.

Lord, R. C., & Yu, N.-T. (1970). Laser-excited Raman spectroscopy of biomolecules: I. Native lysozyme and its constituent amino acids. *Journal of Molecular Biology*, *50*(2), 509–524. https://doi.org/10.1016/0022-2836(70)90208-1.

McCallion, C., Burthem, J., Rees-Unwin, K., Golovanov, A., & Pluen, A. (2016). Graphene in therapeutics delivery: Problems, solutions and future opportunities. *European Journal of Pharmaceutics and Biopharmaceutics*, *104*, 235–250. https://doi.org/10.1016/j.ejpb.2016.04.015.

Novoselov, K. S., Geim, A. K., Morozov, S. V., Jiang, D., Katsnelson, M. I., Grigorieva, I. V., et al. (2005). Two-dimensional gas of massless Dirac fermions in graphene. *Nature*, *438*, 197–200. https://doi.org/10.1038/nature04233.

Novoselov, K. S., Geim, A. K., Morozov, S. V., Jiang, D., Zhang, Y., Dubonos, S. V., et al.Firsov, A. A., (2004). Electric field effect in atomically thin carbon films. *Science*, *306*(5696), 666–669. https://doi.org/10.1126/science.1102896.

Paton, K. R., Varrla, E., Backes, C., Smith, R. J., Khan, U., O'Neill, A., et al.Coleman, J. N., (2014). Scalable production oflarge quantities of defect-free few-layer graphene by shear exfoliation in liquids. *Nature Materials*, *13*, 624. https://doi.org/10.1038/nmat3944.

Pattammattel, A., & Kumar, C. V. (2015). Kitchen chemistry 101: Multigram production of high quality biographene in a blender with edible proteins. *Advanced Functional Materials*, *25*(45), 7088–7098. https://doi.org/10.1002/adfm.201503247.

Pattammattel, A., & Kumar, C. V. (2017). *Introduction to graphene: Chemical and biochemical applications* (1st ed.). Amsterdam: Elsevier.

Pattammattel, A., Pande, P., Kuttappan, D., Puglia, M., Basu, A. K., Amalaradjou, M. A., et al. (2017). Controlling the graphene–bio interface: Dispersions in animal sera for enhanced stability and reduced toxicity. *Langmuir*, *33*(49), 14184–14194. https://doi.org/10.1021/acs.langmuir.7b02854.

Pattammattel, A., Puglia, M., Chakraborty, S., Deshapriya, I. K., Dutta, P. K., & Kumar, C. V. (2013). Tuning the activities and structures of enzymes bound to graphene oxide with a protein glue. *Langmuir*, *29*(50), 15643–15654. https://doi.org/10.1021/la404051c.

Riccardi, C. M., Kasi, R. M., & Kumar, C. V. (2017). Chapter nineteen—Nanoarmoring of enzymes by interlocking in cellulose fibers with poly(acrylic acid). In C. V. Kumar (Ed.), *Vol. 590. Methods in enzymology*. (pp. 475–500): Academic Press.

Riccardi, C. M., Mistri, D., Hart, O., Anuganti, M., Lin, Y., Kasi, R. M., et al. (2016). Covalent interlocking of glucose oxidase and peroxidase in the voids of paper: Enzyme-polymer "spider webs". *Chemical Communications*, *52*(12), 2593–2596. https://doi.org/10.1039/C6CC00037A.

Sridhar, V., Lee, I., Chun, H. H., & Park, H. (2013). Graphene reinforced biodegradable poly(3-hydroxybutyrate-co-4-hydroxybutyrate) nanocomposites. *Express Polymer Letters*, 7, 320–328.

Yang, Y., Asiri, A. M., Tang, Z., Du, D., & Lin, Y. (2013). Graphene based materials for biomedical applications. *Materials Today*, *16*(10), 365–373. https://doi.org/10.1016/j.mattod.2013.09.004.

CHAPTER ELEVEN

Enzyme–Graphene Platforms for Electrochemical Biosensor Design With Biomedical Applications

Luminita Fritea*[,2], Mihaela Tertis†[,2], Robert Sandulescu†, Cecilia Cristea†[,1]

*Department of Preclinical Disciplines, Faculty of Medicine and Pharmacy, University of Oradea, Oradea, Romania

†Department of Analytical Chemistry, Faculty of Pharmacy, "Iuliu Haţieganu" University of Medicine and Pharmacy, Cluj-Napoca, Romania

[1]Corresponding author: e-mail address: ccristea@umfcluj.ro

Contents

Abstract

Due to the growing need for sensitive, reliable, reusable, fast, and cheap devices for the detection of analytes which have an important role in diagnosis of different diseases, in metabolic disorders, in monitoring treatment of serious diseases such as cancers, the sensing field has attracted huge interest from the scientists. The majority of the traditional methods that are currently in use are invasive, expensive, and laborious.

[2] Authors with equal contribution.

Methods in Enzymology, Volume 609
ISSN 0076-6879
https://doi.org/10.1016/bs.mie.2018.05.010

Moreover, highly specialized operators and sophisticated instrumentations are usually required. Taking these into account, the introduction of electrochemical sensors and biosensors avoid a lot of the disadvantages associated with most of the used analytical techniques. The biggest contribution to this development was the use of different nanomaterials as transducers of the analytical signal. The properties, such as high mechanical strength, good electrical conductivity, and ability to serve as efficient signal transducers, make carbon-based nanomaterials, including graphene, ideal materials for biosensor applications. Furthermore, graphene presents high surface area that can be easily modified in different ways to be adapted for the immobilization of various biocompounds for the construction of biosensors. Recent advances regarding the use of graphene and graphene materials for the immobilization of several enzymes for biosensor development and their applications for the detection of chemical and biological species are presented with focus on different enzymes immobilization techniques. In the end, the future trends for the development of graphene-based biosensors in biomedical field are also discussed.

ABBREVIATIONS

ADH	alcohol dehydrogenase
AuNPs	gold nanoparticles
β-CD	β-cyclodextrin
CHER	cholesterol esterase
ChOx	cholesterol oxidase
CLEA	cross-linking enzyme aggregates
CLEC	cross-linking enzyme crystals
CNTs	carbon nanotubes
DA	dopamine
DET	direct electron transfer
EDC	1-ethyl-3-(3-dimethyl-aminopropyl)carbodiimide
FAD	flavin adenine dinucleotide
FDH	D-fructose dehydrogenase
GO	graphene oxide
GCEs	glassy carbon electrodes
GOx	Glucose oxidase
HRP	horseradish peroxide
MET	mediated electron transfer
MNP	magnetic nanoparticles
NHS	*N*-hydroxysuccinimide
PB	Prussian blue
PAMAM	ethylenediamine core polyamidoamine G-4 dendrimers
PtNPs	platinum nanoparticles
rGO	reduced graphene oxide
SPE	screen-printed electrode
Tyr	tyrosinase
XOx	xanthine oxidase

1. INTRODUCTION

There is an enormous interest in the use of carbon-based nanomaterials, especially carbon nanotubes (CNTs) and graphene for the development of biosensors, including the enzymatic ones. This is mainly due to the ability of these nanomaterials to improve the analytical characteristics of macroscale biosensors, as well as to facilitate the development of some completely new nanoscale biosensors. Carbon nanomaterials are among the most commonly used nanomaterials for the development of biosensors since they allow reaching many of the important key issues such as the design of the biosensing interface so that the analyte selectively interacts with it; the achievement of efficient transduction of the biorecognition step; improved sensitivity and selectivity of the biosensor; faster response time in very sensitive systems; good compatibility with biological matrices for testing in complex biological samples and for in vivo applications; fabrication of viable biosensors that can operate inside cells; multiplexing biosensors for testing multiple analytes with only one device.

Electrochemical biosensors are analytical devices capable of converting the biochemical reaction into analytical and measurable signal. They belong to the large class of biosensors which associate to a transducer a biological element (Hulanicki, Glab, & Ingman, 1991). Therefore, biosensors can be categorized according to the biological recognition element (immuno, enzymatic, DNA, and whole-cell biosensors) or the signal transduction method (optical, colorimetric, mass-based, electrochemical, and thermal biosensors) (Fig. 1).

Electrochemical biosensors are attractive analytical devices for biomedical analysis featuring several advantages: detection of the substrate is made without prior separation; the sensitivity could arrive at ppm or ppb levels, higher selectivity (even specificity), high benefit/cost ratio, simple use, and the rapid data collection. Enzyme-based biosensors are the most used biosensing platforms surpassing the immunosensors (using antibodies as recognition element), the genosensors (using nucleic acids), the aptasensors (using aptamers), and the cell-based sensors.

This is due to the fact that enzymes have high affinity toward corresponding substrates, being able to catalyze several biochemical reactions, being relatively easily regenerated several times. The recognition system of a biosensor depends on the enzyme–substrate relation (Kurbanoglu, Ozkan, & Merkoçi, 2017). Enzymes can be immobilized onto the surface of a transducer

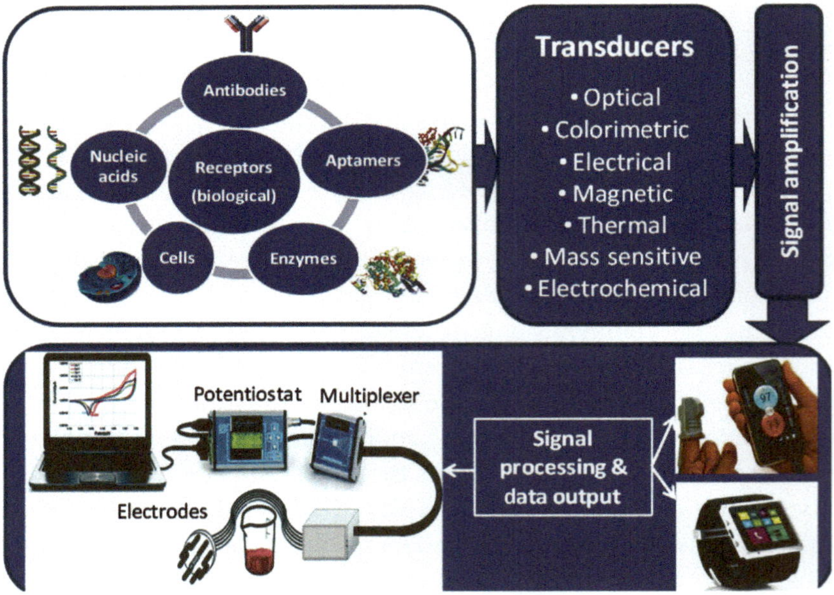

Fig. 1 The schematic representation of a biosensor design.

via several immobilization techniques, each technique having its own limitation regarding the best orientation of the enzymes toward substrate.

The reduced stability, the limited lifetime due to the rapid loss of enzyme activity, the interferences from complex sample matrices, and biofouling in case of in vivo measurements are some of the main disadvantages of the electrochemical biosensors (Castillo et al., 2004; Chaubey & Malhotra, 2002). Nanomaterials are excellent building blocks, generally used in electrochemical biosensors fabrication in order to modify transducers or to enhance their electrochemical signals, also eliminating most of the above-mentioned disadvantages.

The interest of graphene as a nanomaterial employed for biosensors design is due to the success of previous use of CNTs. CNTs are usually referred to as rolled-up graphene sheets, and both allotropes have a meshwork of sp^2-hybridized carbon atoms. Because these nanomaterials have the same composition, it was initially thought that the properties are similar. But practice has proved that this is not always true, and the differences in structure and properties open new ways for further developments in biosensors.

Any biosensor based on graphene should ideally be designed to exploit the unique properties of this nanomaterial. The ideal structure of graphene is completely two-dimensional. It comprises a single layer of sp^2-hybridized carbon atoms joined by covalent bonds to form a flat hexagonal lattice (Castro Neto, Guinea, Peres, Novoselov, & Geim, 2009). Actually, there are different structures unified under the name of graphene. This is the consequence of the difficulty in isolating single layers of graphene in a controlled manner, meaning that what is termed graphene, actually comprises stacks of graphene layers, each containing different numbers of atomic sheets.

Therefore, it is mandatory to accurately assess the number of graphene layers, this being usually done by Raman spectroscopy (Gupta, Chen, Joshi, Tadigadapa, & Eklund, 2006), atomic force microscopy (Novoselov et al., 2004), scanning electron microscopy (SEM) (Grodecki, Jozwik, Baranowski, Teklinska, & Strupinski, 2016), contrast spectroscopy (Ni et al., 2007), and low-energy electron microscopy (Hibino et al., 2008). The above-mentioned methods can be used in order to determine if there are single-layer graphene, bilayer, few-layer graphene, or multilayer graphene. Even if it is demonstrated that the nanomaterial that is intended to be used for the development of biosensors is single-layer graphene, there are other problems that occur and must be taken into consideration. For example, the single-layer graphene is never atomically flat because of its flexibility. Usually the graphene sheets tend to fold, curl, or corrugate (Novoselov, Jiang, et al., 2005).

The mechanism of action of CNTs in electrochemical biosensors is not yet fully elucidated. Unfortunately, in the case of graphene and graphene compounds, the situation is even more unclear. This is mainly due to the fact that the studies dealing with the electrochemical properties of graphene were performed with different forms of graphene (Yang et al., 2010). Despite of this, there is still evidence that graphene and its derivatives can exhibit better electrochemical performance compared with other transducers such as glassy carbon, graphite, or CNTs. It was observed that the excellent electrochemistry of graphene may be attributed to its edges, while the functional groups and defects probably are also important in the case of graphene compounds. The electronic behavior of graphene is the most studied property. This originates in the delocalized π bonds above and below the basal plane, which arise from the sp^2 hybridization responsible for the layered structure of graphene (Castro Neto et al., 2009). The delocalized electrons, and the quality of the graphene, create high electrical conductivity and mobility. The mobility was measured at room temperature and a value of $15{,}000\,cm^2\,V^{-1}\,s^{-1}$ was obtained in graphene with maximum three layers (Novoselov, Geim, et al., 2005), while in the case of single layer graphene, a mobility of 230,000 $cm^2\ V^{-1}\ s^{-1}$ has been measured at temperature around absolute zero (Bolotin et al., 2008).

Graphene also presents an ambipolar electric field effect in which negative gate voltages induce large concentrations of holes, while positive voltages induce large numbers of electrons (Castro Neto et al., 2009). There are other electronic properties and all of these are related to the number of layers in the graphene sheet. For example, a single-layer graphene sheet is a zero-gap semiconductor, while three-layer graphene sheet is a semimetal. The extent of overlap between the conduction and valence bands increases with the number of layers and similar behavior with that of graphite is observed at about 15 layers of graphene (Grüneis et al., 2008).

Raman and infrared spectroscopy can provide useful information about the optical properties of graphene derivatives. Raman spectroscopy is also used in order to identify the number of layers in graphene sheets that is related with the modifications observed in spectra (Ferrari et al., 2006). Thus, there are two highly characteristic Raman bands in graphene: the G band at approximately 1580 cm^{-1} and the D band at about 2700 cm^{-1} (Gupta et al., 2006). Another D band appears at about 1350 cm^{-1}, this being due to the edges, defects, and impurities of the graphene sheets (Ferrari et al., 2006).

Electrochemical enzymatic biosensors based on graphene nanocomposites have been designed for the determination of a wide range of molecules from various fields.

This chapter will focus only on the use of graphene in enzymatic electrochemical biosensors development for the detection of some compounds of biomedical interest. The latest advances in electrochemical biosensors that use graphene will be further discussed and their performances will be critically compared.

2. ELECTROCHEMICAL BIOSENSORS: GENERAL CONSIDERATION

2.1 Definition of Biosensors

Sensors are small or miniaturized devices designed for the monitoring of the physical, chemical, or biochemical properties of specific analytes in order to provide qualitative and quantitative information. Chemical sensors are related to the presence and concentration of analyte by responding selectively to the modification of the properties of the analyte's features such as electrical, optical, thermal, or gravimetric features. Biosensors can be defined as devices with a thin layer serving as support to immobilized biomolecule that contains biorecognition sites. The biomolecules may be enzymes, proteins, receptors, antibodies, oligo- or polynucleotides, microorganisms, cells, or even tissues that display specific interactions with target analyte.

According to IUPAC, an electrochemical biosensor is "a self-contained integrated device that is capable of providing specific quantitative or semiquantitative analytical information by using a biological recognition element (biochemical receptor) in direct spatial contact with a transduction element." Therefore, an electrochemical biosensor contains three parts:

– the biological recognition element
– the transduction element (converting the signal in a quantified result)
– the signal processor (displaying the signal).

However, it must be taken into account that there is a difference between a biosensor and a bioanalytical system. The last one requires additional processing steps, such as reagent addition. According to this definition, there is no size constraint on a biosensor, which can be macroscopic (commercial glucose biosensors), can be composed of a macroscopic device containing a micro or nanoscale-sensing element (field-effect transistor type devices), or the entire device can be on the nanoscale (Yang et al., 2010).

The discovery of biosensors is related to the early 1960s, when Clark and Lyons first reported on amperometric glucose detection using the enzyme glucose oxidase (GOx) (Clark & Lyons, 1962). These sensors were applied for monitoring biological processes, for the detection of biomolecules. The majority of biosensors developed for glucose detection to date are electrochemical ones, these being preferable due to their ease of use, low-cost, small size and relatively fast response, good detection limit, simplicity, rapidity, reliability, less sample requirement, high sensitivity, and less bulky instrumentation. Electrochemical biosensors are potentiometric, amperometric, or conductimetric and are based on the monitoring of the signal generated from electrochemical reactions of the analytes or in response to the presence of different chemical species related to the analytes, thus producing/consuming ions/electrons, leading to the generation of an electrochemically measurable signal (current, potential, charge accumulation, and impedance) that is analyzed by the electrochemical detector. In the case of enzyme-based electrochemical biosensors, the analyte is recognized by enzymes that are immobilized on the working electrode surface, and their catalytic activity causes electron transfer, produces electric current, or contributes to voltage production.

2.2 Nanomaterials: Modifiers for Transducers in Biosensors

Important steps in the development of the biosensors were made with the discovery of nanomaterials and of various nanofabrication tools, such as electron beam lithography, focused ion beam, and nanoimprint lithography (Liu & Guo, 2012). Various nanomaterials coupled with enzymes as biorecognition element led to electrochemical biosensors with high surface area, excellent conductivity, biocompatibility, and (bio)catalytic activity resulting in ultrasensitive and selective biosensing. These nanomaterials include CNTs (Lahiff, Lynam, Gilmartin, O'Kennedy, & Diamond, 2010; Liu et al., 2010; Wang, 2005), nanoparticles (Wilson, 2008), nanowires (Arlett, Myers, & Roukes, 2011; Song et al., 2010), and graphene (He, Wu, Yin, & Zhang, 2012; Huang et al., 2011; Ratinac, Yang, Gooding, Thordarson, & Braet, 2011). A wide range of substances with biomedical importance have been detected by employing graphene-based biosensors (Bollella et al., 2017; Lawal, 2018; Song et al., 2016).

The use of nanosized materials in electrochemical biosensors improved the performance factors such as sensitivity, response time, and limit of detection up to great extent. An ideal nanomaterial must sum up some fundamental properties like ease in preparation, availability, abundance of functional groups, and chemical stability for the use as a suitable matrix in biosensor fabrication. The presence of nanomaterial on the electrode surface facilitates fast electron transfer kinetics, enhances the electroactive surface area, and increases the biomolecule loading. Furthermore, the large surface area increases the binding affinity of immobilized biomolecule via the multivalent immobilization techniques.

Graphene is in the forefront of nanomaterials used for the elaboration of biosensors, and this fact is due to their unique electronic, physical, and chemical properties. The spectacular development of nanomaterial-based electrochemical biosensors is also correlated with their significant advantages compared to conventional optical, biochemical, and biophysical methods. These advantages are high sensitivity, high resolution for localized detection, easy method for integration, and real-time and nondestructive detection ability.

2.3 Enzymatic Biosensors

Enzymes were the first molecular recognition elements used in biosensors technology. Enzymes are macromolecular biological catalysts composed of aminoacids generating globular proteins with 3D structure. Their specific structure (which may also include cofactors) is responsible for the high catalytic activity and specificity of the enzyme for one type of substrate molecule.

An enzymatic biosensor is an analytical device that combines enzyme with transducer and converts signal proportional to target analyte/substrate concentration with high catalytic activity, selectivity, and specificity. Appropriate immobilization of enzyme on the surface of nanomaterials not only enhances the concentration of the biorecognition molecule at the appropriate location but also protects the biomolecules from being inactivated.

The most challenging issue remains the establishment of an efficient and lasting electrical communication between enzymes and the electrode concerning the electrons transfer. Two main strategies were employed for the electric wiring of the active site of the enzyme to the electrode:

- direct electron transfer (DET) (the active site is in direct electronic communication with the electrode surface);
- mediated electron transfer (MET) (using redox mediators).

The most important parameters for DET are the immobilization strategy, the orientation of the enzyme, and the localization of the active site in the protein. Meanwhile, the key parameters for MET envisage: the immobilization strategy for the enzyme and the redox mediator, the redox potentials and structure of the redox mediator, and the electron transport via the redox mediator film.

2.3.1 Strategies for the Immobilization of Enzymes for Biosensors Development

The immobilization of enzymes represents the process through which this bioelement is fixed or localized in a certain position (that is well established) on a special support (electrode in the case of electrochemical biosensors). It is very important to mention that the immobilization process must not affect the catalytic activity of the enzyme, allowing the sensitive, repetitive, and continuous detection of the target analyte. A good biosensor requires optimum amount of enzyme loading together with the preservation of its optimal activity. In order to achieve this, it is important to avoid the use of harsh chemical treatment or exposure to harsh conditions such as high temperature. Immobilization of enzymes on the surface of graphene and graphene compounds provides platforms suitable for the development of biosensors taking advantage of their electronic and optical properties for signal transduction. The best strategy for a successful fabrication of an enzyme biosensor is to elaborate a configuration for the electrode active surface that allows the direct transfer of the electrons from the redox center of the enzyme to the transducer. In this respect, several immobilization strategies are used, such as physical adsorption or covalent immobilization of large biomolecules onto the surface of previously immobilized graphene or graphene derivatives (Gorecka & Jastrzebska, 2011).

The ideal enzyme immobilization strategy must meet several basic criteria, such as:

- employing mild conditions;
- allowing the immobilization of large quantities of enzyme;
- providing substrate with a large surface area vs a small total volume for enzyme;
- removing barriers to mass transport both for the substrate and the product;
- providing a chemically and mechanically robust system.

The immobilization strategy is usually chosen depending on the desired application, and the large number of possible applications was a good reason for the continuous development and improvement of the methods used for immobilization. Another very important issue is represented by the selection of the most appropriate support for the immobilization of a particular biocompound. Selection criteria differ among one another, depending on the biocatalyst of interest, but there are still few basic principles that must be considered for the right choice of the support material such as:

- high chemical, physical, and biological stability during processing and reaction;
- high mechanical strength;
- low or no effect both for the immobilized biocompound and the target analyte that should be detected;
- the presence of adequate functional groups for binding the biocompound;
- adequate physical characteristics such as porosity, swelling, or particle characteristics;
- high loading capacity;
- biodegradability;
- low costs for large-scale production (Gorecka & Jastrzebska, 2011).

There are two categories of immobilization procedures, namely, reversible and irreversible immobilization, respectively. The selection of the immobilization technique and support depends on different factors. The parameters that influence the above selection in case of biosensors are the nature of enzyme, the type of transducer, physicochemical properties of the analyte, and the operating conditions.

Reversible immobilization methods include adsorption, metal chelation, and disulfide bounds formation.

Adsorption represents the simplest and the most commonly used method of reversible immobilization. It is often used for the immobilization of cells and enzymes on different substrates, including graphene-based ones and it is also found in various biotechnological processes. In this case, the immobilization is based on weak forces such as van der Waals forces, ionic, hydrophilic and hydrophobic interactions, hydrogen bonds, and even affinity binding (Bucur, Danet, & Marty, 2004; Mateo et al., 2006).

Disulfide binding is a reversible immobilization method applied especially in the case of enzymes. In fact, this may be considered a covalent bonding, as there are stable covalent bonds formed between activated support and free thiol group (for example, on cysteine) in the biological element. However, this method is classified as a reversible immobilization, while those bonds can be easily broken using suitable agent under mild conditions (Neves-Petersen, Snabe, Klitgaard, Duroux, & Petersen, 2006).

Metal chelation is widely used in chromatography, while it is very important to ensure an easy regeneration of the support without reducing the immobilization yield. The immobilization is based on the ability of the side chains of certain amino acids (histidine, tryptophan, tyrosine, cysteine, and phenylalanine) to substitute weakly bonded ligands in the metal ions that have been immobilized by chelating group covalently bound to a solid support (Afag & Iqbal, 2001).

Irreversible immobilization methods include covalent bonding, cross-linking, entrapment, and encapsulation.

Covalent bonding represents one of the most frequently used methods for enzyme immobilization. This is mainly due to the stability of the bonds formed between the enzyme and the support, which prevents enzyme detachment during the elaboration and detection protocol. There is a wide variety of covalent bounding strategies for enzyme immobilization, all being related on the type of the matrix and substrate. The first strategy involves the generation of functional groups at the surface of the electrode via a polymer matrix without its modifications, while the second one implies the modification of the support matrix in order to generate activated groups. These groups are usually electrophile ones that will react with strong nucleophile ones from the biocompound. There are three different bounding methods that are exploited: the first one involves binding via the side chains of lysine (ε-amino group), the second one involves the thiol group in cysteine, and the third one involves carboxylic group of aspartic and glutamic acids. Furthermore, in this method only the functional groups which are not essential for the catalytic activity of the enzyme are allowed for bonding (Mateo et al., 2006).

Entrapment is an irreversible method where immobilized particles or cells are entrapped in a support matrix or inside fibers. Electropolymerization is an example of entrapment methodology in which the enzyme is mixed with a monomer, followed by electrochemical polymerization, where the enzyme becomes embedded into the polymer matrix. The incorporation of the enzyme into the matrix is often promoted through electrostatic interactions. An important condition for unaltered preservation of the enzyme activity is to avoid the use of harsh chemicals during the polymerization process (Gao, Wang, Diao, Luo, & Dai, 2010).

Encapsulation is another irreversible immobilization method, similar to entrapment. The major difference is that in this process the biocompound molecules are restricted by the membrane walls that have a form of a capsule. However, there are no restrictions for the movement within the capsule space. The membrane is semipermeable, allowing for free flow of substrates and nutrients (when cells are used as a biocatalyst), yet keeping the biocatalyst inside. The factor determining this phenomenon is the proper pore size of the membrane. This method is suitable for multienzyme systems used in sequential enzymatic reactions, as several different bioelements may be incorporated into the same capsule (Cao, 2005a).

Cross-linking represents an irreversible method of enzyme immobilization that does not require a support. It is a useful method to stabilize adsorbed biomaterials, and it generally involves the residues which are not involved in catalysis (e.g., amino group of lysine residues). There are two main strategies used for cross-linking of enzymes, both involving the use of glutaraldehyde which reacts with amino groups from protein surface:

(a) Cross-linking enzyme aggregates (CLEA).
(b) Cross-linking enzyme crystals (CLEC).

An important factor is to use the proper cross-linking agent, glutaraldehyde being the most frequently employed, as it is inexpensive and readily commercial available, but it is unsuitable for immobilization of some enzymes (e.g., nitrilases). Other cross-linking agents are dextran, polysaccharide, bis-isocyanate, bis-diazobenzidine, diazonium salts, and functionally inert proteins, such as bovine serum albumin or ovalbumin (Cao, 2005b; Sheldon, 2007).

The advantages and drawbacks of the above-presented methods for enzyme immobilization are comparatively presented in Table 1.

Table 1 Advantages and Drawbacks of Different Immobilization Methods

Immobilization Method	Advantages	Drawbacks
 Adsorption	– easy in preparation; – low cost; – the possibility to reload the support.	– a very high rate of leakage due to the weak interactions between the support and the biocatalyst; – low reproducibility due to the impossibility to control the enzyme loading; – high susceptibility to changes in ambient conditions (pH, temperature, ionic strength).
 Disulfide bounding	– the possibility to influence the reactivity of thiol groups by alteration of pH; – the possibility to chemically activate the support before the immobilization process.	– the need to establish ideal conditions for fast and reproducible immobilization and the achievement of high coverage rates in biomolecules.
 Metal chelation bounding	– reversibility of immobilization can be attained by exposing the surface to a stronger chelator or by using excess of competitor ligand.	– high difficulty in obtaining reproducible immobilization of enzyme onto the electrodes; – long duration due to the large number of steps involved.

Continued

Table 1 Advantages and Drawbacks of Different Immobilization Methods—cont'd

Immobilization Method	Advantages	Drawbacks
 Covalent bounding	– unlimited contact between the enzyme and the substrate; – increased stability of the enzyme.	– the biosensor usability is determined by the direction of the enzyme binding. The highest activity level is achieved only if the active center of the enzyme is not involved in binding with the support; – the activity of enzyme may be lost during the covalent bonding while this is usually carried out using chemicals and at different pH ranges. – is rarely applied for immobilization of cells due to the fact that agents used for covalent bonds formation are usually cytotoxic.
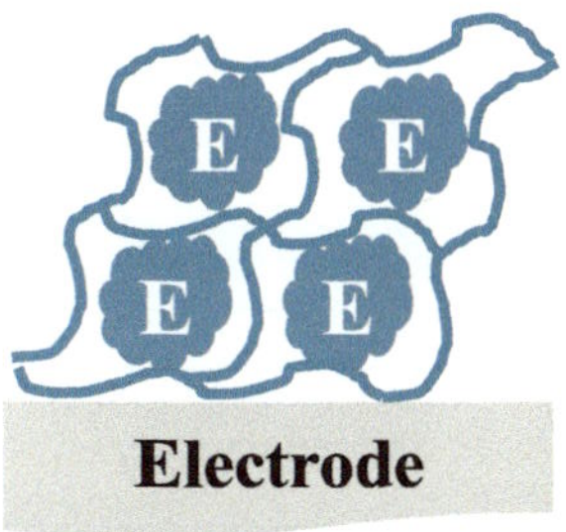 Entrapment	– increased stability of the enzyme at the electrode surface; – great enhancement of sensor response due to the increase in the surface area, as well as the improvement in the electrical communication between the redox centers of the enzyme and the electrode; – simple one-step preparation; – exclusion of electroactive and surface-active interferences; – control of film thickness; – localization of biocatalysts onto tiny electrode surfaces.	– high costs of immobilization; – diffusion limitations; – deactivation during immobilization; – abrasion of support material during usage; – low loading capacity with biocompound as it has to be incorporated into the support matrix: • small pores—low loading while adsorption can be done only on external surface; • big pores—leaking and decreases the loading.

Table 1 Advantages and Drawbacks of Different Immobilization Methods—cont'd

Immobilization Method	Advantages	Drawbacks
 Encapsulation	– limited access to the microcapsule interior protects the bioelement from the harsh environmental conditions; prevents leakage; – increases the efficiency; no chemical modifications of the capsule material are required for the immobilization; – the activity of the bioelement remains intact.	– the necessity for a very strict pore size control is difficult in the case of small enzymes; – this technique cannot be applied for enzymes similar in size to their reaction product.
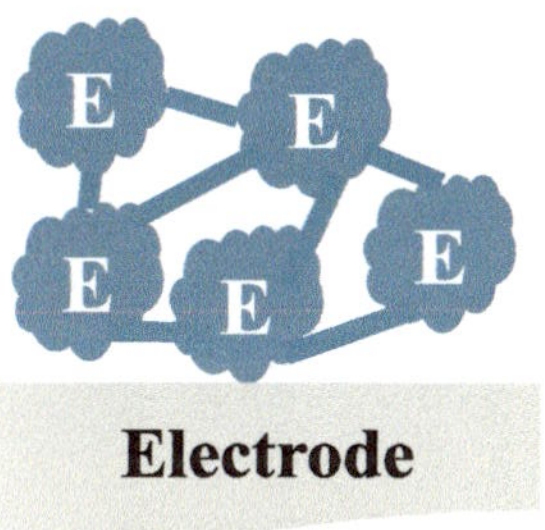 Cross-linking	– low costs; high specificity and activity of the enzyme; – easier recycling; in CLEC method enzymes are much more stable and resistant to denaturation by heat or organic solvents and to proteolysis; – the addition of salts, organic solvents or nonionic polymers determine the aggregation of enzymes which maintain their activity, and increase their stability (CLEA).	– low mechanical stability (CLEA); low reproducibility (CLEA); and low activity retention (CLEA); – the necessity for very high purification of enzyme and crystallization before use is laborious and time consuming (CLEC); – the size of obtained aggregates of enzymes with diameter no greater than 10 μm, because substrate particle may lay in the same size range in a heterogeneous reaction system, and can cause problems for enzymes recovery in continuous processes.

3. GRAPHENE AND GRAPHENE COMPOUNDS USED FOR BIOSENSORS DESIGN

Graphene is the youngest carbon allotrope being discovered in 2004 by Geim and Novoselov who described it as "more than just a flat crystal"; since then, its properties were intensively studied leading to a wide range of nanocomposites (in combination with polymers and nanoparticles) applied in various areas. The use of electrochemical enzymatic biosensors based on graphene in the pharmaceutical and biomedical fields has significantly increased due to their sensitivity, specificity, cost-effective, simplicity, and possibility of miniaturization becoming promising tools for portable devices such as point-of-care (POC) devices (Bo, Zhou, & Guo, 2017).

Graphene is the basic building block for graphitic materials of all other dimensionalities. Graphene is a single layer of sp^2-hybridized carbon atoms that can be folded into fullerenes, rolled into 1D CNTs (Dai, 2002) and can be stacked to obtain 3D graphite. The pristine graphene can be oxidized to graphene oxide (GO) containing oxygen functionalities. It was observed that high amount of oxygen functional groups on the surface of graphene lowers the electron transfer rate. The reduction of GO leads to reduced graphene oxide (rGO) that presents less oxygen functionalities, but allows the recovery of the electrical properties of pristine graphene that was lost after oxidation to GO. All these graphene derivatives can be functionalized, and the composite materials based on graphene are widely used for the immobilization of biomolecules or for the detection of various analytes. The electron transfer rate is high at the edge plane instead of basal plane of graphene. Each carbon atom on the nanocarbon surface is exposed to the environment, and any small changes in the environment can result in drastic changes in the electrical properties of the nanocarbon device, thus forming the basis for ultrasensitive biosensing (Schedin et al., 2007; Wang et al., 2011).

The overall electronic properties of the sensing devices based on graphene and biomolecules can be altered through one of the following mechanisms: surface charge-induced gating, charge transfer, or doping between carbon nanomaterials and biomolecules, a scattering potential across carbon nanomaterials, and Schottky barrier modification between carbon nanomaterials and metal electrodes (Liu & Guo, 2012). The pH, ionic strength, and the type of ions present in the analyzed environment also have a strong influence on the above-mentioned parameters (Heller et al., 2010). Another important feature of graphene is that its dimension is comparable with that of the molecules to be detected and this important property allows the detection of single-molecule events.

Graphene is a molecular chemical entirely composed of carbon atoms which gives high compatibility to biocompounds and allows controlled and selective functionalization of their surface (Liu & Guo, 2012). Despite its unique properties, graphene can stack together damaging its electrochemical performances. In order to minimize the aggregation level of the graphene sheets, some strategies have been employed such as: noncovalent, covalent functionalization, intercalation of nanomaterials, and structure engineering leading to high exposure of active sites and high utilization of catalysts. Graphene metallic nanocomposites containing metal, metal oxides, or metal alloy nanostructures have improved the biosensor performances for different target analytes due to their synergistic effects (Khalil, Rahmati, Julkapli, & Yehye, 2018; Maduraiveeran, Sasidharan, & Ganesan, 2018).

4. ENZYMES USED FOR ELECTROCHEMICAL BIOSENSORS BASED ON GRAPHENE

4.1 Oxidases-Based Biosensors

4.1.1 GOx-Based Biosensors

GOx is a large, dimeric protein having the molecular weight of 160 kDa. It contains one flavin adenine dinucleotide (FAD) unit per monomer, seated deep inside the enzyme. These redox active prosthetic groups are not covalently bound and may be released from the protein during denaturation. The active site where glucose binds is just above the FAD-binding site, in a deep pocket (Fig. 2). The oxidation of glucose to gluconolactone is performed at the FAD level.

In recent years, graphene has been widely used as a high performance two-dimensional material in the development of biosensors based on GOx and the DET between enzyme and electrode in the absence of any mediator is very desirable. The DET property and enzyme activity of GOx on graphene surface without and with mediators were studied. Experimental results showed that the biosensor had no response to glucose in mediator-free solutions, even though the DET of GOx was observed, indicating that the GOx with DET property lacked enzymatically catalytic activity. Two types of GOx adsorbed on rGO/GCE surface (Fig. 2) are described, one with DET property (DET-GOx), other with enzyme catalytic activity (Act-GOx). The DET-GOx might suffer a massive structural rearrangement when

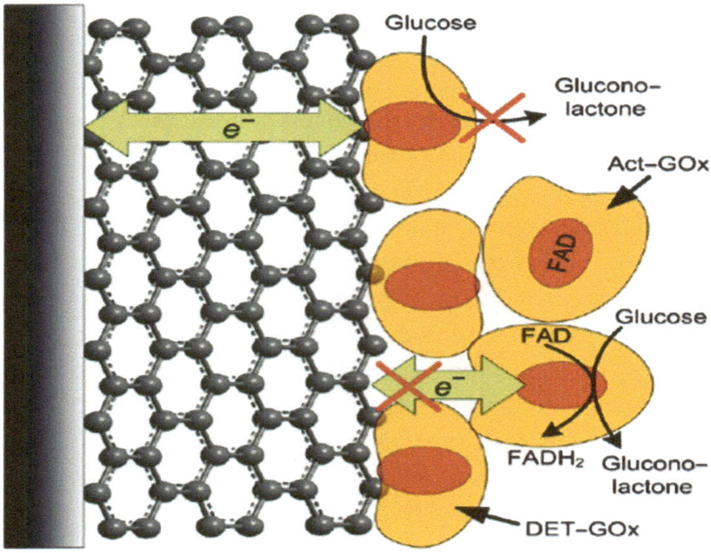

Fig. 2 Illustration of GOx adsorbed on the surface of electrochemically reduced GO-modified GCE (Liang et al., 2015).

adsorbed onto graphene surface, which shortens the distance of FAD to graphene and facilitates the DET with GOx and inactivates the enzyme at the same time (Liang et al., 2015).

4.1.2 Cholesterol Oxidase-Based Biosensors

The determination of cholesterol is of great significance in clinical diagnosis of coronary heart disease, atherosclerosis, myocardial infarction, and brain thrombosis. Among the various biosensors of cholesterol, cholesterol oxidase (ChOx)-based electrochemical biosensors are especially attractive because they offer simple, rapid, and inexpensive means of quantifying cholesterol. Amperometric cholesterol biosensors are based on the ChOx enzyme, which catalyzes the oxidation of cholesterol to 4-cholesten-3-one and generating H_2O_2 in the presence of oxygen. Thus, the quantification of cholesterol can be achieved via electrochemical detection of the enzymatically liberated H_2O_2.

A simple and ultrasensitive cholesterol biosensor based on GO and gold nanoparticles (AuNPs) comediated enzymatic silver deposition was designed by immobilizing ChOx, cholesterol esterase, and GO onto the surface of

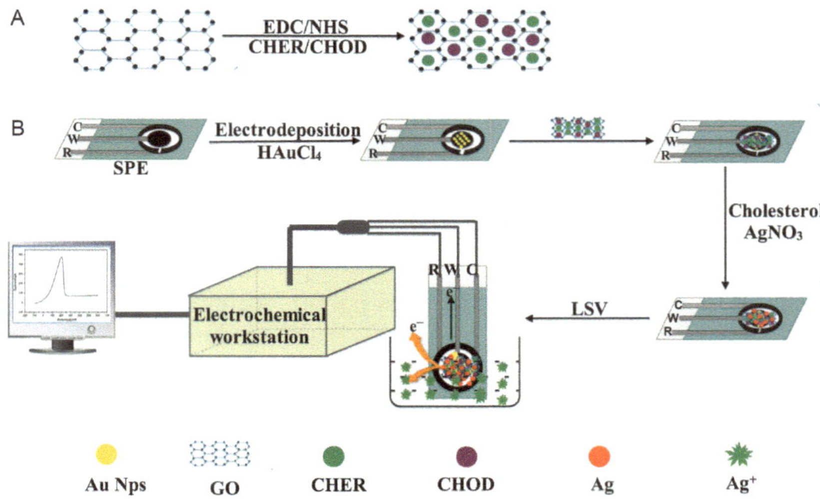

Fig. 3 Electrochemical detection of cholesterol based on GO and AuNPs comediated enzymatic Ag deposition (*LSV*, linear sweep voltammetry; *SPE*, screen-printed electrode) (Huang et al., 2018).

AuNPs modified screen-printed carbon electrode (Fig. 3). First, the carboxyl groups on GO were activated using a mixture of 1-ethyl-3-(3-dimethylaminopropyl)carbodiimide (EDC) and *N*-hydroxysuccinimide (NHS) at room temperature followed by the addition of cholesterol esterase (CHER) and cholesterol oxidase (CHOD) solutions and the mixture was incubated at 4°C for 3 h. First step of the experimental protocol used for the preparation of the cholesterol biosensor was the electrochemical generation of AuNPs onto the activated electrode. Then, the enzyme mixture solution was used in order to increase the specificity of the sensor (see the schematic representation of the experimental protocol in Fig. 3). The cholesterol was hydrolyzed to generate hydrogen peroxide which can reduce the silver ions found in the solution to metallic silver that was quantified by anodic stripping voltammetry measurement. The peak current of Ag increased with the increasing cholesterol concentration in a wide range and a limit of detection of 1 ng mL^{-1} was obtained (Huang et al., 2018).

4.1.3 Xanthine Oxidase-Based Biosensors

Xanthines are a group of alkaloids commonly used for their effects as mild stimulants and as bronchodilators, notably in the treatment of asthma symptoms. Therefore, the detection of these compounds is important, and it has been performed also using electrochemical biosensors. Xanthine oxidases (XOxs) are a class of enzymes that generate reactive oxygen species. These enzymes catalyze the oxidation of hypoxanthine and can further catalyze the oxidation of xanthine to uric acid. These enzymes play an important role in the catabolism of purines in humans.

For example, a novel layer-by-layer biosensor architecture was prepared with poly(dopamine)-modified magnetic nanoparticles (MNP) coated with fourth-generation ethylenediamine core polyamidoamine G-4 dendrimers (PAMAM) through covalent bounding, then decorated with platinum nanoparticles (PtNPs). This nanohybrid was layered on glassy carbon electrodes (GCEs) coated with a GO-carboxymethylcellulose hybrid nanomaterial through electrostatic interactions. The nanostructured surface was then employed as scaffold for the covalent immobilization of the enzyme XOx through a glutaraldehyde-mediated cross-linking. The enzyme electrode allowed the amperometric detection of xanthine in the 50 nM–12 μM range, with a high sensitivity of 140 mA M^{-1} cm^{-2} and low detection limit of 13 nM. The biosensor exhibited high reproducibility and repeatability, and was successfully tested for the quantification of xanthine in fish. Fig. 4

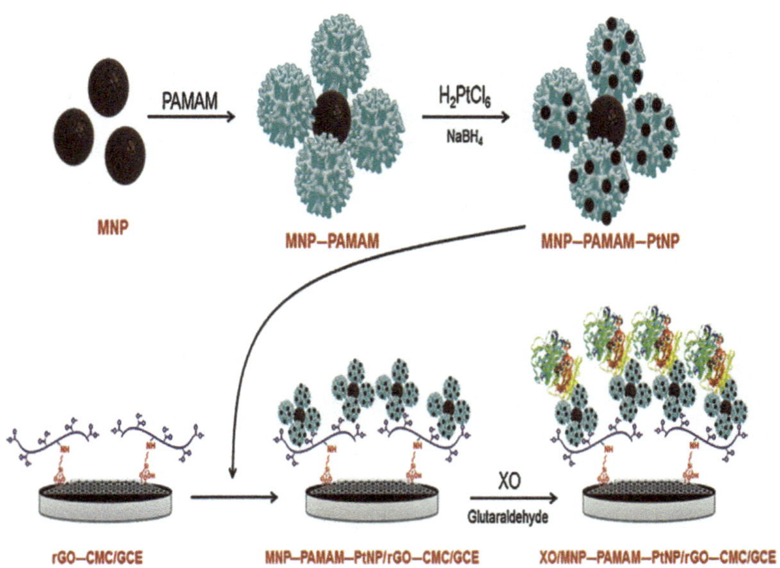

Fig. 4 Preparation of MNP–PAMAM–PtNP and XOx/MNP–PAMAM–PtNP/rGO–CMC/GCE enzyme electrode (Borisova et al., 2016).

shows the steps involved in the preparation of the hybrid nanomaterial and the assembly of the biosensor (Borisova et al., 2016).

Other oxidases used for the development of biosensors with biomedical applications include tyrosinase (Tyr), bilirubin oxidase, lysine oxidase, laccase, and β-galactosidase.

Peroxidases (enzymes that catalyze the substrate oxidization by using H_2O_2 as the oxidizing agent) were also often employed for biosensors development, the most used being horseradish peroxide (HRP) (more details and examples about the uses of HRP in biosensors are presented in Table 3).

4.2 Dehydrogenases-Based Biosensors

Some dehydrogenases including alcohol dehydrogenase (ADH), D-fructose dehydrogenase (FDH), and glutamate dehydrogenase were also employed to develop various electrochemical biosensors. Ethanol biosensors based on ADH employed various platforms based on graphene and Au nanorods and bimetallic nanoparticles. The enzyme was immobilized by physical adsorption and cross-linking with glutaraldehyde (Li, Lu, & Deng, 2013; Tig, 2017).

Electrochemical biosensors based on other types of enzyme such as alkaline phosphatase and urease were also reported.

4.3 Examples of Enzyme-Based Biosensors With Graphene

4.3.1 Biosensors for Pharmaceuticals

The intensive use of drugs in the treatment of various health problems and illness, especially in large cities with many hospitals, can lead to their spreading in the environment and possible contamination especially if released into environment in their complete form, without metabolization. Many of these drugs have a high toxicity and some even mutagenic and carcinogenic potential, posing and causing environmental damage, even at low concentrations (Lutterbeck, Baginska, Machado, & Kümmerer, 2015).

The monitoring of the environmental levels of drugs is therefore essential to minimize the risks associated with these drugs. In addition to environmental and occupational monitoring, the monitoring of plasma concentrations of some drugs is of great importance. This is because ideal treatment in different illness depends on the personalized therapeutic schemes, which are adapted according to the treatment response of each patient. The clinical monitoring of the plasmatic concentration of antineoplastic drugs is also important for the optimization of therapy and management of side effects because at doses that are too high, side effects and toxicity can occur, while at doses that are too low, the desired effects can be reduced or even lost (Florea et al., 2015).

Considering the large number and variety of different electrochemical biosensors based on enzymes and graphene derivatives already described in the literature, it is clear that this subject has been gaining importance. It is a promising scientific area which could bring great advantages to the human health and to the environment.

Some examples related to this subject can be mentioned. The highly sensitive and selective artesunate (an antimalarial drug) detection in pharmaceutical formulation and biological fluids was achieved by using indium tin oxide electrode electropolymerized with GO-polyaniline nanocomposite film on which horseradish peroxidase was immobilized by using glutaraldehyde as cross-linker. The analytical performances of this biosensor included a linear range of 0.05–0.40 ng mL^{-1}, a limit of detection of 0.012 ng mL^{-1}, and a sensitivity of 0.15 mA ng mL^{-1} (Radhapyari, Kotoky, Das, & Khan, 2013).

The same cross-linkage procedure was used for laccase immobilization on a hybrid nanomaterial based on rGO doped with Sb_2O_5 thin film. The electrochemical biosensor was tested for estriol (an endocrine disrupting compound) determination with high selectivity, fast analytical response, high sensitivity (275 mA M^{-1}), low limit of detection (11 n*M*), and a wide linear range (25 n*M*–1.03 m*M*) (Cincotto et al., 2015).

4.3.2 Biosensors for Physiological Compounds and Metabolites

4.3.2.1 Detection of Glucose

Glucose plays an important role in the diagnosis and management of diabetes mellitus; therefore, its accurate determination was the purpose of a wide range of biosensors. Glucose enzymatic biosensors contain the enzyme GOx which catalyzes the oxidation of glucose to gluconolactone by consuming oxygen and liberating H_2O_2, the process being illustrated by the following equations:

$$\text{Glucose} + \text{GOD/FAD} \rightarrow \text{Gluconolactone} + \text{GOD/FADH}_2 \quad (1)$$

$$\text{GOD/FADH}_2 + \text{O}_2 \rightarrow \text{GOD/FAD} + \text{H}_2\text{O}_2 \quad (2)$$

$$\text{FAD} + 2\text{H}^+ + 2\text{e}^- \leftrightarrow \text{FADH}_2 \quad (3)$$

GOx is an oxidoreductase enzyme which has the active center (FAD) deeply embedded in the protein shell, being the reason why the DET between FAD and the electrode surface is difficult. The first generation of glucose biosensors measured the oxygen consumption during the catalytic process, while the second generation quantified the H_2O_2 output. The glucose oxidation by GOx in the absence of oxygen is characteristic for the real third-generation glucose biosensors, the reduced form of the enzyme being oxidized at the electrode surface.

Various types of electrodes (GCE, screen-printed electrode (SPE), Au, Sn, Pt electrodes, porous gold electrode) modified with graphene (GO, rGO, nitrogen-doped graphene, graphene nanodots), polymers (polyethyleneimine, polyvinyl alcohol, polyaniline, poly(3,4-ethylenedioxythiophene), polyamidoamine dendrimer), and metallic nanoparticles (AuNPs, Fe_3O_4NPs, Co_3O_4NPs, ZnO, PtNPs) have been tested for the elaboration of sensors for the determination of glucose with outstanding analytical performances: low limit of detection, high sensitivity, wide linear range, good stability, and improved selectivity toward possible interfering substances (uric acid, ascorbic acid, dopamine (DA), NADH, acetaminophen, cholesterol, glycine, D-galactose, urea, L-phenylalanine, L-tyrosine). The characterization of the hybrid nanocomposites concerning its composition, morphology, enzyme modification, and electrochemical properties was achieved by various techniques such as: UV–vis spectrophotometry, Raman and FTIR spectroscopy, ATR, XRD, EDS, SEM, TEM, CV, EIS, amperommetry, linear sweep voltammetry, and differential pulse voltammetry (Feng, Cheng, Pan, & Zheng, 2015).

The noncovalent immobilization of the enzyme was performed by using polymer entrapment in chitosan, nafion, electrostatic interactions (with Fe_3O_4, rGO), π–π interaction (between the pyrene-modified enzymes and graphene), and physical adsorption. The covalent strategies for enzyme immobilization included the amide bond formation by using glutaraldehyde as cross-linking reagent and succinimidyl ester moiety.

The electrocatalytic oxidation of glucose, lactate, and glutamate was achieved on Au electrode modified with polystyrene beads (500 nm diameter) covered by a layer of Prussian blue (PB); after the template removal, the thiol graphene was deposited in the cavities. The enzymes GOx, glutamate oxidase, and lactate oxidase were immobilized by binding to (3-glycidyloxypropyl) trimethoxysilane. The overall protocol used for sensor elaboration and the SEM images obtained during biosensor elaboration are presented in Fig. 5 (Li et al., 2016).

Table 2 contains some examples of biosensors for glucose based on graphene derivatives.

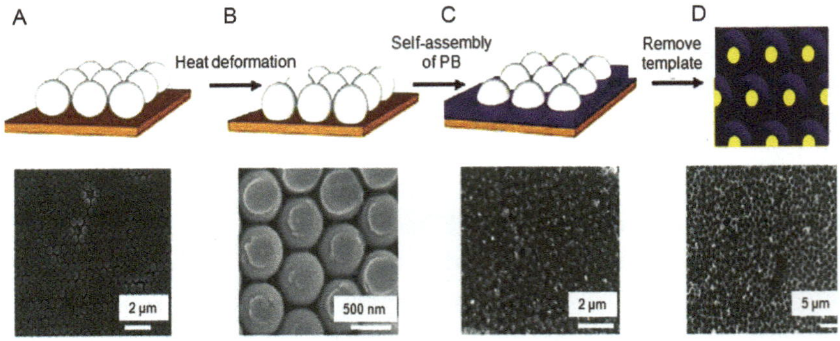

Fig. 5 The Prussian blue film formation on the Au electrode surface by using polystyrene beads (Li, Peng, Chu, Jiang, & Jin, 2016).

4.3.2.2 Detection of Hydrogen Peroxide

Hydrogen peroxide (H_2O_2) is a by-product which is hydrolyzed by oxidases in many biological processes playing an important role as a signaling molecule. Its determination by using electrochemical biosensors based on heme proteins such as hemoglobin (Hb), HRP, cytochrome *c* (Cyt *c*), catalase (CAT), myoglobin (Mb), and microperoxidase-11 (MP-11) is simple, selective, and sensitive. There are two types of heme protein-based H_2O_2 biosensors: mediated biosensors (second generation needs a mediator for the electronic transfer between the redox proteins and the underlying electrodes) and mediated-free biosensors (third generation with DET) (Zhang & Chen, 2017).

Among the heme proteins, HRP is the most used enzyme in H_2O_2 biosensors. HRP catalyzes the activation of peroxides and subsequent redox reaction as an electron acceptor (Eqs. 4–6). This redox enzyme contains 308 amino acid residues, a heme as prosthetic group (iron protoporphyrin type IX) and two calcium ions, being highly stable for chemical modifications (Nandini et al., 2013).

$$\text{HRP } (Fe^{3+}) + H_2O_2 \rightarrow \text{Compound I } (Fe^{4+}{=}O) + H_2O \tag{4}$$

$$\text{Compound I } (Fe^{4+}{=}O) + e^- + H^+ \rightarrow \text{Compound II} \tag{5}$$

$$\text{Compound II} + e^- + H^+ \rightarrow \text{HRP } (Fe^{3+}) + H_2O \tag{6}$$

Graphene-based nanomaterials have been widely employed for H_2O_2 detection because these nanocomposites are a microenvironment which maintains the bioactivity of the redox proteins and also facilitate the electron transfer.

Table 2 Examples of Enzyme-Based Biosensors for Glucose and Related Compounds

Electrode Configuration	Analyte	Linear Range	LOD	Sensitivity	References
rGO-AuNPs/CS-GOx/GCE	Glucose	0.1–1.3 m*M*	76 μ*M*	34 mA M^{-1} cm^{-2}	Bai and Shiu (2014)
rGO/AuNPs/GOx/GCE		0.2–2; 2–20 m*M*	17 μ*M*	56.93; 13.48 μA mM^{-1} cm^{-2}	Wang and Zhang (2013)
rGO-Fe_3O_4/MSPE		0.05–1 m*M*	0.1 μ*M*	5.9 μA mM^{-1}	Pakapongpan and Pooarporn (2017)
rGO-GOx/GCE		0.1–27 m*M*		1.85 mA mM^{-1} cm^{-2}	Unnikrishnan, Palanisamy, and Chen (2013)
AuNPs-PEI-GNS		1–100 μ*M*	0.32 μ*M*	93 μA mM^{-1} cm^{-2}	Rafighi, Tavahodi, and Haghighi (2016)
GOx/PVA-Fe_3O_4/Sn		0.005–30 m*M*	8 μ*M*	9.36 mA mM^{-1}	Sanaeifar et al. (2017)
GF/Co_3O_4-NPs-GOx/GCE		0.5–16.5 m*M*	0.05 m*M*	13.52 μA mM^{-1} cm^{-2}	Karuppiah, Palanisamy, Chen, Veeramani, and Periakaruppan (2014)
CS-GOx/GRA-PANI/Pt		10.0 μ*M*– 1.48 m*M*	2.769 μ*M*	22.1 μA mM^{-1} cm^{-2}	Feng et al. (2015)
GOx/rGO-MWNTs/AuNPs/GCE		10 μ*M*–2 m*M* 2 m*M*– 5.2 m*M*	4.1 μ*M* 0.95 m*M*	0.695 μA mM^{-1} cm^{-2} 0.2380 μA mM^{-1} cm^{-2}	Devasenathipathy et al. (2015)
GOx/rGO/borosilicate glass capillary		10–1000 μ*M*	—	—	Hasan et al. (2015)
GOx/GRA-PEDOT:PSS/SPCE		20–900 μ*M*	0.3 μ*M*	7.23 μA mM^{-1}	Wisitsoraat et al. (2013)
PRGO-AuNPs/GOx/GCE		0.14–4.0 μ*M*	0.06 μ*M*	15.04 mA mM^{-1}	Sabury, Kazemi, and Sharif (2015)
CS-GOX/PTBO/PB/rGO/GCE		20 μ*M*– 1.09 m*M*	8.4 μ*M*	59 mA M^{-1} cm^{-2}	Bai, Chen, and Shiu (2013)

Continued

Table 2 Examples of Enzyme-Based Biosensors for Glucose and Related Compounds—cont'd

Electrode Configuration	Analyte	Linear Range	LOD	Sensitivity	References
GOx/CS/AuNPs/PPy/rGO/GCE		0.2–1.2 m*M*	—	123.8 μA mM^{-1} cm^{-2}	Xue et al. (2014)
rGO/Fe_3O_4/GOx/GCE		0.5–10 m*M*	—	2.645 μA mM^{-1}	Wang et al. (2018)
GOx/Nafion/PANI-TNTs/GCE		10–2500 μ*M*	0.5 μ*M*	11.4 μA mM^{-1} 177.16 μA mM^{-1} cm^{-2}	Zhu, Liu, Wang, Huo, and Yan (2015)
GOx/G-PSE/SPE		0.1–1 m*M*	28.3 μ*M*	32.15 μA mM^{-1}	Chia, Tan, Khiew, Chin, and Siong (2015)
GOx/rGO-CNT-ZnO/GCE		10 μ*M*–6.5 m*M*	4.5 μ*M*	5.36 μA mM^{-1} cm^{-2}	Hwa and Subramani (2014)
GOx/PtNPs/PAMAM-GPTMS-rGO/GCE		10 μ*M*–8.1 m*M*	0.8 μ*M*	24.6 mA *M* cm^{-2}	Araque et al. (2014)
GOx (XOx)/NG/GCE	Glucose Hypoxanthine	Up to 1.2 m*M*	17 μ*M* 180 μ*M*	12.8 μA mM^{-1} cm^{-2} 135.6 μA mM^{-1} cm^{-2}	Barsan, Prathish, Sun, and Brett (2014)
GOx (CAT)/graphene nanodots encaged porous gold electrode	Glucose H_2O_2	0.05–100 m*M* 0.005–4 m*M*	30 μ*M* 1 μ*M*	—	Wang et al. (2015)
GOx (GMOx) (LOx)/tG/PB/Au	Glucose Glutamate Lactate	0.01–0.8 m*M* 0.01–0.09 m*M* 0.01–0.2 m*M*	—	162 μA mM^{-1} cm^{-2} 93 μA mM^{-1} cm^{-2} 70 μA mM^{-1} cm^{-2}	Li et al. (2016)

AuNPs, gold nanoparticles; *CAT*, catalase; *CS*, chitosan; *GCE*, glassy carbon electrode; *GF*, graphene flakes; *GMOx*, glutamate oxidase; *GOx*, glucose oxidase; *GPTMS*, (3-glycidyloxypropyl)trimethoxysilane; *GRA*, graphene; *Lox*, lactate oxidase; *MSPE*, magnetic screen-printed electrode; *MWCNTs*, multiwalled carbon nanotubes; *NG*, nitrogen-doped graphene; *PAMAM*, polyamidoamine; *PANI*, polyaniline; *PB*, Prussian blue; *PEDOT*, poly(3,4-ethylenedioxythiophene); *PEI*, polyethyleneimine; *PRGO*, partially reduced graphene oxide; *PSE*, 1-pyrenebutyric acid *N*-hydroxysuccinimide ester; *PSS*, polystyrene sulfonic acid; *PTBO*, poly(toluidine blue O); *PVA*, polyvinyl alcohol; *rGO*, reduced graphene oxide; *SPCE*, screen-printed carbon electrode; *tG*, thiol graphene; *TNTs*, TiO_2 nanotubes; *Xox*, xanthine oxidase.

Therefore, graphene (as GO, graphene functionalized with COO^-, graphene foam) was used in combination with metallic nanoparticles (such as PdNPs, AuNPs, CeO_2, MoS_2) or polymers (polydopamine, poly(3,4-ethylenedioxythiophene)-poly(styrenesulfonate), polypyrrole) for electrochemical sensing of H_2O_2. These nanocomposites were characterized by SEM/EDS, CV, DPV, and EIS.

The HRP enzyme was immobilized by using galvanostatic electrodeposition, covalent grafting to polydopamine, NHS/EDC system, or physical adsorption, encapsulation in biomimetic graphene capsules, nafion, and chitosan.

Some examples that illustrate the use of graphene for the elaboration of biosensors for H_2O_2 detection are presented in Table 3.

4.3.2.3 Detection of Other Analytes With Biomedical Relevance

4.3.2.3.1 Phenols The oxidation of phenolic compounds such as catechol, phenol, cresol, DA, and epinephrine is catalyzed by Tyr or polyphenol oxidase (PPO), a multifunctional oxidoreductase with two copper atoms in the active site, according to Eqs. (7) and (8):

$$\text{Catechol} + \text{tyrosinase } (O_2) \rightarrow o\text{-quinone} + H_2O \quad (7)$$

$$o\text{-Quinone} + 2H + 2e^- \rightarrow \text{catechol (at electrode)} \quad (8)$$

Catechol can also be oxidized by laccase, a multicopper oxidase, as follows:

$$\text{Catechol} + \text{laccase (oxy)} \rightarrow \text{1,2-benzoquinone} + \text{laccase (deoxy)} + 2H^+ + 2e^- \quad (9)$$

$$\text{Laccase (deoxy)} + O_2 + 4H \rightarrow \text{laccase (oxy)} + 2H_2O \quad (10)$$

Various graphene-based nanocomposites have been used for catechol and DA determination such as graphene nanoplatelets decorated with TiO_2 nanotubes, conducting polymer poly(3,4-ethylenedioxythiophene) doped with GO and Fe_2O_3, palladium–copper alloy nanocages supported on rGO, rGO functionalized with β-cyclodextrin (β-CD), and rGO functionalized with pyrrole-β-cyclodextrin. The enzymes were immobilized by using polymers (Nafion, polyethyleneimine, amphiphilic pyrrole), glutaraldehyde or physical adsorption (see Table 4).

4.3.2.3.2 Cholesterol Cholesterol is a lipid steroid metabolite with serious implication in cardiovascular diseases; therefore, its determination is very important for clinical diagnosis. Various electrochemical biosensors

Table 3 Examples of Enzyme-Based Biosensors for Hydrogen Peroxide

Electrode Configuration	Dynamic Range	LOD	Sensitivity	References
(HRP-Pd)/f-graphene/Gr	25 μ*M*–3.5 m*M*	0.05 μ*M*	92.82 μA mM^{-1} cm^{-2}	Nandini et al. (2013)
HRP/PDA/MB-CNTs/3D-G	0.2 μ*M*–1.1 m*M*	58 n*M*	227.8 μA mM^{-1} cm^{-2}	Liu, Wang, Wang, and Wang (2015)
PEDOT:PSS-rGO-AuNPs-HRP/SPGE	5–400 μ*M*	0.08 μ*M*	677 μA mM^{-1} cm^{-2}	Mercante et al. (2017)
rGO/Ppy nanotube/GCE	—	0.1 μ*M*	—	Cai et al. (2017)
GS-PSS/GRCAPS/ITO	0.01–12 m*M*	3.3 μ*M*	—	Fan et al. (2015)
Nafion/HRP/CeO_2-rGO/GCE	0.1–500 μ*M*	0.021 μ*M*	4.65 μA mM^{-1}	Radhakrishnan and Kim (2015)
HRP-MoS_2–GRA/GCE	0.2–1103 μ*M*	0.049 μ*M*	679.7 μA mM^{-1} cm^{-2}	Song, Ni, and Kokot (2014)
CS/HRP/Fc-rGO/GCE	0.35–150 μ*M*	0.10 μ*M*	—	Deng, Zhou, and Li (2013)

3D-G, 3D-graphene foam freestanding electrode; *AuNPs*, gold nanoparticles; *CS*, chitosan; *f-graphene*, graphene functionalized with COO^-; *Fc*, ferrocene; *GCE*, glassy carbon electrode; *Gr*, graphite; *GRA*, graphene; *GRCAPS*, graphene capsule; *GS-PSS*, graphene-poly(sodium 4-styrenesulfonate); *HRP*, horseradish peroxidase; *ITO*, indium tin oxide; *MB*, methylene blue; *PEDOT*, poly(3,4-ethylenedioxythiophene); *Ppy*, polypyrrole; *PSS*, polystyrene sulfonic acid; *rGO*, reduced graphene oxide; *SPGE*, screen-printed graphite electrode.

Table 4 Examples of Enzyme-Based Biosensors for Other Analytes

Electrode Configuration	Analyte	Dynamic Range	LOD	Sensitivity	References
Tyr/[Demim]Br/Nafion/GNPs–TNTs/GCE	Catechol	0.30–110 μ*M*; 110.5–390 μ*M*	0.055 μ*M*	—	Liu, Yan, Zhu, Zhang, and Liu (2015)
PEDOT-rGO-Fe_2O_3-PPO/GCE		0.04–62 μ*M*	7 n*M*	—	Sethuraman, Muthuraja, Anandha Raj, and Manisankar (2016)
Lac/rGO–PdCu NCs/GCE		0.005–1.155 m*M* 1.655–5.155 m*M*	1.5 μ*M* 2.0 μ*M*	12.65 μA mM^{-1} 5.51 μA mM^{-1}	Mei et al. (2015)
rGO/β-CD/Tyr/PEI/GCE	Dopamine	3.9–328.8 μ*M*	3.9 μ*M*	12.1 nA μM^{-1}	Fritea, Tertiş, Cosnier, Cristea, and Săndulescu (2015)
poly-[NEt^{4+}-Py]/Tyr/poly-[β-CD-Py/rGO]/GCE	Catechol Dopamine	1 n*M*–25 μ*M* 27 n*M*–38.6 μ*M*	1 n*M* 27 n*M*	0.71 A M^{-1} cm^{-2} 0.012 A M^{-1} cm^{-2}	Fritea, Le Goff, et al. (2015)
GO&CHER&ChOx/AuNPs/SPE	Cholesterol	0.01–5000 μg mL^{-1}	0.001 μg mL^{-1}	0.084 μA cm^{-2} μg mL^{-1}	Huang et al. (2018)
ChOx/AuNs/GO-SH/Gr		0.05–11.45 m*M*	0.2 n*M*	273 mA mM^{-1} cm^{-2}	Nandini et al. (2013)
G/Ti(G)-3DNS/CS/ChOx/GCE		0.05–8.0 m*M*	6 μ*M*	3.82 μA cm^{-2} mM^{-1}	Komathi, Muthuchamy, Lee, and Gopalan (2016)
ChOx/AuNPs/TGPHs/GCE		0.05–590 μ*M*	0.017 μ*M*	—	Cao, Zhang, Chai, and Yuan (2013)

Continued

Table 4 Examples of Enzyme-Based Biosensors for Other Analytes—cont'd

Electrode Configuration	Analyte	Dynamic Range	LOD	Sensitivity	References
ADH–AuNRs-GN/GCE	Ethanol	5–377 μ*M*	1.5 μ*M*	102 μA mM^{-1} cm^{-2}	Li et al. (2013)
Naf/ADH/ERGO/P(L-cys)/Au-AgNPs/GCE		0.017–1.845 m*M*	5 μ*M*	0.644 μA mM^{-1}	Tig (2017)
ADH/ERGO–PTH/GCE		0.05–1.0 m*M*	0.3 μ*M*	2.8 μA mM^{-1} cm^{-2}	Li et al. (2013)
Sandwich-type sensor based on AuNPs/MgO nanoflowers and GO-AuNPs/GCE	miRNA	0.0001–100 p*M*	0.05 f*M*	—	Shuai, Huang, Zhang, Cao, and Jia (2017)
HRP-AuNPs–Ab2/CEA/MBs–Ab1/G	CEA	5–60 ng mL^{-1}	5 ng mL^{-1}	—	Jin et al. (2014)
FDH/TRGO	Fructose	0.7–8.8 m*M*	0.7 m*M*	14.5 μA mM^{-1} cm^{-2}	Sakinyte, Barkauskas, Gaidukevic, and Razumiene (2015)
β-Gal-GOx/P(1,5-DAN)/G/Pt	Lactose	0–60 μg mL^{-1}	1.3 μg mL^{-1}	1.33 μA $μg^{-1}$ mL^{-1}	Nguyen et al. (2016)
BOx/GONP/NiNPs/ITO	Bilirubin	0.01–600 μ*M*	0.15 n*M*	—	Rawal, Chauhan, Tomar, and Gupta (2017)
BOx/GONP@Ppy/FTO		0.01–500 μ*M*	0.1 n*M*	—	Chauhan, Rawal, Hooda, and Jain (2016)

ZnPh/GO/Urs/GCE	Urea	0.4–22 μ*M*	0.034 μ*M*	44.53 μA mM^{-1}	Selvarajan, Suganthi, and Rajarajan (2018)
Urs-GLDH/GOS/ITO		3.3–19.9 m*M*	2.1 m*M*	2.6 μA mM^{-1} cm^{-2}	Abraham et al. (2014)
LOx/c-MWCNTs–SnO_2–G–CS/GCE	Lysine	0.99 μ*M*–0.16 m*M*	0.15 μ*M*	55.2 μA mM^{-1} cm^{-2}	Kacar, Erden, and Kilic (2017)
Lac/rGO-RhNP/GCE	17β-Estradiol	0.9–11 p*M*	0.54 p*M*	25.7 μA M^{-1} cm^{-1}	Povedano et al. (2017)

β-CD, β-cyclodextrin; *β-Gal*, galactosidase; *3DNS*, 3D nanostacks; *Ab*, antibody; *ADH*, alcohol dehydrogenase; *AgNPs*, silver nanoparticles; *AuNPs*, gold nanoparticles; *AuNRs*, gold nanorods; *AuNs*, gold nanostructures; *BOx*, bilirubin oxidase; *CEA*, carcinoembryonic antigen; *CHER*, cholesterol esterase; *ChOx*, cholesterol oxidase; *CS*, chitosan; *[Demim]Br*, brominated 1-decyl-3-methyl imid-azole; *ERGO*, electrochemically reduced graphene oxide; *FDH*, D-fructose dehydrogenase; *FTO*, fluorine-doped tin oxide; *G*, graphene; *GCE*, glassy carbon electrode; *GLDH*, glutamate dehydrogenase; *GN*, graphene nanosheets; *GNPs*, graphene nanoplatelets; *GO*, graphene oxide; *GONP*, graphene oxide nanoparticles; *GOS*, mesoporous silica particle embedded graphene oxide; *Gr*, graphite; *GR*, graphene; *HRP*, horseradish peroxidase; *ITO*, indium tin oxide; *Lac*, laccase; *Lox*, lysine oxidase; *MB*, methylene blue; *Naf*, Nafion; *NiNPs*, nickel nanoparticles; *PdCuNCs*, palladium–copper alloyed nanocages; *PEDOT*, poly(3,4-ethylenedioxythiophene); *PEI*, polyethylenimine; *poly-[NEt4+-Py]*, poly[12-(pyrrol-1-yl) dodecyl] triethylammonium tetrafluoroborate; *PPO*, polyphenol oxidase; *Ppy*, poly-pyrrole; *Py*, pyrrole; *P(L-Cys)*, poly(L-cysteine); *PTH*, polythionine; *P(1,5-DAN)*, 1,5-polydiaminonaphthalene film; *rGO*, reduced graphene oxide; *RhNPs*, rhodium nanoparticles; *TGPHs*, TiO_2-graphene-Pt-Pd hybrid materials; *TNTs*, TiO_2 nanotubes; *TRGO*, thermally reduced graphene oxide; *Tyr*, tyrosinase; *Urs*, urease; *ZnPh*, zinc phthalocyanine.

were developed based on nanocomposites containing GO, GO-SH, graphene sheets interconnected graphene decorated with AuNPs, TiO_2, and Pt–Pd hybrid material (see Table 4). The enzymes cholesterol oxidase and cholesterol esterase were immobilized by using NHS/EDC system for covalent bonding to graphene and chitosan as binder, respectively (see Table 4).

4.3.2.3.3 Alcohol Ethanol biosensors employed various platforms such as graphene–Au nanorods hybrid nanosheets, bimetallic nanoparticles onto which poly(L-cysteine) and electrochemically rGO were deposited. Yeast ADH used in biosensors is a Zn-containing metalloenzyme catalyzing the interconversion between alcohols and aldehydes or ketones, oxidation reactions which are coupled with the reduction of NAD^+ coenzyme.

4.3.2.3.4 Other Analytes Other substances with biochemical importance and useful for clinical diagnosis, such as miRNA, carcinoembryonic antigen (CEA), fructose, lactose, bilirubin, urea, lysine, and estradiol, have been determined using various graphene-based biosensors. The employed nanocomposites consisted in MgO nanoflower and GO–AuNPs, magnetic beads and enzyme-labeled antibody-AuNP bioconjugate, thermally rGO, graphene/1,5-polydiaminonaphthalene film, GO nanoparticles decorated with NiNPs/polypyrrole, mesoporous silica particle embedded GO, carboxylated multiwalled CNTs–tin oxide nanoparticles–graphene–chitosan, RhNPs deposited on rGO. The enzymes (alkaline phosphatase, HRP, FDH, β-galactosidase, GOx, bilirubin oxidase, urease, glutamate dehydrogenase, lysine oxidase, laccase) have been immobilized by specific reaction between avidin and biotin, physical adsorption, glutaraldehyde, electrostatic interactions, and amide covalent bind (NHS/EDC).

For the detection of DA a first approach developed by our team consisted in the Tyr immobilization in polyethylenimine (PEI) film. The biosensor was elaborated by layer-by-layer (LBL) deposition of: rGO (1 mg/mL, chemical reduction of GO with ascorbic acid), β-CD (1 mg/mL), Tyr (1 mg/mL), and PEI (1 mg/mL, hydroalcoholic solution 1:1). The optimization process included the variation of rGO and Tyr layers, the most sensitive configuration for DA determination was based on one layer of 5 μL rGO and 50 μg Tyr. Furthermore, the Tyr immobilization onto the graphene layer can be reinforced by the inclusion complexes between hydrophobic residues from Tyr shell with β-CD. This biosensor exhibited good analytical performances such as: linear range from 3.9 to 328.8 μ*M*, limit of detection of 3.9 μ*M*, sensitivity of 12.1 nA/μ*M*, recovery rates of 96.9% and 100.7% for pharmaceutical formulation, 97.25% and 100.19% for urine samples, 98.18% and 101.92% for serum samples, respectively. Interferences studies have been also performed with propylene glycol, ethanol, metabisulfite, glucose, uric acid, and ascorbic acid without any significant modification of the DA signal (Fritea, Tertiş, et al., 2015).

The second strategy was based on Tyr entrapment in an amphiphilic pyrrole derivative [12-(pyrrol-1-yl) dodecyl] triethylammonium tetrafluoroborate] which was electropolymerized at the electrode surface. rGO was noncovalent functionalized with β-CD-pyrrole (β-cyclodextrin-11-pyrrolyl-1-undecyl carboxylic acid amide) leading to an electropolymerizable graphene framework in which the pyrrole derivative bearing a CD moiety acted as an aqueous dispersant (due to the β-CD presence) and also as an electropolymerizable unit (due to the pyrrole presence) generating electrodeposited graphene sheets.

Two possibilities have been tested for the graphene-based biosensor elaboration. The two-step deposition procedure was performed by electrodeposition of poly-(β-CD-pyrrole/rGO) layer followed by the electropolymerization of an adsorbed mixture of amphiphilic pyrrole and Tyr. The one-step technique consisted in the direct electropolymerization of the mixture containing β-CD-pyrrole/rGO, amphiphilic pyrrole, and Tyr. Both configurations (bilayer and mixed layer) generated a maximum current density and sensitivity two times more increased and a lower LOD than in the absence of the electrodeposited graphene layer when the biosensors were tested for catechol and DA amperometric determination. In the case of catechol sensing, the analytical performances recorded on the bilayer configuration were maximum intensity of 171.4 $\mu A\ cm^{-2}$, sensitivity of 0.71 $A\ M^{-1}\ cm^{-2}$, LOD 2.8 n*M*, and linear range 2.8 n*M*–25 μ*M*; meanwhile for the mixed layer were 162.1 $\mu A\ cm^{-2}$, 0.71 $A\ M^{-1}\ cm^{-2}$, 1 n*M* and 1 n*M*–25 μ*M*, respectively. The analytical parameters for DA determination included a maximum intensity of 17.4 $\mu A\ cm^{-2}$, sensitivity of 0.012 $A\ M^{-1}\ cm^{-2}$, LOD 27 n*M*, and linear range 27 n*M*–38.6 μ*M* recorded on the bilayer configuration; meanwhile the parameters obtained on the mixed layer configuration were 13.1 $\mu A\ cm^{-2}$, 0.011 $A\ M^{-1}\ cm^{-2}$, LOD 27 n*M*, linear range 27 n*M*–37 μ*M*, respectively (Fritea, Le Goff, et al., 2015).

In conclusion, the combination of the rGO with CD, Tyr, and PEI enhanced the electroanalytical performances of the biosensor allowing a selective and sensitive detection of DA.

In addition, a new nanostructured graphene framework reinforced by a polypyrrole film was designed and electropolymerized on GCEs leading to the elaboration of rGO-based biosensors applied for catechol and DA sensing.

5. FUTURE TRENDS

The electrochemical biosensors present some advantages compared with other analytical methods including a sensitive, selective, simple, fast, and inexpensive determination of biomolecules. Furthermore, they can be integrated into portable and miniaturized devices such as POC with applications in diagnosis and clinical areas. The great need to bring the laboratory tests much closer to the patient bedside was solved by developing POC diagnostics devices based on electrochemical sensing which offer all the advantages mentioned before, adding also the miniaturized form, portability, easy-to-be used, and disposability.

The main disadvantages of conventional detection methods orientated the researchers to find innovative strategies that will satisfy the demands of POC devices. A set of criteria that an ideal POC device needs to satisfy is currently used by researchers, so-called ASSURED criteria, where the acronym stands for A-affordable, S-sensitive, S-selective, U-user-friendly, R-rapid and robust, *E*-equipment-free, and D-deliverable to end users (Ciui, Jambrec, Sandulescu, & Cristea, 2017; Yu, Maslova, & Hsing, 2017). Electrochemical biosensors are good candidates for future POC devices due to their features and advantages. However, there are more criteria from this list that need to be satisfied. Although several examples of electrochemical biosensors were given through this chapter, their application in the detection of some pharmaceutical compounds and other hazardous compounds in real samples are at the beginning. Those compounds could inhibit the enzymes; therefore, the integration of nanomaterials and other improvements regarding the sensitivity and selectivity of the resulting biosensors must be considered.

Compared to other instrumental methods of chemical analysis, electrochemical biosensors are relatively easy to miniaturize. The main trends in modern electroanalysis include the development of biosensors based on progress in chemical and biochemical methods of molecular recognition, the development of measuring devices of a large integration scale, including sensor arrays, the use of new materials, including nanosized and nanostructured materials, as well as the introduction of flow systems. In order to become ready for the market, a great demand in developing lab-on-a-chip and paper platforms for less assays steps and with interest for *in field* applications are needed, as well as biocompatible devices (for in vitro analysis).

The employment of new materials for easy to use and multidetection devices are also requirements for successfully, environmentally, and human friendly biosensors for biomedical analysis.

6. CONCLUSIONS

The outstanding electrochemical properties of graphene and graphene-based materials determined their intense use as platforms for the construction of a wide range of electrochemical biosensors with improved analytical behavior. Graphene and its composites have shown great potentials in biosensing even that there are issues to be solved in the future: the reproducibility and the stability of the nanocomposites with graphene that must be improved and the possible (and not yet entirely proved) toxicity of the graphene that limits its application to in vivo analysis.

Enzymes, with their high catalytic activity and specificity, have been intensively included in biosensors development using various immobilization methods such as adsorption, disulfide binding, metal chelation, covalent bonding, entrapment, encapsulation, and cross-linking.

Several important physical, chemical, electrical, and optical properties of graphene were summarized, which make them one of the best materials used for the transduction of signals associated with the recognition of analytes. Moreover, there is a huge category of analytes that can be considered for detection using graphene-based platforms such as biomarkers for different diseases, metabolites, and drugs. Besides, graphene can serve as scaffolds for immobilization of biomolecules on their surface. The electrocatalytic activity of graphene-modified transducers and electrodes toward hydrogen peroxide and NADH permits effective low-potential amperometric detection of various important analytes for a wide range of applications with practical relevance. Furthermore, graphene-based biosensors have huge potential for in vivo detection with less cytotoxicity, high sensitivity, and long-term stability for reliable POC diagnostics under physiological conditions. These applications are not restricted to the biomedical field but also have various applications in the food quality assurance, water purification, agriculture, industry, and environment monitoring.

ACKNOWLEDGMENTS

This work was supported by a Grant of the Romanian National Authority for Scientific Research and Innovation, CCCDI–UEFISCDI, project numbers PN-III-P2-2.1-PED-2016-0172 and PN-III-P1-1.2-PCCDI 2017-0407, within PNCDI III.

REFERENCES

Abraham, S., Ciobota, V., Srivastava, S., Srivastava, S. K., Singh, R. K., Dellith, J., et al. (2014). Mesoporous silica particle embedded functional graphene oxide as an efficient platform for urea biosensing. *Analytical Methods*, *6*, 6711–6720.

Afag, S., & Iqbal, J. (2001). Immobilization and stabilization of papain on chelating sepharose: A metal chelate regenerable carrier. *Electronic Journal of Biotechnology*, *4*(3). https://doi.org/10.2225/vol4-issue3-fulltext-1.

Araque, E., Arenas, C. B., Gamella, M., Reviejo, J., Villalonga, R., & Pingarrón, J. M. (2014). Graphene–polyamidoamine dendrimer–Pt nanoparticles hybrid nanomaterial for the preparation of mediatorless enzyme biosensor. *Journal of Electroanalytical Chemistry*, *717–718*, 96–102.

Arlett, J. L., Myers, E. B., & Roukes, M. L. (2011). Comparative advantages of mechanical biosensors. *Nature Nanotechnology*, *6*, 203–215.

Bai, X., Chen, G., & Shiu, K.-K. (2013). Electrochemical biosensor based on reduced graphene oxide modified electrode with Prussian blue and poly(toluidine blue O) coating. *Electrochimica Acta*, *89*, 454–460.

Bai, X., & Shiu, K.-K. (2014). Investigation of the optimal weight contents of reduced graphene oxide–gold nanoparticles composites and theirs application in electrochemical biosensors. *Journal of Electroanalytical Chemistry*, *720-721*, 84–91.

Barsan, M. M., Prathish, K. P., Sun, X., & Brett, C. M. A. (2014). Nitrogen doped graphene and its derivatives as sensors and efficient direct electron transfer platform for enzyme biosensors. *Sensors and Actuators B*, *203*, 579–587.

Bo, X., Zhou, M., & Guo, L. (2017). Electrochemical sensors and biosensors based on less aggregated graphene. *Biosensors and Bioelectronics*, *89*, 167–186.

Bollella, P., Fusco, G., Tortolini, C., Sanzò, G., Favero, G., Gorton, L., et al. (2017). Beyond graphene: Electrochemical sensors and biosensors for biomarkers detection. *Biosensors and Bioelectronics*, *89*, 152–166.

Bolotin, K. I., Sikes, K. J., Jiang, Z., Klima, M., Fudenberg, G., Hone, J., et al. (2008). Ultrahigh electron mobility in suspended graphene. *Solid State Communications*, *146*, 351–355.

Borisova, B., Sánchez, A., Jiménez-Falcao, S., Martín, M., Salazar, P., et al. (2016). Reduced graphene oxide-carboxymethylcellulose layered with platinum nanoparticles/PAMAM dendrimer/magnetic nanoparticles hybrids. Application to the preparation of enzyme electrochemical biosensors. *Sensors and Actuators B*, *232*, 84–90.

Bucur, B., Danet, A. F., & Marty, J. L. (2004). Versatile method of cholinesterase immobilisation via affinity bonds using Concanavalin A applied to the construction of a screen-printed biosensor. *Biosensors and Bioelectronics*, *20*, 217–225.

Cai, Z., Xiong, H., Zhu, Z., Huang, H., Li, L., Huang, Y., et al. (2017). Electrochemical synthesis of graphene/polypyrrole nanotube composites for multifunctional applications. *Synthetic Metals*, *227*, 100–105.

Cao, L. (2005a). *Carrier-bound immobilized enzymes: Principles, applications and design.* New York: John Wiley& Sons.

Cao, L. (2005b). Enzymes: Science or art? *Current Opinion in Chemical Biology*, *9*, 217–226.

Cao, S., Zhang, L., Chai, Y., & Yuan, R. (2013). An integrated sensing system for detection of cholesterol based on TiO_2–graphene–Pt–Pd hybrid nanocomposites. *Biosensors and Bioelectronics*, *42*, 532–538.

Castillo, J., Gáspár, S., Leth, S., Niculescu, M., Mortari, A., Bonţidean, I., et al. (2004). Biosensors for life quality: Design, development and applications. *Sensors and Actuators B: Chemical*, *102*, 179–194.

Castro Neto, A. H., Guinea, F., Peres, N. M. R., Novoselov, K. S., & Geim, A. K. (2009). The electronic properties of graphene. *Reviews of Modern Physics*, *81*(1), 109–162.

Chaubey, A., & Malhotra, B. D. (2002). Mediated biosensors. *Biosensors and Bioelectronics*, *17*, 441–456.

Chauhan, N., Rawal, R., Hooda, V., & Jain, U. (2016). Electrochemical biosensor with graphene oxide nanoparticles and polypyrrole interface for the detection of bilirubin. *RSC Advances*, *6*, 63624–63633.

Chia, J. S. Y., Tan, M. T. T., Khiew, P. S., Chin, J. K., & Siong, C. W. (2015). A bioelectrochemical sensing platform for glucose based on irreversible, non-covalent pi–pi functionalization of graphene produced via a novel, green synthesis method. *Sensors and Actuators B*, *210*, 558–565.

Cincotto, F. H., Canevari, T. C., Machado, S. A. S., Sáncheza, A., Barrioa, M. A. R., Villalonga, R., et al. (2015). Reduced graphene oxide-Sb_2O_5 hybrid nanomaterial for the design of a laccase-based amperometric biosensor for estriol. *Electrochimica Acta*, *174*, 332–339.

Ciui, B., Jambrec, D., Sandulescu, R., & Cristea, C. (2017). Bioelectrochemistry for miRNA detection. *Current Opinion in Electrochemistry*, *5*, 183–192.

Clark, L. C., & Lyons, C. (1962). Electrode systems for continuous monitoring in cardiovascular surgery. *Annals of the New York Academy of Sciences*, *102*(1), 29–45.

Dai, H. J. (2002). Carbon nanotubes: Synthesis, integration, and properties. *Accounts of Chemical Research*, *35*, 1035–1044.

Deng, K. Q., Zhou, J. H., & Li, X. F. (2013). Noncovalent nanohybrid of ferrocene with chemically reduced graphene oxide and its application to dual biosensor for hydrogen peroxide and choline. *Electrochimica Acta*, *95*, 18–23.

Devasenathipathy, R., Mani, V., Chen, S.-M., Huang, S.-T., Huang, T.-T., Lin, C.-M., et al. (2015). Glucose biosensor based on glucose oxidase immobilized at gold nanoparticles decorated graphene-carbon nanotubes. *Enzyme and Microbial Technology*, *78*, 40–45.

Fan, Z., Lin, Q., Gong, P., Liu, B., Wang, J., & Yang, S. (2015). A new enzymatic immobilization carrier based on graphene capsule for hydrogen peroxide biosensors. *Electrochimica Acta*, *151*, 186–194.

Feng, X., Cheng, H., Pan, Y., & Zheng, H. (2015). Development of glucose biosensors based on nanostructured graphene-conducting polyaniline composite. *Biosensors and Bioelectronics*, *70*, 411–417.

Ferrari, A. C., Meyer, J. C., Scardaci, V., Casiraghi, C., Lazzeri, M., Mauri, F., et al. (2006). Raman spectrum of graphene and graphene layers. *Physical Review Letters*, *97*, 187401.

Florea, A., Guo, Z. Z., Cristea, C., Bessueille, F., Sandulescu, R., & Jaffrezic-Renault, N. (2015). Anticancer drug detection using a highly sensitive molecularly imprinted electrochemical sensor based on an electropolymerized microporous metal organic framework. *Talanta*, *138*, 71–76.

Fritea, L., Le Goff, A., Putaux, J.-L., Tertis, M., Cristea, C., Sandulescu, R., et al. (2015). Design of a reduced-graphene-oxide composite electrode from an electropolymerizable graphene aqueous dispersion using a cyclodextrin-pyrrole monomer. Application to dopamine biosensing. *Electrochimica Acta*, *178*, 108–112.

Fritea, L., Tertiş, M., Cosnier, S., Cristea, C., & Săndulescu, R. (2015). A novel reduced graphene oxide/β-cyclodextrin/tyrosinase biosensor for dopamine detection. *International Journal of Electrochemical Science*, *10*, 7292–7302.

Gao, S., Wang, Y., Diao, X., Luo, G., & Dai, Y. (2010). Effect of pore diameter and cross-linking method on the immobilization efficiency of Candida rugosa lipase in SBA-15. *Bioresource Technology*, *101*(11), 3830–3837.

Gorecka, E., & Jastrzebska, M. (2011). Review article: Immobilization techniques and biopolymer carriers. *Biotechnology and Food Sciences*, *75*, 65–86.

Grodecki, K., Jozwik, I., Baranowski, J. M., Teklinska, D., & Strupinski, W. (2016). SEM and Raman analysis of graphene on SiC(0001). *Micron*, *80*, 20–23.

Grüneis, A., Attaccalite, C., Wirtz, L., Shiozawa, H., Saito, R., Pichler, T., et al. (2008). Tight-binding description of the quasiparticle dispersion of graphite and few-layer graphene. *Physical Review B, 78*, 205425.

Gupta, A., Chen, G., Joshi, P., Tadigadapa, S., & Eklund, P. C. (2006). Raman scattering from high-frequency phonons in supported *n*-graphene layer films. *Nano Letters, 6*, 2667–2673.

Hasan, K., Asif, M. H., Hassan, M. U., Sandberg, M. O., Nur, O., Willander, M., et al. (2015). A miniature graphene-based biosensor for intracellular glucose measurements. *Electrochimica Acta, 174*, 574–580.

He, Q. Y., Wu, S. X., Yin, Z. Y., & Zhang, H. (2012). Graphene-based electronic sensors. *Chemical Science, 3*, 1764–1772.

Heller, I., Chatoor, S., Mannik, J., Zevenbergen, M. A. G., Dekker, C., & Lemay, S. G. (2010). Influence of electrolyte composition on liquid-gated carbon nanotube and graphene transistors. *Journal of the American Chemical Society, 132*, 17149–17156.

Hibino, H., Kageshima, H., Maeda, F., Nagase, M., Kobayashi, Y., & Yamaguchi, H. (2008). Microscopic thickness determination of thin graphite films formed on SiC from quantized oscillation in reflectivity of low-energy electrons. *Physical Review B, 77*, 075413.

Huang, Y., Tan, J., Cui, L., Zhou, Z., Zhou, S., Zhang, Z., et al. (2018). Graphene and Au NPs co-mediated enzymatic silver deposition for the ultrasensitive electrochemical detection of cholesterol. *Biosensors and Bioelectronics, 102*, 560–567.

Huang, X., Yin, Z. Y., Wu, S. X., Qi, X. Y., He, Q. Y., Zhang, Q. C., et al. (2011). Graphene-based materials: Synthesis, characterization, properties, and applications. *Small, 7*, 1876–1902.

Hulanicki, A., Glab, S., & Ingman, F. (1991). Chemical sensors. Definitions and classification. *Pure and Applied Chemistry, 63*, 1247–1250.

Hwa, K.-Y., & Subramani, B. (2014). Synthesis of zinc oxide nanoparticles on graphene–carbon nanotube hybrid for glucose biosensor applications. *Biosensors and Bioelectronics, 62*, 127–133.

Jin, B., Wang, P., Mao, H., Hu, B., Zhang, H., Cheng, Z., et al. (2014). Multi-nanomaterial electrochemical biosensor based on label-free graphene for detecting cancer biomarkers. *Biosensors and Bioelectronics, 55*, 464–469.

Kacar, C., Erden, P. E., & Kilic, E. (2017). Amperometric L-lysine biosensor based on carboxylated multiwalled carbon nanotubes-SnO_2 nanoparticles-graphene composite. *Applied Surface Science, 419*, 916–923.

Karuppiah, C., Palanisamy, S., Chen, S.-M., Veeramani, V., & Periakaruppan, P. (2014). A novel enzymatic glucose biosensor and sensitive non-enzymatic hydrogen peroxide sensor based on graphene and cobalt oxide nanoparticles composite modified glassy carbon electrode. *Sensors and Actuators B, 196*, 450–456.

Khalil, I., Rahmati, S., Julkapli, N. M., & Yehye, W. A. (2018). Graphene metal nanocomposites—Recent progress in electrochemical biosensing applications. *Journal of Industrial and Engineering Chemistry, 59*, 425–439.

Komathi, S., Muthuchamy, N., Lee, K.-P., & Gopalan, A.-I. (2016). Fabrication of a novel dual mode cholesterol biosensor using titanium dioxide nanowire bridged 3D graphene nanostacks. *Biosensors and Bioelectronics, 84*, 64–71.

Kurbanoglu, S., Ozkan, S. A., & Merkoçi, A. (2017). Nanomaterials-based enzyme electrochemical biosensors operating through inhibition for biosensing applications. *Biosensors and Bioelectronics, 89*, 886–898.

Lahiff, E., Lynam, C., Gilmartin, N., O'Kennedy, R., & Diamond, D. (2010). The increasing importance of carbon nanotubes and nanostructured conducting polymers in biosensors. *Analytical and Bioanalytical Chemistry, 398*, 1575–1589.

Lawal, A. T. (2018). Progress in utilisation of graphene for electrochemical biosensors. *Biosensors and Bioelectronics, 106*, 149–178.

Li, Z., Huang, Y., Chen, L., Qin, X., Huang, Z., Zhou, Y., et al. (2013). Amperometric biosensor for NADH and ethanol based on electroreduced graphene oxide–polythionine nanocomposite film. *Sensors and Actuators B, 181*, 280–287.

Li, L., Lu, H., & Deng, L. (2013). A sensitive NADH and ethanol biosensor based on graphene–Au nanorods nanocomposites. *Talanta, 113*, 1–6.

Li, L., Peng, J. G., Chu, Z., Jiang, D., & Jin, W. (2016). Single layer of graphene/Prussian blue nano-grid as the low-potential biosensors with high electrocatalysis. *Electrochimica Acta, 217*, 210–217.

Liang, B., Guo, X., Fang, L., Hu, Y., Yang, G., Zhu, Q., et al. (2015). Study of direct electron transfer and enzyme activity of glucose oxidase on graphene surface. *Electrochemistry Communications, 50*, 1–5.

Liu, S., & Guo, X. (2012). Carbon nanomaterials field-effect-transistor-based biosensors. *NPG Asia Materials, 4*, e23.

Liu, S., Shen, Q., Cao, Y., Gan, L., Wang, Z. X., Steigerwald, M. L., et al. (2010). Chemical functionalization of single-walled carbon nanotube field-effect transistors as switches and sensors. *Coordination Chemistry Reviews, 254*, 1101–1116.

Liu, J., Wang, T., Wang, J., & Wang, E. (2015). Mussel-inspired biopolymer modified 3D graphene foam for enzyme immobilization and high performance biosensor. *Electrochimica Acta, 161*, 17–22.

Liu, X., Yan, R., Zhu, J., Zhang, J., & Liu, X. (2015). Growing TiO_2 nanotubes on graphene nanoplatelets and applying the nanocomposite as scaffold of electrochemical tyrosinase biosensor. *Sensors and Actuators B, 209*, 328–335.

Lutterbeck, C. A., Baginska, E., Machado, Ê. L., & Kümmerer, K. (2015). Removal of the anti-cancer drug methotrexate from water by advanced oxidation processes: Aerobic biodegradation and toxicity studies after treatment. *Chemosphere, 141*, 290–296.

Maduraiveeran, G., Sasidharan, M., & Ganesan, V. (2018). Electrochemical sensor and biosensor platforms based on advanced nanomaterials for biological and biomedical applications. *Biosensors and Bioelectronics, 103*, 113–129.

Mateo, C., Pessela, B. C. C., Grazu, V., Torres, R., Lopez-Gallego, F., Guisan, J. M., et al. (2006). One-step purification, immobilization, and stabilization of poly-histidine-tagged enzyme using metal-chelate-epoxy supports. In J. M. Walker (Series Ed.) & J. M. Guisan (Vol. Ed.), *Methods in biotechnology (series): Vol. 11. Immobilization of enzymes and cells* (pp. 117–128, 2nd ed.). Totowa: Humana Press.

Mei, L.-P., Feng, J.-J., Wu, L., Zhou, J.-Y., Chen, J.-R., & Wang, A.-J. (2015). Novel phenol biosensor based on laccase immobilized on reduced graphene oxide supported palladium–copper alloyed nanocages. *Biosensors and Bioelectronics, 74*, 347–352.

Mercante, L. A., Facure, M. H. M., Sanfelice, R. C., Migliorini, F. L., Mattoso, L. H. C., & Correa, D. S. (2017). One-pot preparation of PEDOT:PSS-reduced graphene decorated with Au nanoparticles for enzymatic electrochemical sensing of H_2O_2. *Applied Surface Science, 407*, 162–170.

Nandini, S., Nalini, S., Manjunatha, R., Shanmugam, S. U., Melo, J. S., & Suresh, G. S. (2013). Electrochemical biosensor for the selective determination of hydrogen peroxide based on the co-deposition of palladium, horseradish peroxidase on functionalized-graphene modified graphite electrode as composite. *Journal of Electroanalytical Chemistry, 689*, 233–242.

Neves-Petersen, M. T., Snabe, T., Klitgaard, S., Duroux, M., & Petersen, S. B. (2006). Photonic activation of disulfide bridges achieves oriented protein immobilization on biosensor surfaces. *Protein Science, 15*, 343–351.

Nguyen, B. H., Nguyen, B. T., Vu, H. V., Nguyen, C. V., Nguyen, D. T., Nguyen, L. T., et al. (2016). Development of label-free electrochemical lactose biosensor based on graphene/poly (1,5-diaminonaphthalene) film. *Current Applied Physics, 16*, 135–140.

Ni, Z. H., Wang, H. M., Kasim, J., Fan, H. M., Yu, T., Wu, Y. H., et al. (2007). Graphene thickness determination using reflection and contrast spectroscopy. *Nano Letters*, *7*, 2758–2763.

Novoselov, K. S., Geim, A. K., Morozov, S. V., Jiang, D., Katsnelson, M. I., et al. (2005). Two-dimensional gas of massless Dirac fermions in graphene. *Nature*, *438*, 197–200.

Novoselov, K. S., Geim, A. K., Morozov, S. V., Jiang, D., Zhang, Y., Dubonos, S. V., et al. (2004). Electric field effect in atomically thin carbon films. *Science*, *306*(5696), 666–669.

Novoselov, K. S., Jiang, D., Schedin, F., Booth, T. J., Khotkevich, V. V., Morozov, S. V., et al. (2005). Two-dimensional atomic crystals. *Proceedings of the National Academy of Sciences of the United States of America*, *102*, 10451–10453.

Pakapongpan, S., & Pooarporn, R. P. (2017). Self-assembly of glucose oxidase on reduced graphene oxide-magnetic nanoparticles nanocomposite-based direct electrochemistry for reagentless glucose biosensor. *Materials Science and Engineering C*, *76*, 398–405.

Povedano, E., Cincotto, F. H., Parrado, C., Díez, P., Sánchez, A., Canevari, T. C., et al. (2017). Decoration of reduced graphene oxide with rhodium nanoparticles for the design of a sensitive electrochemical enzyme biosensor for 17β-estradiol. *Biosensors and Bioelectronics*, *89*, 343–351.

Radhakrishnan, S., & Kim, S. J. (2015). An enzymatic biosensor for hydrogen peroxide based on one-pot preparation of CeO_2-reduced graphene oxide nanocomposite. *RSC Advances*, *5*(17), 12937–12943.

Radhapyari, K., Kotoky, P., Das, M. R., & Khan, R. (2013). Graphene-polyaniline nanocomposite based biosensor for detection of antimalarial drug artesunate in pharmaceutical formulation and biological fluids. *Talanta*, *111*, 47–53.

Rafighi, P., Tavahodi, M., & Haghighi, B. (2016). Fabrication of a third-generation glucose biosensor using graphene-polyethyleneimine-gold nanoparticles hybrid. *Sensors and Actuators B*, *232*, 454–461.

Ratinac, K. R., Yang, W. R., Gooding, J. J., Thordarson, P., & Braet, F. (2011). Graphene and related materials in electrochemical sensing. *Electroanalysis*, *23*, 803–826.

Rawal, R., Chauhan, N., Tomar, M., & Gupta, V. (2017). A contrivance based on electrochemical integration of graphene oxide nanoparticles/nickel nanoparticles for bilirubin biosensing. *Biochemical Engineering Journal*, *125*, 238–245.

Sabury, S., Kazemi, S. H., & Sharif, F. (2015). Graphene–gold nanoparticle composite: Application as a good scaffold for construction of glucose oxidase biosensor. *Materials Science and Engineering C*, *49*, 297–304.

Sakinyte, I., Barkauskas, J., Gaidukevic, J., & Razumiene, J. (2015). Thermally reduced graphene oxide: The study and use for reagentless amperometric D-fructose biosensors. *Talanta*, *144*, 1096–1103.

Sanaeifar, N., Rabiee, M., Abdolrahim, M., Tahriri, M., Vashaee, D., & Tayebi, L. (2017). A novel electrochemical biosensor based on Fe_3O_4 nanoparticles polyvinyl alcohol composite for sensitive detection of glucose. *Analytical Biochemistry*, *519*, 19–26.

Schedin, F., Geim, A. K., Morozov, S. V., Hill, E. W., Blake, P., Katsnelson, M. I., et al. (2007). Detection of individual gas molecules adsorbed on graphene. *Nature Materials*, *6*, 652–655.

Selvarajan, S., Suganthi, A., & Rajarajan, M. (2018). A simple sonochemical approach to fabricate a urea biosensor based on zinc phthalocyanine/graphene oxide/urease bioelectrode. *Ultrasonics Sonochemistry*, *42*, 183–192.

Sethuraman, V., Muthuraja, P., Anandha Raj, J., & Manisankar, P. (2016). A highly sensitive electrochemical biosensor for catechol using conducting polymer reduced graphene oxide–metal oxide enzyme modified electrode. *Biosensors and Bioelectronics*, *84*, 112–119.

Sheldon, R. A. (2007). Cross-linked enzyme aggregates (CLEAs): Stable and recyclable biocatalysts. *Biochemical Society Transactions*, *35*(6), 1583–1587.

Shuai, H.-L., Huang, K.-J., Zhang, W.-J., Cao, X., & Jia, M.-P. (2017). Sandwich-type micro RNA biosensor based on magnesium oxide nanoflower and graphene oxide–gold nanoparticles hybrids coupling with enzyme signal amplification. *Sensors and Actuators B*, *243*, 403–411.

Song, Y., Luo, Y., Zhu, C., Li, H., Du, D., & Lin, Y. (2016). Recent advances in electrochemical biosensors based on graphene two-dimensional nanomaterials. *Biosensors and Bioelectronics*, *76*, 195–212.

Song, H. Y., Ni, Y. N., & Kokot, S. (2014). Investigations of an electrochemical platform based on the layered MoS_2–graphene and horseradish peroxidase nanocomposite for direct electrochemistry and electrocatalysis. *Biosensors and Bioelectronics*, *56*, 137–143.

Song, S. P., Qin, Y., He, Y., Huang, Q., Fan, C. H., & Chen, H.-Y. (2010). Functional nanoprobes for ultrasensitive detection of biomolecules. *Chemical Society Reviews*, *39*, 4234–4243.

Tig, G. A. (2017). Highly sensitive amperometric biosensor for determination of NADH and ethanol based on Au-Ag nanoparticles/poly(L-cysteine)/reduced graphene oxide nanocomposite. *Talanta*, *175*, 382–389.

Unnikrishnan, B., Palanisamy, S., & Chen, S.-M. (2013). A simple electrochemical approach to fabricate a glucose biosensor based on graphene–glucose oxidase biocomposite. *Biosensors and Bioelectronics*, *39*, 70–75.

Wang, J. (2005). Nanomaterial-based electrochemical biosensors. *Analyst*, *130*, 421–426.

Wang, Q., Guo, X. F., Cai, L. C., Cao, Y., Gan, L., Liu, S., et al. (2011). TiO_2-decorated graphenes as efficient photoswitches with high oxygen sensitivity. *Chemical Science*, *2*, 1860–1864.

Wang, Y., Liu, X.-Y., Xu, X., Yang, Y., Huang, L. H., He, Z.-Y., et al. (2018). Preparation and characterization of reduced graphene oxide/Fe_3O_4 nanocomposite by a facile in-situ deposition method for glucose biosensor applications. *Materials Research Bulletin*, *101*, 340–346.

Wang, X., & Zhang, X. (2013). Electrochemical co-reduction synthesis of graphene/nanogoldcomposites and its application to electrochemical glucose biosensor. *Electrochimica Acta*, *112*, 774–782.

Wang, J., Zhu, H., Xu, Y., Yang, W., Liu, A., Shan, F., et al. (2015). Graphene nanodots encaged 3-D gold substrate as enzyme loading platform for the fabrication of high performance biosensors. *Sensors and Actuators B*, *220*, 1186–1195.

Wilson, R. (2008). The use of gold nanoparticles in diagnostics and detection. *Chemical Society Reviews*, *37*, 2028–2045.

Wisitsoraat, A., Pakapongpan, S., Sriprachuabwong, C., Phokharatkul, D., Sritongkham, P., Lomas, T., et al. (2013). Graphene–PEDOT:PSS on screen printed carbon electrode for enzymatic biosensing. *Journal of Electroanalytical Chemistry*, *704*, 208–213.

Xue, K., Zhou, S., Shi, H., Feng, X., Xin, H., & Song, W. (2014). A novel amperometric glucose biosensor based on ternary goldnanoparticles/polypyrrole/reduced graphene oxide nanocomposite. *Sensors and Actuators B*, *203*, 412–416.

Yang, W., Ratinac, K. R., Ringer, S. P., Thordarson, P., Gooding, J. J., & Braet, F. (2010). Carbon nanomaterials in biosensors: Should you use nanotubes or graphene? *Angewandte Chemie International Edition*, *49*, 2114–2138.

Yu, H. L. L., Maslova, A., & Hsing, I. M. (2017). Rational design of electrochemical DNA biosensors for point-of-care applications. *ChemElectroChem*, *4*, 795–805.

Zhang, R., & Chen, W. (2017). Recent advances in graphene-based nanomaterials for fabricating electrochemical hydrogen peroxide sensors. *Biosensors and Bioelectronics*, *89*, 249–268.

Zhu, J., Liu, X., Wang, X., Huo, X., & Yan, R. (2015). Preparation of polyaniline–TiO_2 nanotube composite for the development of electrochemical biosensors. *Sensors and Actuators B*, *221*, 450–457.

CHAPTER TWELVE

Graphene Quantum Dots: Synthesis and Applications

Ankarao Kalluri*, Debika Debnath*, Bhushan Dharmadhikari†, Prabir Patra*,‡,1

*Department of Biomedical Engineering, University of Bridgeport, Bridgeport, CT, United States
†Department of Computer Science and Engineering, University of Bridgeport, Bridgeport, CT, United States
‡Department of Mechanical Engineering, University of Bridgeport, Bridgeport, CT, United States
[1]Corresponding author: e-mail address: ppatra@bridgeport.edu

Contents

Abstract

Graphene and its derivatives having at least one dimension in nanoscale range have attracted tremendous attention in recent years due to their unique electronic, optical, chemical, and mechanical properties. This chapter is about graphene quantum dots (GQDs) that are zero-dimensional graphene derivatives with one to few layers of graphene sheet having size range less than 20nm. This chapter is an overview of synthesis of GQDs by top-down and bottom-up approaches, as well as detailed methods of synthesis of GQDs by acidic oxidation of carbon fibers. Owing to their extremely small size, quantum confinement, edge effect, biocompatibility, low toxicity, photostability as well as water solubility they are excellent candidates for understanding biological

Methods in Enzymology, Volume 609
ISSN 0076-6879
https://doi.org/10.1016/bs.mie.2018.07.002

systems and cellular processes at the molecular scale. These are also suitable nanomaterials to replace inorganic semiconducting nanoparticles (e.g., CdS, CdSe, ZnS, and Si) which are toxic to biological systems.

1. INTRODUCTION

Carbon nanomaterials have a wide variety of promising applications in energy, environmental, electronic, optical, and biomedical fields. Graphene, a 2D nanomaterial with sp^2 hybridized carbon atoms in a single layer, has received much attention because of its unique physical and chemical properties. These include high surface area, superior mechanical flexibility, high electronic mobility of $10^6\,cm^2\,V^{-1}\,s^{-1}$ (Long et al., 2009), excellent thermal, and chemical stability. This is very promising and highly attractive for studies in physics, chemistry, and materials science. However, graphene is a zero-bandgap semiconductor, which limits its electronic and optoelectronic applications. Due to the lack of the bandgap, there is an absence of luminescence in pristine graphene. However, bandgap can be introduced in graphene quantum dots (GQDs) and graphene nanoribbons (GNRs) due to the quantum confinement and edge effects (Ritter & Lyding, 2009) of these graphene derivatives.

Recently, among all graphene derivatives, GQDs were considered as a newly emerging material, which represent a class of zero-dimensional material with a typical dimensions less than 20 nm in width (Bacon, Bradley, & Nann, 2014; Li & Yan, 2010; Pan, Zhang, Li, & Wu, 2010). These show unique quantum confinement and edge effects (Ponomarenko et al., 2008), which increase its potential for applications for nanoscale optoelectronics. GQDs also show biocompatibility (Nurunnabi et al., 2013), low toxicity, excellent photostability, tunable photoluminescence, nanosecond lifetime, multicolored emission, excellent dispersibility, high-water solubility, and easy surface grafting which are promising for nanotechnology applications (Farka et al., 2017).

Furthermore, edge-functionalized GQDs contain carboxyl, carbonyl, epoxy, and hydroxyl groups that can interact with various biological molecules like proteins or enzymes or antibodies (Sapsford et al., 2013). GQDs have been successfully conjugated to antibody and DNA molecules through amide formation (Kim et al., 2013), which is useful for making high sensitivity diagnostic nanodevices. Hence, high quality colloidal GQDs show tremendous potential in the fields of biolabeling (Zhu et al., 2011) and bioimaging (Zheng et al., 2013), biomolecule sensing (Al-Ogaidi et al., 2014), electrocatalysis (Pan et al., 2013), light-emitting diodes (Gupta et al., 2011), solar cells (Kolay et al., 2017), supercapacitors, and fuel cells. Moreover, photoluminescent GQDs have gained great attention as GQDs are ecofriendly, less toxicity, biocompatible (Nurunnabi et al., 2013), and more photostable (Wang et al., 2015). Inorganic semiconductor quantum dots attracted more attention over the past two decades, owing to their unique electronic, optical, and bioimaging applications. However, the inorganic colloidal QDs have drawbacks like intrinsic toxicity (CdSe) due to the leaching of the metal ions, as well as it is necessary to modify their surfaces coupling chemistries.

2. GRAPHENE QUANTUM DOTS

GQDs are synthesized either by top-down or bottom-up approaches. In the top-down approach, the zero-dimensional GQDs are synthesized and fabricated from particular carbon source materials (CF, GO, GNR) by a variety of methods. These include electron beam lithography (Ponomarenko et al., 2008), chemical oxidation (Peng et al., 2012), solvothermal, microwave-assisted hydrothermal, microfluidization (Buzaglo, Shtein, & Regev, 2016), and electrochemical methods. The bottom-up approach involves condensation of smaller units of benzene derivatives forming bigger entities of GQDs, and these include stepwise solution chemistry, pyrolysis and oxidation, and hydrothermal heating.

2.1 Top-Down Approach

In "top-down" method, GQDs have been processed by cutting sp^2 or sp^3 carbon allotropes of graphene, GO, carbon fiber, carbon black by acidic oxidation, microfluidization, exfoliation, hydrothermal process, electrochemical oxidation. These approaches have the advantages of having abundant raw materials, simple straightforward processes and large-scale production of bulk reagents and precursors. In addition, the GQDs synthesized by top-down methods usually produce oxygen functional groups on the surface, because of these, the edge-functionalized GQDs show excellent water solubility and passivation of their surfaces. However, this top-down method has some drawbacks such as requiring special equipment, particular reaction conditions, and nonselective destruction of the aromatic carbon framework. These lead to the formation of disordered and defective structures, poorly crystalline nanomaterials, GQDs with relatively low luminescence quantum yields, and give mixtures of particles with various diameters (3–25 nm). Controlling their morphology by this method is difficult. Some of these methods are impractical for bulk production due to their high-cost, difficult scalability, the use of acid, high-temperatures, and environmental issues.

2.1.1 Acidic Oxidation

GQDs are fabricated by a facile, one-step, exfoliated, acidic treatment of the carbon fibers (Fig. 1A). As-synthesized GQDs are more polar in nature due to the edge-functionalized groups such as OH, COOH, epoxy, and carbonyl groups, and hence, these GQDs are highly soluble in water and other organic solvents such as DMF and DMSO. Interestingly, the band gaps of nanoparticles are controlled by varying the reaction temperature and conditions. GQDs formed at 120°C showed relatively a narrow size distribution between 1 and 4 nm, high crystallinity, with a lattice parameter of 0.242 nm, (1120), the lattice fringes of graphene. The topographic morphology measured by AFM indicated heights in the range between 0.4 and 2 nm. Further, formation of the GQDs depended upon the domain of the sp^2 carbons, which are present in the pitched carbon fiber structure. The photoluminescence of these GQDs is tailored through bandgap engineering by varying the process parameters like temperature and carbon precursors.

Ye et al. proposed an inexpensive, facile, one-step, wet-chemistry method to fabricate GQDs from the bituminous coal, which is the most abundant and readily available. GQDs were made by chemical oxidation. In this method, the coal was sonicated in a mixture of concentrated sulfuric acid and nitric acid, followed by heat treatment at 80/100/120°C for 24 h. The bituminous coal GQDs (b-GQDs) showed uniform size distribution and size ranged from 2.96 ± 0.96 nm in diameter, as shown by transmission electron microscopy (TEM). The topological heights of b-GQDs were in the range of 1.5–3 nm, indicating that these have 2–3 layers of graphene oxide-like structures. The I_D/I_G ratio for bituminous coal is 1.06 ± 0.12, which increased to 1.55 ± 0.19 after oxidative cutting of b-GQDs owing to the introduction of defects to the basal planes and the edges.

Ultrasmall GQDs were prepared by the cage-opening fullerene C_{60} molecules having well-defined carbon nanomaterial with a diameter of ~1 nm by top-down wet chemistry approach (Fig. 1C). By using modified Hummers' method, C_{60} was treated with a mixture of concentrated H_2SO_4 and HNO_3 (1:3 V/V ratio) and potassium permanganate which yielded GQDs with well-defined diameters of 2–3 nm (Chua et al., 2015).

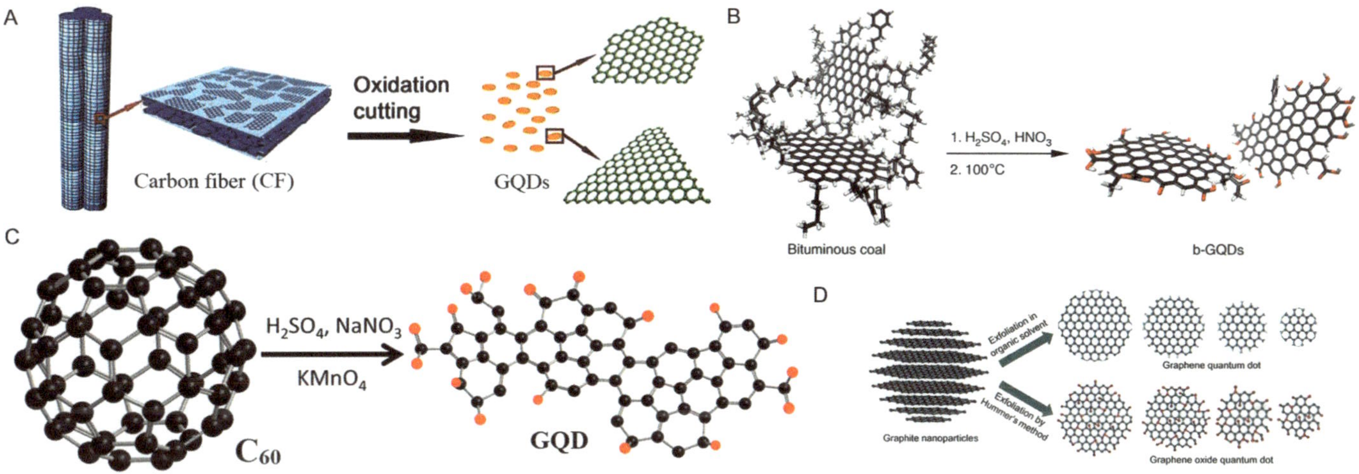

Fig. 1 Schematic illustration of GQD synthesis by: (A) Oxidative cutting of the carbon fibers (CF). (B) Acidic oxidation of coal. Scale bar: 10 nm. (C) Oxidation and cage-opening of fullerene C60 with treatment of strong acid and chemical oxidant. (D) Synthesis of GQDs and GOQDs by chemical exfoliation of GNPs (oxygen sites are shown as *red dots*). *Panels (A and B) reproduced with permission. Copyright 2012, American Chemical Society. Panel (C) reproduced with permission. Copyright 2015, American Chemical Society. Panel (D) reproduced with permission. Copyright 2013, Wiley-VCH.*

2.1.2 Hydrothermal and Exfoliation Method

Carbon precursor is crucial to predetermine the shape, size, and exfoliation of the GQDs. Graphite nanoparticle (GNP) is an ideal precursor to produce both homogeneous GQDs without any oxygen functional groups and GOQDs after an oxidation process. The spherical GNP has many layers of graphene stacked together through the van der Waals and π–π interactions. GQD and GOQD with regular shape and size are produce by a simple exfoliation process. Liu et al. proposed GOQDs fabrication by modified Hummers' method and GQDs by exfoliation (Fig. 1D) of GNPs in ethanol/H_2O (1:1 v/v ratio) mixture on a vortex mixing at 2700 rpm for 1 h. After centrifugation at 2000 rpm for 30 min, the GQD suspension was collected from the supernatant. The resulting GQDs showed a diameter less than 4 nm and a height profile of 0.5–1 nm, and these indicated the presence of monolayered graphene sheets (Liu et al., 2013).

GQDs have extraordinary properties due to quantum confinement and edge effects. Notably, the magnetism of GQDs, due to their spin-polarized edge states and high edge-to-area ratio and possible substantial spin-polarized edge states at the zigzag segments show potential applications in spintronic devices. Recently, Sun et al. reported the high yield and high purity of GQDs produced by annealing a single layers of GOQDs in argon atmosphere at 1010°C (Fig. 2D). These had intrinsic magnetism (ca. 2.04 nm) with mostly pristine edges and nearly perfect basal planes. They showed pure Curie-like paramagnetism at 2 K and the magnetism may mainly arise from the edge functional states of passivated hydroxyl groups (Sun et al., 2017).

2.1.3 Nanolithography

Ponomarenko and coworkers used ultrahigh-resolution beam lithography to breakdown the 2D graphene into zero-dimensional GQDs to a size as small as 10 nm. Even though this work reported a very high level of precision, this nanolithography technique required very specialized equipment, difficult to obtain smooth edges and lead to material with low luminescence quantum yields (Ponomarenko et al., 2008).

2.2 Bottom-Up Approaches

The bottom-up approach involves the synthesis of the GQDs from small molecular precursors, and benzene derivatives are condensed to form bigger entities, having well-defined size, shape, and desirable properties. However, these methods involve stepwise and complex synthetic routes, which require

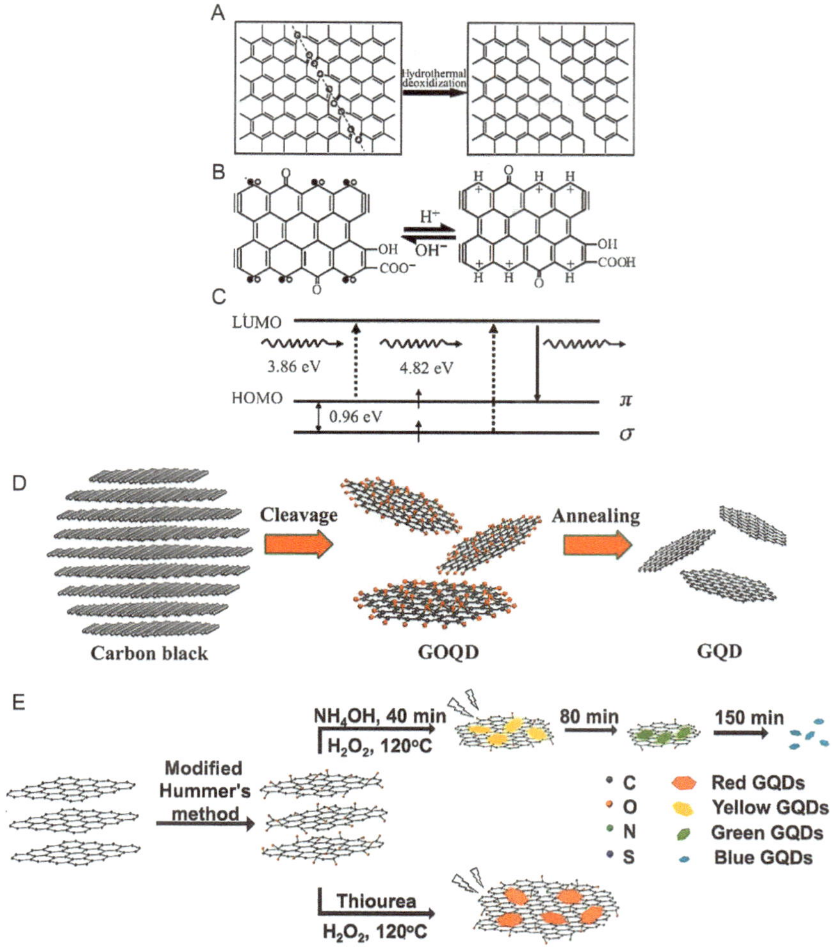

Fig. 2 (A) Mechanism for the hydrothermal cutting of oxidized GSs into GQDs: a mixed epoxy chain composed of epoxy and carbonyl pair groups (*left*) is converted into a complete cut (*right*) under the hydrothermal treatment. (B) Models of the GQDs in acidic (*right*) and alkali (*left*) media. The two models can be converted reversibly depending on pH. The pairing of σ (●) and π (○) localized electrons at carbene-like zigzag sites and the presence of triple bonds at the carbyne-like armchair sites are represented. (C) Typical electronic transitions of triple carbenes at zigzag sites observed in the optical spectra. (D) Thermal annealing of GQDs from carbon black. The *black and red balls* denote carbon and oxygen atoms, respectively. (E) Modified Hummer's method to prepare red-, blue-, and yellow-colored GQDs by radical-induced decomposition of graphene oxide. *Panel (C) reproduced with permission. Copyright 2010, Wiley-VCH. Panel (D) reproduced with permission. Copyright 2017, Springer Nature. Panel (E) reproduced with permission. Copyright 2015, Wiley-VCH.*

special reagents and conditions. Alternatively, uniform GQDs were obtained by solution phase cyclodehydrogenation of polypropylene precursors (Wu, Pisula, & Müllen, 2007).

2.2.1 Microwave–Hydrothermal Synthesis

The GQDs were prepared by a microwave-assisted hydrothermal (MAH) method that combined both the advantages of hydrothermal and microwave techniques. The microwave heating provides simultaneous, homogeneous, fast heating leading to uniform size distribution of GQDs.

The glucose and fructose were converted to functionalized GQDs by microwave and hydrothermal treatment. Tang et al. first report the deep ultraviolet emission at 303 nm (~4.1 eV) from the GQDs when excited by a 197 nm laser, which is a shortest emission wavelength reported from solution-based inorganic or organic QDs. The emission wavelength of glucose-derived nanoparticles is independent of the size and size has been tuned from 1.65 to 21 nm by increasing the heating time from 1 to 9 min, respectively. The proposed mechanism is schematically shown in Fig. 3B (Tang et al., 2012). Glucose molecules are dehydrated to form the nucleus of GQDs that is primarily composed of C=C bonds, as confirmed by FTIR, HRTEM, EELS, and XPS. With increasing the heating time, the growth of GQD occurs at the peripheral surface (edge growth) by formation of C=C bonds between glucose molecules. The source molecules reach the surface of the GQDs and generate new C=C by dehydration. By this way GQD becomes larger as heating time increases. Many parameters may affect the growth of the GQDs which include microwave power, source concentration, and heating time.

2.2.2 Pyrolysis and Oxidation

Gram-scale mass production and highly luminescent water-soluble GQDs are fabricated using the cheap precursor, pyrene (Ranjit et al., 2016). The main synthesis involved the nitration of pyrene in HNO_3 at 80°C followed by the hydrothermal treatment in 0.2 M NaOH aqueous solutions at 200°C for 10 h. Post hydrothermal treatment the solution is dialyzed to get an overall yield of 63% black OH-GQD powder. The OH-GQDs in aqueous solution emit bright green fluorescence by UV irradiation. The thickness of the GQDs from atomic force microscope indicated a size range of 1–2 nm, with an average thickness of 1.47 ± 0.86 nm, which is equal to ~5 layers. The TEM images showed GQDs well dispersed with uniform lateral sizes. Their average lateral size determined by TEM was 3.5 ± 0.6 nm (Fig. 4C).

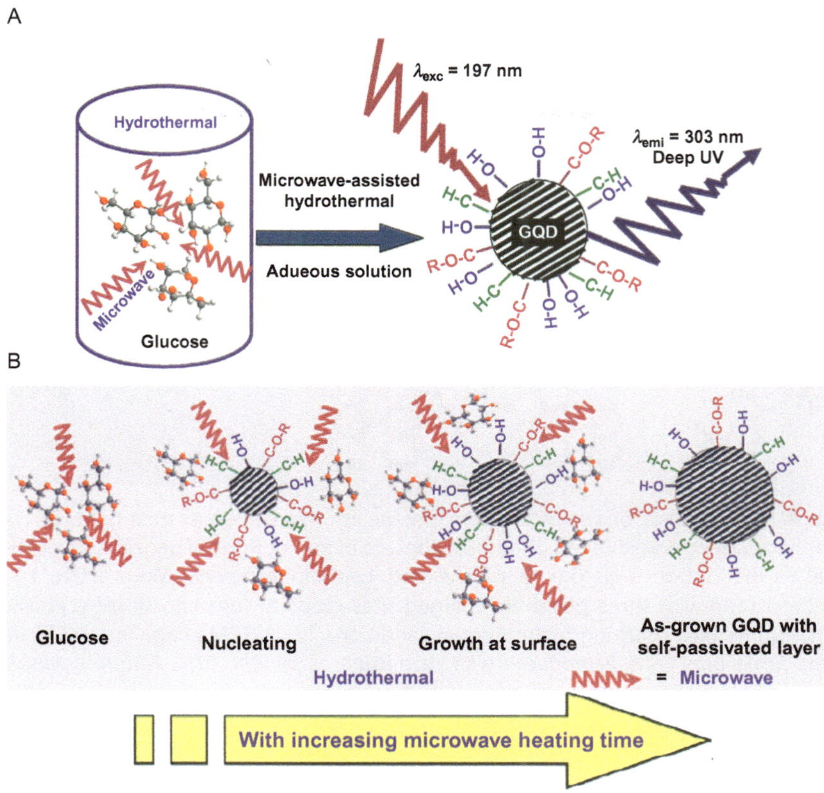

Fig. 3 Schematic representation of fabrication of GQDs by microwave-assisted hydrothermal (MAH) reaction (A) from glucose. (B) Mechanism and growth of GQDs by MAH method. *Reproduced with permission. Copyright 2012, American Chemical Society.*

3. METHODS

3.1 Synthesis of GQDs by Acid Oxidation

GQDs are synthesized by simple oxidation of carbon fibers in the presence of acids at various temperatures. These GQDs had a size range of 1–10 nm, 2D morphology, with zigzag edge structure, consisting of 1–3 atomic layers. They have high stability, good biocompatibility, low toxicity, and high-water solubility, which are excellent for bioimaging and biosensing applications. By this chemical oxidation process, Peng et al. synthesized three types of GQDs of varying sizes and can be tailored by simply choosing different reaction temperatures (see Table 1). Here are brief methods of making these nanomaterials (Fig. 5).

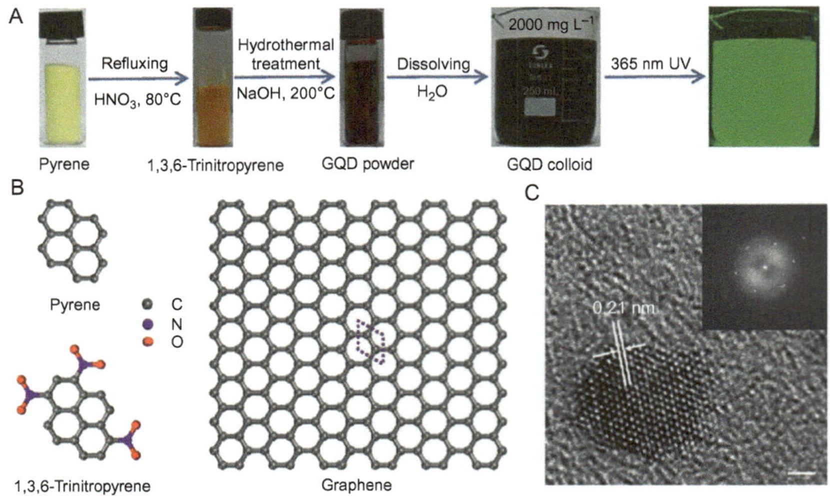

Fig. 4 (A) Synthesis of OH-GQDs from pyrene by hydrothermal treatment. (B) The graphene lattice primitive cell (rhombus) consists of four peri-fused benzene rings identical to the carbon skeletons of pyrene and 1,3,6-trinitropyrene. More active 1,3,6-trinitropyrene with three positively charged sites tends to fuse into single-crystalline GQDs under base-assisting hydrothermal conditions. (C) HRTEM image (Inset: FFT pattern). Scale bar: 1 nm. *Reproduced with permission. Copyright 2010, Nature Publishing Group.*

Table 1 Distinct Types of GQDs With Size, Temperature, and Optical Parameters

No.	GQD Types	Temperature (°C)	Size Range (nm)	Excitation λ_{max} (nm)	Emission λ_{max} (nm)	Color
1.	GQD-80	80	8–10	429	564	Yellow
2.	GQD-100	100	4–7	331	500	Green
3.	GQD-120	120	1–4	318	434	Blue

3.1.1 General Equipment

1. *Safety:* safety goggles, nitrile gloves, lab coat, ventilation hood, nose mask
2. Round-bottom flask (500-mL)
3. Beakers, conical flasks, pipettes of different volumes
4. Heating mantle and oil bath (Sigma-Aldrich)
5. Temperature controller
6. Reflux condenser
7. Analytical balance
8. Stirrer and hot plate

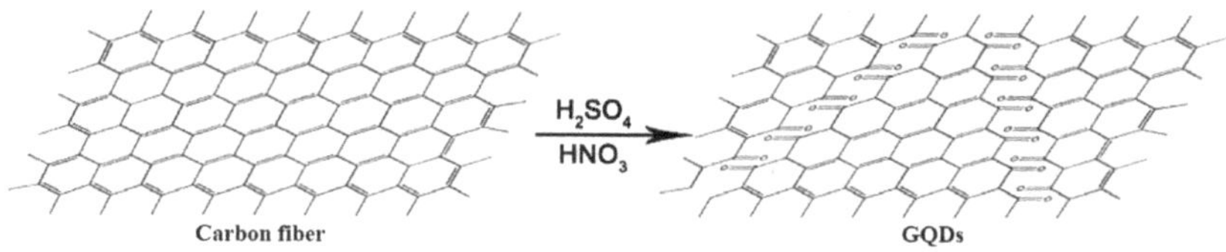

Fig. 5 Proposed mechanism for the chemical oxidation of carbon fibers (CF) into GQDs. *Reproduced with permission from Peng, J., et al. (2012). Graphene quantum dots derived from carbon fibers.* Nano Letters, 12*(2), 844–849. Copyright 2012, American Chemical Society.*

9. Ultrasonic bath
10. pH meter
11. Filter membrane: 0.22-μm PES filter (Corning)
12. Dialysis tubeing MWCO 2000 (Sigma-Aldrich)

3.1.2 Materials and Reagents

1. Pitch carbon fibers (Fiber Glast Development Corporation)
2. Sulfuric acid (Sigma-Aldrich)
3. Nitric acid (Sigma-Aldrich)
4. Sodium carbonate (Sigma-Aldrich)
5. Paraffin wax (Sigma-Aldrich)

3.1.3 Synthesis

1. Weigh 0.3 g of pitch carbon fibers and add them to 3:1 mixture of conc. H_2SO_4 and conc. HNO_3.
2. Sonicate the solution for 2 h.
3. Transfer the sonicated solution onto a heating mantle and stir for 24 h at 80°C, 100°C, or 120°C.
4. Cool the reaction mixture with 500 mL of deionized water.
5. Adjust the pH to 8 with Na_2CO_3.
6. Filter the solution by using 0.22 μm PES membrane and collect the solution.
7. Dialyze the final product for 3 days in dialysis bags (MWCO 2000)

4. CHARACTERIZATION OF GQDs

4.1 Transmission Electron Microscopy

These studies were carried out to determine the morphology of the GQDs that are synthesized by the acidic oxidation method described earlier.

4.1.1 Equipment

JEOL 2100 Field Emission Gun TEM.

4.1.2 Materials and Reagents

1. TEM copper grid (TED Pella).
2. JEOL 2100 Field Emission Gun TEM at accelerating voltage of 120 keV

4.1.3 Procedure

1. Prepare dilute solutions (5 n*M*) of GQDs
2. Use micropipette to drop-cast 5 μL of water solubilized GQDs on the copper grid.
3. Blot the excess solution using filter paper, and air dry the sample
4. Image the sample using JEOL 2100 Field Emission Gun TEM.

4.1.4 Data Analysis

The GQDs are synthesized by chemical oxidation of micrometer-sized pitch-based carbon fibers (CF). Fig. 6A shows the TEM image of the fiber-derived GQDs (synthesized reaction temperature at 120°C), showing relatively a narrow size distribution between 1 and 4 nm (Fig. 6C). The high-resolution TEM (HRTEM) image (inset of Fig. 6B) indicated high crystallinity, with a lattice parameter of 0.242 nm, and (1120) lattice fringes of graphene.

4.2 Characterization by UV–Visible

4.2.1 Equipment

Perkin Elmer UV–visible spectrophotometer.

4.2.2 Materials and Reagents

1. Silica Quartz cuvettes (1.4 mL, 1 cm path length)
2. Pipettes (1000 μL)

4.2.3 Procedure

1. Prior to use, triple rinse all the glassware used in this step with purified water and allow to dry.
2. Record the reading for blank by filling the cuvette with deionized water.
3. Vortex and sonicate the GQD solution up to 15 min.
4. Transfer 1 mL of GQD solution into quartz cuvettes using pipette and record the absorbance values.

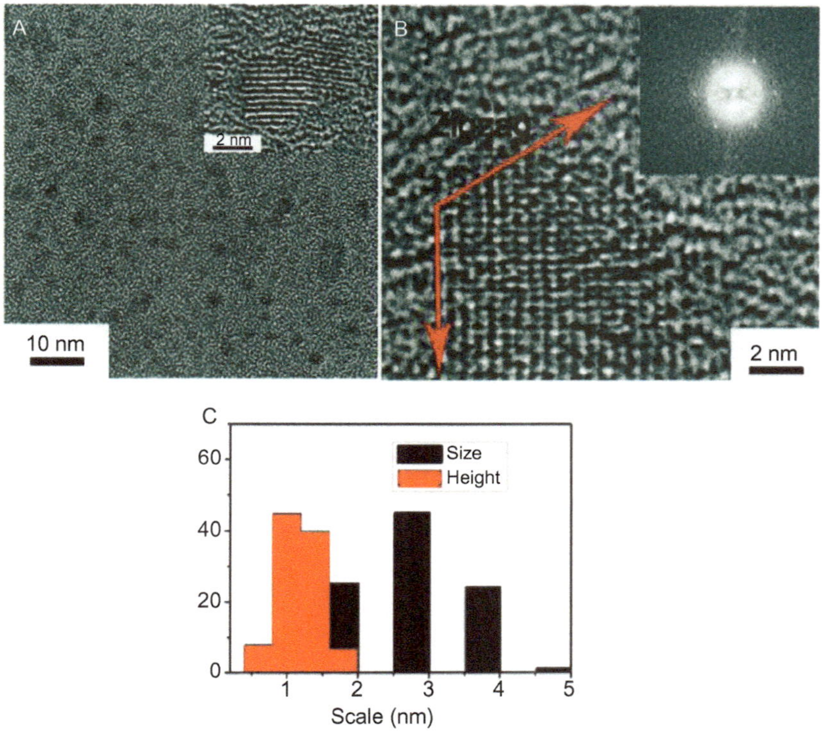

Fig. 6 (A) TEM images of GQDs (synthesized reaction temperature at 120°C), inset is the HRTEM of GQDs. (B) HRTEM image of the edge of GQD, inset is the 2D FFT of the edge. (C) Size and height distribution of GQDs. *Reproduced with permission from Peng, J., et al. (2012). Graphene quantum dots derived from carbon fibers.* Nano Letters, 12*(2), 844–849. Copyright 2012, American Chemical Society.*

4.2.4 Data Analysis

Fig. 7A shows the UV–visible absorbance spectra of three distinct types of GQDs synthesized at 120, 100, or 80°C, respectively. The shift in the absorbance wavelengths from 330 to 270nm with increase in temperature is observed, it reveals a reaction temperature affect on the absorption properties of GQDs. Samples made at lower temperatures absorb at longer wavelengths. The inset of Fig. 3A shows the optical images of GQDs under UV light as blue, green, and yellow depending on the synthesis temperature.

4.3 Characterization by Photoluminescence Spectroscopy

4.3.1 Equipment

1. Photoluminescence spectrometer (Nanolog, HORIBA, Jobin Yvon)
2. Xenon lamp

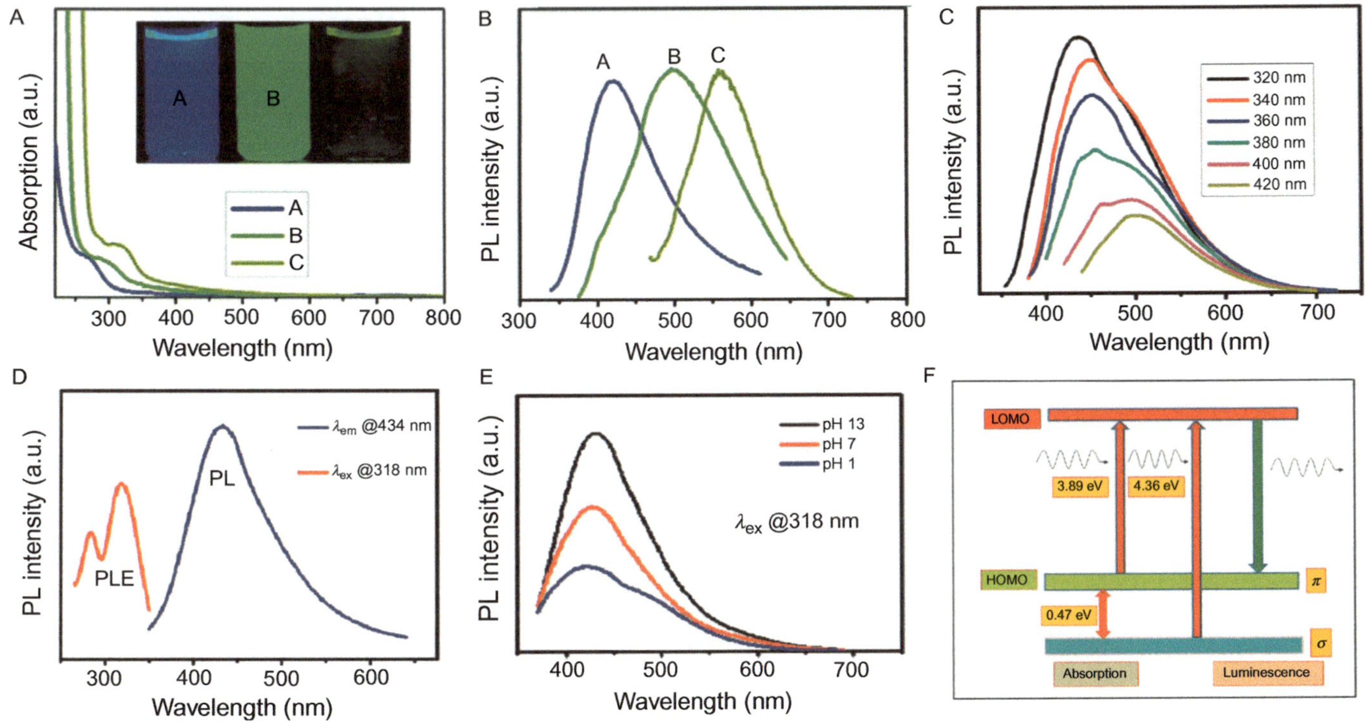

Fig. 7 Optical properties of the GQDs. (A) UV–vis spectra of GQDs A, B, and C, correspond to synthesis temperatures of 120, 100, and 80°C, respectively. Inset of panel (A) is a photograph of the corresponding GQDs under UV light with 365 nm excitation. (B) PL spectra of GQDs with different emission color excited at 318, 331, and 429 nm, respectively. (C) PL spectra of the blue GQDs at different excitation wavelengths. (D) PLE spectrum of the blue GQDs with emission at 434 nm and PL spectrum of the blue GQD excited at 318 nm registered at room temperature. (E) PL spectra at 318 nm excitation of blue GQDs at different pH values. (F) Typical electronic transitions of triple carbenes at zigzag sites observed in the optical spectra for blue emission GQDs. *Reproduced with permission from Peng, J., et al. (2012). Graphene quantum dots derived from carbon fibers.* Nano Letters, 12*(2), 844–849. Copyright 2012, American Chemical Society.*

4.3.2 Materials and Reagents

Silica Quartz cuvettes (1.4 mL, 1 cm path length).

Pipettes (1000 μL).

4.3.3 Procedure

1. Prior to use, triple-rinse all the glassware to be used with purified water and dry.
2. Record the absorbance reading for the blank by filling the cuvette with deionized water.
3. Vortex and sonicate the GQD solution for upto 15 min.
4. Transfer 1 mL of GQD solution into quartz cuvettes, using pipette and record the absorbance data.

4.3.4 Data Analysis

Photoluminescence (PL) spectra of GQDs shown in Fig. 7B had emission wavelengths at 318, 331, and 429 nm of the three different samples made at different temperatures. Thus, the synthesis temperature can be used to tune the emission wavelength of as-synthesized GQDs. The emission color depends on the size, shape, and defects. Hence, the energy gap of GQDs can be tuned by changing the size of GQDs, by simply controlling the synthesis temperature. Fig. 7C shows the pH dependence of PL behavior from pH 1 to 13. The blue GQDs were excited at 318 to 420 nm, and the corresponding data shown in Fig. 7D. The carbene ground state multiplicity is related to energy difference between the sigma and π orbitals and it should be below 1.5 eV for a triplet ground state carbine. The corresponding gaps for the GQDs of blue, green, and yellow emission are 0.47, 0.82, and 1.25 eV, respectively (Fig. 7F).

4.4 Characterization by Raman Spectroscopy

4.4.1 Equipment

1. Renishaw instrument via Raman microscope
2. Laser wavelength: 515.5 nm
3. Laser power: 20 mW
4. Laser beam size (diameter): 0.41 μm
5. Grating: 1800 g/mm
6. Objective lens: 50 ×
7. Step size = 200 nm for Raman mapping

4.4.2 Materials and Reagents

PTFE nanoporous membrane.

4.4.3 Procedure

1. Calibrate the Raman instrument by using silicon wafer, prior to each acquisition.
2. Prepare the GQD samples by filtering a freshly sonicated dispersion on PTFE nanoporous membrane.
3. Place this PTFE porous membrane in the sample holder and record the data.

4.4.4 Data Analysis

Raman spectroscopy is used for characterizing the GQDs properties such as disorder, edge chemistry, doping, and thermal conductivity. In Fig. 8, the 2D peak of GQDs is at 2700 cm^{-1} and the G band had a blue shift when compared to that of the CF. The relative intensity of I_D/I_G ratio is 0.91, which indicated low crystalline nature and more defects in GQDs, likely due to the incorporation of oxygen functional groups at the edges.

4.5 Characterization by FTIR

FTIR analysis is often used to identify the functional groups that were introduced at the edges of the GQDs.

4.5.1 Equipment

1. Nicolet FTIR microscope with an MCT/A detector
2. Mercury cadmium telluride (MCT/B) detector for FTIR measurements

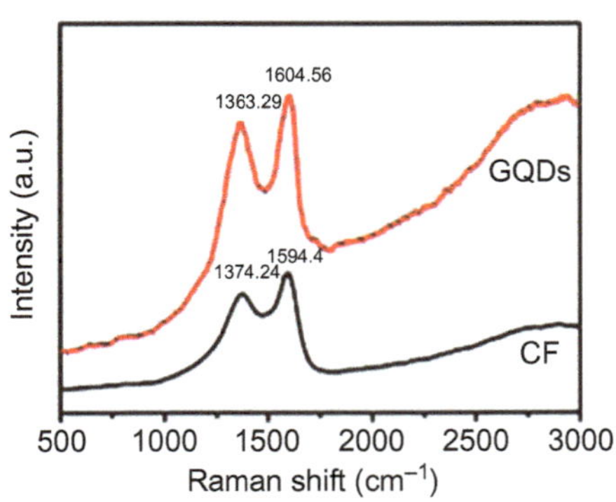

Fig. 8 Raman spectra of CF and GQDs. FT-IR spectra of CF and GQDs. *Reproduced with permission from Peng, J., et al. (2012). Graphene quantum dots derived from carbon fibers.* Nano Letters, 12*(2), 844–849. Copyright 2012, American Chemical Society.*

4.5.2 Materials and Reagents

1. Traceable polystyrene film standard
2. Vacuum drying oven
3. KBr pellets
4. Pellet holder
5. Microcompression cell with diamond window

4.5.3 Procedure

1. Using the vacuum drying oven, dry the GQD suspension for FTIR analysis.
2. Place the powdered GQD sample into the cavity of the pellet holder.
3. Place the pellet holder in a plastic sample holder and mount it in the spectrometer.
4. On a Nicolet FTIR microscope with MCT/A detector, obtain the FTIR spectra.
5. Use MCT/B detector for FTIR measurements.

4.5.4 Data Analysis

FTIR spectra (Fig. 9) indicated the oxidation producing the oxygen containing functional groups such as carbonyl (1724 cm^{-1}), hydroxyl, carboxyl, epoxy groups were introduced at the edges of the GQDs and the basal planes.

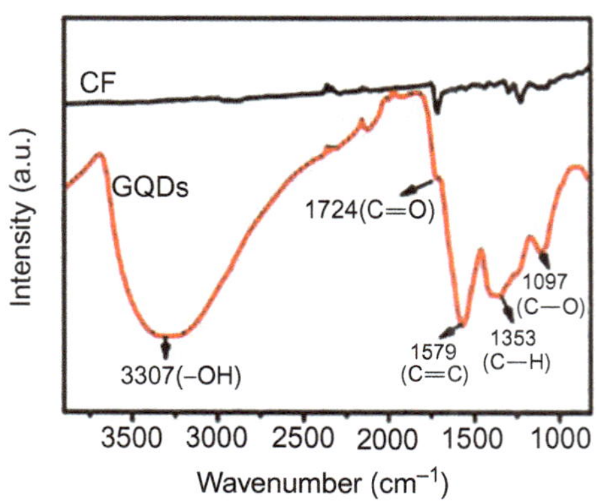

Fig. 9 FT-IR spectra of CF and GQDs. *Reproduced with permission from Peng, J., et al. (2012). Graphene quantum dots derived from carbon fibers.* Nano Letters, 12*(2), 844–849. Copyright 2012, American Chemical Society.*

5. APPLICATIONS OF THE GQDs

The green GQDs derived from the carbon fibers were used for bioimaging application by staining the nucleus with DAPI and the surroundings show the green color (Fig. 10) (Peng et al., 2012) from human breast cancer cell lines T47D. The images clearly show the nucleus stained with blue DAPI and the surroundings by the green GQDs, and this demonstrates that GQDs can be used for bioimaging and biomedical applications.

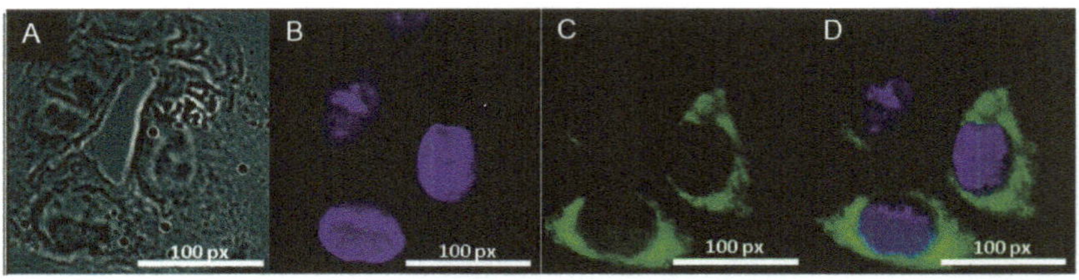

Fig. 10 Fluorescence images of human breast cancer cells T47D after incubation with green GQDs for 4h (A) phase contrast picture. (B) Individual nuclei stained blue with DAPI. (C) Agglomerated green GQDs surrounding each nucleus. (D) The overlay of high-contrast image of DAPI and GQD stained cell images. *Reproduced with permission from Peng, J., et al. (2012). Graphene quantum dots derived from carbon fibers.* Nano Letters, 12*(2), 844–849. Copyright 2012, American Chemical Society.*

6. SUMMARY AND CONCLUSIONS

GQDs, owing to their quantum confinement, edge effect, chemical stability, high surface to volume ratio, and biocompatibility, have potential in bioimaging, sensing and drug delivery. However, the research is still at its early stages and their full potential is yet to be exploited. Several issues need to be resolved which include but not limited to low PL quantum yield, low product yield, inability to control lateral dimensions, the surface chemistry, and cytotoxicity. Therefore, there is potential for the development of GQDs to revolutionize biomedical research as well as for optical, electronics, and imaging applications.

ACKNOWLEDGMENTS

A.K., D.D., B.P., and P.P. acknowledge financial support from the University of Bridgeport.

REFERENCES

Al-Ogaidi, I., et al. (2014). Detection of the ovarian cancer biomarker CA-125 using chemiluminescence resonance energy transfer to graphene quantum dots. *Chemical Communications*, *50*(11), 1344–1346. http://xlink.rsc.org/?DOI=C3CC47701K. March 31, 2018.

Bacon, M., Bradley, S. J., & Nann, T. (2014). Graphene quantum dots. *Particle & Particle Systems Characterization*, *31*(4), 415–428. http://doi.wiley.com/10.1002/ppsc.201300252. March 30, 2018.

Buzaglo, M., Shtein, M., & Regev, O. (2016). Graphene quantum dots produced by microfluidization. *Chemistry of Materials*, *28*(1), 21–24. http://pubs.acs.org/doi/10.1021/acs.chemmater.5b03301. March 31, 2018.

Chua, C. K., et al. (2015). Synthesis of strongly fluorescent graphene quantum dots by cage-opening buckminsterfullerene. *ACS Nano*, *9*(3), 2548–2555. http://pubs.acs.org/doi/10.1021/nn505639q. March 31, 2018.

Farka, Z., et al. (2017). Nanoparticle-based immunochemical biosensors and assays: Recent advances and challenges. *Chemical Reviews*, *117*(15), 9973–10042. http://pubs.acs.org/doi/10.1021/acs.chemrev.7b00037. June 13, 2018.

Gupta, V., et al. (2011). Luminscent graphene quantum dots for organic photovoltaic devices. *Journal of the American Chemical Society*, *133*(26), 9960–9963. http://pubs.acs.org/doi/abs/10.1021/ja2036749. March 31, 2018.

Kim, J. K., et al. (2013). Balancing light absorptivity and carrier conductivity of graphene quantum dots for high-efficiency bulk heterojunction solar cells. *ACS Nano*, *7*(8), 7207–7212. http://pubs.acs.org/doi/10.1021/nn402606v. March 30, 2018.

Kolay, A., et al. (2017). New antimony selenide/nickel oxide photocathode boosts the efficiency of graphene quantum-dot co-sensitized solar cells. *ACS Applied Materials & Interfaces*, *9*(40), 34915–34926. http://pubs.acs.org/doi/10.1021/acsami.7b09754. November 29, 2017.

Li, L.-s., & Yan, X. (2010). Colloidal graphene quantum dots. *The Journal of Physical Chemistry Letters*, *1*(17), 2572–2576. http://pubs.acs.org/doi/10.1021/jz100862f. March 30, 2018.

Liu, F., et al. (2013). Facile synthetic method for pristine graphene quantum dots and graphene oxide quantum dots: Origin of blue and green luminescence. *Advanced Materials*, *25*(27), 3657–3662.

Long, M.-Q., et al. (2009). Theoretical predictions of size-dependent carrier mobility and polarity in graphene. *Journal of the American Chemical Society*, *131*(49), 17728–17729. http://pubs.acs.org/doi/abs/10.1021/ja907528a. June 13, 2018.

Nurunnabi, M., et al. (2013). In vivo biodistribution and toxicology of carboxylated graphene quantum dots. *ACS Nano*, *7*(8), 6858–6867. http://pubs.acs.org/doi/10.1021/nn402043c. March 31, 2018.

Pan, D., Zhang, J., Li, Z., & Wu, M. (2010). Hydrothermal route for cutting graphene sheets into blue-luminescent graphene quantum dots. *Advanced Materials*, *22*(6), 734–738.

Pan, D., et al. (2013). Electrophoretic fabrication of highly robust, efficient, and benign heterojunction photoelectrocatalysts based on graphene-quantum-dot sensitized TiO_2 nanotube arrays. *Journal of Materials Chemistry A*, *1*(11), 3551. http://xlink.rsc.org/?DOI=c3ta00059a. March 31, 2018.

Peng, J., et al. (2012). Graphene quantum dots derived from carbon fibers. *Nano Letters*, *12*(2), 844–849. http://pubs.acs.org/doi/10.1021/nl2038979. June 13, 2018.

Ponomarenko, L. A., et al. (2008). Chaotic Dirac billiard in graphene quantum dots. *Science*, *320*(5874), 356–358. http://www.sciencemag.org/cgi/doi/10.1126/science.1154663.

Ranjit, S., et al. (2016). Effect of polyaryl hydrocarbons on cytotoxicity in monocytic cells: Potential role of cytochromes P450 and oxidative stress pathways. ed. Sanjay B.

Maggirwar, *PLoS One*, *11*(9), e0163827. http://dx.plos.org/10.1371/journal.pone.0163827. June 13, 2018.

Ritter, K. A., & Lyding, J. W. (2009). The influence of edge structure on the electronic properties of graphene quantum dots and nanoribbons. *Nature Materials*, *8*(3), 235–242. http://www.nature.com/articles/nmat2378. March 30, 2018.

Sapsford, K. E., et al. (2013). Functionalizing nanoparticles with biological molecules: Developing chemistries that facilitate nanotechnology. *Chemical Reviews*, *113*(3), 1904–2074. http://pubs.acs.org/doi/10.1021/cr300143v. June 13, 2018.

Sun, Y., et al. (2017). Magnetism of graphene quantum dots. *NPJ Quantum Materials*, *2*(1), 5. http://www.nature.com/articles/s41535-017-0010-2. April 5, 2018.

Tang, L., et al. (2012). Deep ultraviolet photoluminescence of water-soluble self-passivated graphene quantum dots. *ACS Nano*, *6*(6), 5102–5110. http://pubs.acs.org/doi/10.1021/nn300760g. March 31, 2018.

Wang, H., et al. (2015). A biomimetic hybrid nanoplatform for encapsulation and precisely controlled delivery of therasnostic agents. *Nature Communications*, *6*, 10081. http://www.nature.com/doifinder/10.1038/ncomms10081.

Wu, J., Pisula, W., & Müllen, K. (2007). Graphenes as potential material for electronics. *Chemical Reviews*, *107*(3), 718–747. https://pubs-acs-org.libproxy.bridgeport.edu/doi/full/10.1021/cr068010r. June 13, 2018.

Zheng, X. T., et al. (2013). Graphene quantum dots as universal fluorophores and their use in revealing regulated trafficking of insulin receptors in adipocytes. *ACS Nano*, 7(7), 6278–6286. http://pubs.acs.org/doi/10.1021/nn4023137. March 31, 2018.

Zhu, S., et al. (2011). Strongly green-photoluminescent graphene quantum dots for bioimaging applications. *Chemical Communications*, *47*(24), 6858. http://xlink.rsc.org/?DOI=c1cc11122a. March 31, 2018.

CHAPTER THIRTEEN

Shielding of Enzymes on the Surface of Graphene-Based Composite Cellular Foams Through Bioinspired Mineralization

Yiying Sun*[,‡], Jingjing Zhao*[,‡], Jiafu Shi[†,§,1], Shaohua Zhang*[,‡], Zhongyi Jiang*[,‡,1]

*Key Laboratory for Green Chemical Technology of Ministry of Education, School of Chemical Engineering and Technology, Tianjin University, Tianjin, China
[†]School of Environmental Science and Engineering, Tianjin University, Tianjin, China
[‡]Collaborative Innovation Center of Chemical Science and Engineering (Tianjin), Tianjin, China
[§]Guangdong Provincial Key Laboratory of New and Renewable Energy Research and Development, Guangzhou, China
[1]Corresponding authors: e-mail address: shijiafu@tju.edu.cn; zhyjiang@tju.edu.cn

Contents

Abstract

Immobilization of enzyme on the surface of graphene-based composite cellular foams (GCCFs) is commonly prone to acquire stable and ultrahigh loading of enzymes and fast transport of substrates during the catalytic process. In this chapter, we reported a method of preparing GCCFs through combination of redox assembly and biomimetic mineralization with in situ enzyme immobilization. Briefly, GCCFs were first prepared through redox assembly of graphene oxide (GO) nanosheets enabled by

Methods in Enzymology, Volume 609
ISSN 0076-6879
https://doi.org/10.1016/bs.mie.2018.05.011

polyethyleneimine (PEI). The cationic PEI in the resultant reduced GO/PEI (rGO/PEI) cellular foams acted as the mineralization-inducing agent could catalyze the condensation of silicate to form silica (biomimetic silicification) on the reduced graphene oxide (rGO) surface, where enzyme (with penicillin G acylase as model enzyme) is in situ entrapped and shielded within the silica network. Enzymes could be stably resided on the surface of GCCFs without any leaching against a broad range of pH values (3.5–10.0). GCCFs show a three-dimensional (3D) porous structure, which facilitates the fast transfer of substrate and, thereby, leads to desirable catalytic activity. Combined with the monolithic feature, GCCFs exhibit ease of recyclability and superior thermal/recycling stabilities during the catalytic synthesis of 6-aminopenicillanic acid (6-APA, an important pharmaceutical intermediate) compared to free enzyme and enzyme adsorbed on rGO/PEI cellular foams.

1. INTRODUCTION

Graphene, a two-dimensional (2D) carbon nanomaterial, has exhibited superior properties in biological-related applications, e.g., biocatalysis, biomedical, etc. Because of the high surface area, high mechanical and thermal stabilities, scalable production and low cost, graphene oxide (GO), and reduced graphene oxide (rGO) have been widely explored as carriers to immobilize enzymes for the preparation of nanobiocatalysts (Bi, Chen, Lin, & Huang, 2015; Jiang & Fan, 2014; Liu, Dong, & Chen, 2012; Yang, Feng, Shi, & Liu, 2013). Assembly of 2D graphene into three-dimensional (3D) cellular foams with monolithic feature offers a feasible way for the preparation of easily recovered immobilized enzymes, which could not only inherit the high surface area of graphene but also retain large percentage of water entrapped in the porous structure that implied an adequate environment for enzymes (Wang, Li, Wang, Li, & Lin, 2011).

Currently, physical adsorption and chemical covalent binding are two commonly utilized methods to immobilize enzymes on GCCFs (Chen, Wei, Zhang, & Li, 2013; Ormategui, Veloso, Leal, Rodriguez-Couto, & Tomovska, 2015; Song, Chen, & Wang, 2015; Zhang et al., 2012). In this chapter, we would like to present an alternative method to in situ immobilize enzyme on the surface of GCCFs during the formation process (Scheme 1). Cationic polymers, like polyethyleneimine (PEI), are adopted as the reducing agents to trigger the redox assembly of GO into rGO/PEI cellular foams under conventional hydrothermal conditions (Sui, Cui, Zhu, & Han, 2013). The occurrence of such assembly process is primarily based on the multiple interactions (hydrogen bonding, electrostatic interaction, chemical bonding, etc.) between PEI and GO (Cong, Ren, Wang, & Yu, 2012).

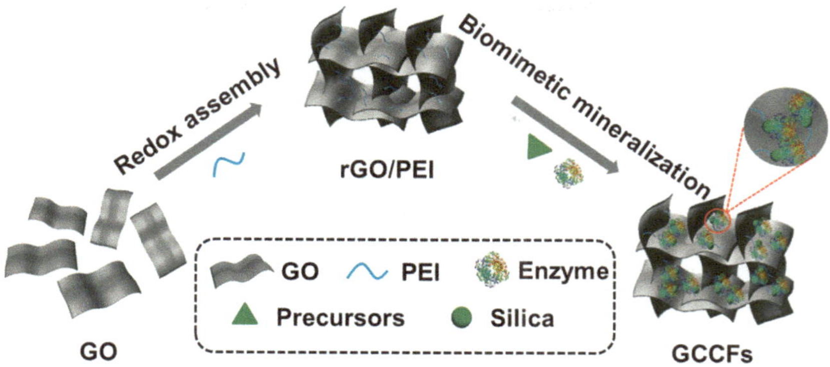

Scheme 1 Preparation of graphene-based composite cellular foams (GCCFs) with enzyme shielded by the silica network on the surface (Yang et al., 2017a, 2017b).

Then, free amine groups in PEI, as the mineralization-inducing agent, catalyze the condensation of silicate to form silica on the rGO surface. Since enzyme is dissolved in the silicate aqueous solution, enzyme is then in situ entrapped by the as-formed silica networks during the silicification, thus forming the final GCCFs (Kou et al., 2015; Ramanathan et al., 2009; Wang, Zhang, Huang, Zhang, & Dong, 2017). Both PEI and silica could enhance the structural stability of GCCFs against external stress, whereas the structure of GCCFs is strongly dominated by the mass ratios of PEI to GO that needed to be accurately controlled. Enzymes could be hardly leached from GCCFs against extreme pH values due to the robust shielding of enzyme by the mineralized silica network. Combined with the easy-recycling behavior, GCCFs could be recycled for many times with minor activity changes.

2. PREPARATION OF GCCFs

As mentioned earlier, GCCFs are formed through the combination of redox assembly and biomimetic mineralization, which includes the formation of rGO/PEI cellular foams through hydrothermal process followed by bioinspired silicification process with in situ enzyme immobilization (Tang, Shen, Zhuang, & Wang, 2010; Vahidi, Yang, Ngo, & Li, 2015). Penicillin G acylase (PGA, EC 3.5.1.11, $7 \times 5 \times 5.5\,nm^3$), a class of N-terminal nucleophile hydrolases with broad applications in pharmaceutical fields, was adopted as a typical enzyme to prepare GCCFs for the catalytic hydrolysis of penicillin G potassium (PGK) to 6-aminopenicillanic acid (6-APA) (Mateo, Palomo, Fernandez-Lorente, Guisan, & Fernandez-Lafuente, 2007).

2.1 Equipments

- Electric-heated thermostatic water bath (DF-101S) (Xingke Instrument Co. Ltd., China).
- Centrifugal machine (3–16 PK) (Sigma-Aldrich Chemical Co. Ltd., USA).
- Freeze dryer (Alpha 1–2) (Martin Christ Freeze Dryers Co. Ltd., Germany).
- Ultrasonic sound apparatus (KQ2200E) (Kunshan ultrasonic instrument Co. Ltd., China).
- UV–vis spectrophotometer (U-3010) (Hitachi Ltd., Japan).
- Purification system (Milli-Q) (Millipore Ltd., USA).

2.2 Buffers and Reagents

- Penicillin G acylase (PGA, EC 3.5.1.11, 1000 U mL^{-1}) (Junfeng Bioengineering Co. Ltd., China).
- Potassium permanganate ($KMnO_4$) (Sigma-Aldrich Chemical Co. Ltd., USA).
- Graphite powder (99.9995%) (Sigma-Aldrich Chemical Co. Ltd., USA).
- Sodium metasilicate ($Na_2SiO_3 \cdot 9H_2O$) (Guangfu Co. Ltd., Tianjin, China). *Tip*: Na_2SiO_3 aqueous solution should be prepared freshly.
- Polyethylenimine (PEI, $M_w \approx 600$ Da, 99 wt% aqueous solution) (Aladdin Co. Ltd., China).
- Penicillin G potassium salt (Penicillin G, 98%) (Hwrk Chem Co. Ltd., China).
- 6-Aminopenicillanic acid (6-APA, 95%) (Hwrk Chem Co. Ltd., China).
- *P*-dimethylaminobenzaldehyde (PDAB) (Hwrk Chem Co. Ltd., China).
- Tris (hydroxymethyl) aminomethane (Tris) (Sigma-Aldrich Chemical Co. Ltd., USA).
- Fluorescein isothiocyanate (FITC) (Sigma-Aldrich Chemical Co. Ltd., USA). *Tip*: FITC should be stored under dark condition before use.
- Water (purified by a Millipore Milli-Q purification system with a resistivity higher than 15.0 MΩ cm).
- Other reagents were analytical grade and used without further purification.

2.3 Procedure

2.3.1 Preparation of GO Nanosheets

- 5 g graphite, 115 mL H_2SO_4, and 2.5 g $NaNO_3$ were mixed and stirred under ice bath until completely dissolved.
- 15 g $KMnO_4$ was slowly added into the above solution, then stirred for 2 h.

- The mixed solution was transferred to 308.15 ± 2 K water bath and stirred for 30 min.
- 230 mL water was slowly added. Keep the temperature of the mixture lower than 373.15 K.
- The reaction solution was stirred for 3 h at 371.15 K. 1 L deionized water was added to dilute the solution.
- 30 mL H_2O_2 was added into the mixture to remove the residual $KMnO_4$ and MnO_2.
- The mixture need to be repeatedly washed by hydrochloric acid (5%) until a yellow dispersion was obtained.
- The yellow dispersion was centrifuged followed by washing by deionized water until the pH of mixture turned to be 7.0.
- The GO powder was obtained after freeze drying for 24 h.

2.3.2 Preparation of rGO/PEI Cellular Foams

- 4 mg GO was sonicated for 30 min in 2 mL water until a homogeneous solution was acquired, while the pH value of the solution was adjusted to 8.0 by NaOH.
- 2 mL PEI aqueous solution (2 mg mL^{-1}) was added to the GO aqueous solution.
- The above mixture was transferred to a 10 mL cylindrical glass vial and then placed into a 363.15 K oil bath for 6 h without stirring.
- rGO/PEI cellular foams were then formed and taken out, washed with deionized water.

2.3.3 Preparation of GCCFs

- 4 mL sodium silicate aqueous solution (10 mmol L^{-1}) containing 3.4 mg PGA was added to the suspension of rGO/PEI cellular forms, followed by the continuous reaction for another 15 min to form GCCFs.
- GCCFs (immobilized PGA) were taken out, and washed with Tris–HCl buffer solution (pH 7.8) for four times to remove free and loosely adsorbed PGA.

2.3.4 Activity Assay of GCCFs

- Penicillin G can be hydrolyzed into 6-APA by the free or immobilized PGA (Scheme 2).
- *P*-dimethylaminobenzaldehyde (PDAB, 0.5 w/v%) was used as the colored indicator to measure the amount of 6-APA.

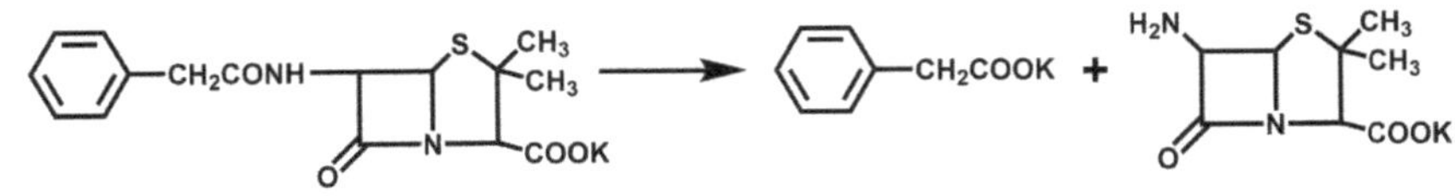

Scheme 2 Reaction diagram of PGA-catalyzed synthesis of 6-APA.

- The penicillin G (4 w/v%) solution was prepared by dissolving certain amount of penicillin G into Tris–HCl buffer solution (50 mmol L^{-1}, pH 7.8).
- The reaction solution was shaken for 5 min and then 1.0 mL solution was taken out and added into the mixture of PDAB methanol solution (1.0 mL, 0.5 w/v%) and NaAc–HAc buffer solution (3.0 mL, 50 mmol L^{-1}, pH 2.5) to inactivate the enzymes.
- The solution was protected from light for 10 min at room temperature. The absorbance at 415 nm was determined by a UV–vis spectrophotometer (Hitachi U-3010).
- 6-APA concentration was acquired through comparing the absorbance with the standard curve, and then the conversion rate of penicillin G was calculated.
- The relative activity was calculated through comparing the activity of immobilized PGA and free PGA.

2.3.5 The Kinetic Parameters Test of GCCFs

- The kinetic parameters were evaluated with different concentrations of penicillin G solution (0.1–0.6 w/v%), using the same free and immobilized PGA (15 U).
- The reaction was performed in the Tris–HCl buffer solution (50 mmol L^{-1}, pH 7.8) at the room temperature.
- The initial reaction rate at different substrate concentration was measured and the Michaelis–Menten equation (Eq. 1) was used to calculate the K_m and V_{max}.

$$\frac{1}{V} = \frac{K_m}{V_{max}} \times \frac{1}{[S]} + \frac{1}{V_{max}} \quad (1)$$

where V (mmol (L min)$^{-1}$) was the initial reaction rate for different substrate concentrations; $[S]$ (mmol L^{-1}) was the initial concentration of the reactants; V_{max} was the maximum reaction rate; and K_m was the Michaelis–Menten constant.

2.3.6 Thermal and Recycling Stabilities Test of GCCFs

- The thermal stability was evaluated by measuring the activities of residual enzyme, after the immobilized PGA and free PGA were incubated in water bath at different temperatures (278.15, 293.15, 303.15, 313.15, 323.15, and 328.15 K) for 3 h.
- 1.0 mL solution was taken out and added into the mixture of PDAB methanol solution (1.0 mL, 0.5 w/v%) and NaAc–HAc buffer solution (3.0 mL, 50 mmol L^{-1}, pH 2.5).
- The solution was protected from light for 10 min at room temperature. The absorbance at 415 nm was determined by a UV–vis spectrophotometer (Hitachi U-3010).
- The initial activities of the free and immobilized PGA used in all these stability evaluations were 15 U and taken to be 100%. The relative activity (%) represented the ratio of residual activity to initial activity of each sample.
- The recycling stability of the immobilized PGA was measured through taking out the immobilized PGA by using tweezers and washed with Tris–HCl buffer solution (50 mmol L^{-1}, pH 7.8) for the next-cycled reuse.
- In each cycle, the immobilized PGA collected from the last cycle was added to fresh penicillin G (4 w/v%) solution. The activity detection method was the same as that of the thermal stability. The recycling activity of immobilized PGA was defined as the ratio of the enzyme activity at the *n*th recycling to that at the first recycling.

2.4 Some Notes

- During the preparation of GO nanosheets, $KMnO_4$ must be slowly added into the solution in batches to guarantee safety.
- The GO powder must be sonicated for at least 30 min in water thoroughly to form a homogeneous solution.
- During the forming process of rGO/PEI cellular foams, the glass vial need to keep sealed to avoid liquid enter it.
- GCCFs need to be washed with deionized water for three or more times.
- Na_2SiO_3 aqueous solution should be prepared freshly.
- FITC should be stored under dark condition before use.
- During the recycling process, the sample must be fully centrifuged to avoid the loss of enzyme and minimize the artificial error.

3. STRUCTURE CONTROL OF GCCFs

3.1 Influence of GO/PEI Ratios

The mechanism behind the formation of monolithic 3D porous structure of rGO/PEI cellular foams is the electrostatic interaction, π–π stacking and chemical bonding forces during chemical/thermal reduction of GO and physical/chemical crosslinking of PEI and GO. With the increase of GO/PEI ratios, π–π stacking became stronger, which is the only driving force to astringe the structure (Fig. 1A). As a result, the structural stability

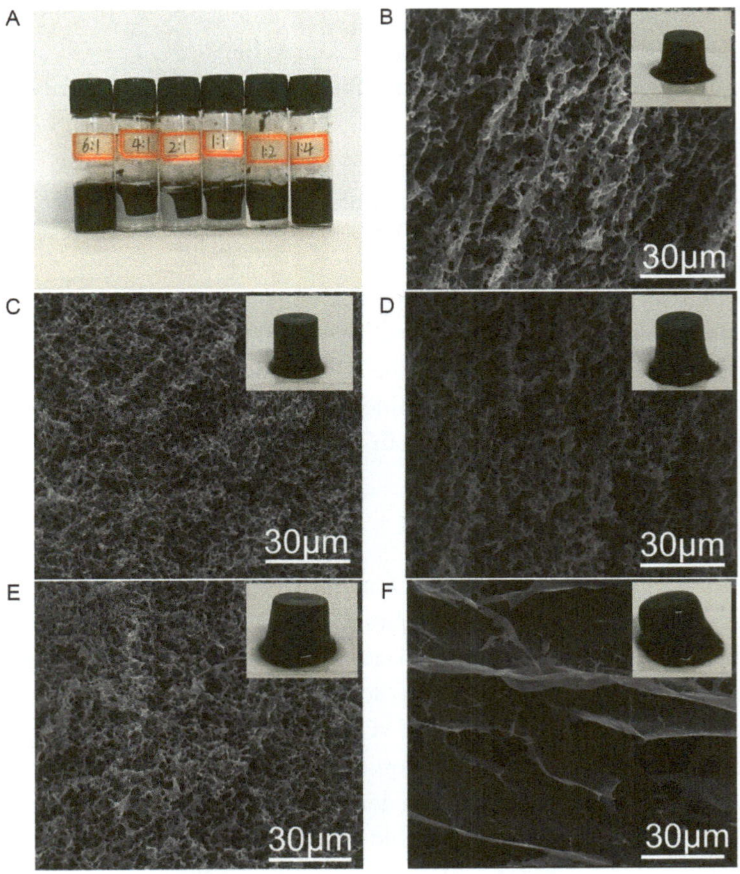

Fig. 1 (A, *insets* of (B–F)) Optical images and (B–F) scanning electron microscopy (SEM) images of rGO/PEI cellular foams with different weight ratios of GO/PEI (B: 4:1; C: 2:1; D: 1:1; E: 1:2; and F: 1:4). Notably, the weight of GO was unaltered and foams could not be formed in the absence of PEI (Yang et al., 2017a).

and porosity of rGO/PEI cellular foams were influenced by the GO/PEI ratios, so we regulate the GO/PEI ratios with 4:1, 2:1, 1:1, 1:2, and 1:4 to find the rules (Yang et al., 2017a). According to the results, lower GO/PEI ratios with PEI higher than that of rGO in rGO/PEI results in softer cellular foams with more flexible structures (insets of Fig. 1B–F). For the microstructure, rGO/PEI with GO/PEI ratios from 4:1 to 1:2 exhibited 3D interconnected porous structure with pore diameters ranging from several nanometers to micrometers (Fig. 1B – F). rGO/PEI cellular foams with GO/PEI ratio of 1:1 would have the smallest pore size but highest porosity. Interestingly, layered structure was formed when GO/PEI ratio was down to 1:4, the reason of which was still under investigation. Softer GCCFs that prepared with GO/PEI ratios of 1:2 and 1:4 were not quite suitable for enzyme catalysis as the immobilized enzyme should have the relatively strong mechanical stability against external stress.

3.2 Influence of Silica Concentration

The rGO/PEI with a GO/PEI ratio of 1:1, which would have higher porosity and appropriate structural stability, were adopted for triggering the silica mineralization. Stability test showed that rGO/PEI/silica with ~8 mm in diameter could support 10 g weight without any apparent deformation (Fig. 2A). These graphene-based cellular composite foams (GCCFs) still retained 3D interconnected porous structure with high surface area and large pore volume, where numerous nanoparticles were distributed on the rGO surface (Fig. 2B, *red circle*). The energy spectrum analysis (energy-dispersive spectroscopy, EDS) further indicated that the nanoparticles were probably to be silica with a silicon content of ~5.11 wt% (Fig. 2C).

Then, TGA was conducted to examine the thermal property of rGO/PEI and rGO/PEI/Silica. As demonstrated in Fig. 3A, both samples exhibited a three-step weight loss: the first step (room temperature ~513.15 K) was due to the thermal desorption of water physically adsorbed on the rGO surface; the second step (513.15–673.15 K) was due to the decomposition of PEI and oxygen-containing groups on the rGO surface; the third step (673.15–793.15/823.15 K resulted from the decomposition of carbon skeleton in graphene (for both samples) and the further condensation of silica (only for rGO/PEI/Silica). When the temperature exceeded 793.15 K for rGO/PEI and 823.15 K for rGO/PEI/Silica, minor thermosteresis was found. By contrast with rGO/PEI, the shift of TGA curve in rGO/PEI/Silica implied the enhanced thermal stability

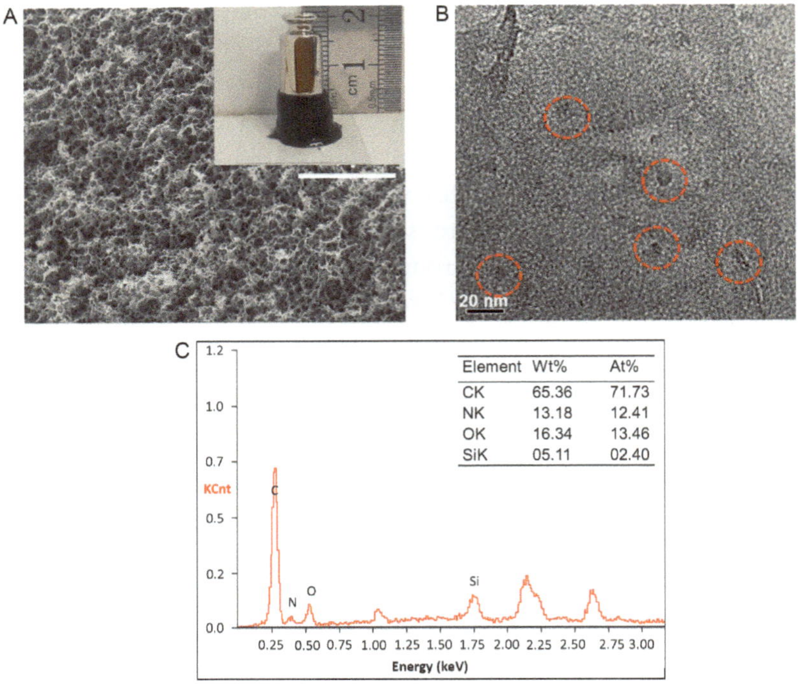

Fig. 2 (A) SEM, (B) transmission electron microscopy (TEM) and (*inset* of A) optical images of rGO/PEI/Silica. (C) EDS curve of rGO/PEI/Silica (Yang et al., 2017a).

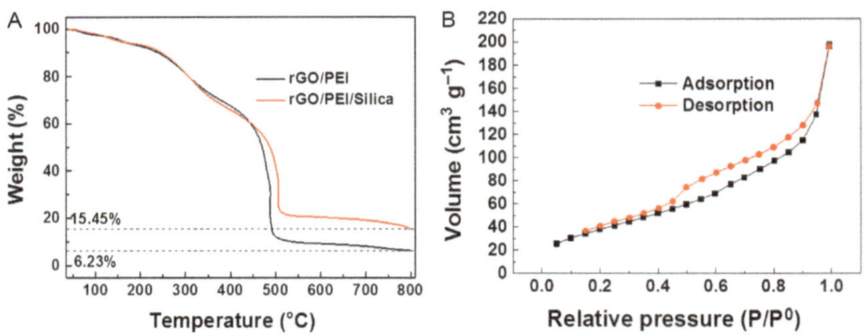

Fig. 3 (A) Thermogravimetric analysis (TGA) curves of rGO/PEI and rGO/PEI/Silica. (B) N_2 adsorption–desorption isotherm of rGO/PEI/Silica (Yang et al., 2017a).

after silica deposition. The final silica content in rGO/PEI/Silica was about 10.80 wt%, which was almost equal to the result (10.95%) calculated from the EDS result. We further conducted the N_2 adsorption–desorption experiment (Fig. 3B), according to which the BET surface area of rGO/PEI/Silica was calculated to be $174.513\,m^2\,g^{-1}$ and the average pore size was over 270 nm. Considering the size of the enzyme was about $5.3 \times 6.1 \times 6.9\,\text{nm}^3$, all fractions (100%) of the exposed surface was available for enzyme loading theoretically.

To intensively study the immobilization and catalytic process, three samples of GCCFs with identical amounts of rGO/PEI but different concentration of Na_2SiO_3 (10, 30–50 mmol L^{-1}) were investigated under the same PGA solution concentration (Yang et al., 2017a). Usually, enzyme entrapped within the silica matrix could effectively inhibit the enzyme leaching. However, much higher content of silica on the rGO/PEI cellular foams would lead to a thicker "barrier" round enzyme molecules, which lowered the mass transfer rate of substrates, and finally resulted in a decrease in activity recovery from 31.6% to 18.6%. As a result, PGA immobilized on the rGO/PEI/Silica (10 m*M*) would be better selected as GCCFs for evaluating the catalytic activities and stabilities in the subsequent experiments.

4. ACTIVITY AND STABILITIES OF GCCFs

GCCFs were then adopted to immobilize PGA (an enzyme that catalyzed the conversion of 6-APA). The position of PGA in GCCFs was shown by confocal laser scanning microscope (CLSM). The CLSM result (Fig. 4A) confirmed the uniform distribution of PGA in/on GCCFs. In order to rule out the possibility of physical adsorption of PGA on the rGO surface, GCCFs were washed with 0.5 M NaCl solution. The enzyme leaching behavior of rGO/PEI/PGA (PGA directly adsorbed on rGO/PEI) and GCCFs were examined under different pH conditions. As shown in Fig. 4B, rGO/PEI/PGA showed relatively high leaching ratios under acidic (>75%), neutral (>45%), and alkaline (>3%) pH values. The leaching from rGO/PEI/PGA would be due to the weakened electrostatic interaction between PGA and PEI under extreme pH values (Wan et al., 2016). The changes in pH would affect the free state and the charges of polar groups in the PGA, which in turn affected the combination of PGA and carriers. The isoelectric point of the PGA was 6.5–6.7, thus PGA was negatively charged under alkaline pH values, which promoted its binding to rGO/PEI with positive charges. When the pH of the solution was 3.5, PGA was positively charged and the electrostatic interaction between PGA and carrier disappeared, causing the decrease of the enzyme loading. When the pH was 7.0, PGA was basically neutral electrically and the interaction weakened and the loading still decreased with the time. On the contrary, nearly no enzyme came out from GCCFs in the pH values from 3.5 to 10.0.

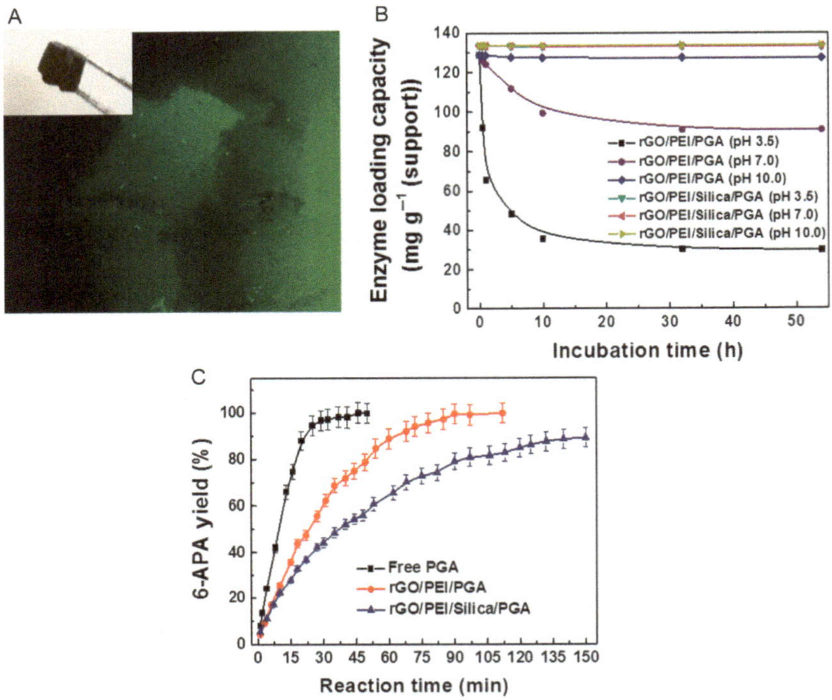

Fig. 4 (A) Fluorescence microscope and (*inset* of A) optical images of GCCFs. (B) The enzyme loading capacity of rGO/PEI/PGA and GCCFs vs incubation time under different pH conditions. (C) 6-APA yield with reaction time (298.15 K, pH 7.8) (Yang et al., 2017a).

This phenomenon was mainly attributed to the formation of mineralized silica network that inhibited the structural changes and leaching of PGA (Pchelintsev, Neville, & Millner, 2008). The yield of 6-APA along with the reaction time catalyzed by free PGA and immobilized PGA was then measured to assess the catalytic property of GCCFs-bearing PGA (Fig. 4C). The reaction catalyzed by free PGA could reach equilibrium in 30 min with a 6-APA yield of 98.3%. By contrast, rGO/PEI/PGA and GCCFs required 90 and 150 min to reach reaction equilibrium, respectively. rGO/PEI/PGA showed a 6-APA yield of 99.5%, which was nearly equal to free PGA (98.3%) but slightly higher than GCCFs (89.1%). The decreased yield of 6-APA catalyzed by the immobilized PGA would be ascribed to the slower diffusion of the substrate (penicillin G potassium) into GCCFs and the product (6-APA) out of GCCFs (Frenkel-Mullerad & Avnir, 2005).

Herein, the kinetic parameters of free PGA, rGO/PEI/PGA, and rGO/PEI/Silica/PGA were calculated by Linerweave–Burk method.

Table 1 Kinetic Parameters of Free and Immobilized PGA

	K_m (mmol L^{-1})	V_{max} (mmol (L min)$^{-1}$)
Free PGA	1.25	1.34
rGO/PEI/PGA	12.93	1.07
rGO/PEI/Silica/PGA	13.92	1.01

As shown in Table 1, the Michaelis–Menten constant K_m of free PGA, rGO/PEI/PGA, and rGO/PEI/Silica/PGA were 1.25, 12.93, and 13.92 mmol L^{-1}, respectively. The remarkable increase of K_m after immobilization indicated that the affinity between the PGA and the substrate in the gel network was decreased. However, the K_m of rGO/PEI/PGA and rGO/PEI/Silica/PGA only had a little difference, indicating that embedding PGA through biomimetic mineralization did not affect its affinity with the substrate. Besides, compared with free PGA, rGO/PEI/PGA, and rGO/PEI/Silica/PGA had a slightly lower V_{max}, which was mainly due to the larger substrate/product diffusion resistance caused by the foams (for both samples) and silica layer (for rGO/PEI/Silica/PGA).

Besides activity, the thermal and recycling stabilities were two other critical parameters to evaluate the noncatalytic property of immobilized PGA. In Fig. 5A, GCCFs showed higher thermal stability than free PGA and rGO/PEI/PGA. All the immobilized PGA could keep 100% of relative activity at the temperature ranging from 4 to 313.15 K. When the temperature reached 323.15 K, the activity of free PGA and rGO/PEI/PGA decreased quickly (60% off) until fully denatured at 328.15 K. The increase of temperature could accelerate the thermal motion of enzyme, which was then prone to destroy the intramolecular hydrogen bonds in enzyme and cause the denaturation of enzyme. Notably, GCCFs showed only 20% off in catalytic activity under 323.15 K. Even better, GCCFs could retain 35.7% of its initial activity when the temperature was increased up to 328.15 K. This was mainly because the silica coating provided a gentler microenvironment for the enzyme and reduced its thermal movement, which reduced the possibility of conformational changes, thereby increasing the thermal stability of enzyme. In addition to thermal stability, our GCCFs could be taken out directly for reuse, indicating a rather ease of recycling behavior. GCCFs could retain 89.2% of the initial enzyme activity after over 16th reuses, indicating a superior recycling stability compared to rGO/PEI/PGA (77%) and PEI/Silica/PGA (55%) (Fig. 5B).

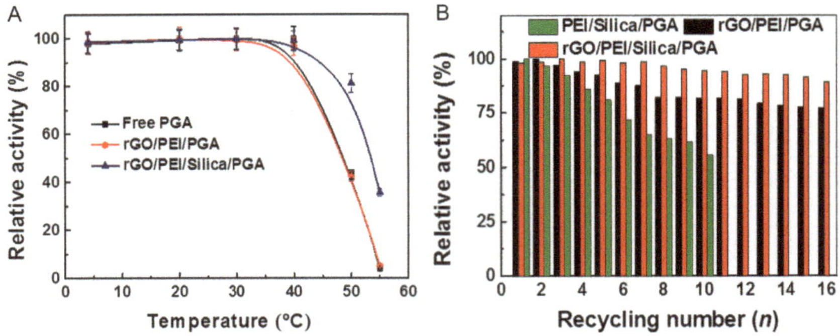

Fig. 5 (A) Thermal stability of free PGA, rGO/PEI/PGA, and GCCFs. (B) Recycling stability of PEI/Silica/PGA, rGO/PEI/PGA, and GCCFs (Yang et al., 2017a).

As for rGO/PEI/PGA, the enzyme molecules were immobilized on/in the foams through relatively strong electrostatic interaction, which benefited for retaining the enzyme conformation and elevating the stability of PGA under the reaction conditions. Further mineralization of silica coating on the foams not only reinforced the mechanical stability of the foams but also entrapped the enzymes in a silica network, which then lead to little leakage and conformation changes of PGA. In short, although showing a bit lower catalytic activity than free PGA and rGO/PEI/PGA, our GCCFs exhibited remarkably enhanced thermal and recycling stabilities.

5. CONCLUDING REMARKS

A versatile method of combining redox assembly and biomimetic mineralization is developed to prepare GCCFs along with in situ enzyme immobilization. The feature of 3D interconnected porous structure endows GCCFs with easy recovery and rapid transport of substrates/products. More importantly, the space-confined property of silica network on the rGO surface significantly inhibits the enzyme leaching, thus conferring enzyme better recycling and storage stabilities. Actually, the formation of GCCFs is not only restricted to PGA immobilization but also applicable to a broad range of other enzymes, like catalase, dehydrogenase, etc. Meanwhile, due to the versatility of bioinspired mineralization in forming other different kinds of metals/metal oxides besides silica, enzymes could be then composited with other active metals/metal oxides and, thereby, applied for chemobiocoupled catalysis.

ACKNOWLEDGMENTS

The authors thank financial support from National Natural Science Foundation of China (21621004, 21776213, 91534126), Tianjin Research Program of Application Foundation and Advanced Technology (15JCQNJC10000), Open Funding Project of the National Key Laboratory of Biochemical Engineering (2015KF-03), Open funding supported by Guangdong Provincial Key Laboratory of New and Renewable Energy Research and Development (Y707s61001), and the Program of Introducing Talents of Discipline to Universities (B06006).

REFERENCES

Bi, H., Chen, I. W., Lin, T., & Huang, F. (2015). A new tubular graphene form of a tetrahedrally connected cellular structure. *Advanced Materials*, *27*, 5943–5949.

Chen, L., Wei, B., Zhang, X. T., & Li, C. (2013). Bifunctional graphene/gamma-Fe_2O_3 hybrid aerogels with double nanocrystalline networks for enzyme immobilization. *Small*, *9*, 2331–2340.

Cong, H. P., Ren, X. C., Wang, P., & Yu, S. H. (2012). Macroscopic multifunctional graphene-based hydrogels and aerogels by a metal ion induced self-assembly process. *ACS Nano*, *6*, 2693–2703.

Frenkel-Mullerad, H., & Avnir, D. (2005). Sol–gel materials as efficient enzyme protectors: Preserving the activity of phosphatases under extreme pH conditions. *Journal of the American Chemical Society*, *127*, 8077–8081.

Jiang, L., & Fan, Z. (2014). Design of advanced porous graphene materials: From graphene nanomesh to 3D architectures. *Nanoscale*, *6*, 1922–1945.

Kou, L., Liu, Z., Huang, T., Zheng, B., Tian, Z., Deng, Z., et al. (2015). Wet-spun, porous, orientational graphene hydrogel films for high-performance supercapacitor electrodes. *Nanoscale*, *7*, 4080–4087.

Liu, Y. X., Dong, X. C., & Chen, P. (2012). Biological and chemical sensors based on graphene materials. *Chemical Society Reviews*, *41*, 2283–2307.

Mateo, C., Palomo, J. M., Fernandez-Lorente, G., Guisan, J. M., & Fernandez-Lafuente, R. (2007). Improvement of enzyme activity, stability and selectivity via immobilization techniques. *Enzyme and Microbial Technology*, *40*, 1451–1463.

Ormategui, N., Veloso, A., Leal, G. P., Rodriguez-Couto, S., & Tomovska, R. (2015). Design of stable and powerful nanobiocatalysts, based on enzyme laccase immobilized on self-assembled 3D graphene/polymer composite hydrogels. *ACS Applied Materials & Interfaces*, *7*, 14104–14112.

Pchelintsev, N. A., Neville, F., & Millner, P. A. (2008). Biomimetic silication of surfaces and its application to preventing leaching of electrostatically immobilized enzymes. *Sensors and Actuators B: Chemical*, *135*, 21–26.

Ramanathan, M., Luckarift, H. R., Sarsenova, A., Wild, J. R., Ramanculov, E. K., Olsen, E. V., et al. (2009). Lysozyme-mediated formation of protein-silica nanocomposites for biosensing applications. *Colloids and Surfaces B: Biointerfaces*, *73*, 58–64.

Song, Y., Chen, C. H., & Wang, C. L. (2015). Graphene/enzyme-encrusted three-dimensional carbon micropillar arrays for mediatorless microbiofuel cells. *Nanoscale*, *7*, 7084–7090.

Sui, Z. Y., Cui, Y., Zhu, J. H., & Han, B. H. (2013). Preparation of three-dimensional graphene oxide-polyethylenimine porous materials as dye and gas adsorbents. *ACS Applied Materials & Interfaces*, *5*, 9172–9179.

Tang, Z., Shen, S., Zhuang, J., & Wang, X. (2010). Noble-metal-promoted three-dimensional macroassembly of single-layered graphene oxide. *Angewandte Chemie International Edition*, *49*, 4603–4607.

Vahidi, A. K., Yang, Y., Ngo, T. P. N., & Li, Z. (2015). Simple and efficient immobilization of extra cellular his-tagged enzyme directly from cell culture supernatant as active and recyclable nanobiocatalyst: High-performance production of biodiesel from waste grease. *ACS Catalysis*, *5*, 3157–3161.

Wan, Y. J., Yang, W. H., Yu, S. H., Sun, R., Wong, C. P., & Liao, W. H. (2016). Covalent polymer functionalization of graphene for improved dielectric properties and thermal stability of epoxy composites. *Composites Science and Technology*, *122*, 27–35.

Wang, Y., Li, Z. H., Wang, J., Li, J. H., & Lin, Y. H. (2011). Graphene and graphene oxide: Biofunctionalization and applications in biotechnology. *Trends in Biotechnology*, *29*, 205–212.

Wang, Q., Zhang, X., Huang, L., Zhang, Z., & Dong, S. (2017). One-pot synthesis of Fe_3O_4 nanoparticle loaded 3D porous graphene nanocomposites with enhanced nanozyme activity for glucose detection. *ACS Applied Materials & Interfaces*, *9*, 7465–7471.

Yang, K., Feng, L. Z., Shi, X. Z., & Liu, Z. (2013). Nano-graphene in biomedicine: Theranostic applications. *Chemical Society Reviews*, *42*, 530–547.

Yang, D., Zhao, J. J., Wang, X. Y., Shi, J. F., Zhang, S. H., & Jiang, Z. Y. (2017a). Combination of redox assembly and biomimetic mineralization to prepare graphene-based composite cellular foams for versatile catalysis. *ACS Applied Materials & Interfaces*, *9*, 43950–43958.

Yang, D., Zhao, J. J., Wang, X. Y., Shi, J. F., Zhang, S. H., & Jiang, Z. Y. (2017b). Monolithic biocatalytic systems with enhanced stabilities constructed through biomimetic silicification-induced enzyme immobilization on rGO/FeOOH hydrogel. *Biochemical Engineering Journal*, *117*, 52–61.

Zhang, Y., Zhang, J. Y., Huang, X. L., Zhou, X. J., Wu, H. X., & Guo, S. W. (2012). Assembly of graphene oxide–enzyme conjugates through hydrophobic interaction. *Small*, *8*, 154–159.

CHAPTER FOURTEEN

Enzyme Immobilization on Functionalized Graphene Oxide Nanosheets: Efficient and Robust Biocatalysts

Asieh Soozanipour, Asghar Taheri-Kafrani[1]
Department of Biotechnology, Faculty of Advanced Sciences and Technologies, University of Isfahan, Isfahan, Iran
[1]Corresponding author: e-mail address: a.taheri@ast.ui.ac.ir

Contents

Methods in Enzymology, Volume 609
ISSN 0076-6879
https://doi.org/10.1016/bs.mie.2018.06.010

Abstract

Enzymes are used as biocatalysts for analytical purposes in diagnostics and preparative purposes in large-scale industrial processes. Despite perfect catalytic properties of enzymes, their industrial applications are limited due to the drawbacks regarding the lack of long-term stability under process conditions. The difficulties associated with recycling have to be resolved before enzyme implementation at industrial scale. Enzyme immobilization, as a novel approach, can improve the half-life, stability, catalytic activity, and reusability of enzymes. Graphene-based nanomaterials, as nanoscaled and thermostable inorganic carriers, are nontoxic materials and selective modulators for enzyme activity. Herein, we have concentrated on strategies for preparing graphene-based nanocomposites for enzyme immobilization. Nanostructures of graphene, hybrid graphene, and their derivatives with adjustable surface chemistry, caused them to be excellent candidates for immobilization of enzymes. For instance, the synthesis and functionalization of Fe_3O_4–graphene oxide (GO) hybrids were improved recently, in our research group, using cyanuric chloride and polyethylene glycol bis-amine for the immobilization of xylanase and glucoamylase enzymes, via physical and covalent attachments. Decorating GO nanosheets with Fe_3O_4 nanoparticles has facilitated the reusability of enzymes and increased the surface area for enzyme loading. The use of these hydrophilic crosslinkers may change the microenvironment of the immobilized enzymes that could result in the enhancement of their catalytic activity. As a result of the fascinating properties of graphene-based nanocarriers, with respect to structures that can be oriented and surfaces that can be modified, in our opinion, they offer some important advantages for biotechnological applications, especially in the areas of enzyme immobilization and medicine.

ABBREVIATIONS

BSA	bovine serum albumin
CC (TCT)	cyanuric chloride
CVD	chemical vapor deposition
GBP	gold binding polypeptide
GFNs	graphene family nanomaterials
GO	graphene oxide
GOD	glucose oxidase
HA	hydroxyapatite
HRP	horseradish peroxidase
OPH	organophosphorus hydrolase
OxOx	oxalate oxidase
PEGA	polyethylene glycol bis-amine
PNFs	peptide nanofibers
rGO	reduced graphene oxide
SPGO	superparamagnetic graphene oxide

1. INTRODUCTION

Nowadays, significant advancements in nanotechnology have led to the utilization of nanostructured materials for a variety of scientific research due to their specific physical and chemical properties, particularly quantum-size effects, compared to bulk materials (Sagadevan & Periasamy, 2014). During the recent years, the study of unique structure of carbon-based layered materials such as graphene and its derivatives, as new-found 2D nanosystems, has become one of the interesting topics in the fields of biochemical and biomaterials sciences (Pavlidis, Patila, Bornscheuer, Gournis, & Stamatis, 2014; Zare-Zardini et al., 2015). Compared with other carbon materials, graphene-based systems are younger in development. Although the initial theory of graphene was explored in 1974, the first single-atom-thick crystal was extracted from bulk graphite in 2004 and has been the subject of 2010 Nobel Prize for Physics (Gerstner, 2010; Sassolas, Blum, & Leca-Bouvier, 2012).

Graphene is a two-dimensional sheet and defined by IUPAC as "A single carbon layer of graphite structure, describing its nature by analogy to a polycyclic aromatic hydrocarbon of quasi-infinite size" (Carbone, Gorton, & Antiochia, 2015). This allotrope of carbon is composed of sp^2 hybridized carbons. They combine with each other to form a sheet of hexagonal arranged lattice with a thickness of about 0.34–1.6 nm. What distinguishes graphene from most conventional three-dimensional materials is its unique physicochemical properties. Graphene is one of the strongest materials ever tested with Young's modulus of 0.5–1 TPa, the ultimate tensile strength of 130 GPa, that is 200 times greater than steel (Kim, Abdala, & Macosko, 2010). While the electron mobility of graphene at room temperature is very significant, its value at temperatures between 10 and 100 K is almost independent of temperature. Under ambient conditions, the charge mobility of graphene can be higher than 15,000 $cm^2\ V^{-1}\ S^{-1}$, which is 10 times higher than the mobility of silicon. Furthermore, it is the best-known conductor till now, and the maximum current density is 100 and 60 times greater than copper and carbon nanotubes, respectively. These high thermal conductivity and excellent electronic properties can be due to the presence of π electrons in its structure (Neto, Guinea, Peres, Novoselov, & Geim, 2009;

Yi & Shen, 2015). High surface area to weight ratio on both sides of the planar axis (theoretical limit: 2600 m^2/g, which is two times higher than the specific surface area of carbon nanotubes), measured specific capacitance of 135 F/g, optical transparency, and saturation absorption under strong excitation, gas impermeability, good dispersibility in most solvents and most importantly biocompatibility and good planar support for biomaterials, can be considered as some of the advantages of graphene-based materials (Li, Jiang, et al., 2009). Regarding these features, there is the notable tendency in the application of graphene-based nanomaterials in modern biotechnology.

In the last decade, the synergistic interactions between nanotechnology and biotechnology have led to developing new nanobiosystems. Graphene-related nanoscaled materials have been illustrated as ideal enzyme immobilization scaffolds. Enzyme immobilization can improve the application of enzyme technology in various fields (Misson, Zhang, & Jin, 2015).

Enzymes, as homogeneous biocatalysts, are used in modern industries including biosensors, health care and pharmaceuticals, green chemical, food, water, etc. (Bosio, Islan, Martínez, Durán, & Castro, 2016; Bustamante-Vargas et al., 2015; Shahrestani, Taheri-Kafrani, Soozanipour, & Tavakoli, 2016). Despite perfect catalytic properties of enzymes such as selectivity, use of mild reaction conditions, operation in shorter synthetic routes compared with the chemical catalysts and absence of secondary reactions, their industrial applications are limited due to the drawbacks regarding the lack of long-term stability under process conditions and difficulties associated with recycling (Amirbandeh, Taheri-Kafrani, Soozanipour, & Gaillard, 2017; Soozanipour, Taheri-Kafrani, & Isfahani, 2015). These issues have to be resolved for enzymes before their implementation at industrial scale. Immobilization process provides a way for heterogeneous biocatalyst preparation with advantages including reuse over multiple cycles, sequester enzyme from product stream, improve stability over soluble enzyme forms, favorable alterations in pH and temperature optima, and retain activity for longer periods (Defaei, Taheri-Kafrani, Miroliaei, & Yaghmaei, 2018; Wu, Yang, & Ge, 2017).

The best enzymatic performance depends on the stability of both enzyme and support and their interactions. The effectiveness of immobilized enzyme is often less than its soluble equivalent (DiCosimo, McAuliffe, Poulose, & Bohlmann, 2013; Klein et al., 2012). Given the significant degree to which the physical and chemical properties of the support influence catalytic performance, an appreciation of materials science is required for the successful development of an immobilized enzyme catalyst. Thus, sufficient knowledge of enzyme chemistry, material science, and surface engineering is required for choosing a good methodology for enzyme immobilization (Klein et al., 2016; Mohamad, Marzuki, Buang, Huyop, & Wahab, 2015).

This chapter presents an updated and confederated collection of the synthesis methods and functionalization of suitable graphene-based nanomaterials for enzyme immobilization. The application of immobilized enzymes based on graphene nanomaterials in certain fields has also been summarized.

2. SYNTHESIS OF GRAPHENE-BASED NANOMATERIALS

The graphene family nanomaterials (GFNs) are mainly classified based on layer number, lateral size, composition, chemical properties, stiffness, defect density, quality, and purity. GFNs involve monolayer graphene, ultrathin graphite, few-layer graphene, graphene nanosheets, graphene oxide (GO), and reduced graphene oxide (rGO). They have different physicochemical properties and probably can affect the biological responses (Goenka, Sant, & Sant, 2014; Sanchez, Jachak, Hurt, & Kane, 2011). Among graphene-based nanomaterials, GO and rGO are extensively developed for biological applications. GO is a highly oxidized form of chemically modified graphene. It consists of an amphiphilic one-atom-thick sheet of carbon with abundant oxygen-containing groups (such as hydroxyl, epoxy, carbonyl, and carboxyl groups). rGO is usually prepared by the reduction of GO sheets. This procedure introduces defects in the carbon network and reduces the surface charge and water dispersibility, but it retains the structure and electrical conductivity to a large extent (Ke et al., 2014; Ke & Wang, 2016). Depending on the synthesis method, the type, density, and distribution of functional groups can be developed on the surface of GO and rGO.

Various methods have been established for fabrication of GFNs. The mechanical exfoliation (Novoselov et al., 2004), chemical synthesis (Park & Ruoff, 2009), electrochemical exfoliation (Allen, Tung, & Kaner, 2009; Viculis, Mack, & Kaner, 2003), chemical vapor deposition (CVD) (Reina et al., 2008), and plasma-enhanced CVD (PECVD) (Reina et al., 2008) are much more convenient methods. As shown in Fig. 1, two basic principles are used in all methods: cleaving multilayer graphite into single layers (exfoliation) and depositing one layer of carbon onto another material (epitaxy), which are described below.

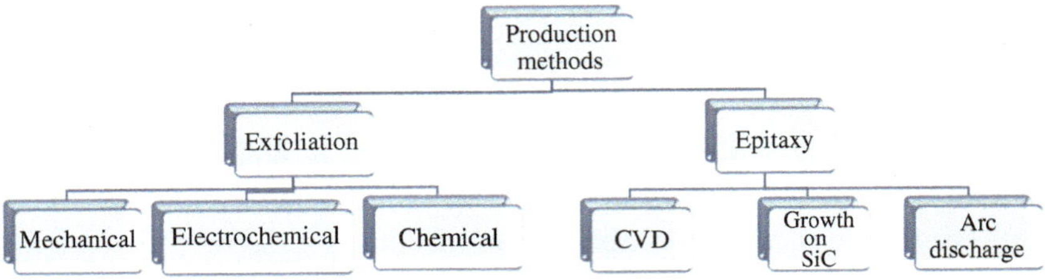

Fig. 1 Flow chart of the methods for synthesizing GFNs.

2.1 Mechanical Exfoliation

Mechanical exfoliation is the safe and simple process employed to prepare graphene while preserving its structure and properties. However, the probability of impurities being present in the graphene obtained is less, and the sample preparation is simplified. This method is not suitable for large-scale synthesis of graphene, due to its low yield production (Chang, Kim, Lee, & Song, 2010).

2.2 Chemical Exfoliation

The chemical exfoliation methods are suitable approaches for preparing graphene-based materials on the industrial scale. Brodie (1859), Staudenmaier (1898), and Hummers and Offeman (1958) methods are most commonly used chemical synthesis methods. In recent years, much attention has been paid to Hummers' method due to short reaction time, reaction safety, and high production yields. This method is usually applied for producing graphene by two consequent steps. First, graphite is oxidized to GO, using suitable oxidizing agents such as $KMnO_4$ and $NaNO_3$, and subsequently, the synthesized GO reduced by chemical agents to obtain graphene-like material (Stankovich et al., 2007). According to the recent studies, Hummers' method was improved by excluding $NaNO_3$ and increasing the amount of $KMnO_4$. The modified Hummers' method introduces a way to obtain more stable GO colloidal solution and to reduce toxic gas evolution. Despite these advantages, the thickness control of the produced graphene sheet is not as promising as in the epitaxial graphene growth or the CVD process (Chen, Yao, Li, & Shi, 2013).

2.3 Electrochemical Exfoliation

The electrochemical exfoliation method is an eco-friendly method for producing high-quality graphene. This is performed using two simultaneous procedures of reduction and deposition. The graphene obtained can be functionalized depending upon electrolyte and hence, can be more compatible with certain organic compounds or polymers. The high negative voltage applied can provide a condition for the reduction of oxygen functionalities on the surface of GO, and therefore, rGO can be efficiently produced. Although this method has important advantages including fast, and easy processing, controllable and scalable to industrial level, but the ionic liquid used in the synthesis is expensive, and impurities may be present in the form of unwashed salts in between the graphene layers that may affect the conductivity of graphene (Fang, Shen, & Chen, 2015; Rao, Senthilnathan, Liu, & Yoshimura, 2014).

2.4 Chemical Vapor Deposition

Catalytic CVD is used extensively for large-scale graphene fabrication. CVD growth is a way of depositing gaseous reactants on transition metal substrates (Ni, Cu, and Pd) or on some ceramic such as glass. The selection of substrate depends on the feasibility of transferring the graphene onto the required material. The formation of high-quality graphene from CVD process can usually be controlled by optimizing the various reaction parameters, such as the type and concentration of carbon precursor, temperature, time, and cooling rate (Liu, Chai, Mohamed, & Hashim, 2014). When the CVD process is done in a resistive heating furnace, it is known as thermally enhanced CVD, and when the process consists of plasma-assisted growth, it is called PECVD (Bhuyan, Uddin, Islam, Bipasha, & Hossain, 2016). Thermally enhanced CVD method yields harder, crystallized graphene film with uniform morphology and good reproducibility, compared with mechanical exfoliation method. However, production of corrosive and toxic gases and high temperatures needed limit the use of this method. High temperatures (higher than 900°C) lead to wrinkled graphene due to the difference in coefficient of thermal expansion (Chae et al., 2009; Li, Cai, et al., 2009). PECVD method is known as efficient and low-cost method with two advantages compared to CVD, including less deposition time and lower growth temperature, because of having additional high-density reactive gas atoms and radicals generating by plasma sources (Nandamuri, Roumimov, & Solanki, 2010).

2.5 Arc Discharge

One of the eco-friendly techniques for the fabrication of graphene nanosheets is arc discharge. Compared to chemical exfoliation methods, this rapid method without the need of metal catalyst can produce good crystallinity and high thermal stability graphene on small-scale with low cost. Arc discharge method involves two electrodes, a buffer gas containing carbon dioxide in the atmospheres of an inert gas, helium, hydrogen, and NH_3. The type of gas, its pressure, rate of heating and cooling affect size, number of graphene layers, and yield. Plasmas of the arc discharge can also provide a convenient way to add functional groups in situ during the synthesis (Fang, Shashurin, & Keidar, 2015; Li, Wang, & Shi, 2011).

2.6 Epitaxial Growth of Graphene on Silicon Carbide (SiC)

Graphene can be produced on a single crystalline SiC by annealing SiC surface using ultra-high vacuum (UHV). The main concept behind the process is that the vapor pressure of silicon is higher, as a result of heating the SiC wafer, the Si evaporates leaving behind the graphene layers on the SiC. Multilayer, bilayer, or single layer graphene can be grown on the SiC substrate by controlling various parameters such as SiC temperature, pressure, and annealing time. With the removal of silicon atoms, the arrangement of carbon atoms will take place to form graphene layers. Although patterning of graphene is easier due to the use of insulating SiC substrate, and lattice constant mismatch and difference in coefficient of thermal expansion of SiC–C can lead to various defects. Conditions such as high temperature, difficulty in growing uniform multilayer graphene on different SiC polar face (Si-face or C-face) in UHV, limit this method for usage in large-scale practical applications (Deng et al., 2010).

All of the above-mentioned synthesis techniques have some advantages and limitations, depending upon the final application of products that are summarized in Table 1.

3. MODIFICATION OF GRAPHENE-BASED NANOMATERIALS

In the last decade, GFNs with unique properties have attracted attention as valuable supports to interact with various materials. However, pure GFNs are limited for many applications. It is well known that many properties of graphene and its derivatives depend on their structure and surface functional groups. So, surface modification not only should protect structure but also provide conditions for surface optimization or second functionalization that can improve their biocompatibility, adsorption capacity, surface functionality, selectivity, dispersibility, and solubility. Functionalization of GFNs with polymers, biomolecules, and metal nanoparticles may increase

Table 1 A Comparison of Approaches for Synthesis of GFNs

Synthesis Method	Precursors	Advantages	Disadvantages
Mechanical exfoliation	Graphite	Safe and simple, low cost, impurity free	Low yield, time consuming
Electrochemical exfoliation	Graphite	Fast, controllable and hazardous solvent-free process, impurity free	Cost of ionic liquids, yield
CVD	Hydrocarbon gases	High quality, impervious, good reproducibility, high growth rates	High temperatures, difficulty in controlling number of layers, cost
Chemical synthesis (Hummer method)	Graphite	High yield, scalable to industrial level, efficient technique	Existence of defects, quality
Epitaxial growth on SiC	SiC wafer	Easy, high quality, controllable	High temperature, high vacuum, existence of defects
Arc discharge	Graphite	Small size	Low yield of graphene; carbonaceous impurities

their activity and reactivity and provide a promising nanocomposite for biological applications (Wang, Sun, Ang, & Tadé, 2013). The modification can be carried out through either noncovalent or covalent binding. Covalent attachment involves the chemical transformation of sp^2 carbon to sp^3 hybridization by covalent addition which alters some of its properties. While similar to covalent modification, changes in physicochemical properties are also made with noncovalent modification routes; their effects are weaker. One of these two methods will be preferred depending on whether the stability and mechanical properties are important or maintain conductivity and physical properties (Karimi, Othman, Uzunoglu, Stanciu, & Andreescu, 2015; Liu, Tang, & Gooding, 2012).

3.1 Modification of Graphene-Based Nanomaterials With Polymers

The use of polymers as precursors for functionalization of GFNs can provide a multifunctional composite with good mechanical properties. The number of polymers including both insulating polymers (e.g., Nafion, chitosan, polyethylene glycol, poly(vinyl alcohol), polyesters) and conducting polymers (e.g., polyaniline, polypyrrole, poly(phenylene ethynylene)) are used for functionalization of GFNs (Chen, Wu, Jiang, Wang, & Chen, 2011).

Polymerization process often is performed based on covalent interaction or physical adsorption via van der Waals, hydrogen bond, and π–π stacking interactions. The homogenous composites produced using van der Waals and π–π stacking are reversible and preserve electronic properties of graphene. The formation of multiple hydrogen bonds between hydrophilic polymers and graphene endows composite with appropriate tensile strength and glass transition temperature (Wang, Hao, Yang, Lu, & Wang, 2010).

Graphene polymerization with covalent attachment is usually performed via residual functionalities. It is noteworthy that covalent modification may compromise some graphene properties, but thermal and mechanical stability and also degradation time can be enhanced owning to stronger bonds between graphene and the modifier. For instance, GO has negatively charged surface which helps good solubility in water. On the other hand, electrostatic charge on the surface of GO may lead to its aggregation. Grafting GO with polymers such as chitosan, dextran not only prevents its aggregation but also enhances stability and surface functional groups for interaction with other molecules (Boujakhrout et al., 2015; Li et al., 2013; Wang, Zhao, Jing, Peng, & Li, 2015).

3.2 Modification of Graphene-Based Nanomaterials With Biomolecules

The biofunctionalization of GFNs with biomolecules and cells via various interactions (Fig. 2) has built-up a promising biosystem with appropriate biocompatibility and selectivity. Decorating of surface graphene layers using nucleic acid, protein, aptamer, peptide, and small molecules through self-assembly makes it a terrific support for immobilizing desired macromolecules (Patil, Vickery, Scott, & Mann, 2009; Zare-Zardini, Taheri-Kafrani, Amiri, & Bordbar, 2018). The formation of a monolayer of ssDNA molecules on graphene sheets via π–π staking enhances their hole density and is used as either platform for certain molecules analysis or as two-dimensional bionanointerfaces for attachment of other molecules (Guo, Wang, Qian, Wang, & Xia, 2009; Jia, Liu, & Wang, 2015).

The conjugation of GFNs with protein/peptide facilitates their applications in biological fields. For example, the improvement in the solubility of

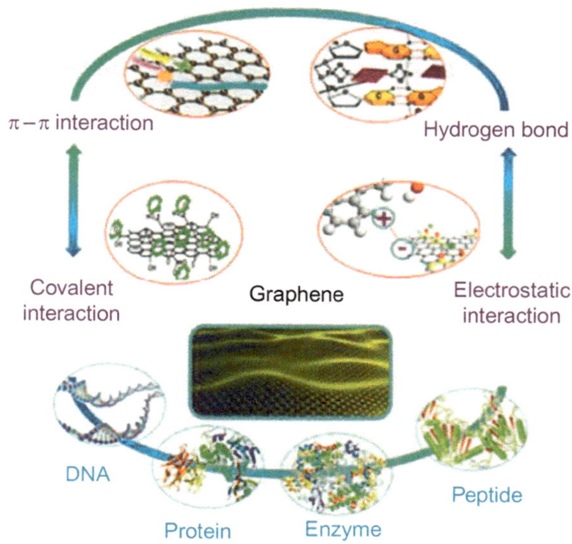

Fig. 2 Different interactions of biomolecules with graphene (Li et al., 2016).

graphene layers functionalized with hydrophobic protein hydrophobins via hydrophobic interaction creates a suitable biomimetic nanocomposite for using in aqueous solutions (Laaksonen et al., 2010).

Additionally, a biotin–streptavidin complex as the linking molecule was assembled on GO by forming stable amide linkages between biotinylated molecules and streptavidin functionalized GO. The support has excellent loading capacity and can be used as a substrate for biomolecular affinity chromatography (Liu, Jiang, et al., 2010).

Besides the direct chemical interaction between the surface functional groups of GO and biomolecules, the crosslinkers can also be used for covalent attachment of biomolecules onto GO. The bovine serum albumin (BSA) was covalently immobilized on GO nanosheets through diimide-activated amidation reaction of the amine groups of BSA with the carboxyl groups of GO. The GO–BSA complex has high biocompatibility and aqueous solubility for biomedical applications (Shen et al., 2010). On the other hand, the amphiphilic feature of BSA is tunable according to the pH of the solution. Metallic nanoparticles (NPs) can interact with the hydrophobic or hydrophilic residues on the BSA based on their charge. The product, NP–BSA–GO nanocomposite, has high capability for various ligands (Liu, Fu, et al., 2010).

The combination of various peptide assemblies such as nanofibers, nanotubes, and nanowires with GFNs is a promising approach for designing new nanobiosystems. For instance, self-assembled peptide nanofibers (PNFs) conjugated with graphene, can be applied to achieve GO–PNFs nanohybrids as a scaffold for biomimetic mineralization of hydroxyapatite (HA). The created GO–PNF–HA has good biocompatibility and renewability, in order to improve the proliferation and differentiation of cells (Li, Zhang, et al., 2016).

3.3 Modification of Graphene-Based Nanomaterials With Metal Nanoparticles

In accordance with two previous modification routes, the synergistic effects of graphene derivatives conjugated with inorganic nanoparticles can promote catalytic and electrochemical properties. Graphene–NP hybrids have high sensitivity, large surface area, and enhanced binding characteristics with biomolecules (Georgakilas et al., 2012). The interaction of metal adatoms with graphene layers has been studied extensively (Wu, Liu, Ge, & Jiang, 2009). The consensus is that there are three different positions (the center of hexagons, C—C bonds, and C atoms) for the adsorption of metal adatoms. Adatoms will bind to graphene at one of these three symmetry positions. Depending on the bonding energies of the adatom, the adsorption may take place via either chemisorption or physisorption (graphene-based materials) (Hu et al., 2010).

Most of inorganic materials that are usually used for decorating graphene–nanoparticle hybrids contain metal NPs (Au, Ag, Ti, Co, Cu, Pd, and Pt) (Liu, Guo, Duan, Dong, & Gao, 2017; Mani et al., 2015; Wu, Li, et al., 2017), metal oxide NPs (Fe_3O_4, ZrO_2, Co_3O_4) (Pakapongpan & Poo-Arporn, 2017; Puangjan & Chaiyasith, 2016), and quantum dots (ZnS and CdSe) (Huo et al., 2017). The preparation of these hybrids is based on two conventional methods, namely, in situ growth and solution mixing following a bottom-up approach, in which metal ions are reduced to produce metal NPs. In reality, the presence of surface functional groups, especially the oxygenated groups on the surface of GO, facilitates the reduction of metal ions and subsequently provides active centers for NPs nucleation. Simultaneous with nucleation, the size, and shape of NPs can be controlled by adding capping agents into synthesis system (Chen et al., 2011). Furthermore, graphene–NP hybrids provide a hydrophilic microenvironment to protect the activities of immobilized biomolecules. Furthermore, they can improve the interlayer electrical communications and fast conductivity for biosensor applications. The incorporation of NPs can also impart interesting properties

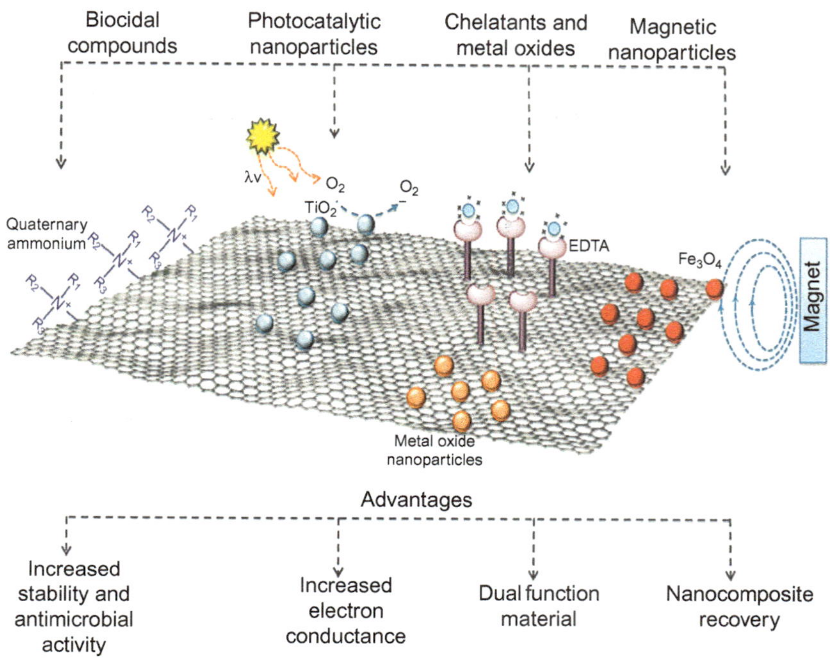

Fig. 3 Incorporation of some nanoparticles onto graphene nanosheets (Perreault, De Faria, & Elimelech, 2015).

(Fig. 3). For instance, Ag and TiO_2 NPs promote antibacterial and photocatalytic activity (Perreault et al., 2015; Qu, Ma, Hui, Chen, & Gao, 2017), and Pd and Pt NPs enhance thermal stability (Araque et al., 2014; Song et al., 2017). Also, GO–Fe_3O_4 NP hybrids that are functionalized with silane derivatives not only reveal magnetic responses but also can act as an active site for easy and stable interactions with biomolecules (Pavlidis et al., 2014).

4. APPLICATION OF GRAPHENE-BASED NANOMATERIALS FOR ENZYME IMMOBILIZATION

In green biotechnology, the alternation of chemical catalysts with enzymes is a turning point in the development of biocatalytic processes. The main challenges in the usage of enzymes are short-term operational stability and lack of reusability. Enzyme immobilization on appropriate supports not only can overcome these drawbacks but also makes a balance between costs and benefits of industrial enzyme applications (Landarani-Isfahani et al., 2015). The prominent properties of support chosen for immobilization are low synthesis cost, stability, high binding capacity, high available surface area, feasibility of functionalization, and above all, maintaining the biological function of enzyme (Ahmad & Sardar, 2015).

Recently, graphene derivatives with exceptional physicochemical features mentioned earlier are increasingly used as particular nanoscaffolds to facilitate enzymes immobilization and their applications. The surface chemistry of graphene derivatives can determine the kind of interactions with enzymes as well as stability. Besides, the surface engineering for desirable grafting of functional groups can influence enzyme catalytic behavior (Jin et al., 2012). Therefore, the performance of immobilized enzyme is affected by the structure and chemistry of the support and also by the immobilization method. The optimum method depends on the nature of the enzyme, the type of support, and the practical applicability. Based on the charge and the presence of specific groups on the surface of the support, the attachment of enzyme on either pristine graphene or functionalized graphene can be done through a variety of interactions including physical adsorption, site specific, and covalent binding (Fig. 4) (Shen et al., 2010). Also, some functionalized graphenes function similar to enzymes. For instance, folic acid functionalized graphene and carboxyl-modified GO possess intrinsic peroxidase-like activity, as candidates for fabrication of the composites in order to cancer cell detection and glucose monitoring, respectively (Song, Chen, Feng, Ren, & Qu, 2011; Song, Qu, Zhao, Ren, & Qu, 2010).

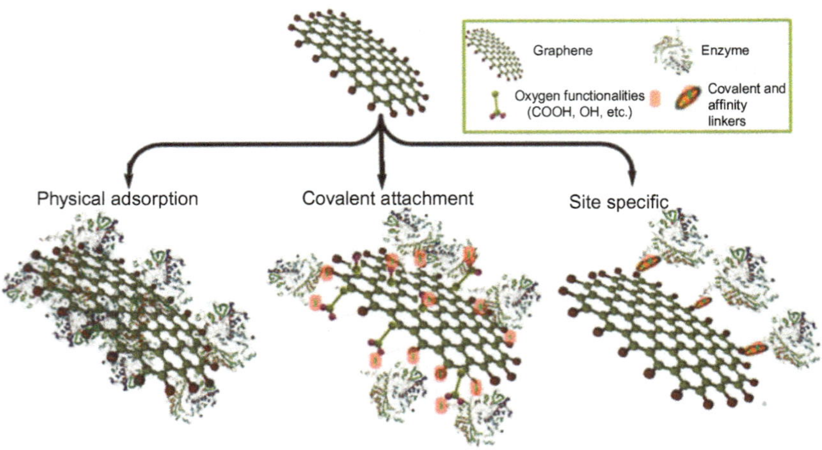

Fig. 4 Illustration of enzyme immobilization methods onto graphene nanosheets (Karimi et al., 2015).

4.1 Physical Adsorption

The physical binding approach is considered to be a more effective method for the immobilization of enzymes on graphene derivatives, because the absence of additional functional reagents and complicated process often preserves conformational changes of enzymes. Although electrostatic and hydrophobic interactions are weak, it is expected that extreme environmental changes lead to the leakage of the enzyme from the surface of the support and decrease the shelf life (Jesionowski, Zdarta, & Krajewska, 2014).

Among GFNs, GO is widely used for physical adsorption. The abundant oxygen-containing groups on GO facilitate enzyme attachment via van der Waals, ionic, and π–π stacking interactions. For example, lysozyme immobilized onto GO nanosheets may improve antibacterial activity based on synergistic effects of GO and lysozyme (Zhu et al., 2017). In another study, it has been found that coimmobilization of horseradish peroxidase (HRP) and lysozyme onto GO nanosheets without any surface modification caused higher catalytic performance compared to that of each free enzyme (Zhang et al., 2010).

PEGylation of Fe_3O_4–GO nanocomposite caused strong ionic interactions between xylanase and nanocomposite modified polyethylene glycol bis-amine. This enhanced the stability and accessibility of active sites of the immobilized xylanases to the substrate (Mehnati-Najafabadi, Taheri-Kafrani, & Bordbar, 2018).

GO has lower electrical conductivity than rGO. Therefore, it is better to recover its conductivity by a chemical reduction for electrochemical applications and biosensors (Karimi et al., 2015). The most interactions between rGO and enzymes can be done via hydrophobic interactions. A new artificial biocatalyst system was fabricated by coimmobilization of glucose oxidase (GOD) and glucoamylase on rGO via hydrophobic interactions. This biosystem could accomplish starch to gluconic acid in one-pot conversion, in a short time. Also, these amphiphilic enzymes were also immobilized on GO, and the results showed less loading capacity and lower reusability. The obvious difference between results of GO and rGO in the reusability experiments could probably be attributed to the different interactions between the enzymes–GO (electrostatic interaction) and enzymes–rGO (hydrophobic interaction). The optimal pH value of the catalytic reaction is higher for the immobilized enzyme, compared to that of the free one, and the electrostatic interactions are more sensitive to the pH and thermal changes (Zhao, Li, Jiang, Wang, & Mu, 2014).

Simultaneous immobilization of HRP and oxalate oxidase (OxOx) on GO and rGO indicated that the rGO–HRP/OxOx conjugates exhibit higher enzyme activity and stability than GO–HRP/OxOx (Zhang et al., 2012).

Inspired by nature, an H_2O_2 biosensor was constructed using biomimetic graphene capsule. This scaffold contains porous calcium carbonate microspheres as temporary templates for HRP and GO. After deletion of calcium carbonate microspheres, HRP was encapsulated in graphene capsule to mimic existence form of HRP in living cells (Fan et al., 2015).

4.2 Covalent Attachment

Some enzymes cannot be stably adsorbed on the surface of pristine graphene and then it is required to functionalize the surface. The durable attachment of enzyme can be achieved via covalent conjugation which involves two main steps, including surface activation by adding bifunctional components or spacers, and the binding of enzymes onto functionalized graphene. The spacer molecules not only provide high loading capacity as well as the greater degree of mobility of enzyme but also decline mass transfer resistance and good availability to the substrate (Feng & Ji, 2011). The common crosslinkers used for surface functionalization of graphene involved glutaraldehyde for aminofunctionalized graphene and carbodiimide/succinamide for carboxylated graphene. The usage of the novel and multifunctional crosslinkers can improve the efficacy and stability of enzyme–graphene conjugates.

Recently, a suitable Fe_3O_4–GO hybrid for covalent immobilization of xylanase and glucoamylase was designed. The functionalization of this solid support with cyanuric chloride (CC (TCT)) as a biocompatible linker has provided nucleophilic groups for covalent attachment of enzymes. Furthermore, decorating GO nanosheets with superparamagnetic NPs has facilitated the reusability of enzymes of the reaction products without complex equipment (Fig. 5) (Amirbandeh & Taheri-Kafrani, 2016; Royvaran et al., 2016).

The development of durable and high-sensitivity enzymatic biosensors is highly essential for the detection of nitrite in additive food preservatives. A biocompatible nanocomposite based on core–shell heterostructured multiwalled carbon nanotubes@rGO nanoribbons (MWCNTs@rGONRs)/chitosan (CHIT) was fabricated for the immobilization of myoglobin (Mb) as an important heme-containing redox enzyme for biosensing nitrite and H_2O_2. The features of MWCNTs@rGONRs/CHIT are its high edge defects which provide sites and functional groups for the covalent and π–π stacking interactions with Mb and then fast electron transfer for fast

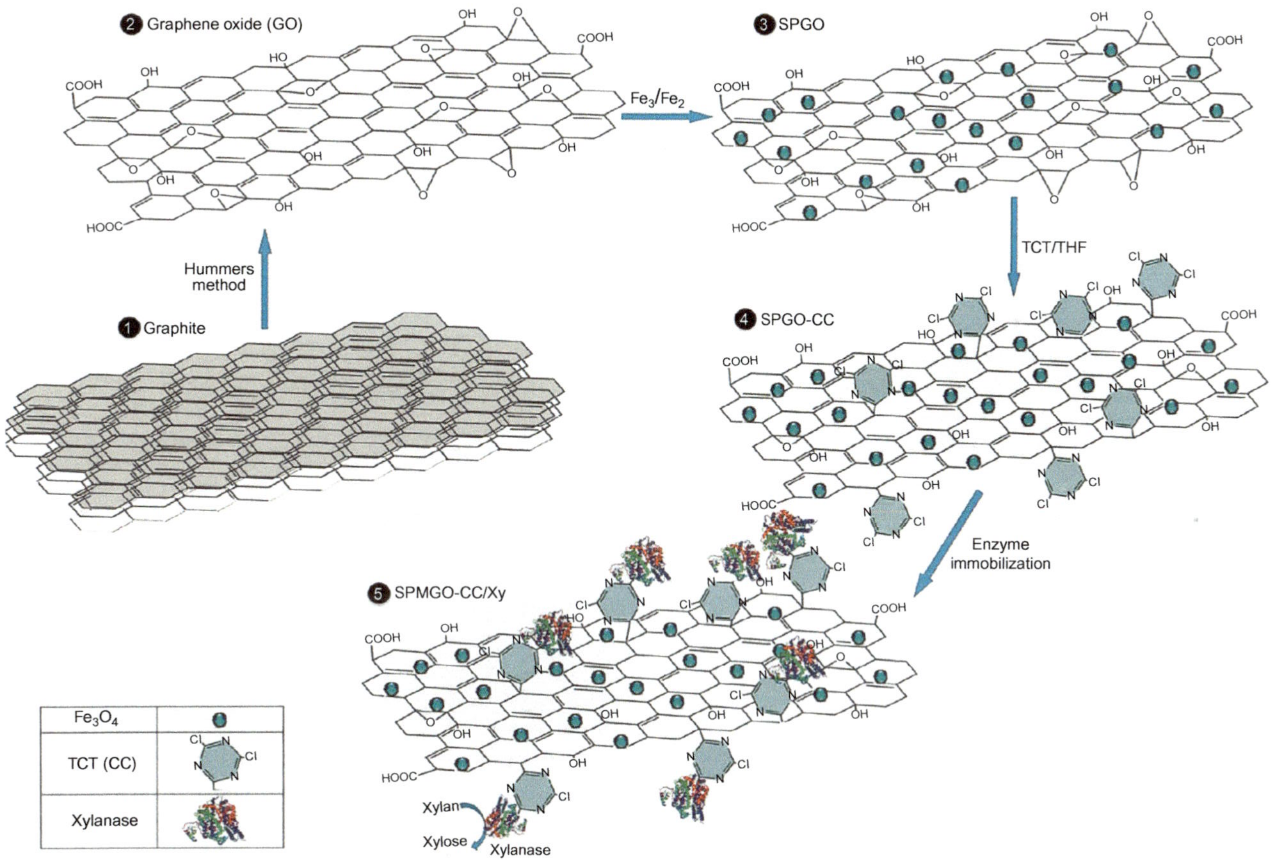

Fig. 5 Schematic of xylanase immobilization onto Fe_3O_4–GO hybrid (Royvaran, Taheri-Kafrani, Isfahani, & Mohammadi, 2016).

monitoring target analytes in nanomolar concentrations. The effectiveness of the application of this biosensor for the detection of H_2O_2 and nitrite was confirmed in contact lens cleaning solution and meat samples (Mani et al., 2017).

4.3 Site-Specific Immobilization

An ideal immobilization process for the best enzymatic performance is supposed to be particularly based on specific interactions between the enzyme and modified supports, with appropriate orientation for high catalytic activity while preserving conformation and stability of enzymes. The reasons why some immobilized enzymes lost their activities are explained by random attachment, different orientation, structural changes, inaccessibility of active site to the substrate, and different loadings of enzymes. Enzymes may require special orientation (Abad, Mertens, Pita, Fernández, & Schiffrin, 2005; Blank, Morfill, & Gaub, 2006). Therefore, site-specific covalent immobilization can overcome some of these drawbacks by the formation of affinity bonds to allow the enzymes to be arranged in a definite, controlled fashion. In reality, the chemoselective reaction between enzyme and support requires the presence of two unique and mutually reactive groups on the protein and the support surface. Gene fusion technology, site-directed mutagenesis, and posttranslational modification methods are employed for site-specific immobilization. The selection of these approaches relies on accommodating site-specific immobilization of enzymes with different structural characteristics (Butterfield, Bhattacharyya, Daunert, & Bachas, 2001). The incorporation of a peptidic affinity tag at the N- or C-terminus of the enzyme (far from active site) through gene fusion can attach enzyme with high affinity for NPs on graphene–NPs composites. Posttranslational modification can incorporate a single biotin moiety on a specific lysine residue of enzymes with avidin–graphene complex. In addition, sugar residues on glycoenzymes can provide strong interactions with lectin inserted onto graphene. If enzyme contains no cysteine residues, with cooperation of site-directed mutagenesis approach, sulfhydryl group on the side chain of the introduced the cysteine can be coupled with thiol-reactive groups on graphene or Au NPs–graphene composites (Ganesana, Istarnboulie, Marty, Noguer, & Andreescu, 2011; Loo, Bonanni, & Pumera, 2013).

GOD is a glycoenzyme with high-mannose type carbohydrate content. Functionalization of GO with a lectin (concanavalin A) is a suitable support for oriented immobilization of GOD through lectin–sugar bioaffinity interactions between sugar residues of GOD and concanavalin A. Compared with free and randomly immobilized GOD, oriented immobilized GOD exhibited better pH and thermal stability, higher affinity to glucose, and long-term storage stability. The precursors of modification, random, and site-specific immobilization of GOD are depicted in Fig. 6. (Zhou et al., 2012).

With the fine advances created in the fields of designed polypeptides (GDPs) and nanotechnology, the development of high precision electrochemical biosensors based on immobilization of organophosphorus hydrolase (OPH) onto Au–graphene hybrid is one of the most effective methods for pesticide detection. In this work, gold binding polypeptide (GBP) as a GDPs has been employed for the preparation of GBP–OPH fusion protein. Au NPs decorated on graphene exposed sites were used for oriented immobilization of GBP–OPH fusion protein on the support surface. This biosensor showed high sensitivity for detection of paraoxon as a pesticide compound model (Yang et al., 2011).

Several enzymes have been successfully immobilized onto a variety of graphene derivatives based on physical, covalent, ionic, and metal binding interaction methods which some of them are reported in Table 2.

5. OVERVIEW OF OUR STRATEGY TO FABRICATE A NOVEL GO NANOSHEET FOR ENZYME IMMOBILIZATION

Although we have shortly introduced, in previous sections, our recent efforts on enzyme immobilization onto functionalized GO, in this section we describe the detailed protocols used in the preparation of cyanuric chloride functionalized superparamagnetic GO (SPGO-CC) hybrids for xylanase immobilization. Among the various types of GFN nanocarriers, GOs, and their magnetic derivatives have been the focus of our laboratory for a variety of reasons that were previously mentioned. The superparamagnetic GO (SPGO) has received considerable attention in enzyme immobilization because of the unique combination of favorable properties. In order to improve the immobilization of enzymes on SPGO, the latter must first be functionalized with linkers or adsorptive surface coats. Therefore, CC was chosen as a novel linker for providing covalent immobilization of xylanase as a model of industrial enzymes. Herein, we describe synthesis, functionalization, and covalent immobilization procedures that were used in our laboratory.

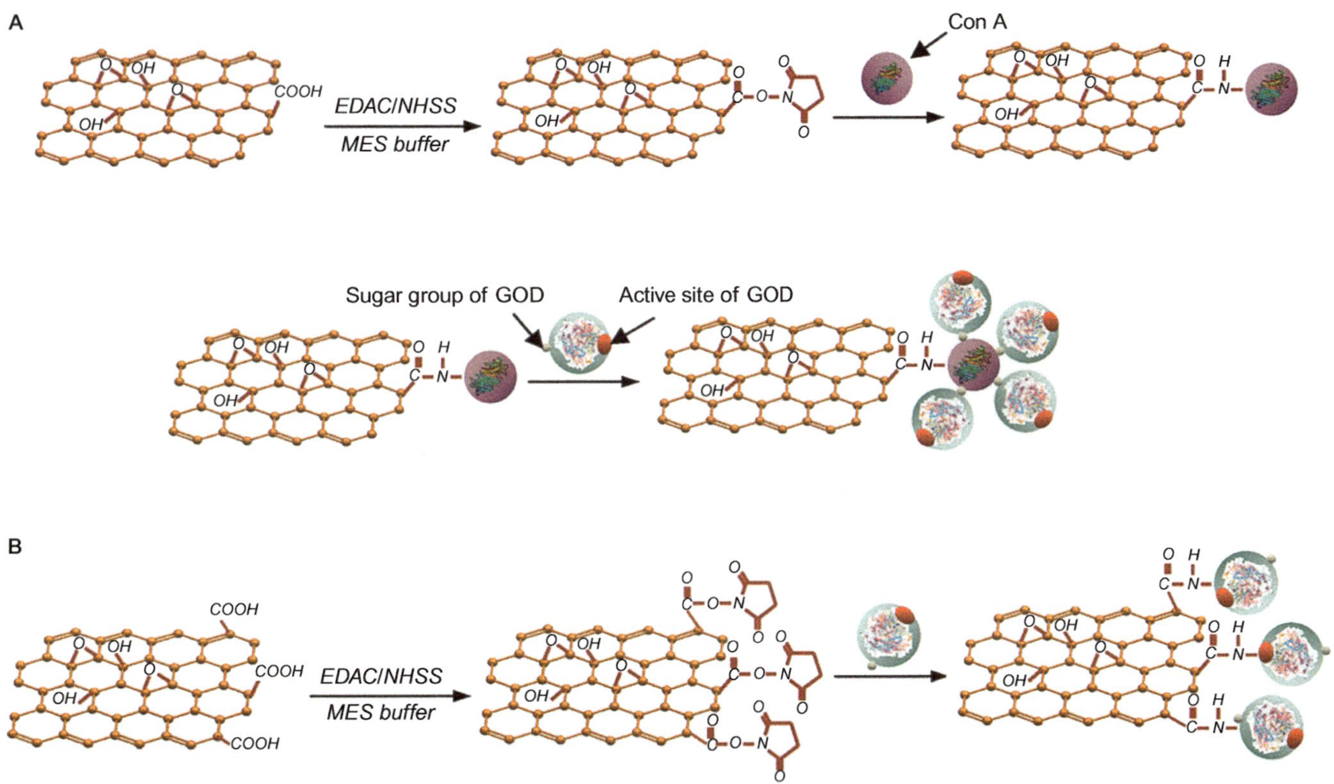

Fig. 6 The schematic representation the oriented (A) and random (B) immobilization of GOD onto GO nanosheets (Zhou et al., 2012).

Table 2 The Characteristics of Enzyme Immobilization on Various Graphene Derivatives

Graphene Derivatives	Modification	Enzyme	Technique Immobilization	Application	References
Graphene	Cystamine–GA	α-Galactosidase	Covalent	Enzymatic hydrolysis of RFOs in soymilk	Singh et al. (2014)
GO	Dextran–AuNPs	Tyrosinase	Covalent	Amperometric enzyme biosensor for catechol detection	Boujakhrout et al. (2015)
GO	PEG–NH_2	Nuclease P1	Ionic interaction	Designing a new biocatalyst for production technology of nucleotides	Zhuang et al. (2016)
	PEG–NH_2–GA		Covalent		
Graphene	3D nitrogen-doped graphene–chitosan	GOx	Electrostatic and hydrogen bonds	Glucose biosensing	Guo, Zhang, Hu, and Fu (2015)
rGO	AgNPs	GOx–laccase	Electrostatic interaction	Preparation bioelectrode for glucose–O_2 biofuel cell	Qu et al. (2017)
rGO	PCHNSs	HRP	Electrostatic interaction	Nitrite biosensing	Liu et al. (2015)
Graphene	PSE	GOx	π–π stacking	Glucose biosensing	Chia, Tan, Khiew, Chin, and Siong (2015)
GO	ZIFs	HRP	Entrapment	H_2O_2 detection	Fan et al. (2016)
rGO	RhNPs–diglycidyl ether–GA	Laccase	Covalent	17ß-Estradiol biosensing	Povedano et al. (2017)

Continued

Table 2 The Characteristics of Enzyme Immobilization on Various Graphene Derivatives—cont'd

Graphene Derivatives	Modification	Enzyme	Technique Immobilization	Application	References
GO	Fe_3O_4 NPs–chitosan	Lipase	Electrostatic interaction	Hydrolysis of olive oil	Wang et al. (2015)
	Fe_3O_4–chitosan–GA		Covalent		
	Fe_3O_4–chitosan–IDU–Cu		Metal ion affinity		
GRNs	PEI–PAA	DAAO	Covalent	An electrochemical microfluidic strategy for the separation of D-Met and D-Leu	Batalla, Martín, López, González, and Escarpa (2015)
GO	Epoxy chloropropane	γ-Lactamase	Covalent	Designing stable biosystem for production of lactam	Li, Wen, Shi, and Zheng (2016)

DAAO, D-Amino acid oxidase; *D-Leu*, D-Leucine; *D-Met*, D-Methionine; *GA*, glutaraldehyde; *GOx*, glucose oxidase; *GRNs*, graphene-reduced nanoribbons; *PAA*, polyallylamine; *PCHNSs*, porous Co_3O_4 hexagonal nanosheets; *PEG*, polyethylene glycol; *PEI*, polyethileneimine; *PSE*, pyrenebutyric acid hydroxyl succinimide ester; *RFOs*, raffinose family oligosaccharides; *ZIFs*, zeolitic imidazolate frameworks.

5.1 Synthesis of GO

An easy modified Hummers' method is developed for the preparation of GO from graphite powder according to the following protocol:

1. Add 2 g of graphite powder and 2 g of $NaNO_3$ in conc. H_2SO_4 (50 mL) at 0°C.
2. Add 4.0 g $KMnO_4$ slowly to the mixture while maintaining vigorous stirring.
   Note: The color of the suspension turns into purple–green immediately.
3. Stir the mixture at 40°C until it becomes pasty brownish.
4. Dilute the brown-colored paste with 50 mL of deionized water (DI) and allow to stir for 10 min.
5. Add slowly 10 mL of H_2O_2 (30%wt) solution into the mixture.
   Note: The color of the mixture changes to golden brown.
6. Centrifuge the golden brown sol acquired in the previous step.
7. Wash final mixture of bottom product with HCl solution (1 *M*) and warm DI water several times to remove the residual metal ions.
8. Dry the powder at room temperature, under vacuum.

5.2 Preparation of SPGO Nanocomposite

1. Add 400 mg of GO to 100 mL of 0.1 *M* NaOH solution and sonicate at room temperature for 45 min.
2. Separate the desired product by centrifuging and wash with warm DI water, repeatedly, to adjust the pH to 6.
3. Add 20 mL of degassed water to the isolated powder and disperse by an ultrasonic bath for 30 min.
4. Add a solution of $FeCl_3 \cdot 6H_2O$ (480 mg) and $FeCl_2 \cdot 4H_2O$ (176 mg) in degassed water (10 mL) dropwise to the suspension under N_2 atmosphere for 30 min, while the system is kept for 24 h under vigorous agitation at room temperature.
5. Add ammonium hydroxide solution (23%) to the mixture to adjust pH to 9–12.
6. Separate the product by a magnet and wash several times with methanol, under sonication.
7. Dry the resulting solid (SPGO nanocomposite), overnight, under vacuum.

5.3 Preparation of SPGO-CC

1. Add a mixture of SPGO (100 mg) to 10 mL anhydrous tetrahydrofuran (THF), while being exposed to ultrasonic waves at 0°C for 15 min.

2. Add slowly a solution of CC (TCT) (18 mg, 0.1 mmol) and *n,n*-diisopropylethylamine (DIPEA) (100 μL) in THF, and sonicate for 10 min.
3. Stir the final mixture under N_2 atmosphere at 0°C for 5 h.
4. Collect the product by a magnet, and wash with anhydrous THF several times.
5. Dry the resulting solid under vacuum, at room temperature.

5.4 Conjugation of Xylanase and SPGO Hybrid for Enzyme Immobilization

1. Disperse 10 mg of SPGO-CC in 4 mL of phosphate buffer (20 m*M*, pH 6.5), in a round bottom flask.
2. Add 2 mL of the xylanase solution (4.0 mg/mL) into the suspension.
3. Shake the mixture at room temperature for 16 h.
4. Separate the immobilized xylanase from free xylanases by magnetic separation, and wash three times with phosphate buffer (20 m*M*, pH 6.5).

   Note 1: Determine the amounts of the enzyme immobilized on SPGO-CC nanocarriers by measuring the initial and final concentrations of xylanase, in the immobilization medium, using the Bradford protein assay. In this method, the amount of protein in the supernatant is determined colorimetrically (595 nm) with the Bio-Rad protein assay reagent concentrate, using BSA as the standard protein.

   Note 2: Determine the percentage of total xylanase immobilized on SPGO-CC nanocarriers indirectly by measuring the concentration of total xylanase before immobilization (C_i) and the concentration of free xylanase in the supernatant after completion of the reaction (C_s), according to Eq. (1):

$$\text{Immobilization efficiency}\ (\%) = \left(\frac{C_i - C_s}{C_i}\right) \times 100 \quad (1)$$

5.5 Characterization of Nanocomposites and Immobilized Xylanase

The size and morphology of the as-prepared nanocomposites are determined using transmission electron microscopy (Philips CM10 HT-100KV). X-ray photoelectron spectroscopy is employed to analyze the surface composition of synthesized nanoparticles and xylanase/SPGO-CC. Presence of surface functional groups and the binding of xylanase onto SPGO-CC are analyzed by Fourier transform infrared (FTIR) spectroscopy (Perkin-Elmer, USA). Thermogravimetric analysis is performed on powdered samples, with a heating rate of 10°C/min from room temperature to 650°C, under a high purity nitrogen atmosphere. X-ray powder diffraction spectra are taken on a Bruker D8-advance X-ray diffractometer with Cu K_{α} radiation to show the X-ray diffraction patterns of the synthesized SPGO. Elemental analysis was performed on a LECO CHNS-932 analyzer to confirm the presence of specific elements. Vibrating sample magnetometer system is used to measure the magnetic properties of MNPs, as a function of the applied magnetic field.

5.6 Determination of Xylanase Activity

The activity of immobilized xylanase is assayed according to the method of Bailey et al. with minor modification. In this method, the rate of the reducing sugar (xylose equivalent) production was measured under standard conditions, using 3,5-dinitrosalicylic acid reagent, according to the protocol:

1. Prepare xylan solution (1%, w/v) as substrate by dissolving beechwood xylan, in 20 m*M* phosphate buffer (pH 6.5) as the substrate, under the standard conditions.
2. Incubate the reaction mixture containing substrate and the appropriate amount of either immobilized xylanase or free xylanase at 60°C for 15 min, under shaking conditions.
3. Separate the immobilized xylanase from the solution, using an external magnetic field.
4. Add 2 mL of DNS reagent to the separated supernatant, and then terminate it at 100°C for 5 min.
5. At the end of the process, measure the absorbance of reducing sugar released in the reaction at 530 nm and calculate the enzyme activity using standard curve of reducing xylose.

   Note 1: The control experiments without any enzyme are performed, no xylose is generated when just SPGO-CC nanocarriers are used as catalysts.

   Note 2: All experiments are carried out three times, and the experimental error was less than 3%.

   Note 3: One unit (IU) of xylanase activity is defined as the amount of enzyme that can catalyze 1 μmol of reducing sugar (xylose) equivalent per minute, under the specified assay conditions. The activity recovery of the immobilized enzyme is calculated from Eq. (2):

$$R(\%) = \frac{A}{A_0} \times 100\% \qquad (2)$$

where R is the activity recovery of the immobilized enzyme (%), A is the activity of the immobilized enzyme, and A_0 is the activity of the free enzyme in solution, before immobilization.

6. CHALLENGES AND CONCLUDING REMARKS

In the last decade, the application of enzymes has created beneficial alteration in various industries. Although free enzymes often have poor performance toward environmental changes or even increase the probability of contamination of the productions in the reaction mixture, immobilization process has been introduced as a fascinating route for improving efficiency and reusability of enzymes.

One of the major issues of practical implementation of immobilized enzymes is the effort for enhancement of stability and yield, and following that, minimizing the inconsistency and decreasing enzyme activity. The problem of traditional immobilization routes is that most enzymes are randomly immobilized either directly or using spacer molecules on the supports through their amino acid residues. On the other hand, the presence of multiple surface lysine or glutamate residues causes multi-point attachment of enzyme onto the support. The various orientations of the enzymes lead to their immobilization in potentially inactive conformations and consequently limited substrate access to active sites. Because of the random orientation, uncontrolled conformational changes, and different loading of enzymes, it is difficult to achieve similar activity from immobilized enzymes in various repetitions.

In order to reach a successful immobilization, it should be considered that the selected methodology and type of support should have only a slight effect on the nature of enzyme and maintain its active conformation. Besides, the support should have a large surface area, for high loading capacity. Recently, the use of nanomaterials, especially carbon-based nanomaterials, has emerged as a versatile tool for generating excellent supports for enzyme stabilization. As a result of the fascinating properties of graphene, it is supposed to display reactive groups on the support for ideal immobilization of enzymes in a controlled manner.

Despite all the unique features, there are some challenges for the application of graphene as the building block of functionalized biosystems such as purity, dispersibility status, control of the size. The number of single layers of graphene separated from bulk graphite can be adjusted by choosing suitable synthesis method. Considering all the drawbacks and benefits together, more efforts are still required to promote the potential of the application of immobilized enzymes in industries.

ACKNOWLEDGMENT

The authors are grateful to the Center of Excellence of Research Council of the University of Isfahan for the financial support of this work.

REFERENCES

Abad, J. M., Mertens, S. F., Pita, M., Fernández, V. M., & Schiffrin, D. J. (2005). Functionalization of thioctic acid-capped gold nanoparticles for specific immobilization of histidine-tagged proteins. *Journal of the American Chemical Society*, *127*(15), 5689–5694.

Ahmad, R., & Sardar, M. (2015). Enzyme immobilization: An overview on nanoparticles as immobilization matrix. *Biochemistry & Analytical Biochemistry*, *4*(2), 1.

Allen, M. J., Tung, V. C., & Kaner, R. B. (2009). Honeycomb carbon: A review of graphene. *Chemical Reviews*, *110*(1), 132–145.

Amirbandeh, M., & Taheri-Kafrani, A. (2016). Immobilization of glucoamylase on triazine-functionalized Fe_3O_4/graphene oxide nanocomposite: Improved stability and reusability. *International Journal of Biological Macromolecules*, *93*, 1183–1191.

Amirbandeh, M., Taheri-Kafrani, A., Soozanipour, A., & Gaillard, C. (2017). Triazine-functionalized chitosan-encapsulated superparamagnetic nanoparticles as reusable and robust nanocarrier for glucoamylase immobilization. *Biochemical Engineering Journal*, *127*, 119–127.

Araque, E., Arenas, C. B., Gamella, M., Reviejo, J., Villalonga, R., & Pingarrón, J. M. (2014). Graphene–polyamidoamine dendrimer–Pt nanoparticles hybrid nanomaterial for the preparation of mediatorless enzyme biosensor. *Journal of Electroanalytical Chemistry*, *717*, 96–102.

Batalla, P., Martín, A., López, M. A., González, M. C., & Escarpa, A. (2015). Enzyme-based microfluidic chip coupled to graphene electrodes for the detection of D-amino acid enantiomer-biomarkers. *Analytical Chemistry*, *87*(10), 5074–5078.

Bhuyan, M. S. A., Uddin, M. N., Islam, M. M., Bipasha, F. A., & Hossain, S. S. (2016). Synthesis of graphene. *International Nano Letters*, *6*(2), 65–83.

Blank, K., Morfill, J., & Gaub, H. E. (2006). Site-specific immobilization of genetically engineered variants of Candida antarctica lipase B. *Chembiochem*, *7*(9), 1349–1351.

Bosio, V. E., Islan, G. A., Martínez, Y. N., Durán, N., & Castro, G. R. (2016). Nanodevices for the immobilization of therapeutic enzymes. *Critical Reviews in Biotechnology*, *36*(3), 447–464.

Boujakhrout, A., Sánchez, A., Díez, P., Jiménez-Falcao, S., Martínez-Ruiz, P., Peña-Álvarez, M., et al. (2015). Decorating graphene oxide/nanogold with dextran-based polymer brushes for the construction of ultrasensitive electrochemical enzyme biosensors. *Journal of Materials Chemistry B*, *3*(17), 3518–3524.

Brodie, B. C. (1859). XIII. On the atomic weight of graphite. *Philosophical Transactions of the Royal Society of London*, *149*, 249–259.

Bustamante-Vargas, C. E., de Oliveira, D., Nyari, N. L., Valduga, E., Soares, M. B. A., Backes, G. T., et al. (2015). In situ immobilization of commercial pectinase in rigid polyurethane foam and application in the hydrolysis of pectic oligosaccharides. *Journal of Molecular Catalysis B: Enzymatic*, *122*, 35–43.

Butterfield, D. A., Bhattacharyya, D., Daunert, S., & Bachas, L. (2001). Catalytic biofunctional membranes containing site-specifically immobilized enzyme arrays: A review. *Journal of Membrane Science*, *181*(1), 29–37.

Carbone, M., Gorton, L., & Antiochia, R. (2015). An overview of the latest graphene-based sensors for glucose detection: The effects of graphene defects. *Electroanalysis*, *27*(1), 16–31.

Chae, S. J., Güneş, F., Kim, K. K., Kim, E. S., Han, G. H., Kim, S. M., et al. (2009). Synthesis of large-area graphene layers on poly-nickel substrate by chemical vapor deposition: Wrinkle formation. *Advanced Materials*, *21*(22), 2328–2333.

Chang, Y. M., Kim, H., Lee, J. H., & Song, Y.-W. (2010). Multilayered graphene efficiently formed by mechanical exfoliation for nonlinear saturable absorbers in fiber mode-locked lasers. *Applied Physics Letters*, *97*(21), 211102.

Chen, X.-m., Wu, G.-h., Jiang, Y.-q., Wang, Y.-r., & Chen, X. (2011). Graphene and graphene-based nanomaterials: The promising materials for bright future of electroanalytical chemistry. *Analyst*, *136*(22), 4631–4640.

Chen, J., Yao, B., Li, C., & Shi, G. (2013). An improved Hummers method for eco-friendly synthesis of graphene oxide. *Carbon*, *64*, 225–229.

Chia, J. S. Y., Tan, M. T., Khiew, P. S., Chin, J. K., & Siong, C. W. (2015). A bioelectrochemical sensing platform for glucose based on irreversible, non-covalent pi–pi functionalization of graphene produced via a novel, green synthesis method. *Sensors and Actuators B: Chemical*, *210*, 558–565.

Defaei, M., Taheri-Kafrani, A., Miroliaei, M., & Yaghmaei, P. (2018). Improvement of stability and reusability of α-amylase immobilized on naringin functionalized magnetic nanoparticles: A robust nanobiocatalyst. *International Journal of Biological Macromolecules*, *113*, 354–360.

Deng, D., Pan, X., Zhang, H., Fu, Q., Tan, D., & Bao, X. (2010). Freestanding graphene by thermal splitting of silicon carbide granules. *Advanced Materials*, *22*(19), 2168–2171.

DiCosimo, R., McAuliffe, J., Poulose, A. J., & Bohlmann, G. (2013). Industrial use of immobilized enzymes. *Chemical Society Reviews*, *42*(15), 6437–6474.

Fan, Z., Lin, Q., Gong, P., Liu, B., Wang, J., & Yang, S. (2015). A new enzymatic immobilization carrier based on graphene capsule for hydrogen peroxide biosensors. *Electrochimica Acta*, *151*, 186–194.

Fan, Z., Wang, J., Nie, Y., Ren, L., Liu, B., & Liu, G. (2016). Metal-organic frameworks/graphene oxide composite: A new enzymatic immobilization carrier for hydrogen peroxide biosensors. *Journal of the Electrochemical Society*, *163*(3), B32–B37.

Fang, X., Shashurin, A., & Keidar, M. (2015). Role of substrate temperature at graphene synthesis in an arc discharge. *Journal of Applied Physics*, *118*(10), 103304.

Fang, Q., Shen, Y., & Chen, B. (2015). Synthesis, decoration and properties of three-dimensional graphene-based macrostructures: A review. *Chemical Engineering Journal*, *264*, 753–771.

Feng, W., & Ji, P. (2011). Enzymes immobilized on carbon nanotubes. *Biotechnology Advances*, *29*(6), 889–895.

Ganesana, M., Istarnboulie, G., Marty, J.-L., Noguer, T., & Andreescu, S. (2011). Site-specific immobilization of a (His) 6-tagged acetylcholinesterase on nickel nanoparticles for highly sensitive toxicity biosensors. *Biosensors and Bioelectronics*, *30*(1), 43–48.

Georgakilas, V., Otyepka, M., Bourlinos, A. B., Chandra, V., Kim, N., Kemp, K. C., et al. (2012). Functionalization of graphene: Covalent and non-covalent approaches, derivatives and applications. *Chemical Reviews*, *112*(11), 6156–6214.

Gerstner, E. (2010). Nobel Prize 2010: Andre Geim & Konstantin Novoselov. *Nature Physics*, *6*(11), 836.

Goenka, S., Sant, V., & Sant, S. (2014). Graphene-based nanomaterials for drug delivery and tissue engineering. *Journal of Controlled Release*, *173*, 75–88.

Guo, H.-L., Wang, X.-F., Qian, Q.-Y., Wang, F.-B., & Xia, X.-H. (2009). A green approach to the synthesis of graphene nanosheets. *ACS Nano*, *3*(9), 2653–2659.

Guo, J., Zhang, T., Hu, C., & Fu, L. (2015). A three-dimensional nitrogen-doped graphene structure: A highly efficient carrier of enzymes for biosensors. *Nanoscale*, 7(4), 1290–1295.

Hu, L., Hu, X., Wu, X., Du, C., Dai, Y., & Deng, J. (2010). Density functional calculation of transition metal adatom adsorption on graphene. *Physica B: Condensed Matter*, *405*(16), 3337–3341.

Hummers, W. S., Jr., & Offeman, R. E. (1958). Preparation of graphitic oxide. *Journal of the American Chemical Society*, *80*(6), 1339.

Huo, P., Guan, J., Zhou, M., Ma, C., Liu, X., Yan, Y., et al. (2017). Carbon quantum dots modified CdSe loaded reduced graphene oxide for enhancing photocatalytic activity. *Journal of Industrial and Engineering Chemistry*, *50*, 147–154.

Jesionowski, T., Zdarta, J., & Krajewska, B. (2014). Enzyme immobilization by adsorption: A review. *Adsorption*, *20*(5–6), 801–821.

Jia, L.-P., Liu, J.-F., & Wang, H.-S. (2015). Electrochemical performance and detection of 8-hydroxy-2′-deoxyguanosine at single-stranded DNA functionalized graphene modified glassy carbon electrode. *Biosensors and Bioelectronics*, *67*, 139–145.

Jin, L., Yang, K., Yao, K., Zhang, S., Tao, H., Lee, S.-T., et al. (2012). Functionalized graphene oxide in enzyme engineering: A selective modulator for enzyme activity and thermostability. *ACS Nano*, *6*(6), 4864–4875.

Karimi, A., Othman, A., Uzunoglu, A., Stanciu, L., & Andreescu, S. (2015). Graphene based enzymatic bioelectrodes and biofuel cells. *Nanoscale*, 7(16), 6909–6923.

Ke, Q., Liu, Y., Liu, H., Zhang, Y., Hu, Y., & Wang, J. (2014). Surfactant-modified chemically reduced graphene oxide for electrochemical supercapacitors. *RSC Advances*, *4*(50), 26398–26406.

Ke, Q., & Wang, J. (2016). Graphene-based materials for supercapacitor electrodes—A review. *Journal of Materiomics*, *2*(1), 37–54.

Kim, H., Abdala, A. A., & Macosko, C. W. (2010). Graphene/polymer nanocomposites. *Macromolecules*, *43*(16), 6515–6530.

Klein, M. P., Hackenhaar, C. R., Lorenzoni, A. S., Rodrigues, R. C., Costa, T. M., Ninow, J. L., et al. (2016). Chitosan crosslinked with genipin as support matrix for application in food process: Support characterization and β-d-galactosidase immobilization. *Carbohydrate Polymers*, *137*, 184–190.

Klein, M. P., Nunes, M. R., Rodrigues, R. C., Benvenutti, E. V., Costa, T. M., Hertz, P. F., et al. (2012). Effect of the support size on the properties of β-galactosidase immobilized on chitosan: Advantages and disadvantages of macro and nanoparticles. *Biomacromolecules*, *13*(8), 2456–2464.

Laaksonen, P., Kainlauri, M., Laaksonen, T., Shchepetov, A., Jiang, H., Ahopelto, J., et al. (2010). Interfacial engineering by proteins: Exfoliation and functionalization of graphene by hydrophobins. *Angewandte Chemie International Edition*, *49*(29), 4946–4949.

Landarani-Isfahani, A., Taheri-Kafrani, A., Amini, M., Mirkhani, V., Moghadam, M., Soozanipour, A., et al. (2015). Xylanase immobilized on novel multifunctional hyperbranched polyglycerol-grafted magnetic nanoparticles: An efficient and robust biocatalyst. *Langmuir*, *31*(33), 9219–9227.

Li, X., Cai, W., An, J., Kim, S., Nah, J., Yang, D., et al. (2009). Large-area synthesis of high-quality and uniform graphene films on copper foils. *Science*, *324*(5932), 1312–1314.

Li, L., Fan, L., Sun, M., Qiu, H., Li, X., Duan, H., et al. (2013). Adsorbent for chromium removal based on graphene oxide functionalized with magnetic cyclodextrin–chitosan. *Colloids and Surfaces B: Biointerfaces*, *107*, 76–83.

Li, J., Jiang, Z., Wu, H., Long, L., Jiang, Y., & Zhang, L. (2009). Improving the recycling and storage stability of enzyme by encapsulation in mesoporous $CaCO_3$–alginate composite gel. *Composites Science and Technology*, *69*(3–4), 539–544.

Li, N., Wang, Z., & Shi, Z. (2011). Synthesis of graphenes with arc-discharge method. In *Physics and applications of graphene-experiments*. InTech.

Li, W., Wen, H., Shi, Q., & Zheng, G. (2016). Study on immobilization of (+) γ-lactamase using a new type of epoxy graphene oxide carrier. *Process Biochemistry*, *51*(2), 270–276.

Li, D., Zhang, W., Yu, X., Wang, Z., Su, Z., & Wei, G. (2016). When biomolecules meet graphene: From molecular level interactions to material design and applications. *Nanoscale*, *8*(47), 19491–19509.

Liu, W.-W., Chai, S.-P., Mohamed, A. R., & Hashim, U. (2014). Synthesis and characterization of graphene and carbon nanotubes: A review on the past and recent developments. *Journal of Industrial and Engineering Chemistry*, *20*(4), 1171–1185.

Liu, H., Duan, C., Yang, C., Shen, W., Wang, F., & Zhu, Z. (2015). A novel nitrite biosensor based on the direct electrochemistry of hemoglobin immobilized on MXene–Ti_3C_2. *Sensors and Actuators B: Chemical*, *218*, 60–66.
Liu, J., Fu, S., Yuan, B., Li, Y., & Deng, Z. (2010). Toward a universal "adhesive nanosheet" for the assembly of multiple nanoparticles based on a protein-induced reduction/decoration of graphene oxide. *Journal of the American Chemical Society*, *132*(21), 7279–7281.
Liu, H., Guo, K., Duan, C., Dong, X., & Gao, J. (2017). Hollow TiO_2 modified reduced graphene oxide microspheres encapsulating hemoglobin for a mediator-free biosensor. *Biosensors and Bioelectronics*, *87*, 473–479.
Liu, Z., Jiang, L., Galli, F., Nederlof, I., Olsthoorn, R. C., Lamers, G. E., et al. (2010). A graphene oxide streptavidin complex for biorecognition—Towards affinity purification. *Advanced Functional Materials*, *20*(17), 2857–2865.
Liu, J., Tang, J., & Gooding, J. J. (2012). Strategies for chemical modification of graphene and applications of chemically modified graphene. *Journal of Materials Chemistry*, *22*(25), 12435–12452.
Loo, A. H., Bonanni, A., & Pumera, M. (2013). Biorecognition on graphene: Physical, covalent, and affinity immobilization methods exhibiting dramatic differences. *Chemistry—An Asian Journal*, *8*(1), 198–203.
Mani, V., Devasenathipathy, R., Chen, S.-M., Wang, S.-F., Devi, P., & Tai, Y. (2015). Electrodeposition of copper nanoparticles using pectin scaffold at graphene nanosheets for electrochemical sensing of glucose and hydrogen peroxide. *Electrochimica Acta*, *176*, 804–810.
Mani, V., Govindasamy, M., Chen, S.-M., Chen, T.-W., Kumar, A. S., & Huang, S.-T. (2017). Core-shell heterostructured multiwalled carbon nanotubes@ reduced graphene oxide nanoribbons/chitosan, a robust nanobiocomposite for enzymatic biosensing of hydrogen peroxide and nitrite. *Scientific Reports*, 7(1), 11910.
Mehnati-Najafabadi, V., Taheri-Kafrani, A., & Bordbar, A.-K. (2018). Xylanase immobilization on modified superparamagnetic graphene oxide nanocomposite: Effect of PEGylation on activity and stability. *International Journal of Biological Macromolecules*, *107*, 418–425.
Misson, M., Zhang, H., & Jin, B. (2015). Nanobiocatalyst advancements and bioprocessing applications. *Journal of the Royal Society Interface*, *12*(102), 20140891.
Mohamad, N. R., Marzuki, N. H. C., Buang, N. A., Huyop, F., & Wahab, R. A. (2015). An overview of technologies for immobilization of enzymes and surface analysis techniques for immobilized enzymes. *Biotechnology & Biotechnological Equipment*, *29*(2), 205–220.
Nandamuri, G., Roumimov, S., & Solanki, R. (2010). Remote plasma assisted growth of graphene films. *Applied Physics Letters*, *96*(15), 154101.
Neto, A. C., Guinea, F., Peres, N. M., Novoselov, K. S., & Geim, A. K. (2009). The electronic properties of graphene. *Reviews of Modern Physics*, *81*(1), 109.
Novoselov, K. S., Geim, A. K., Morozov, S. V., Jiang, D., Zhang, Y., Dubonos, S. V., et al. (2004). Electric field effect in atomically thin carbon films. *Science*, *306*(5696), 666–669.
Pakapongpan, S., & Poo-Arporn, R. P. (2017). Self-assembly of glucose oxidase on reduced graphene oxide-magnetic nanoparticles nanocomposite-based direct electrochemistry for reagentless glucose biosensor. *Materials Science and Engineering C*, *76*, 398–405.
Park, S., & Ruoff, R. S. (2009). Chemical methods for the production of graphenes. *Nature Nanotechnology*, *4*(4), 217.
Patil, A. J., Vickery, J. L., Scott, T. B., & Mann, S. (2009). Aqueous stabilization and self-assembly of graphene sheets into layered bio-nanocomposites using DNA. *Advanced Materials*, *21*(31), 3159–3164.
Pavlidis, I. V., Patila, M., Bornscheuer, U. T., Gournis, D., & Stamatis, H. (2014). Graphene-based nanobiocatalytic systems: Recent advances and future prospects. *Trends in Biotechnology*, *32*(6), 312–320.

Perreault, F., De Faria, A. F., & Elimelech, M. (2015). Environmental applications of graphene-based nanomaterials. *Chemical Society Reviews*, *44*(16), 5861–5896.

Povedano, E., Cincotto, F. H., Parrado, C., Díez, P., Sánchez, A., Canevari, T. C., et al. (2017). Decoration of reduced graphene oxide with rhodium nanoparticles for the design of a sensitive electrochemical enzyme biosensor for 17β-estradiol. *Biosensors and Bioelectronics*, *89*, 343–351.

Puangjan, A., & Chaiyasith, S. (2016). An efficient ZrO_2/Co_3O_4/reduced graphene oxide nanocomposite electrochemical sensor for simultaneous determination of gallic acid, caffeic acid and protocatechuic acid natural antioxidants. *Electrochimica Acta*, *211*, 273–288.

Qu, F., Ma, X., Hui, Y., Chen, F., & Gao, Y. (2017). Preparation of close-packed silver nanoparticles on graphene to improve the enzyme immobilization and electron transfer at electrode in glucose/O_2 biofuel cell. *Chinese Journal of Chemistry*, *35*(7), 1098–1108.

Rao, K. S., Senthilnathan, J., Liu, Y.-F., & Yoshimura, M. (2014). Role of peroxide ions in formation of graphene nanosheets by electrochemical exfoliation of graphite. *Scientific Reports*, *4*, 4237.

Reina, A., Jia, X., Ho, J., Nezich, D., Son, H., Bulovic, V., et al. (2008). Large area, few-layer graphene films on arbitrary substrates by chemical vapor deposition. *Nano Letters*, *9*(1), 30–35.

Royvaran, M., Taheri-Kafrani, A., Isfahani, A. L., & Mohammadi, S. (2016). Functionalized superparamagnetic graphene oxide nanosheet in enzyme engineering: A highly dispersive, stable and robust biocatalyst. *Chemical Engineering Journal*, *288*, 414–422.

Sagadevan, S., & Periasamy, M. (2014). Recent trends in nanobiosensors and their applications—A review. *Reviews on Advanced Materials Science*, *36*, 62–69.

Sanchez, V. C., Jachak, A., Hurt, R. H., & Kane, A. B. (2011). Biological interactions of graphene-family nanomaterials: An interdisciplinary review. *Chemical Research in Toxicology*, *25*(1), 15–34.

Sassolas, A., Blum, L. J., & Leca-Bouvier, B. D. (2012). Immobilization strategies to develop enzymatic biosensors. *Biotechnology Advances*, *30*(3), 489–511.

Shahrestani, H., Taheri-Kafrani, A., Soozanipour, A., & Tavakoli, O. (2016). Enzymatic clarification of fruit juices using xylanase immobilized on 1,3,5-triazine-functionalized silica-encapsulated magnetic nanoparticles. *Biochemical Engineering Journal*, *109*, 51–58.

Shen, J., Shi, M., Yan, B., Ma, H., Li, N., Hu, Y., et al. (2010). Covalent attaching protein to graphene oxide via diimide-activated amidation. *Colloids and Surfaces B: Biointerfaces*, *81*(2), 434–438.

Singh, N., Srivastava, G., Talat, M., Raghubanshi, H., Srivastava, O. N., & Kayastha, A. M. (2014). Cicer α-galactosidase immobilization onto functionalized graphene nanosheets using response surface method and its applications. *Food Chemistry*, *142*, 430–438.

Song, Y., Chen, Y., Feng, L., Ren, J., & Qu, X. (2011). Selective and quantitative cancer cell detection using target-directed functionalized graphene and its synergetic peroxidase-like activity. *Chemical Communications*, *47*(15), 4436–4438.

Song, Y., Qu, K., Zhao, C., Ren, J., & Qu, X. (2010). Graphene oxide: Intrinsic peroxidase catalytic activity and its application to glucose detection. *Advanced Materials*, *22*(19), 2206–2210.

Song, X., Shi, Q., Wang, H., Liu, S., Tai, C., & Bian, Z. (2017). Preparation of Pd-Fe/graphene catalysts by photocatalytic reduction with enhanced electrochemical oxidation–reduction properties for chlorophenols. *Applied Catalysis B: Environmental*, *203*, 442–451.

Soozanipour, A., Taheri-Kafrani, A., & Isfahani, A. L. (2015). Covalent attachment of xylanase on functionalized magnetic nanoparticles and determination of its activity and stability. *Chemical Engineering Journal*, *270*, 235–243.

Stankovich, S., Dikin, D. A., Piner, R. D., Kohlhaas, K. A., Kleinhammes, A., Jia, Y., et al. (2007). Synthesis of graphene-based nanosheets via chemical reduction of exfoliated graphite oxide. *Carbon*, *45*(7), 1558–1565.

Staudenmaier, L. (1898). Verfahren zur darstellung der graphitsäure. *European Journal of Inorganic Chemistry*, *31*(2), 1481–1487.

Viculis, L. M., Mack, J. J., & Kaner, R. B. (2003). A chemical route to carbon nanoscrolls. *Science*, *299*(5611), 1361.

Wang, H., Hao, Q., Yang, X., Lu, L., & Wang, X. (2010). Effect of graphene oxide on the properties of its composite with polyaniline. *ACS Applied Materials & Interfaces*, *2*(3), 821–828.

Wang, S., Sun, H., Ang, H.-M., & Tadé, M. (2013). Adsorptive remediation of environmental pollutants using novel graphene-based nanomaterials. *Chemical Engineering Journal*, *226*, 336–347.

Wang, J., Zhao, G., Jing, L., Peng, X., & Li, Y. (2015). Facile self-assembly of magnetite nanoparticles on three-dimensional graphene oxide–chitosan composite for lipase immobilization. *Biochemical Engineering Journal*, *98*, 75–83.

Wu, X., Li, M., Li, Z., Lv, L., Zhang, Y., & Li, C. (2017). Amyloid-graphene oxide as immobilization platform of Au nanocatalysts and enzymes for improved glucose-sensing activity. *Journal of Colloid and Interface Science*, *490*, 336–342.

Wu, M., Liu, E.-Z., Ge, M., & Jiang, J. (2009). Stability, electronic, and magnetic behaviors of Cu adsorbed graphene: A first-principles study. *Applied Physics Letters*, *94*(10), 102505.

Wu, X., Yang, C., & Ge, J. (2017). Green synthesis of enzyme/metal-organic framework composites with high stability in protein denaturing solvents. *Bioresources and Bioprocessing*, *4*(1), 24.

Yang, M., Choi, B. G., Park, T. J., Heo, N. S., Hong, W. H., & Lee, S. Y. (2011). Site-specific immobilization of gold binding polypeptide on gold nanoparticle-coated graphene sheet for biosensor application. *Nanoscale*, *3*(7), 2950–2956.

Yi, M., & Shen, Z. (2015). A review on mechanical exfoliation for the scalable production of graphene. *Journal of Materials Chemistry A*, *3*(22), 11700–11715.

Zare-Zardini, H., Amiri, A., Shanbedi, M., Taheri-Kafrani, A., Kazi, S., Chew, B., et al. (2015). In vitro and in vivo study of hazardous effects of Ag nanoparticles and Arginine-treated multi walled carbon nanotubes on blood cells: Application in hemodialysis membranes. *Journal of Biomedical Materials Research. Part A*, *103*(9), 2959–2965.

Zare-Zardini, H., Taheri-Kafrani, A., Amiri, A., & Bordbar, A.-K. (2018). New generation of drug delivery systems based on ginsenoside Rh2-, lysine- and arginine-treated highly porous graphene for improving anticancer activity. *Scientific Reports*, *8*(1), 586.

Zhang, Y., Zhang, J., Huang, X., Zhou, X., Wu, H., & Guo, S. (2012). Assembly of graphene oxide–enzyme conjugates through hydrophobic interaction. *Small*, *8*(1), 154–159.

Zhang, J., Zhang, F., Yang, H., Huang, X., Liu, H., Zhang, J., et al. (2010). Graphene oxide as a matrix for enzyme immobilization. *Langmuir*, *26*(9), 6083–6085.

Zhao, F., Li, H., Jiang, Y., Wang, X., & Mu, X. (2014). Co-immobilization of multi-enzyme on control-reduced graphene oxide by non-covalent bonds: An artificial biocatalytic system for the one-pot production of gluconic acid from starch. *Green Chemistry*, *16*(5), 2558–2565.

Zhou, L., Jiang, Y., Gao, J., Zhao, X., Ma, L., & Zhou, Q. (2012). Oriented immobilization of glucose oxidase on graphene oxide. *Biochemical Engineering Journal*, *69*, 28–31.

Zhu, J., Wang, J., Hou, J., Zhang, Y., Liu, J., & Van der Bruggen, B. (2017). Graphene-based antimicrobial polymeric membranes: A review. *Journal of Materials Chemistry A*, *5*(15), 6776–6793.

Zhuang, W., He, L., Zhu, J., Zheng, J., Liu, X., Dong, Y., et al. (2016). Efficient nanobiocatalytic systems of nuclease P1 immobilized on PEG-NH_2 modified graphene oxide: Effects of interface property heterogeneity. *Colloids and Surfaces B: Biointerfaces*, *145*, 785–794.

FURTHER READING

Wang, P., Jiang, T., Zhu, C., Zhai, Y., Wang, D., & Dong, S. (2010). One-step, solvothermal synthesis of graphene–CdS and graphene–ZnS quantum dot nanocomposites and their interesting photovoltaic properties. *Nano Research*, *3*(11), 794–799.

FSC
www.fsc.org
MIX
Papier | Fördert gute Waldnutzung
FSC® C083411

Kolibri 360 GmbH
Ferdinand-Jühlke-Straße 7
99095 Erfurt, Deutschland
produktsicherheit@kolibri360.de